The International Library of Bioethics

Volume 104

The *International Library of Bioethics* comprises international and interdisciplinary volumes on foundational and applied issues in bioethics. The series focuses on perennial topics of aging, mental health, preventive medicine, medical research issues, end of life, biolaw, and other areas of bioethics, whilst expanding into other current and future questions involving the living world. The *International Library of Bioethics* examines the role of bioethics against the background of increasing globalization and interdependency of the world's cultures and government.

The series explores meaningful ways to answer the questions of how to direct biotechnology to human and other living things' ends, how to deal with changed values in the areas of religion, society, and culture, and how to formulate a new way of thinking. That is, to create a new, practical bioethics.

The international Library of Bioethics welcomes book proposals representing the broad interest of the series' focus. We especially encourage proposals addressing aspects of changes in biological and medical research and clinical health care, health policy, medical and biotechnology, and other applied ethical areas involving living things, with an emphasis on those interventions and alterations that force us to re-examine foundational issues.

Jaro Kotalik · David W. Shannon
Editors

Medical Assistance in Dying (MAID) in Canada

Key Multidisciplinary Perspectives

Springer

Editors
Jaro Kotalik
Centre for Health Care Ethics
Lakehead University
Thunder Bay, ON, Canada

Northern Ontario School of Medicine
University
Thunder Bay, ON, Canada

David W. Shannon
Shannon Law Office
Thunder Bay, ON, Canada

ISSN 2662-9186 ISSN 2662-9194 (electronic)
The International Library of Bioethics
ISBN 978-3-031-30001-1 ISBN 978-3-031-30002-8 (eBook)
https://doi.org/10.1007/978-3-031-30002-8

This Springer imprint is published by the registered company Springer Nature Switzerland AG
The registered company address is: Gewerbestrasse 11, 6330 Cham, Switzerland

Foreword

Under the penumbra of the Canadian Charter of Rights and Freedoms, the Supreme Court of Canada and the Canadian Parliament have created a veritable milestone in the history of Canada with respect to identifying a new 'foundational right,' autonomy. Nothing could be more definitive than proclaiming the elevated value of an individual right which in this volume and related commentaries has been described as taking on a position of 'radical autonomy.' The critique of 'radical autonomy' in this volume has meant to expose the importance of defending related rights which should not be subsumed if there are accompanying violations of both individual and social rights.

This volume is co-edited by two significant authors in the field, one a lawyer and the other a medical doctor, both having had elaborate careers in defending the rights of dying patients, the medically and mentally handicapped, and those with physical or mental disabilities.

Leading clinicians and legal authorities have been assembled to constitute chapters which give a cautionary and in-depth look at the ambiguities and potential failures of the enacted legislation and its forthcoming consolidations. There are grave concerns in anticipation of March 2024, when a much-debated issue will again come to the fore, led by citizens and professionals alike, in contesting whether an open-door policy should be put in place for persons with mental illness to have a shortcut through this legislation to end their lives. This is connected to a serious polarization among professionals whether a mental illness should ever be summarily regarded as irremediable.

So far tens of thousands of cases have emerged with many controversies already in place about whether to define MAID as having a criminal dimension beyond the provisions of health services. This volume in effect argues that the MAID legislation violates international covenants and treaties to which Canada is bound. It goes further to claim that the legislation is in violation of UN law.

'Medical Assistance in Dying (MAID) in Canada: Key Multidisciplinary Perspectives' clarifies the areas where serious monitoring and reporting is needed. There is less surveillance than in comparable legislation already in place for many years such as in Holland and Belgium. Are there already serious transgressions in violation of what the mandate currently demands? Little is known about the efficacy and comprehensiveness of recommendations made or options given to the claimants, most importantly in the area of palliative care. If physicians refuse to comply with a request as a matter of conscience, are their recommendations streamlined as either pro or against? Is the level of disclosure ethically justified in regard to MAID in the light of credible alternatives?

In a number of chapters, the MAID issue is expanded into the domain of systemic shortcomings beyond the delivery of health services, in order to deal with challenges of institutional injustice, maltreatment of certain populations, endemic poverty, and an unequal distribution of services based upon gender or race.

The editors assert that longitudinal studies are not yet available and that some arguments will remain unresolved until the appropriate research is conducted across Canada. Nevertheless, the chapters offer an up-to-date picture on legislation and judicial decisions which are under the surveillance of an extensive international community that understand the far-reaching implications of the most aggressive legislation extant.

The core question that this volume raises is whether the Canadian authorities have in haste put significant communities in jeopardy, including Indigenous populations and ethnic minorities, and all variances of beliefs and commitments about the roles of families and community groups to participate in decisions regarding the finality of life.

Canada is likely the most respected country in the world on multicultural protections and recognition of differences. Therefore, it is a worthy pursuit to have investigated all possible directions that the MAID legislation might take. It is a major accomplishment to have listed solid policy recommendations contained in many of the chapters about how to rectify the possibility of mishaps and injustices, those which have already occurred and likely yet to come.

'Medical Assistance in Dying (MAID) in Canada: Key Multidisciplinary Perspectives' is timely and should be celebrated as representing the voices that must surround the final enactments and applications of the law. In sum, the volume is not opposed to having MAID for a select minority of those who have gone through the process of adequate assessment. In spirit, the chapters make a series of strong suggestions that MAID legislation must minimally be attached to a surveillance structure led by concerned and professional representatives of Canadian fundamental values, namely the preserving and protecting of human life among vulnerable populations. This volume has charted an admirable course of action.

Montreal, QC, Canada

David N. Weisstub
Honorary Life President
International Academy of Law and Mental Health
Co-president
International Academy of Medical Ethics and Public Health

Preface

We are writing in a time of great change. The Canadian government, when amending the *Criminal Code* to allow medical assistance in dying (MAID), made one of the most significant legal, medical, cultural, and social shifts in recent Canadian history. Since this time, tens of thousands of Canadians have availed themselves of this liberalized access to MAID. With this monumental development, and subsequent rush to utilize MAID services, there has been limited serious study into the ramifications of this decision, the substantive interpretation of its eligibility criteria, and the procedural safeguards in place to prevent vulnerable persons from choosing MAID in a time of vulnerability. With the thoughtful, and hopefully intellectually provocative, insights of this work's contributing authors, we hope to lay the foundation for this broad and needed discussion.

Until 2015, the Criminal Code of Canada held that, without exception, anyone who aids or abets a person in committing suicide commits an indictable offence and that no person may consent to death being inflicted upon them. Four individuals and a provincial civil liberties association challenged the constitutionality of these provisions. In 2015, the dispute reached the Supreme Court of Canada, which ruled that such absolute prohibition of physician-assisted death conflicts with the Canadian Charter of Rights and Freedoms. In response to the direction of the Supreme Court, the Canadian Parliament in 2016 enacted Bill C-14 to amend the Criminal Code and to create an exemption for 'medical assistance in dying' (MAID) when provided by a medical practitioner or nurse practitioner. The *Criminal Code* was further amended in 2021 by Bill C-7, which resulted in major changes in safeguards and an expansion of eligibility for MAID.

The objectives of the program were most ambitious. As described in the Preamble to C-14, the MAID law set out to provide, on request, medical assistance in dying to competent, informed persons who are not under external pressure, who have an incurable medical condition and are experiencing intolerable suffering. At the same time, the law aimed to protect vulnerable persons who could choose to end their life in a moment of weakness, for example, crushed by a serious medical diagnosis or overwhelmed by symptoms that medical care as yet did not have a chance to control. It is not mentioned in the Preamble, but justice and compassion require that the law and practice also protect from selecting premature death the people who are vulnerable because of their physical or mental disability or are under the duress of difficult social and economic conditions. Safeguards were established in Bill C-14, intendent to prevent errors and abuse in the provision of MAID. The law also wanted to ensure that the availability of MAID would not encourage 'negative perceptions of the quality of life of persons who are elderly, ill or disabled.' Finally, the Preamble to the law recognized that 'suicide is a significant public health issue that can have lasting and harmful effects on individuals, families, and communities.'

Has the MAID program, which is now five years old, been successful in meeting those demanding objectives that were annunciated at the program's inauguration? The purpose of this book is to attempt to answer this question. We are very grateful that we were able to enlist co-operation of very competent individuals who also thought this question important. Our team of lawyers, philosophers, social scientists, bioethicists, community activists, psychiatrists, palliative care specialists, and other medical practitioners, most of them with strong academic credentials, examined various aspects of the MAID program, based on publicly available evidence. This multidisciplinary approach allowed us to construct a comprehensive picture of the MAID program, something which has not been attempted before. We hope that our work will contribute to the unavoidable and ongoing development and refinement of the program that is unique because of its size (more than 33,000 Canadians have already received MAID) and its ground-breaking, radical nature. MAID program has been very successful in managerial terms, attracting unexpectedly large fraction of the population (about 3.3% of all death), but it has also been very worrisome, raising concerns of many individuals and groups. If the book should be helpful in shaping the future of the program as well as related elements of public policy, we had to focus on apparent and potential weaknesses, on tensions, disagreement, and conflicts. The authors of the chapters often made recommendations that aim to control undesirable aspects of program and support its positive features, aiming for a balance between personal and community interests. We must remember that the MAID program was launched extremely rapidly as a high political priority because it was understood as a demand of the *Canadian Charter of Rights and Freedoms*. But now, after five years of experience, it is time to look back and recognize what changes are needed, so we can look to the future.

One limitation of the book stems from a lack of good quality longitudinal research concerning MAID due to its recent implementation. However, we all feel that this book is an urgent project, and that rapidly collecting and commenting on the existing evidence and hearing urgent contemporary voice should be first priority, without waiting for more research. Many chapters also suggest research projects that are desirable for a number of reasons, for example, to better understand the situation of fellow Canadians who resigned themselves to apply for MAID because their suffering was not ameliorated in our health care system.

Among limitations of the book is that the reader will learn little about medical assistance in dying in Quebec. The reason is that this second most populous province of Canada is in many ways different from the rest of the country in its history, and legal and administrative structure. Quebec became the first province to legalize medically assisted deaths in 2014. It was accomplished by a law passed by the Quebec National Assembly, and the provision of death was defined as a medical service, thus bypassing the federal *Criminal Code*. There are so many differences in the origin, governance, and delivery of assisted deaths between Quebec and the rest of Canada that to give Quebec justice would require a book of its own.

The initial idea for the book was to avoid all overlaps; however, in the end, we decided that each chapter must make sense on its own and must provide enough background to observations and arguments that would make it understandable to anyone, being in Canada or abroad. And so, we invite readers to pick and choose, skip or come back and read those chapters that you are most keenly interested in, and pursue them in any order.

It may seem that medical assistance in dying involves just two persons—one who wants to have it done and the other who is willing to do it. The reality is much more complicated. The pages that follow show that each of these acts involve much more than two people; it affects the delicate mesh of mutual relationships that we call human society.

We hope that this work will stimulate ideas, provoke debate and additional discourse, and lay the foundation for new and advanced research that will provide the answers many Canadians are seeking. We urgently advocate for thoughtful evidence, sound empirical research, and human sensitivity that will advise future MAID related amendments, regulations, policy, and the inevitable review by the courts.

Thunder Bay, ON, Canada
June 2022

Jaro Kotalik
David W. Shannon

Acknowledgements

We greatly appreciate the kind support and contribution of each author. We hope you find the same thoughtful intellectual probity that we did. Your collegiality and commitment to your peers has shown us that your work is foundational to elevating the discourse in this highly charged subject. We also express deep gratitude to Louisa Pedri, Darlene Bruzzese, David N. Weisstub, and Ashley Bénard-Legris, as well as the generous support of the staff at Springer Nature publishing. Above all, it is with gravity and sensitivity that we acknowledge the thousands of Canadians and their families for whom MAID has been a central question in their lives and for those who may one day confront this decision.

Jaro Kotalik
David W. Shannon

Acknowledgements

Contents

Contributors

Kathleen Absolon Department of Theological Studies, Concordia University, Montreal, Canada

S. Zaki Ahmed Department of Theological Studies, Concordia University, Montreal, Canada

Alireza Bagheri Research Affiliate Center for Healthcare Ethics, Lakehead University, Thunder Bay, ON, Canada

Jonas-Sébastien Beaudry McGill University, Montreal, QC, Canada

Rabbi Catharine Clark Department of Theological Studies, Concordia University, Montreal, Canada

Simon Czajkowski Division of Urology, Department of Surgery, University Health Network, Toronto, Canada

Saumil Dholakia Department of Mental Health, The Ottawa Hospital, Ottawa, Canada;
Department of Psychiatry, University of Ottawa, Ottawa, ON, Canada

Travis Dumsday Concordia University of Edmonton, Edmonton, AB, Canada

K. Sonu Gaind University of Toronto and Humber River Hospital, Toronto, ON, Canada

Alexandra Giancarlo Faculty of Kinesiology, University of Calgary, Calgary, AB, Canada

Ewan C. Goligher Interdepartmental Division of Critical Care Medicine, University of Toronto, Toronto, Canada;
Department of Medicine, Division of Respirology, University Health Network, Toronto, Canada

Mara Grunau Centre for Suicide Prevention, Calgary, AB, Canada

Amy Hendricks Dalhousie University, Halifax, NS, Canada

John Heng The King's University College of Western University, London, ON, Canada

Leonie Herx Queen's University, Kingston, ON, Canada

Christine Jamieson Department of Theological Studies, Concordia University, Montreal, Canada

Heidi Janz Associate Adjunct Professor, University of Alberta, Edmonton, AB, Canada

Kerri Joffe ARCH Disability Law Centre, Toronto, Canada

Ebru Kaya University of Toronto, Toronto, ON, Canada

Sister Nuala Kenny Department of Theological Studies, Concordia University, Montreal, Canada

Tom Koch University of British Columbia, Vancouver, BC, Canada;
Alton Medical Centre, Toronto, Canada

Jaro Kotalik Centre for Health Care Ethics Lakehead University and Northern Ontario School of Medicine University, Thunder Bay, ON, Canada

Nita Kotiuga Department of Theological Studies, Concordia University, Montreal, Canada

Roberto Lattanzio ARCH Disability Law Centre, Toronto, Canada

Timothy Lau University of Ottawa, Ottawa, ON, Canada

John Maher Barrie & South Georgian Bay ACT Teams, Canadian Mental Health Association, Barrie, ON, Canada

Irmo Marini The University of Texas Rio Grande Valley, Edinburg, TX, USA

Dylan McGuinty Jr. McGuinty Law, Ottawa, Canada

Sean Murphy Protection of Conscience Project, Powell River, Canada

Dwight Newman University of Saskatchewan, Saskatoon, Canada

Robert Olson Centre for Suicide Prevention, Calgary, AB, Canada

Louisa Pedri Centre for Health Care Ethics, Lakehead University, Thunder Bay, Ontario, Canada

Gabrielle Robitaille University of Saskatchewan, Saskatoon, Canada

Derek Ross Christian Legal Fellowship, London, Canada

Harvey Schipper University of Toronto, Toronto, Canada

John F. Scott University of Ottawa, Ottawa, ON, Canada

Mary M. Scott The Ottawa Hospital Research Institute, ON, Ottawa, Canada

David W. Shannon Barrister-and-Solicitor, Shannon Law, Toronto, ON, Canada

Mary J. Shariff Faculty of Law, University of Manitoba, Winnipeg, Manitoba, Canada;
Centre on Aging, University of Manitoba, Winnipeg, Manitoba, Canada

Tim Stainton University of British Columbia, Vancouver, BC, Canada

William F. Sullivan University of Toronto, Toronto, ON, Canada

Sephora Tang University of Ottawa, Ottawa, ON, Canada

Ryan Tonkens Centre for Health Care Ethics, Lakehead University, Thunder Bay, Canada;
Northern Ontario School of Medicine University, Thunder Bay, ON, Canada

Elaine Toombs Clinical Psychologist, Dilico Anishnawbe Family Care, Toronto, ON, Canada

Laura Villarreal The University of Texas Rio Grande Valley, Edinburg, TX, USA

Crystal Walker Centre for Suicide Prevention, Calgary, AB, Canada

Deina Warren Canadian Centre for Christian Charities, Elmira, Canada

Part I
MAID in Law and Regulations

Chapter 1
Radical Autonomy in Supreme Court of Canada Jurisprudence

Dylan McGuinty Jr.

Abstract Since the adoption of the Canadian *Charter of Rights and Freedoms* in 1982, there has been a movement afoot at the Supreme Court of Canada that advances a vision of radical autonomy. This essay uses the term radical autonomy to refer to a view of autonomy that is defined by freedom from any comprehensive moral anthropology. This chapter will outline the movement towards radical autonomy at the Supreme Court of Canada, which arguably culminated in the 2015 *Carter* (*Carter v. Canada (Attorney General)* (2015).) decision. This movement arguably supplants a commitment to objective truth with a preference for subjectivity. This chapter will argue that radical autonomy is an obstacle to collective action that can be taken through legislative bodies for particular communities, by which reasonable limits may be placed on members of society to set up the law as a pedagogical tool to advance demonstrable universal goods.

Keywords Freedom · Moral anthropology · Radical autonomy · Virtue · Ethics

1.1 Introduction

Since the adoption of the Canadian *Charter of Rights and Freedoms* in 1982, there has been a movement afoot at the Supreme Court of Canada that advances a vision of radical autonomy. This essay uses the term radical autonomy to refer to a view of autonomy that is defined by freedom from any comprehensive moral anthropology[1]

[1] "Michael Klenk notes that in recent years anthropology has taken what he terms an "ethical turn," recognizing moral systems, and ethics more generally, as a distinct object of anthropological study. This is a move away from the Durkheimian paradigm, and includes the study of self-development, virtues, habits, and the role of explicit deliberation when moral breakdowns occur." See Klenk (2019). See also: Gert and Gert (2020). For Durkheim, morality was simply a matter of how a given society enforces whatever social rules it happens to have (*ibid.*).

D. McGuinty Jr. (✉)
McGuinty Law, Ottawa, Canada
e-mail: dylanjr@mcguintylaw.ca

J. Kotalik and D. W. Shannon (eds.), *Medical Assistance in Dying (MAID) in Canada*, The International Library of Bioethics 104,
https://doi.org/10.1007/978-3-031-30002-8_1

that would articulate a shared and substantive view of human nature, of ultimate human goods and the common good; a moral anthropology that could, in turn, be the basis for "reasonable limits prescribed by law as can be demonstrably justified in a free and democratic society."[2] The author is not aware of any use of the term "radical autonomy" in legal or philosophical literature, except for the following passage by Robert Cardinal Sarah:

> The modern mistrust of all dependance explains many ills. Alas, it has not finished displaying its deleterious effects… Now only a radically autonomous and independent man, a man who is alone, without any bond, can be a free man.[3]

Radical autonomy is characterized by existential freedom in the tradition of Jean-Paul Sartre, that is, the idea that no general account of what it means to be human can be given, and "the idea of existence as self-making-in-a-situation".[4] In the context of euthanasia, the term radical autonomy refers to the prioritization of the real or perceived autonomy of those who wish to die, over and against the well-being of vulnerable persons who would not express a wish to die but for the hardship of their circumstances, including a tragic lack of social supports. Thus, radical autonomy is contrasted with a view of human beings as interconnected and as owing to each other basic obligations arising out of a shared human nature and natural law, which obligations are breached by the provision of euthanasia, whether or not a desire for a hastened death is expressed.

The movement towards radical autonomy arguably culminated in the Supreme Court of Canada's 2015 *Carter*[5] decision. This movement supplants a commitment to objective truth with a preference for subjectivity and has little, if any, commitment to the possibility of ultimate goods and the common good within an understanding of a demonstrably objective human nature. This movement may be called post-modern in its apparent preference for subjectivity, or at least in its implied skepticism about the possibility of knowable objective human nature.[6]

Radical autonomy is an obstacle to collective action that can be taken through legislative bodies for particular communities, by which reasonable limits may be placed on members of society to set up the law as a pedagogical tool to advance demonstrable universal goods, such as life, knowledge and marriage that is based on the unique complementarity of the male/female relationship. Such reasonable limits would be made on the basis of a comprehensive and collectively held understanding of human nature and would be conducive towards human well-being within specific political communities in Canada. By analyzing underlying philosophical assumptions in Supreme Court *Charter* and non-*Charter* jurisprudence about human nature, this article seeks to demonstrate the advancement of radical autonomy in Canadian Supreme Court jurisprudence.

---

[2] *Canadian Charter of Rights and Freedoms* (1982).

[3] Diat and Sarah (2019), p. 153.

[4] Crowell (2020), citing Fackenheim (1961).

[5] *Carter v. Canada (Attorney General)* (2015).

[6] The author relies here on the description of modernism and postmodernism offered by Hicks (2011).

1.2 Constitutional Liberal Democracies and Moral Anthropology

The law is prescriptive; it sets an arch over the building of civilization, as it were, to give it both stability and proper limits. It is therefore important for constitutional liberal democratic societies to have and to heed to a shared conception of the human person, for he is the foundation of liberal, constitutional law. Otherwise, questions of law and rights become irresolvable. That is because rights protect fundamental human goods. If lawmakers are precluded from using objective human nature, which points to fundamental human goods and the common good, as the basis to develop law, then the task becomes exceedingly difficult, if not impossible, to formulate policies that promote certain rights and limit others. For example, if it cannot be assumed that the desire for a hastened death is contrary to the human good, and needs to be addressed primarily, if not exclusively, through psychiatric treatment and social determinants of health, and not through demonstrable harm to the patient; and if, further, it cannot be assumed that euthanasia is itself a harm to the patient, then it is unlikely that policy-makers will be able to pass laws that limit the aiding of suicide. Indeed, *Carter* struck down Canadian Criminal Code s.241(1)(b) prohibiting the aiding of a person to die by suicide on the basis that the desire for a hastened death can be properly resolved through demonstrable harm to the patient, namely, by ending her life prematurely, and worse, in the absence of adequate social supports.

1.3 Radical Autonomy, Euthanasia and the Supreme Court of Canada

As mentioned, Supreme Court *Charter* jurisprudence shows a movement towards radical autonomy in specific judgments. By advancing radical autonomy, the Supreme Court has limited Parliament's ability to constrain the individual in society on the basis of a robust or "thick" moral anthropology, in which reasonable limits to freedom advance a collective understanding of the individual good and the common good, including dignity and the sacredness of life. The movement in favour of radical autonomy reached new dramatic heights in the Supreme Court's *Carter* decision. There, the Court decided that the choice of euthanasia by apparently consenting, autonomous individuals was ultimately worth the risk that the same option would be used to end the lives of patients who are under duress on account of the tragedy of their hardships, be it lack of housing, money, social supports or existential distress[7] and yet who, by all appearances, choose euthanasia.

---

[7] According to Dr. James Downar, Head of the Division of Palliative Care at the University of Ottawa, existential distress is "the patient's feeling of distress that their role in the world has changed and their identity has been undermined by their illness." See Laucius (2020). See also: McGuinty and Lau (2020).

The radicality of such autonomy for apparently consenting, autonomous individuals is highlighted by the fact that the euphemistically termed "medical assistance in dying" has turned the act of killing into a state-regulated economic good, used primarily by wealthy dying people[8] in the comfort of their homes. The moral accountability of the act of killing is effectively outsourced to a vast and impersonal Weberian administrative state, in conjunction with the medical profession and pharmaceutical industry.[9] This outsourcing of the act of killing arguably creates a neo-liberal marketplace that disregards the negative externalities it creates, namely, the killing of those who lack the emotional and social supports to live until a natural death; and who live under the weight of tragic existential distress and the well-documented loneliness[10] that has arguably resulted from the (also well-documented) breakdown of the family and communities in modern times.[11]

The decisions of such vulnerable persons to choose euthanasia arguably cannot be prevented by appropriate controls, because the philosophical lens that is applied to their decision is one of "autonomy". In other words, persons "choosing euthanasia" are not seen as vulnerable and acting under psychological distress, but are narrowly conceived of as autonomous. Patients' words expressing a desire for a hastened death become a showstopper at the hospital bedside, as the writer has witnessed firsthand, thus precluding the most robust efforts possible to heal the root causes of such a desire, and opening up the very real possibility of euthanasia. And this seems to be because the situational factors that lead to the expression of the desire for a hastened death are excluded from the sanctuary, as it were, of the expression itself, the expression being fiercely guarded from the moral offence of paternalism. Therefore, whatever criteria a patient meets in order to demonstrate consent, the criteria see the vulnerable person suffering from existential distress through the one-dimensional lens of self-determination. This is clear from the fact that s.241(2)(c) of the *Criminal Code of Canada* sets out psychological suffering not as grounds to enhance the patient's social supports and mental healthcare in order to *prevent* euthanasia, but as a criterion to *qualify* for it. What is more, s.241.2(1)(e) of the *Criminal Code of Canada* merely requires that the patient be informed of alternative means to relieving their pain and suffering, but imposes no obligation on the state to actually provide those alternatives to its members. Meanwhile, the law imposes on the state the obligation to provide euthanasia. What is more, less than 15% of adult Canadians who died in 2016–2017 received palliative home care services,[12] and in the same time periods, most long-term care residents with less than 6 months to live did not have a record of palliative care.[13] Finally, according to an analysis by the Canadian Institute for Health Information, 94% of Canadians who died in a

---

[8] Joanne Laucius, *ibid*.

[9] Indeed, even the *Carter* decision's authorship was depersonalized by being published under the authorship of "The Court," rather than setting out the names of its individual authors.

[10] See for example Korzinski (2022).

[11] See for example Putnam (2000). See also Riesman et al. (2001).

[12] Canadian Institute for Health Information (2018) at p. 14.

[13] *ibid*, p. 17.

hospital in 2016–2017 could potentially have benefitted from palliative care during their stay.[14] This is to say nothing of the well-documented lack of social supports and affordable housing, as well as loneliness, suicide rates and untreated depression among many Canadians.

As a result, the criteria for euthanasia in Canada do not seem to leave any room, in the calculus of whether to permit euthanasia, for a rich, multifaceted and holistic view of the patient seeking euthanasia, as a person situated within a particular place and time, with particular social relations and needs; as a person often trapped by a devastating absence of social supports. The Criminal Code might have considered psychological suffering to be a showstopper that ceases any discussion of euthanasia and triggers the implementation of social supports and mental health care. Instead, the Criminal Code seems to consider psychological suffering to be a showstopper that ceases social supports and mental health care and triggers the implementation of euthanasia. And it would appear that this is so because the Criminal Code provision permitting euthanasia seems to prioritize autonomy, real or perceived, to such an extent that psychological suffering is not seen as the impetus for better care, but as a justification for ending human life. The Criminal Code in this regard thus seems to be based on the apparently inverted value that death is good because suffering essentially has no meaning, being that there is no shared comprehensive moral anthropology that would give meaning to suffering. Thus, the criteria for euthanasia in Canada seem to screen out, as it were, a patient's psychological suffering from the factors that would preclude euthanasia. The criteria seem to focus instead on the patient's agency, real or constructed, however compromised by tragedy and human frailty.

Furthermore, the criteria for euthanasia seem blind to the normative principle that euthanasia offends the sacredness and inviolability of human life, principles that were affirmed by the majority of the Supreme Court in *Rodriguez*.[15] This situation is a dramatic departure from the natural law against the taking of innocent human life, as well as from the principle, expressed by the majority of the Supreme Court in *Rodriguez*, that the state has an interest in preserving the lives of its members, even when the individual expresses a desire for a hastened death:

> Section 241(b) (which criminalised assisted suicide) has as its purpose the protection of the vulnerable who might be induced in moments of weakness to commit suicide. This purpose is grounded in the state interest in protecting life and reflects the policy of the state that human life should not be depreciated by allowing life to be taken (emphasis added)[16]

Indeed, according to a 2020 study in the Canadian Medical Association Journal (**CMAJ**),[17] 96% of Ontarians who received euthanasia in 2016–2018 reported psychological suffering. Yet merely 6.2% of "medically assisted death" cases were given psychiatric consultations. According to the study's lead author, Dr. James Downar of the University of Ottawa and The Ottawa Hospital, "the motivation behind

[14] *ibid*, pp. 18–19.

[15] *Rodriguez v British Columbia (Attorney General)* (1993).

[16] *ibid*.

[17] Downar et al. (2020).

most requests (for euthanasia) is existential distress"[18] and the psychological cause at the root of euthanasia is existential distress.[19] Dr. Downar describes existential distress as "the patient's feeling of distress that their role in the world has changed and their identity has been undermined by their illness."[20] And yet, existential distress is arguably treatable.[21] Similarly, treatment resistant depression ("TRD") has been observed to improve significantly over time, and clinicians cannot make accurate predictions of irremediability for particular patients with TRD.[22] Therefore, one might reasonably ask: if existential distress is a basis for many if not most requests for euthanasia, and if it can be treated, then why would society see fit to euthanize its members rather than advance their continued treatment? Does human agency, real or constructed, override all other considerations absolutely? What are the limits of agency? What are the limits of consent, especially in the presence of existential distress? What is the role of medicine? What is the meaning of suffering? What is the meaning of life? These seem to be the questions whose answers have been turned upside down, first in the culture of the meritocratic elite who are making pro-euthanasia policy decisions; second, among those who mobilize politically to advocate for and sustain those decisions; and third, in the law itself.

a. Radical Autonomy by Misrecognition in Carter

i. Freedom, Recognition, and "Release" in Hegel

In the *Phenomenology of Spirit* G. W. F. Hegel argues that adequate recognition from others is the foundation of just social relations. Further, Hegel has been read as expressing the idea that freedom becomes actual through the process of reciprocal recognition.[23] Hegel's concept of freedom includes the notion of "release", that is, affirming the other as she determines herself. The process of release, for Hegel, advances a community of freedom that does not reduce individuals to some homogeneity.[24]

Charles Taylor, the pre-eminent Canadian philosopher with a background in Hegelian philosophy, and co-author of the *Bouchard-Taylor Commission* on reasonable accommodation in the Province of Quebec, has written:

---

[18] Downar (2019).

[19] Laucius, *supra* note 7. See also: McGuinty and Lau, *supra* note 7.

[20] *ibid*.

[21] Cooper et al. (2015). See in particular pp. 123–124: "In general, clients seem to benefit from meaning group therapy interventions as compared with participating in a social support group, being on a waiting list, or receiving care as usual. In particular, they seemed to find greater meaning or purpose in life, their level of psychopathology decreases to a moderate extent, and their self-efficacy was strengthened moderately. Thus, this type of existential therapy seems to be promising as a means of addressing meaning-orientated and existential concerns in people with serious and life-threatening illnesses and deserves further investigation and development as an evidence-based treatment in this area."

[22] Gastmans et al. (2022).

[23] Mullender (2011), pp. 46–47.

[24] *ibid*, p. 47, citing Siep (1992).

> our identity is partly shaped by recognition or its absence, often by the misrecognition of others, and so a person or group of people can suffer real damage, real distortion, if the people or society around them mirror back a confining or demeaning or contemptible picture of themselves. Nonrecognition or misrecognition can inflict harm, can be a form of oppression, imprisoning someone in a false, distorted, and reduced mode of being.[25]

It is one thing to affirm a person's identity; it is quite another to affirm that they can be something other than their objective human nature. And, it is one thing to impose a monolithic identity on others; it is quite another to insist that we all have uniqueness within the bounds of our shared human nature, that is, within the natural limits of being that are set, as it were, in everyone's nature, which is in-laid with normative, transcendent principles. If truth gives way to self-creation as the central organizing principle prescribing social and political arrangements, as Nietzche predicted,[26] then universal human nature and the ultimate good of the human person are rejected in favour of self-determined identity *beyond* human nature. And if this is so, then it no longer makes sense to speak of standard deviations from human nature. All standard deviations become arbitrary data points in a void of radical existential freedom. With no anchor in universal human nature, all values are relative to the perspectives of those who pronounce them, all experience is subjective, and there can be no ascertainable right or wrong answer on pressing moral questions, such as whether it is acceptable to set up an administrative apparatus to end the lives of apparently consenting, but psychologically suffering, persons.

In such a world, the possibility of objective mental health becomes highly questionable, if not suspect, and the Diagnostic and Statistical Manual of Mental Disorders risks becoming meaningless, if not hopelessly patronizing. The rights-claimant insists that his compatriots respect, enable and finance his right to things that demonstrably harm him, because the compass showing the way towards his well-being is cast to sea from the ship of society. A patient may invoke his constitutional rights in demanding thus and such from his physician, including death, even against his physician's better judgment. The marker shifts from what is true and right, upon reflection and against a well-founded and longstanding understanding of human nature, to what a person demands as a matter of her own self-directed autonomy, often whilst caught in the stranglehold of mental illness. Society thus risks descending into a Hobbesian struggle of Nietzschean will to power. Strife becomes the animating principle of such a state of affairs, in keeping with the pre-Socratic philosopher Heraclitus' first principle. Meanwhile, the compass washes up on the shore of a distant land, fossilized and sand swept, now speaking a foreign tongue, replaced by the wanderings of the human heart. The tragedy is that the North Star remains, but it is covered by the storm clouds of human passions, and the same human heart, now without a compass, still longs to move in its direction.

Rather than measuring claims to rights on the basis of whether they promote or harm the human person, properly understood, claims to rights risk being decided on the basis of mere individual preferences. The desire for a hastened death is thus

[25] Taylor (1995), p. 225. See also: Bates (2002), p. 31.

[26] See Nietzsche (1973, 2022).

seen through the lens of autonomy, not mental health. The bystander is encouraged to shout "stop!" to the desperate man whose toes hang off the ledge, but the same bystander, now working as a physician, is required by law to tell the same man, now in a hospital bed, about his right to a lethal injection. On this rudderless ship, it becomes possible for physicians and families to be inclined, or required by law, to encourage an individual to misrecognize herself as something she is not: as someone whose life is not worth living, whose suffering has no meaning, and whose suffering does not warrant the co-suffering (*com-passion*) of those who would form a nest of mutually supportive relationships. As the palliative care physician A. J. Donkin writes: "an expression of a wish to die may be more a question than a statement: "Am I of any worth? Am I in the way? Patients are looking for a picture of themselves in the eyes of their beholder. If the answer is returned that they should be helped to die, it is an affirmation of what they suspected – that others now see them as worthless."[27]

While the Supreme Court dismissed concerns about the possibility of erecting adequate safeguards to protect vulnerable populations within a legalised euthanasia regime, the question is not whether such safeguards can be erected, but rather whether they can be effective at all if they screen out psychological suffering as grounds to preclude euthanasia, when psychological suffering arguably accompanies every expression of desire for a hastened death. The argument can be succinctly put this way: the desire for a hastened death itself is evidence of grave psychological suffering; consent to a matter as grave as death cannot adequately be obtained in the grip of grave psychological suffering; therefore, consent to euthanasia or assisted suicide can never be adequately obtained. The safeguards are powerless that are put in place to prevent the harm of a lack of adequate consent, for the lack of adequate consent is there right when the words expressing a desire for a hastened death are first uttered. To the allegation that this argument is paternalistic, it is useful to point to the soundness of an argument by the fruit that it bears. In pre-*Carter* Canada, those not truly consenting to euthanasia were not euthanized; whether they should have had better supports is another and a valid question. But according to an investigation by *The New Atlantis*[28] into Canada's euthanasia regime published in December 2022, in a post-*Carter* world:

a. Poor people lacking adequate social supports, such as housing, healthcare and long-term care, are requesting euthanasia because they lack adequate social supports;
b. At least one such person has been euthanized;
c. Some euthanasia providers in Canada are aware of this risk;
d. At least one prolific euthanasia provider in Canada seems to have suggested that this is OK.

It is worth remembering that Canada abolished Capital Punishment because it was believed that one wrongful death is too many.

b. Two Scapegoats: Rodriguez v. Carter

---

[27] A. J. Donkin, cited in Hartling (2021).

[28] Raikin (2022).

Justices McLachlin's and L'Heureux-Dubé's dissent in *Rodriguez v. British Columbia*,[29] which were delivered by Justice McLachlin, can be read as an argument in favour of radical autonomy. In *Rodriguez*, Justices McLachlin and L'Heureux-Dubé favoured striking down the ban on aiding suicide in Criminal Code s.241(b), in order to allow Sue Rodriguez to elect to have assisted suicide at a point in time when her degenerative disease would prevent her from taking her own life "by her own hand". For, if Sue Rodriguez were to commit suicide on her own, she would be ending her life sooner than if assisted suicide were an option. Assisted suicide, then, was a way to prolong Sue Rodriguez's life. Justice McLachlin put forward the "scapegoat argument" to defend Sue Rodriguez' putative right to assisted suicide. She wrote:

> It is argued that the denial to Sue Rodriguez of the capacity to treat her body in a way available to the physically able is justified because to permit assisted suicide will open the doors, if not the floodgates, to the killing of disabled persons who may not truly consent to death. The argument is essentially this. There may be no reason on the facts of Sue Rodriguez's case for denying to her the choice to end her life, a choice that those physically able have available to them. Nevertheless, she must be denied that choice because of the danger that other people may wrongfully abuse the power they have over the weak and ill, and may end the lives of these persons against their consent. Thus, Sue Rodriguez is asked to bear the burden of the chance that other people in other situations may act criminally to kill others or improperly sway them to suicide. She is asked to serve as a scapegoat.

For our purposes, the difference between euthanasia and assisted suicide is immaterial. There are effectively two possible scapegoats, to use Justice McLachlin's terminology. On one view, Sue Rodriguez should be prevented from using assisted suicide for the sake protecting less autonomous vulnerable persons, who might be unduly influenced by the duress of their circumstances to choose assisted suicide, and thus do "not truly consent to death." Here, the scapegoat is Sue Rodriguez,the one, we are told, who autonomously desires assisted suicide but who is being denied it in order to protect against the death of those who do not autonomously wish to die by assisted suicide. But on another view, Sue Rodriguez should be allowed to die by assisted suicide. Here, the one autonomously desiring to die should be allowed to die by assisted suicide,even if it means that other, less autonomous vulnerable persons "who may not truly consent to death" will be scapegoats who die by assisted suicide, by "choosing" it on account of, for instance, lack of psychological and social supports. On Justices McLachlin's and L'Heureux-Dubé's preferred view, Sue Rodriguez' autonomy ought to prevail.

Can it then be surmised that, between Sue Rodriguez wishing to die and the less autonomous vulnerable person not truly wishing to die, Justices McLachlin's and L'Heureux-Dubé's preferred scapegoat is the less autonomous vulnerable person not truly wishing to die, in order to clear the way for other people to die, who are either more autonomous or conceived of as such?

Is the sin of the preferred scapegoat then that he lacks autonomy when compared with his peers, in addition to his being dependent upon others for ongoing support?

---

[29] [1993] 3 SCR 519, [1993] SCJ No 94 (QL).

Is his lack of autonomy the price he pays with his life, for is not the only material difference between he and the one who "truly wishes" to die the fact that he lacks autonomy? Does our society display its values by loading up this scapegoat with the sin of non-autonomy and sending him out to his death? Autonomy arguably becomes radical when, being the overriding priority, it puts underfoot the non-autonomous, who are prone to being unduly influenced by the duress of their circumstances to "choose" euthanasia.

And so, are the non-autonomous sacrificed at the Altar of Autonomy, in what we may dare to call the Worship of the Hero-Self, or the Reign of the *Übermensch*? If one were to argue in favour of such radical autonomy, then would the final backstop to such an argument not be the view that those who choose euthanasia, on account of the duress of their circumstances, actually lead lives that are not worth living? By prioritizing the autonomous over the non-autonomous, the law seems to exalt John Galt, the paragon of hyper-individualism in Ayn Rand's 1957 novel *Atlas Shrugged*. In *Carter*, the Supreme Court of Canada seems to have made its choice abundantly clear: it preferred to give the choice for euthanasia to those conceived of as autonomous, at the risk of creating new scapegoats, namely those who are less autonomous and who, in their vulnerability, do not truly consent to death. Is *Carter* not then truly a sign of the times?

1.4 Radical Autonomy in Supreme Court of Canada Jurisprudence

a. The Harm Principle

i. The Moral Anthropology Underlying the Harm Principle

In *On Liberty*, John Stuart Mill famously advanced the Harm Principle, according to which restrictions to liberty can only be justified to prevent harm to others.[30] Further, to constitute a harm, an action must be injurious to important interests in which others have rights. Whereas Mill appears to reject the regulation of mere offense, the Harm Principle appears to be the one justification he recognizes for restricting liberty.[31]

Even though Mill's Utilitarianism can be seen as primarily social by nature, Mill's Harm Principle is in fact predicated on a particular moral anthropology, i.e. of the individual, the general outline of which can be detected in *On Liberty*.[32] For Mill, man is fundamentally:

1. Self-expressive
2. Individualist

---

[30] Brink (2022) at 3.6.

[31] *ibid*.

[32] Mill (1960).

3. Sovereign
4. Self-determining
5. The final judge of what is true
6. Free to err and free to be right
7. Self-reliant
8. To be free, must be of full age, uncoerced and undeceived to decide for himself.

ii. Radical Autonomy and the Harm Principle

Mill's moral anthropology contrasts with a moral anthropology in which man is dependent on his family and community for his identity and well-being, as an organic part of a society based on a shared past and traditions:

> the individual should seek and find his completion in society, and… he should find himself as part of an order that is greater than himself, in the sense of transcending anything that could have been brought about through his own enactment. He must see himself as the inheritor, not the creator, of the order in which he participates, so that he may derive from it (from the picture of its "objectivity") the conceptions and values which determine self-identity. He will see his extension in time from birth to death as taking on significance from civil stability: his world was not born with him, nor does it die when he departs from it.[33]

Professor Alan Hutchinson and Professor Patrick Monahan (as he then was) similarly wrote of community in their well-known essay Democracy and Rule of Law. They charged that "the Rule of Law's language of rights reinforces the assumption that communities are nothing more than aggregations of private interests." Further, "[r]ightholders are defined in contradistinction to the community rather than as integral components of it."[34]

Further,

> The difficulty with this individualist ideology is that it ignores and suppresses actual human experience. Individuals are located in history, within a context of allegiances. They are not abstract or bloodless, but are in part constituted by their social context. To divorce individuals from this structure of allegiances is to rob them of the "railings to which [individuals] can cling as they walk into the [midst] of their social lives."[35]

And finally,

> Values such as justice and equality are the products of politics, not its antecedents. They take root in a public that engages in debate and argument and that is given the opportunity to nurture notions of reasonableness and commonality.[36]

The individual for Mill is arguably conceived of in opposition to community and its conception of the good life, including values of justice and equality, or what Rawls described as "comprehensive doctrines."[37] As such, it can reasonably be asked

---

[33] Scruton (1980), at p. 66. See also: Stewart (1999) at p. 1–13.

[34] Hutchinson and Monahan (1987).

[35] *ibid.*

[36] *ibid.*

[37] Rawls (1993).

whether Mill's Harm Principle, being predicated on what may be called an "individualist" moral anthropology, sows the seeds of radical autonomy, which stands in opposition to, rather than in harmony with, community.

Georgia du Plessis shows that the Harm Principle ousts the application of what may be called normative principles arising out of one's community, which are richer than the Harm Principle's threadbare approach to intervention in human conduct.

Writing on Mill's Harm Principle, du Plessis notes

> [N]ot all conduct that causes injury or offence to society or other individuals is to be restricted. Mill agrees that an action that a person does to himself may seriously affect the sympathies, feelings or interests of those closely connected to him, but these will not be definite damage or a definite risk of damage. These will be mere "contingent" or "constructive" injuries.[38]

And, Mill does not allow that the state may proscribe contingent injuries. He writes:

> But with regard to the merely contingent...injury...by conduct which neither violates any specific duty to the public, nor occasions perceptible hurt to any... individual except himself; the inconvenience is one which society can afford to bear, for the sake of the greater good of human freedom.[39]

du Plessis concludes:

> Thus, if a person watches pornography in the privacy of his home, and the only external effect or injury thereof is that his neighbours are offended or appalled by this action, this will be considered a contingent injury (not part of the three instances of definite damage), and therefore, such action cannot be limited.[40]

If Mill does not allow that a community can express its vision of the human good through law in a way that bars certain actions such as the consumption of pornography, it can reasonably be asked to what extent community and a shared comprehensive moral anthropology are possible under the Harm Principle, and to what extent the Harm Principle gives license to radical autonomy.

Indeed, in the realm of sexual morality from 1962 to the present, the Supreme Court of Canada moved the fulcrum of criminal law from substantive, moral organizing principles with a claim to truth, towards a liberal, tolerant (one might say relativist), harms-based logic that is not characterised by a substantive moral anthropology centred upon a shared conception of human nature and the common good (other than freedom from harm, restrictively conceived). In particular, the Supreme Court moved away from (a) the notion that obscene and indecent acts and images can corrupt human nature and throw well-ordered society into disorder, towards (b) a metric of harm.[41] This movement occurred in four well-defined stages: the *Hicklin*

---

[38] Du Plessis (2016), p. 354.

[39] Du Plessis, citing Mill *On Liberty*, p. 356.

[40] Du Plessis, p. 354.

[41] Jochelson and Kramar (2011).

Era (1868–1962), named after the 1868 British case of *R. v. Hicklin*[42]; the Community Standards Era (1962–1992); the Community Standards of Tolerance for Harm Era (1992–2005); and the "Political Harm" Era (2005 onward).

During the *Hicklin* Era, the legal test for obscenity was "whether the tendency of the matter charged as obscenity is to deprave and corrupt those whose minds are open to such immoral influences and into whose hands the publication of this sort is likely to fall."[43] The *Hicklin* test can arguably be characterized by a commitment to the notion of a fixed human nature that is capable of corruption.

The Community Standards Era was ushered in by the case of *R. v. Brodie.*[44] According to Jochelson and Kramar, "[h]ere the Court began to appeal to positivist research in the human sciences as a technique for determining [a community's] tolerance for 'undue exploitation'."[45] Whereas in the *Hicklin* Era the role of government was arguably to act as the guardian of the proper ordering of society on the basis of substantive moral principles centred upon human nature, in the Community Standards Era, the Court can be seen as attempting to assess a community's views "as arbiter of sexual morality". With the arrival of the Community Standards Era (1962–1992), there seems to have been a movement away from objective morality, towards subjective attitudes and preferences, notwithstanding Jochelson and Kramar's contention that the Community Standards Era sought to use an "objective technique". That is because in the movement from the *Hicklin Era* to the Community Standards Era, the guardrails of objective morality seem to have given way to the Court's understanding of subjective community assessments about harm done by obscenity.

It is noteworthy that one of the justifications put forward by the Supreme Court in *Carter* for removing the criminal prohibition on euthanasia was that the "substantial consensus" of Western countries in favour of the prohibition had changed since the Supreme Court had upheld the democratically legislated prohibition on aiding suicide in the 1993 *Rodriguez* case.[46] It should also be noted that the principle of Parliamentary Supremacy in Canada justifies the position that the Canadian consensus should be formalised in law through the legislature, not the courts. Therefore, it is interesting to note that nine unelected Supreme Court justices relied on their assessment of societal consensus to render their decision, seemingly arrogating to themselves the position of a representative body. This, in contradistinction to the 338 Members of Parliament who hit the hustings every four years, if not more frequently, to win re-election, and who have their ear to their constituents' concerns year-round, and yet who had not chosen to amend social policy on the life and death issues of assisted suicide and euthanasia since 1993, when the Rodriguez case was decided.

---

[42] *R. v. Hicklin* (1868).

[43] *R. v. Hicklin* at p. 373.

[44] *R. v. Brodie* (1962).

[45] Jochelson and Kramar, *supra* note 41, p. 291.

[46] *Carter*, *supra* note 5 at para 47. See also para 24 for an argument by appeal to public opinion.

The Community Standards of Tolerance for Harm Era (1992–2005) was ushered in by the Supreme Court in the case of *R. v. Butler*.[47] This Era merely extended the preference for subjectivity inaugurated in the Community Standards Era. Here, the Court "refined" the community standards test of tolerance of undue exploitation of sex causing harm.[48] During this era, the Supreme Court in the 1993 case of *R. v. Tremblay*[49] allowed common bawdy houses in which women perform erotic dances in rooms equipped with peep-holes. It is true that the Supreme Court proscribed public sexual touching between dancers and patrons in the 1997 case of *R. v. Mara*,[50] but the Supreme Court allowed, less than a decade later, in *R. v. Labaye*, a swingers club in which members engaged in group sex. In these ways, the Court pushed the legal parameters even further away from moral principles centred on a comprehensive conception of human nature. This movement seems to further validate radical autonomy by loosing the individual from moral principles centred upon a defensible moral anthropology that had previously undergirded Canadian law. Finally, in *R. v. Labaye*,[51] the Supreme Court ushered in the Political Harm Era by reconfiguring the community standards of tolerance test so that the risk of harm, rather than mere harm, could be regulated.

Through the aforementioned four stages of jurisprudence, the Supreme Court of Canada seems to have replaced substantive, moral organizing principles that have a claim to truth, towards a liberal, tolerant, harms-based, and one might even say relativist, logic that is agnostic about morality based on comprehensive moral anthropology. The Court's decisions throughout these eras arguably show a removal of the guardrails around human action that had been preserved through the democratic process, and a move towards radical autonomy. One can see in this short history a movement at the Supreme Court of Canada away from a shared moral anthropology towards a view of man as radically autonomous and constrained only by the most threadbare limits of the Harm Principle.

b. Radical Autonomy as Antithesis to Collective Action

i. R. v. Big M Drug Man Ltd. and the Charter Right to Freedom of Religion

In 1906, Parliament passed the *Lord's Day Act*, which prohibited most business transactions on Sundays. In 1985, the Supreme Court struck down the Act in *R. v. Big M Drug Man Ltd*.[52] Chief Justice Dickson, writing for the majority wrote:

> A truly free society is one which can accommodate a wide variety of beliefs, diversity of tastes and pursuits, customs and codes of conduct... Freedom must surely be founded in respect for the inherent dignity and the inviolable rights of the human person.[53]

---

[47] *R. v. Butler* (1992).

[48] Jochelson and Kramar, *supra* note 41, p. 293.

[49] *R. v. Tremblay* (1993).

[50] *R. v. Mara* (1997).

[51] *R. v. Labaye* (2005).

[52] *R. v. Big M Drug Mart Ltd.* (1985).

[53] *ibid*, p. 33.

It can reasonably be asked what role a commitment to the truth, through the assertion of reasonably ascertainable human nature, social realities and fundamental values, plays in a 'free society'. In other words, what are the guardrails to a wide variety of beliefs, diversity of tastes and pursuits, customs, and codes of conduct? When does the exercise of freedom become undignified, both for the free individual and the society in which he lives? Given Dickson CJ's expansive prescription for the toleration of different "beliefs," "tastes and pursuits", "customs and codes of conduct", it is worth asking what, if any, objective view of human nature can possibly ground limits to human behaviour in society and pass *Charter* scrutiny. *Big M* goes some way towards answering these questions by asserting certain philosophical views of the individual in society.

Big M shows a conception of the individual that is arguably antagonistic towards the claims of community. Further, by not leaving it to the legislature within Canadian political culture to find the right balance between the claims of community and those of the individual, the Court arguably saw itself as the guardian of the claims of autonomy against community, thus ironically contributing to the atrophy of the individual's ability to exercise his autonomy within his political community.[54] Further, in *Big M*, the Court merely considered the purpose of the *Lord's Day Act*, not its effect. The purpose of the Act was to compel sabbatical observance, and this purpose was enough to strike it down, even if the "effects [of the Act] were found inoffensive". In other words, it didn't matter that the legislature did nothing that actually compelled sabbatical observance. So sensitive to offence was the Court's notion of autonomy that it was enough to offend autonomy for the Act to merely *mean* to limit it on the basis of a longstanding societal tradition. Thus, Dickson CJ wrote in relation to the purpose of the Act using the following terms:

c. "tyranny of the majority"
d. "sectarian Christian ideal"
e. "coercion"
f. "contrary to… dignity"
g. "hostile to non-Christians"
h. "appearing discriminatory against non-Christians"
i. "constant reminder to religious minorities… of alienation from the dominant religious culture".

Such views in relation to the purpose of the Act arguably reveal a view of the individual as hyper-sensitive to the claims of his or her political community, and as unable, without the guardianship of the court, to change the rules by which he is bound within a free and democratic society.

ii. Andrews v. Law Society of British Columbia and the Charter Right to Equality

[54] Alan Hutchinson and Patrick Monahan make a similar argument in "Democracy and Rule of Law", *supra* note 34.

In *Andrews*,[55] the respondent, a British subject permanently resident in Canada, met all the requirements for admission to the British Columbia bar except that of Canadian citizenship. Becoming a Canadian citizen would take three years. The Supreme Court of Canada held that the citizenship requirement was not justified in a free and democratic society.

According to Justin Bates, Justice Bertha Wilson

> clearly understood that section 15(1) operated to prevent 'otherness' and promote the 'right to belong'. As she made clear in her judgement in *Andrews*, groups lacking in political power (in this case, non-citizens who, as they lack the vote, make up a group "whose needs and wishes elected officials have no apparent interest in attending") are vulnerable to abuse and being excluded, either intentionally or by oversight.[56]

A common theme of radical autonomy seems to be the jettisoning of the claims of community in favour of the individual. One may ask, as McIntyre and Lamer JJ. did in their dissenting opinion in *Andrews*, whether it was unreasonable for the community of BC lawyers to insist that one must be adopted into Canadian society, through processes set out in law, in order to have the privilege of being an officer of the court in the administration of justice. Instead of prioritizing the importance of citizenship as a collective value, the Supreme Court seems to have prioritized the individual's desire to practise a profession, and to have freed him from the social and political requirements imposed by the collective in which he lives. Though *Andrews* was limited to the legal profession, the normative claim of the Court seems clear: the individual's desire trumps the democratic right to impose a citizenship requirement to obtain a privilege as important to the functioning of society as the practice of law.

c. *Radical Autonomy in Non-Charter Criminal Law Cases*

Radical autonomy advances a thin moral anthropology because it reserves judgment about the objects of one's freedom. On a thin moral anthropology, respect for autonomy runs the gamut from non-interference with another's choice, to the aiding and abetting of that choice, however discordant the choice is with fundamental values, or biological and social realities as revealed by human nature.

The Supreme Court has arguably endorsed a thin moral anthropology in *R. v. Mabior*[57] and *R. v. DLW*,[58] a case that followed on the heels of *Carter*. In *R. v. DLW*, the Court held that bestiality within the Criminal Code does not include sexual touching. Thus, the Court held that the Criminal Code allowed sexual acts with animals, such as sexual touching, as long as such acts were not penetrative. In this way, the Court allowed sexual acts that are not in harmony with the objective nature

[55] *Andrews v. Law Society of British Columbia* (1989).

[56] Bates, *supra* note 25.

[57] *R. v. Mabior* (2012).

[58] *R. v. D.L.W.* (2016).

and purpose of the sexual act. The Court jettisoned the trial judge's effort to interpret bestiality as including sexual touching, and instead took a restrictive view of bestiality which maximized sexual freedom. An interesting research project would be to correlate the Court's restrictive vs. expansive statutory interpretations with its decisions that either enhance or restrict radical autonomy, in order to determine whether the Court's chosen method of statutory interpretation is based on a preference for restrictive or expansive statutory interpretation as such, or a preference for radical autonomy.

R. v. Mabior is a non-*Charter* case that relied on *Charter* values. Relying on *R. v. Cuerrier*,[59] the Court allowed persons living with HIV to keep secret their HIV status from their sexual partners if there is no realistic possibility of transmission of HIV. But this effectively overrides people's ability to choose for themselves the scope and risk of their sexual activities.[60] Further, the Court assumed a model of sexual relations that is not in keeping with authentic human relationships, of which disclosure of HIV status with no risk of transmission is a constitutive part. Instead, the Court endorsed a view of human sexuality in which keeping such information secret is acceptable, however difficult it may be to disclose. On that account, the Court seems to have endorsed radical autonomy by prioritizing one person's freedom over another's well-being, here the ability to choose the scope and risk of his or her sexual activities; and by prioritizing freedom instead of authentic human relationships within a stable and objective understanding of human nature.

1.5 Conclusion: Democracy and Parliamentary Supremacy by Subsidiarity

This essay has endeavoured to articulate the problem of radical autonomy, to show its presence in Canadian Supreme Court jurisprudence, and to show its culmination in the *Carter* decision. It is hoped that this essay has demonstrated the demonstrable harm to vulnerable persons and the public good posed by euthanasia in Canada.

This essay has also discussed the imposition of radical autonomy across Canadian civil society by the Supreme Court of Canada. The solution to this imposition is arguably to follow the prescription laid down by the politicians who long ago understood that the possibility of the *polis* depends on a wise division of powers among communities. The Fathers of Confederation crafted the Canadian project in no small measure around the principle of subsidiarity, which permits individual communities to decide social and political questions within the Rule of Law. Thus, the federal order of government should be given a "subsidiary" function, performing only the functions that more local orders of government are unable to perform, and which the more local orders of government delegate to central authority. Power thus begins in communities, and is delegated outward, so that the balance of power remains within

---

[59] *R. v. Cuerrier* (1998).

[60] See Mathen (2012); Mathen and Plaxton (2011).

communities themselves, rather than the other way around. This avoids the alternative in which power begins at the centre and is imposed over communities, except to the extent that those at the centre decide to delegate power to communities. Power is rarely given up voluntarily, therefore a central authority is unlikely to give it up to smaller communities; whereas smaller communities cannot give up power they do not have, therefore that notional power is delegated to a central authority for practical reasons, such as the laying of transmission lines or the building of trade networks. Furthermore, when power is centralized, the harm done by power wrongly used is greater when a central authority wields it over many communities, than when a single community wrongly uses it against its own, for its members can more readily effect change within their own political community than over a central authority. And, in the last resort, members of a political community have the option to exit, but they cannot exit as easily from the territory covered by a central authority as they can from their own, smaller, political constituency.

Canadian law ought arguably to permit, more than it does now, for local laws that are based on a comprehensive moral anthropology firmly anchored in Western thought. Diversity of creed and culture require that communities be permitted to live out their own ethics and create boundaries that are animated by their respective moral anthropologies. Canada's success as a coherent conception and country is arguably enriched by the inheritance of cultures and creeds, for example, Christianity and the Western tradition, which are capable of advancing the common good in local constituencies.

Section 1 of the *Charter* allows rights to be limited on the basis of reasonable limits that are prescribed by law, and that can be demonstrably justified in a free and democratic society. Claims of community can and have been used as justifications for limiting individual rights under s. 1 of the *Charter* in order to advance the common good. Furthermore, s. 33 of the *Charter*, known as the 'notwithstanding clause', can be invoked by provincial legislatures to opt-out of the application of sections 2 and 7–15 of the *Charter*. This means that on questions of deep moral significance, it is understandable for communities to seek to invoke either s. 1 or s. 33 of the *Charter* in order to avoid the experience of being asked to give up too much of their collective ethics in the name of what they perceive as libertinism and moral degradation, including the unjustifiable taking of innocent human life. Thus, there should be a conversation about what types and sizes of communities in Canada should be allowed to use sections 1 and 33 to uphold their comprehensive conceptions of human nature and the public good. And, while it is true that criminal law in Canada is under exclusive federal jurisdiction, non-federal jurisdictions can and should have a say on their provision of certain public services, such as euthanasia, within their jurisdictions. Further, since modern municipalities are more populous than provinces were at Confederation, and since many modern municipalities, and even townships, are arguably more socially cohesive than provinces are today, and are arguably as socially cohesive as provinces were at Confederation, there is a strong arguable case that smaller communities, including townships and cities, should be recognized under the Constitution as polities that have the right to invoke sections 1 and 33 of the *Charter*.

In conclusion, sections 1 and 33 of the *Charter* should be seen as avenues for communities to build up laws and institutions in which they can enjoy the kind of freedom Hegel described, by truly seeing themselves reflected in their laws and institutions.

Acknowledgements The author wishes to thank Savannah Harlan for her diligent work assisting with formatting bibliographic citations.

References

Andrews v. Law Society of British Columbia. 1989. 1 SCR 143, [1989] SCJ No 6 (QL).

Bates, Justin. 2002. *Canada's hidden destiny: The charter and the right to belong.* Master's dissertation. University of Toronto.

Brink, David. Spring 2022 Edition. *The stanford encyclopedia of philosophy*, s.v. "Mill's moral and political philosophy". In ed. Edward N. Zalta. https://plato.stanford.edu/archives/spr2022/entries/mill-moral-political/.

Canadian Charter of Rights and Freedoms, Part I of the Constitution Act, 1982, being Schedule B to the Canada Act 1982 (UK). 1982. c 11, s 1.

Canadian Institute for Health Information. 2018. *Access to palliative care in Canada.* Ottawa, ON: CIHI. https://www.cihi.ca/sites/default/files/document/access-palliative-care-2018-en-web.pdf.

Carter v. Canada (Attorney General). 2015. SCC 5.

Cooper, Mick, Meghan Craig, and Joël Vos. 2015. A meta-analysis of their effects on psychological outcomes. *Journal of Consulting and Clinical Psychology* 83 (1): 115–128.

Crowell, Steven. Summer 2020 Edition. *The stanford encyclopedia of philosophy*, s.v. "Existentialism". In ed. Edward N. Zalta. https://plato.stanford.edu/archives/sum2020/entries/existentialism/.

Diat, Nicolas, and Robert Cardinal Sarah. 2019.*The day is now far spent.* San Francisco: Ignatius Press.

Downar, James. June 2019. *Maid: Palliative care team perspective. TOHCP grand rounds.* Ottawa: Ottawa Hospital General Campus.

Downar, James, Robert A. Fowler, Roxanne Halko, Larkin Davenport Huyer, Andrea D. Hill, and Jennifer L. Gibson. 2020. Early experience with medical assistance in dying in Ontario, Canada: A cohort study. *Canadian Medical Association Journal* 192 (8): 173–181.

Du Plessis, Georgia. 17 June 2016. The legitimacy of using the harm principle in cases of religious freedom within education. *Human Rights Review* 17: 349–370.

Fackenheim, Emil L. 1961. *Metaphysics and historicity.* Milwaukee, Wisconsin: Marquette University Press.

Gastmans, Chris, E. J. Jardas, Scott Y. H. Kim, Marie E. Nicolini, and, Carlos A. Zarate. 28 October 2022. Irremediability in psychiatric euthanasia: Examining the objective standard. *Psychology Medicine* 1–19.

Gert, Bernard, and Joshua Gert. Fall 2020 Edition. *The stanford encyclopedia of philosophy*, s.v. "The definition of morality,". In ed. Edward N. Zalta. https://plato.stanford.edu/archives/fall2020/entries/morality-definition/.

Hartling, Ole. 29 November 2021. We need to rethink the idea of 'dying with dignity. *Mercatornet.*

Hicks, Stephen. 2011. *Explaining postmodernism: Skepticism and socialism from Rousseau to Foucault, expanded edition.* Brisbane, Queensland: Connor Court Publishing.

Hutchinson, A.C., and Patrick Monahan. 1987. Democracy and rule of law. In *The rule of law: Ideal or ideology*, eds. A.C. Hutchinson and Patrick Monahan. Toronto: Carswell.

Jochelson, Richard, and Kirsten Kramar. 2011. Governing through precaution to protect equality and freedom: Obscenity and indecency law in Canada After *R. v. Labaye* [2005]. *Canadian Journal of Sociology* 36 (4): 283–312.

Klenk, Michael. 2019. Moral philosophy and the 'ethical turn' in anthropology. *Zeitschrift Für Ethik Und Moralphilosophie* 2 (2): 331–353. https://doi.org/10.1007/s42048-019-00040-9.

Korzinski, Dave. 2022. *A portrait of social isolation and loneliness in Canada today*. Angus Reid Institute. Last accessed 13/06/2022. https://angusreid.org/social-isolation-loneliness-canada/.

Laucius, Joanne. 12 February 2020. MAiD recipients more likely to be high-income, married, living at home, study says. *Ottawa Citizen*. https://ottawacitizen.com/health/maid-recipients-more-likely-to-be-high-income-married-living-at-home-study-says.

Mathen, Carissima. 2012. "R v. Mabior; R v. DC." *Slaw*. 57.https://www.slaw.ca/2012/10/05/r-v-mabior-r-v-dc/.

Mathen, Carissima and Platon, Michael. 2011. *Criminal Law Quarterly*. In ed. Carswell, A Thomson Reuters Business, 57(4). https://www.academia.edu/1119440/HIV_Consent_and_Criminal_Wrongs_Mathen_and_Plaxton_2011_Final_with_Copyright_Notice_1_.

McGuinty, Dylan, and Timothy Lau. 24 February 2020. Assisted dying study suggests we should put much more focus on palliative care. *Ottawa Citizen*. https://ottawacitizen.com/opinion/columnists/mcguinty-and-lau-assisted-dying-study-suggests-we-should-put-much-more-focus-on-palliative-care.

Mill, John Stuart. 1960. *On liberty; representative government; the subjection of women: Three essays*, Introduction by M.G. Fawcett, London, England: Oxford University Press.

Mullender, Richard. 08 September 2011. Hate speech, human rights, and G. W. F. Hegel. In *Handbook of human rights*, ed. Thomas Cushman. Abingdon: Routledge.

Nietzsche, Frederich. 1973. *Beyond good and evil: Prelude to a philosophy of the future*, translated by R. J. Hollingdale, Harmondsworth: Penguin Books.

Nietzsche, Frederich. 2022. *Thus spoke Zarathustra: A book for all and none*, translated by Michael Hulse, New York: New York Review Books.

Putnam, Robert D. 2000. *Bowling alone: The collapse and revival of American community*. New York: Simon and Schuster.

R. v. Big M Drug Mart Ltd. 1985. 1 SCR 295, [1985] SCJ No 17 (QL).

R. v. Brodie. 1962. SCR 681, 32 DLR (2d) 507.

R. v. Butler. 1992. 1 SCR 452, [1992] SCJ No 15 (QL).

R. v. Cuerrier. 1998. 2 SCR 371, [1998] SCJ No 64 (QL).

R. v. D.L.W. 2016. SCC 22, [2016] 1 SCR 402.

R. v. Hicklin. 1868. LR. 3 Q.B. 360.

R. v. Labaye. 2005. SCC 80.

R. v. Mabior. 2012. SCC 47, [2012] 2 SCR 584.

R. v. Mara. 1997. 2 SCR 630, [1997] SCJ No 29 (QL).

R. v. Tremblay. 1993. 2 SCR 932, 23 CR (4th) 98.

Raikin, Alexander. 16 December 2022. No other option. *The New Atlantis*. https://www.thenewatlantis.com/publications/no-other-options.

Rawls, John. 1993. *Political liberalism*. New York: Columbia University Press.

Riesman, David, Nathan Glazer, and Reuel Denney. 2001. *The lonely crowd: A study of the changing American character*. New Haven, CT: Yale University Press.

Rodriguez v British Columbia (Attorney General). 1993. 3 SCR 519, [1993] SCJ No. 94 (QL).

Scruton, Roger. 1980. *The meaning of conservatism*, 1st ed. Markham, Ontario: Penguin Books Canada.

Siep, Ludwig. 1992. *Praktische hilosophie im Deutschen Idealismus*, Frankfurt: Suhrkamp.

Stewart, Alan Glen. 1999. *Moral legislation and the conservative constitutional tradition*. Ph.D. dissertation. University of Toronto.

Taylor, Charles. 1995. The politics of recognition. In*Philosophical arguments*. Cambridge, Massachusetts: Harvard University Press.

Dylan McGuinty Jr, BA, MA, JD is a lawyer who serves families in the areas of consent, capacity, substitute decision-making, estate planning and estate administration. Dylan regularly acts as counsel for elderly and vulnerable people living at home, in hospital settings and in palliative, hospice and long-term care.

Dylan earned his Bachelor of Arts degree with Majors in Philosophy and Public Administration from the University of Ottawa in 2009, studying both ethics and the implementation of public policy. He earned his Master's in Philosophy from the University of Ottawa on an Admission Scholarship in 2011 with a focus on ethics, before graduating from Osgoode Hall Law School with a Juris Doctor in Law in 2014, with particular interests in Administrative and Constitutional Law.

Dylan has represented clients at the Supreme Court of Canada in the area of fiduciary obligations and trust law, and has represented clients at the Human Rights Tribunal of Ontario, the Landlord Tenant Board and the Ontario Labour Relations Board. He currently serves on the Board of Directors of Bruyere Continuing Care where he seeks to advance the care of patients in need of sub-acute, geriatric, and palliative care. Dylan has taught Administrative and Constitutional Law at Carleton University.

Chapter 2
Carter v Canada: Exploring the Ebb and Flow of "Competing" Societal Values Through Sections 7 and 1 of the *Canadian Charter of Human Rights and Freedoms*

Mary J. Sharif

Abstract This chapter explores *Carter v Canada*, the case that led to the decriminalization of termination of life practice in Canada, otherwise known as "Medical Assistance in Dying" or "MAID". The preliminary objective of this chapter is to assist those trying to better understand what the Supreme Court of Canada actually decided by providing the reader with a concise summary of the Court's Sections 7 and 1 analysis pursuant to the *Canadian Charter of Rights and Freedoms*. The secondary objective however is to scrutinize within this summary, how the analysis identifies and balances "competing" rights and values as between persons *seeking* physician-assisted death (the Plaintiff in *Carter*) and the protection of vulnerable persons from being *induced* to commit suicide in times of weakness (the target of the *Criminal Code* provision under scrutiny). The main purpose of this chapter is not to provide a general critique of the Sections 7 and 1 analyses *per se*. Rather the purpose is to demonstrate how the overall *Charter* analysis ultimately operates to sideline "competing" societal values and obscure the critical concepts of *seeking* versus *inducing* MAID. As more information regarding the lived experiences of "vulnerable" Canadians begins to come to light, these concepts are of utmost significance to ongoing discussions around MAID reform.

Keywords Ableism · Assisted suicide · Assisted death · *Bill C-14* · *Bill C-7* · *Carter v. Canada* · *Canadian Charter of Rights and Freedoms* · *Convention on the Rights of Persons with Disabilities* · *Criminal Code* · Death · Disability · Disability · End of life · Euthanasia · Human rights · MAID · MAiD · MAIL · Medical assistance in dying · Medical assistance in living · *Michaud* · *Oakes* · Physician-assisted suicide · Physician-assisted dying · Right to die · Right to life · Right to liberty · Right to security · *Rodriguez* · Section 7 · Section 1 · Slippery slope · Suicide · Termination of life · *Truchon* · Vulnerable persons

---

M. J. Shariff (✉)
Faculty of Law, University of Manitoba, Winnipeg, Manitoba, Canada
e-mail: mary.shariff@umanitoba.ca

Centre on Aging, University of Manitoba, Winnipeg, Manitoba, Canada

J. Kotalik and D. W. Shannon (eds.), *Medical Assistance in Dying (MAID) in Canada*, The International Library of Bioethics 104,
https://doi.org/10.1007/978-3-031-30002-8_2

2.1 Introduction

In February 2015, the Supreme Court of Canada (SCC or the "Court") issued its historic decision, *Carter v Canada* (*Carter*).[1] In *Carter*, the SCC held that the *Criminal Code's* absolute prohibition[2] against "physician-assisted dying"[3] was an infringement of the rights to life, liberty and security pursuant to Section 7 of the *Canadian Charter of Rights and Freedoms* (*Charter*)[4] and furthermore could not be saved by Section 1.[5] The SCC suspended its "declaration of invalidity" of the infringing provisions for 12 months allowing the Canadian Parliament (and provincial legislatures) 12 months to respond with legislative changes.[6] Following the Court's granting of an additional 4 month extension,[7] Parliament passed *Bill C-14* in June 2016.[8]

Bill C-14 amended the *Criminal Code* to create an exemption to allow physicians and nurse practitioners to provide "medical assistance in dying" (MAID)[9] to eligible persons, namely, capable adults who have made a voluntary request for MAID, are suffering from a "grievous and irremediable medical condition",[10] and who have given informed consent to receive MAID.

---

[1] *Carter v. Canada (Attorney General)* (2015) [hereinafter "*Carter SCC*"].

[2] *Criminal Code of Canada* R.S.C. (1985), c. C-46 [hereinafter "*Criminal Code*"] at Sections 241(b) and 14.

[3] The terms "physician-assisted dying" and "physician-assisted death" and are used interchangeably throughout the *Carter* decision and were defined by the trial court to include both the acts of: "physician-assisted suicide" or PAS (whereby the patient intentionally kills themselves with assistance of a physician or someone acting under a physician's direction); and "voluntary euthanasia" (whereby the patient's life is intentionally terminated by a physician or someone acting under a physician's direction). See *Carter v Canada (Attorney General)*, 2012 BCSC 886 [hereinafter "*Carter BCSC*"] at paras 23 and 37–39; In Canada, these terms have been superseded by the term "medical assistance in dying" or "MAID" when the *Criminal Code of Canada* was amended in 2016 to allow these practices by physicians and nurse practitioners with the passing of *Bill C-14, Act to amend the Criminal Code and to make related amendments to other Acts (medical assistance in dying)*, S.C. 2016, c. 3 [hereinafter "Bill C-14"]. Bill C-14 received Royal Assent June 17, 2016. Online: https://www.parl.ca/DocumentViewer/en/42-1/bill/C-14/royal-assent.

[4] *Canadian Charter of Rights and Freedoms*, Part I of the *Constitution* Act, 1982, being Schedule B to the *Canada* Act, 1982 (U.K.), 1982, c. 11 [hereinafter "*Charter*"].

[5] *Carter SCC*, Footnote 1 at para 123.

[6] *Carter SCC*, Footnote 1 at paras 126–128.

[7] *Carter v. Canada (Attorney General)*, 2016 SCC 4, [2016] 1 SCR 13 [hereinafter "*Carter SCC 2016*"]. During this extension period, the SCC granted an exemption "for those who wish to seek assistance in ending their life" to apply to the superior court of their jurisdiction to demonstrate eligibility on basis of the "criteria" set out in "para 127" of *Carter*. *Carter SCC 2016*, Footnote 7 at para 7. Approximately 15 cases were heard by provincial courts during this period.

[8] Bill C-14, Footnote 3. For Parliamentary debates leading to Bill C-14 see Parliament of Canada (2015).

[9] *Criminal Code*, Footnote 2 at Section 241.1.

[10] Pursuant to the Bill C-14 amendments to the *Criminal Code*, a person would have a "grievous and irremediable medical condition" if they met four (4) specific criteria which included the criterion that their "natural death has become reasonably foreseeable". Following subsequent legislative

Controversy over the amendments quickly erupted with some arguing that *Bill C-14* was too restrictive and not in compliance with the SCC's decision particularly with respect to the "grievous and irremediable medical condition" requirement that natural death be "reasonably foreseeable".[11] When a subsequent 2019 court judgment from the Province of Quebec found the "reasonably foreseeable" natural death criterion to be unconstitutional,[12] Parliament again amended the *Criminal Code* in 2021 through Bill C-7[13] to create two "tracks" to MAID: track one for suffering persons whose natural death is reasonably foreseeable and track two for suffering persons whose natural death is *not* reasonably foreseeable (with different procedural safeguards attaching to each track).[14]

Rather than being stable law, MAID reform discussions in Canada have been virtually non-stop and most recently have involved debate over expanding eligibility to requests by persons with mental illness as the sole underlying condition, mature minors, as well as pursuant to advance directives.[15] These potential expansions of MAID law are presently situated as questions for Parliament and its attendant processes. The debate however has also been taking place in other public spaces where MAID expansion is discussed with recourse to what at times seems like a tortuous war of words and logic related to what the Supreme Court of Canada did or did not establish in the 2015 *Carter* case.[16]

Given that MAID has been decriminalized and established as a medicalized service in Canada,[17] the unending *Carter* interpretation scrum stands to detract from

---

amendment (discussed further below at Footnote 11 and accompanying text), the "reasonably foreseeable" natural death eligibility criterion was removed, leaving the remaining three, namely that: (a) they have a serious and incurable illness, disease or disability; (b) they are in an advanced state of irreversible decline in capability; and (c) that illness, disease or disability or that state of decline causes them enduring physical or psychological suffering that is intolerable to them and that cannot be relieved under conditions that they consider acceptable. See *Criminal Code*, Footnote 2 at Sections 241.2(1)(c) and 241.2(2).

[11] For brief discussion see Tiedemann (2019) at pp. 4–5; For an example of media commentary see Coyne (2020).

[12] *Truchon v Canada (AG)*, 2019 QCCS 3792 [hereinafter "*Truchon*"]. The Attorney General of Canada decision not to appeal this trial court decision was and remains controversial. See for example, commentary by Sikkema (2019); For a technical review of the government's response to *Truchon*, see Government of Canada, *Legislative Background: Bill C-7: Government of Canada's Legislative Response to the Superior Court of Québec Truchon Decision* (2021). Online: https://justice.gc.ca/eng/csj-sjc/pl/ad-am/c7/c7-eng.pdf.

[13] Bill C-7, *An Act to amend the Criminal Code (medical assistance in dying)*, S.C. 2021, c. 2 [hereinafter "Bill C-7"] which received Royal Assent March 17, 2021. Online: https://www.parl.ca/DocumentViewer/en/43-2/bill/C-7/royal-assent.

[14] See *Criminal Code*, Footnote 2 at Sections 241.2(3) and 241.2(3.1).

[15] Independent review of these areas was mandated by Bill C-14. Bill C-14 Footnote 3 at Sections 9.1(1); See also Bill C-7, Footnote 13 at Sections 3.1(1) and 5(1).

[16] For an example of argumentation with recourse to the Supreme Court of Canada's direction or non-direction in *Carter* SCC, see Downie (2022); Compare to response: Lemmens (2022).

[17] In 2017 it was estimated MAID "could reduce annual health care spending across Canada by between \$34.7 million and \$138.8 million". Traachtenberg and Manns (2017) [hereinafter "Traachtenberg"]; With Bill C-7, the predicted net cost reduction for provincial health care

what arguably should be the government's central focus in current debate and discussion[18]: how to appropriately address the rights and interests of Canadians *not seeking* MAID,[19] particularly those who require support and whose personal attributes match MAID eligibility criteria such that they are at greater risk of vulnerability within a MAID-permissive society.[20] For the author, this would, at minimum, require attention to the identification and development of prudent systems and safeguards to shore up and guard against *non-seeking* persons from being *induced* to end their lives in times of "weakness". Indeed, despite being at the centre of the constitutional balancing act undertaken by the Court in *Carter*, the specific concepts of the MAID *seeking* person versus the *inducement* of vulnerable persons have not been given sufficient scrutiny to date. In addition to the complexity of the *Charter* analysis, this apparent oversight has contributed to the myriad of conflicting accounts of what the SCC said or meant in *Carter* and subverts critical considerations that could assist in protecting and advancing the rights of all Canadian citizens who find themselves at the interface of federal criminal law and provincial health care.[21]

The preliminary objective of this chapter is therefore to assist those trying to better understand what the Supreme Court of Canada actually decided in *Carter* by providing the reader with a concise summary of the SCC's Sections 7 and 1 *Charter* analysis. The secondary objective, however, is to scrutinize within this summary, how the analysis "balances" competing rights and values as between individual persons *seeking* death (the Plaintiff in *Carter*) and the prevention or protection of others from being *induced* to commit suicide in times of weakness (the target of the challenged *Criminal Code* provision[22]).

The purpose here is not to provide a general critique of the *Charter* analysis but rather to demonstrate how the analysis operates within the *Carter* case to sideline "competing" societal interests and values and ultimately obscure the most critical concepts of *seeking*, and *inducing*.

Part 2 below presents a brief description of the *Carter* claim and the SCC's opening statement of the competing values at issue which establishes the overall context for

spending is $86.9 million. Office of the Parliamentary Budget Officer(2020) [hereinafter "OPBO-BDPB"]. Online: https://www.pbo-dpb.gc.ca/web/default/files/Documents/Reports/RP-2021-025-M/RP-2021-025-M_en.pdf.

[18] See *Carter SCC*, Footnote 1 at para 95.

[19] See *Carter SCC*, Footnote 1 at para 2.

[20] See for example, discussion in United Nations (2021); See also recent discussions in Canada where persons have described that it is not their medical condition (which makes them eligible for MAID) but rather their experiences of poverty, homelessness or lack of supports to live and engage in society that is motivating them to request or proceed with MAID. For example see: Alberga (2022a); See also Favaro (2022); See also Leffler and Dimain (2022); See also LeGal (2022); See also Alberga (2022b); See also Mulligan and Bond (2022); See also Phillips ((2022); See also Kruger (2022); See also Farquhar (2022).

Canadian Association of MAID Assessors and Providers (2022).

[21] For example, what should or should not be said to a patient whose medical condition would appear to make them eligible for MAID. See discussion in Ho et al. (2021); Compare to guidance document by the Canadian Association of MAiD Assessors and Providers (2022) [hereinafter "CAMAP"].

[22] See *Carter SCC*, Footnote 1 at paras 74–78.

the Court's Sections 7 and 1 analysis. Section 7 is specifically considered in Part 3 while Section 1 is examined in Part 4.

The chapter concludes in Part 5 with a brief comment on the ongoing role of the *Carter* case in Parliamentary MAID reform discussions along with some recommendations in Part 6. In addition to recommending that the criminal law (which permits this termination of life practice) ought to expressly distinguish between persons *seeking* MAID versus those who are not, the chapter closes with the suggestion that perhaps it is time to seriously reflect upon whether judicial interpretation and application of the *Charter* (as it is currently interpreted and applied) is appropriate for resolving questions involving fundamental and highly complex social issues such as MAID where "competing" individual and societal interests may themselves be protected under the *Charter*. If this latter recommendation is taken by the reader to be unduly bold or unduly naïve, then it is at least hoped that this chapter has provided a decent foundation to assist the reader in determining for themselves, "what did *Carter* say?".

2.2 The "Competing Values"

2.2.1 Brief Overview of the Claim

The main plaintiff in the *Carter* case was Gloria Taylor,[23] a 64 year old woman suffering from a "fatal, neurodegenerative disease, amyotrophic lateral sclerosis (or ALS)".[24] Ms. Taylor "did not want to die slowly, piece by piece" or "wracked with pain"[25] and brought a claim before the British Columbia court challenging the constitutionality of several *Criminal Code* provisions[26] that prohibited "physician-assisted dying".[27]

Although Ms. Taylor (the "Claimant") challenged the prohibition as being an unjustified infringement of both her Section 15 (equality) and Section 7 (life, liberty and security) *Charter* rights,[28] the SCC found it unnecessary to consider

[23] In addition to Ms. Taylor, the claim was filed by the British Columbia Civil Liberties Association, a physician willing to participate in PAS or euthanasia, and the daughter and son-in-law of Kay Carter who assisted Ms. Carter in receiving assisted death in Switzerland. While the trial court found that the prohibition of physician-assisted dying was an unjustified infringement of Ms. Taylor's Sections 15 right and the Section 7 rights of both Ms. Taylor and those who would assist, the Supreme Court of Canada (SCC) confined its decision "to the rights of those who seek assistance in dying". *Carter SCC*, Footnote 1 at para 69.

[24] *Carter SCC*, Footnote 1 at para 11.

[25] *Carter SCC*, Footnote 1 at paras 11–12.

[26] *Criminal Code* Footnote 2 at sections 14, 21, 22, 222 and 241. See *Carter SCC*, Footnote 1 at para 11.

[27] *Carter SCC*, Footnote 1 at para 11.

[28] Note: Both the Sections 15 and 7 arguments succeeded at trial. *Carter BCSC*, Footnote 3 at paras 16–19. The trial decision however was successfully appealed to the BC Court of Appeal on the

the Section 15 argument once it concluded that the absolute prohibition violated Section 7.[29]

2.2.2 *The "Competing Values"*

The SCC opens its judgment with a clear statement of the "competing values" at issue:

> It is a crime in Canada to assist another person in ending her own life. As a result, people who are grievously and irremediably ill **cannot seek** a physician's assistance in dying and may be **condemned** to a life of severe and intolerable suffering. **A person facing this prospect has two options:** she can take her own life prematurely, often by violent or dangerous means, or she can suffer until she dies from natural causes. **The choice is cruel.**
>
> The question on this appeal is whether the criminal prohibition **that puts a person to this choice** violates her *Charter* rights to life, liberty and security of the person (s. 7) and to equal treatment by and under the law (s. 15). This is a question that asks us to balance **competing values** of great importance. On the one hand stands the **autonomy and dignity of a competent adult who seeks death** as a response to a grievous and irremediable medical condition. On the other stands the **sanctity of life and the need to protect the vulnerable.**[30]

These first two paragraphs establish the overall context for the SCC's decision which makes it clear that the violation of rights to be assessed relates to competent adults who are *seeking* death. In addition to the Court's unequivocal language in this regard, this interpretation is supported by the Court's description of the "cruel choice": by prohibiting that which the Claimant is specifically *seeking* (i.e. physician-assisted death) the Claimant is left with two options, of which the choice between them is described as cruel: suffer until natural death or end her life prematurely and potentially dangerously or violently. In other words, the "cruel choice" as constructed by the SCC manifests when the suffering person is *seeking* and then, through operation of the criminal law, is *prevented from* obtaining death with assistance.

It should also be noted that the Claimant herself had described the "cruel choice" slightly differently: a choice between "killing herself while still physically capable of doing so" (i.e. a premature and potentially dangerous or violent death) or "giving up the ability to exercise any control over the manner and timing of her death" (i.e. impacting on suffering, quality of life and the exercise of autonomy).[31] So although there may be other ways to address suffering arising from or associated with a medical condition (evidence of which was extensively explored at trial), the specific question

ground that the trial judge was bound to follow *Rodriguez v. British Columbia (Attorney General)*, [1993] 3 SCR 519 [hereinafter "*Rodriguez*"], an earlier Supreme Court of Canada decision which upheld the blanket prohibition on assisted suicide. *Carter v. Canada (Attorney General)* (2013) [hereinafter "*Carter BCCA*"]. The case was then appealed to the Supreme Court of Canada. *Carter SCC*, Footnote 1.

[29] *Carter SCC*, Footnote 1 at para 93.

[30] *Carter SCC*, Footnote 1 at paras 1 and 2. Emphasis added.

[31] *Carter SCC*, Footnote 1 at para 13. Emphasis added.

provoked is whether the denial of an assisted death to those individuals *seeking* it violates rights protected by the *Charter*.

Accordingly, in order to determine whether there was an unjustifiable violation of the Claimant's rights (and of other persons in her position), the SCC set out to balance what it described as the competing "values": "the autonomy and dignity" of a competent adult *seeking* physician-assisted death (PAD) versus "the sanctity of life and the need to protect the vulnerable".[32]

2.2.3 *The Relevant Statutory Provision: Criminal Code Section 241(b)*

Consistent with the trial decision, the SCC identified Sections 241(b) and 14 of the *Criminal Code* as being most relevant to the claim:

> In our view, two of these provisions are at the core of the constitutional challenge: s. 241(*b*), which says that everyone who aids or abets a person in committing suicide commits an indictable offence, and s. 14, which says that no person may consent to death being inflicted on them. It is these two provisions that prohibit the provision of assistance in dying.[33]

Note that the Court scoping the claim to particular *Criminal Code* provision(s) is a most critical step because it is the *object* or *target* of the identified provision(s) upon which the balancing of competing values and interests turns.

2.3 Section 7

2.3.1 *The Analytical Framework*

Section 7 of the *Charter* states that:

> Everyone has the right to life, liberty and security of the person and the right not to be deprived thereof except in accordance with the principles of fundamental justice.

In order to demonstrate a violation of Section 7, the courts have established that claimants:

> must first show that the law interferes with, or deprives them of, their life, liberty or security of the person. Once they have established that s. 7 is engaged, they must then show that the deprivation in question is not in accordance with the principles of fundamental justice.[34]

---

[32] *Carter SCC*, Footnote 1 at para 2.

[33] *Carter SCC*, Footnote 1 at para 20. Emphasis added.

[34] *Carter SCC*, Footnote 1 at para 55.

Thus under Section 7, the onus is on the claimant to demonstrate the interference or deprivation. Furthermore, the analysis is to be conducted entirely from the perspective of the claimant. Indeed the focus of Section 7 has been described as "relentlessly on the individual claimant".[35] Consequently, any balancing between the individual infringement and competing societal interests and moral claims (including potential public benefits conferred by the impugned law)[36] are to be left for later consideration under Section 1.[37] Having said that, to date, the Supreme Court of Canada has never found a limit of Section 7 rights to be justified under Section 1.[38]

2.3.2 Section 7 Analysis

2.3.2.1 Infringement on Life, Liberty and Security

The SCC held that the prohibition on physician-assisted death infringed upon the right to life, liberty and security of the Claimant "and of persons in her position".[39] Specifically, the prohibition:

- **deprived individuals of life**[40] by having the effect of "forcing" some individuals to take their own lives earlier than they would otherwise if they had access to a physician-assisted death,[41] i.e. the right to life was engaged because of the "threat" of premature death.[42]
- **impinged upon individuals' right to liberty**[43]**:** denying the "right to request" a "physician's assistance in dying" interferes with fundamentally important and

[35] *R v. Michaud* (2015)127 O.R. (3d) 81, 2015 ONCA 585 at para 79 [hereinafter "*Michaud*"]; See also *Canada (Attorney General) v. Bedford* 2013 SCC 72, [2013] 3 SCR 1101 at paras 123–127 [*Bedford*]; For further discussion see *Michaud*, Footnote 35 at paras 146–154.

[36] *Carter SCC*, Footnote 1 at para 79.

[37] The separation of the individual deprivation from societal interests is repeated in various different ways throughout the *Carter SCC* judgment. See for example, *Carter SCC*, Footnote 1 at paras 85, 89; See also *Bedford*, Footnote 35 at para 123.

[38] For review of case law and academic commentary regarding the limited role, insignificance and/or redundancy of the *Charter* Section 1 analysis once the court has found a Section 7 violation see Haak (2022) [hereinafter "Haak"]; See also Carter (2015) at 210. See also generally Carter (2017) [hereinafter "Carter, Different Questions"]. Compare to analysis in *Michaud*, Footnote 35, at paras 62, 83, 144–154.

[39] *Carter SCC*, Footnote 1 at para 56.

[40] *Carter SCC*, Footnote 1 at paras 57–58.

[41] Note that the effect of the prohibition on premature death was not challenged before the court. *Carter SCC*, Footnote 1 at para 58.

[42] *Carter SCC*, Footnote 1 at paras 57; See also *Carter SCC*, Footnote 1 at para 63 where Court discusses interpretation of the "right to life".

[43] "Liberty protects "the right to make fundamental personal choices free from state interference' ". *Carter SCC*, Footnote 1 at para 64.

personal medical decision-making,[44] i.e. the "ability to make decisions concerning bodily integrity and medical care".[45]

- **impinged upon individuals' right to security[46]:** the unavailability of physician-assisted dying leaves "people like" the Claimant "to endure intolerable suffering".[47]

With respect to the rights to liberty and security, the SCC further stated that:

> ss. 241(*b*) and 14 of the *Criminal Code*, insofar as they prohibit physician-assisted dying for competent adults who **seek** such assistance as a result of a grievous and irremediable medical condition that causes enduring and intolerable suffering, infringe the rights to liberty and security of the person.[48]

Observations/Comment

Note that in addition to the Court's express use of the word "seek",[49] the element of *seeking* is indeed critical to establishing the infringement on the rights to liberty and security which are constructed on choice and the denial of choice. The *seeking* element is also critical to the establishment of the infringement of the right to life, in that for the prohibition to have the effect of "forcing" the Claimant towards premature death (and thus poses a "threat" to life), the Claimant must actually be seeking death. Accordingly, per *Carter*, there is no infringement of Section 7 rights with respect to persons who are *not* seeking death.

2.3.2.2 Infringement not in Accordance with the Principles of Fundamental Justice

The next stage in the Section 7 analysis is to determine whether the infringement or deprivation is or is not in accordance with the "principles of fundamental justice" (PFJ). If not in accordance with the PFJ, there will be a violation of Section 7.

Over time, and contrary to the expectations of many key drafters of the *Charter*,[50] the judiciary has interpreted the PFJ assessment as being broader than determining

---

[44] *Carter SCC*, Footnote 1 at para 64.

[45] *Carter SCC*, Footnote 1 at para 66.

[46] "Security of the person encompasses "a notion of personal autonomy involving ... control over one's bodily integrity free from state interference"... ". *Carter SCC*, Footnote 1 at para 64.

[47] *Carter SCC*, Footnote 1 at para 66.

[48] *Carter SCC*, Footnote 1 at para 68. Emphasis added.

[49] In addition to aforementioned paragraphs 1, 2 and 68 see *Carter SCC*, Footnote 1 at paras 5, 12, 30, 42, 69 and 98. See also description of plaintiffs at Footnote 23 above.

[50] See presentation by Chief Justice Glenn Joyal, "The Charter and Canada's New Political Culture: Are We All Ambassadors Now?", (January 6, 2017) Canadian Constitution Foundation, 2017 Law and Freedom Conference. Online: https://theccf.ca/2207-2/ [hereinafter "Chief Justice Joyal"].

procedural fairness (akin to the American concept of "due process").[51] This is particularly interesting because *if* the PFJ assessment was restricted to simply ruling out improper process, this would no doubt operate to afford significant deference to Parliament and its policy choices.[52] Regardless of this broader interpretation, the court has maintained that the principles of fundamental justice do not arise out of public policy considerations but rather are to be found in the "basic tenets of our legal system" and lie "in the inherent domain of the judiciary as guardian of the justice system".[53]

Thus, at the PFJ stage of analysis, the "object" of the law is to be compared against what the court has identified to date as "vices" that run *contrary* to the principles of fundamental justice, namely[54]: *arbitrariness*; *overbreadth*; and *gross disproportionality*.[55] In *Carter*, the SCC further explains that the PFJ are derived from essential elements of the system of justice and founded on "belief in the dignity and worth of every human person" and that the above-mentioned vices "diminish" that worth and dignity.[56] Practically speaking, within the Section 7 analysis however, the focus is not on the dignity and worth of *all* persons, but rather the dignity and worth of the individual claimant. Citing *Rodriguez*,[57] the SCC in *Carter* describes that a law asks a claimant to serve as a "scapegoat" if that law "imposes a deprivation via a process that is fundamentally unfair to the rights claimant".[58]

To summarize, Section 7 has come to be concerned with the individual claimant and the impact or effect of the provision at issue on the claimant (and people like the claimant) divorced from other societal concerns, values or benefits that contextualize the particular law and reflect (or impart) its narrower and broader socio-political targets. Taking the Section 7 methodology at face value, and knowing the Court's analytical disposition towards the impugned law at this stage (as reflected in the word "scapegoat"), it would seem incumbent upon a court to identify the specific object of the law as clearly and accurately as possible and apply that object consistently throughout its analysis (recognizing that the object of the law is relevant to applying the PFJs within the Section 7 analysis, as well as to the government's prospective justification for limits on the claimant's right within the subsequent Section 1 analysis). Any variation or deviation by the court from the identified object of the law as the court's *Charter* analysis proceeds should be made as transparent as possible.

---

[51] For discussion see *Carter BCCA*, Footnote 28 at paras 283–290; See also discussion in *Bedford*, Footnote 35 at paras 93–123; See also *Re B.C. Motor Vehicle Act* (1985) [1985] 2 SCR 486 [hereinafter "*Re B.C. Motor Vehicle Act*]; See also Chief Justice Joyal, Footnote 50.

[52] For further discussion on history and critique, see Chief Justice Joyal, Footnote 50; See also discussion in Dufraimont (2014).

[53] *Re B.C. Motor Vehicle Act*, Footnote 51 at page 503; See also discussion in *Carter BCCA*, Footnote 28 at para 283.

[54] *Carter SCC*, Footnote 1 at para 73. For fuller description and discussion of the principles of fundamental justice see *Michaud*, Footnote 35 at paras 66 et seq.

[55] *Carter SCC*, Footnote 1 at para 72.

[56] *Carter SCC*, Footnote 1 at para 81.

[57] See *Rodriguez*, Footnote 28.

[58] *Carter SCC*, Footnote 1 at para 81. Emphasis added.

Step One: The Object of the Provision

The SCC identified the objective or target of Section 241(b) as follows:

> The direct target of the measure is the narrow goal of preventing vulnerable persons from being induced to commit suicide at a time of weakness.[59]

Although the Respondent Attorney General of Canada (Canada) agreed that the prohibition was intended to protect vulnerable persons, Canada also argued that the object of the prohibition should be defined more broadly and simply as "the preservation of life".[60] The SCC rejected this proposition describing that the object must be defined "precisely", further describing that[61]:

- the "preservation of life" should be understood as an "animating social value";
- if the object of a prohibition is defined too broadly it will "short-circuit" the analysis and effectively "immunize" the law from *Charter* challenge; and
- the "object of the prohibition should be confined to measures directly targeted by the law".

Observations/Comment

In addition to the preliminary observation that the Court's stated object or target of the prohibition includes the word, concept and mischief of *induced*, three additional comments might be made. First, recourse to the notion of the "belief in the dignity and worth of every human person" (which arguably could also be described as an "animating social value") prompts a sense of intensity with respect to the individual infringement and would support more expansive interpretations of Section 7 rights based on the values of autonomy and self-determination.[62] Second, in the rejection of "the preservation of life" as an object of the prohibition, it can be again observed how the sentiment of "the dignity and worth of every human person" applies more to one side of the scale (the individual claim) than to the other (the object or target of the provision), at least at this stage of analysis. Third, notwithstanding that the purpose at this step is to identify the specific "target" of the prohibition, it remains difficult not to perceive this judicial exercise as touching on public policy considerations in that this step simply affords the court the discretion to demonstrate its

[59] *Carter SCC*, Footnote 1 at para 78. Emphasis added; See also *Carter SCC*, Footnote 1 at paras 29, 74, 76, 78.

[60] *Carter SCC*, Footnote 1 at para 75.

[61] *Carter SCC*, Footnote 1 at paras 76–78. Note that Sopinka J in *Rodriguez*, Footnote 28, described the "purpose" of protecting vulnerable persons who might be induced in moments of weakness to commit suicide as being "grounded in the state interest in protecting life and reflects the policy of the state that human life should not be depreciated by allowing life to be taken". See *Carter* SCC, Footnote 1 at para 76.

[62] For example see, the Court's interpretation of the prohibition's infringement on the Section 7 right to life. *Carter SCC*, Footnote 1 at paras 57–63; See also analysis in *Michaud* drawing on *Law Society of British Columbia v. Andrews* (1989), [1989] S.C.J. No. 6: "The broader the reach given to s. 7, the more likely it is that it will be deprived of any real content". *Michaud*, Footnote 35 at para 43.

preference for a particular objective; in this case, preference for a narrower legislative objective than the broader one.[63] The possibility that this narrowing step comprises a judicial public policy exercise is potentially supported by the *Charter* analysis later dropping what the Court has identified as the specific "target" of the prohibition (i.e. "preventing vulnerable persons from being *induced* to commit suicide) and restating or reasserting the target as a broader and more nebulous societal goal of "protection" of the vulnerable. These contracting and expanding judicial machinations with respect to naming. Identifying and applying "objectives", "objects", "targets", "measures" and "social values", along with the impacts on the unfolding legal analysis are discussed further below.

Step Two: Application of the Principles of Fundamental Justice

Arbitrariness

The concept of *arbitrariness* is that there is no "rational connection" between the *object* of the law and the *limit* it imposes on life, liberty and security. "An arbitrary law … exacts a constitutional price in terms of rights, without furthering the public good that is said to be the object of the law."[64]

The SCC readily found that the prohibition did not limit rights in an arbitrary way:

> The object of the prohibition on physician-assisted dying is to protect the vulnerable from ending their life in times of weakness. A total ban on assisted suicide clearly helps achieve this object.[65]

Observations/Comment

Within this step, the SCC drops the word "induced" from its original description of the law's object. When the means taken (the prohibition on aiding suicide) is re-conceptualized from the right-holder's perspective into a *limit* of their right (absolute prohibition on physician-assisted dying), it appears to provoke a lens which operates to focus on the beneficiaries or class of persons "protected" by the prohibition (i.e. vulnerable persons) rather than the actual *mischief* the prohibition seeks to target: the *inducing* of vulnerable persons to commit suicide. In so doing, and at this early stage of analysis, the impugned behaviour targeted by the law (*inducement*) is essentially dropped from scrutiny. Note also the Court's analytical fluidity in re-conceptualizing the "object" of the law as a "public good" rather than the "measure directly targeted by the law" which arguably leans the Court into the public policy domain.

Public policy considerations aside, whether an absolute prohibition against PAD is necessary *to prevent persons from inducing vulnerable people to end their lives in a*

[63] Note also that in identifying the conduct at issue as an issue of suicide or self-killing, the Court had already removed from consideration, the other *Criminal Code* provisions and attendant policy considerations relevant to the actions of third parties who intentionally end an individual's life. See Footnotes 26 and 33 and associated text. See also *Carter SCC*, Footnote 1 at paras 19–20.

[64] *Carter SCC*, Footnote 1 at para 83. Emphasis added.

[65] *Carter SCC*, Footnote 1 at para 84.

moment of weakness asks a substantively different question than whether an absolute prohibition against PAD is *necessary to prevent vulnerable persons from ending their lives in a moment of weakness*. Abandoning the word *induce* paradoxically broadens and dilutes the target of the prohibition and downplays or sidelines the mischief of *inducement* as a risk or concept for possible later consideration. Retaining focus on *inducement* (the "mischief") and not only on vulnerable persons (the "who") would have also helped retain focus on the nature of the infringement: the denial of choice to persons who are *seeking*. Increased clarity on the act of *seeking* (which stands in contrast to the act of *induced*), in the author's view, would have helped to guard against analytical slippage in relation to the SCC's eventual "declaration of invalidity"[66] and its attendant remedy whereby, as will be discussed, the word "seeking" is absent and seemingly substituted with the words "clearly consents".[67]

Overbreadth

Overbreadth involves evaluating whether the law denies the rights of some individuals in a way that bears no relation to the object.[68] As described by the SCC:

> A law that is drawn broadly to target **conduct** that bears no relation to its purpose "in order to make enforcement more practical" may therefore be overbroad... .
>
> The question is not whether Parliament has chosen the least restrictive means, but whether the chosen means infringe life, liberty or security of the person in a way that has no connection with **the mischief contemplated** by the legislature. The focus is not on broad social impacts, but on the **impact of the measure** on the individuals whose life, liberty or security of the person is trammelled.[69]

Here the SCC again states the object of the law:

> The object of the law, as discussed, is to protect vulnerable persons from being induced to commit suicide at a moment of weakness.[70]

Although the word *induced* reappears and would clearly be the "mischief contemplated by the legislature", the SCC continues to focus on "vulnerable persons" to describe that, like the Claimant, not every person who wishes to commit suicide is vulnerable.[71] The SCC thus continues:

> It follows that the limitation on their rights is in at least some cases not connected to the objective of *protecting vulnerable persons*.[72]

---

[66] *Carter SCC*, Footnote 1 at para 126.

[67] *Carter SCC*, Footnote 1 at para 127.

[68] *Carter SCC*, Footnote 1 at para 87 et seq.

[69] *Carter SCC*, Footnote 1 at para 85. Emphasis added. Also note that it appears that the Court *may* be utilizing the word "measure" to indicate the law or prohibition at issue, rather than narrower "object" of the law or prohibition (as in measures taken or "targeted by the law"). See Footnote 59 and associated text and compare to Footnote 61 and associated text.

[70] *Carter SCC*, Footnote 1 at para 86. Emphasis added.

[71] Note that the finding that not everyone who wishes to commit suicide is vulnerable was conceded by the Respondent AG Canada. See *Carter SCC*, Footnote 1 at para 86.

[72] *Carter SCC*, Footnote 1 at para 86. Court's emphasis.

Accordingly, the SCC found the blanket prohibition on assisted dying *overbroad* because it catches persons outside of the "class" of vulnerable persons and "sweeps conduct into its ambit that is unrelated to the law's objective".[73]

Observations/Comment

In this part of the analysis, the SCC does not refer to the Claimant "seeking" PAD, but rather refers to persons, like the Claimant, who have a "considered, rational and persistent wish to end their own lives" and are "competent, fully informed, and free from coercion or duress".[74] While this profile could imply the "conduct" of *seeking* PAD, it is curious that this was not expressly stated or juxtaposed against the objective of "protecting vulnerable persons from being induced". On the other hand, perhaps the SCC was not actually contemplating "seeking" as the unrelated conduct being swept into the ambit of the blanket prohibition. Either way, by not clarifying what "conduct" it is actually referencing, the SCC provokes a grey area.

When Canada attempts to argue against *overbreadth* by describing that it is difficult to conclusively identify who is "vulnerable" and that "every person is *potentially* vulnerable"[75] and therefore the prohibition is not overbroad (an argument that seems logically tied to the "mischief" of *inducing* someone to commit suicide as contemplated by the prohibition), the SCC intimates that this is precisely why the provision is overbroad for purposes of the Section 7 analysis: the law is flawed because it does not distinguish between those who would exploit and those who would not[76]; and although the line between exploitative and non-exploitative relationships may be "blurry" such that the provision is drawn broadly to capture "targets", according to the SCC this argument is more appropriately addressed in Section 1,[77] where the infringing law can potentially be saved (discussed further below).

At this point there is a serious socio-political question to be raised: how can protecting vulnerable persons be truly understood as a "competing" societal interest? Yet this is the message carried by the *Charter* analysis.[78] By focusing on the "who" (i.e. the "vulnerable"), the "mischief" targeted by the legislation (*inducement*) is stripped away with the effect that the law as a means to protect vulnerable persons from the conduct of bad actors is recast as a generalized, competing idea of *protection* of the vulnerable and lobbed over for consideration in Section 1. Put a different way, the overbreadth step appears to select for analysis, the "conduct" of *protecting* (a public good associated with the prohibition) rather than the "mischief" of *inducement*. Similarly, by focusing on the rights-holding "who" (the non-vulnerable, competent,

---

[73] *Carter SCC*, Footnote 1 at para 86. Emphasis added.

[74] See *Carter SCC*, Footnote 1 at para 86. Emphasis added.

[75] *Carter SCC*, Footnote 1 at para 87. Court's emphasis.

[76] *Carter SCC*, Footnote 1 at para 88.

[77] *Carter SCC*, Footnote 1 at para 88.

[78] For a discussion of proposal to use language of rights "in tension" as opposed to "competing rights" in order to permit "a more nuanced assessment" see Hughes (2012): 171; See also discussion in Haak, Footnote 38 at footnote 4.

fully informed Claimant who is free of coercion and duress), the "conduct" wrongfully swept up by the prohibition i.e. *seeking* is also obscured, unless again, this is not the conduct the SCC had in mind.

In the result, by disregarding the mischief of *inducement*, the Court's overbreadth analysis encourages an overall impression that the prohibition is antiquated and paternalistic, and thus repugnant; trammelling on the individual Claimant's Section 7 rights.[79]

Gross Disproportionality

The *gross disproportionality* inquiry "compares the law's purpose, 'taken at face value', with its negative effects on the rights of the claimant, and asks if this impact is completely out of sync with the object of the law".[80] The standard in terms of finding a law disproportionate is high.[81]

Here the SCC, in agreement with the trial judge, found that the impact of the prohibition on "affected individuals" was "severe" imposing unnecessary suffering and depriving choice, potentially causing persons to take their lives sooner than they would if PAD was available.[82] When restating the "object" of the prohibition, the SCC reverts back to including the word *induced* (i.e. "to protect vulnerable persons from being induced to commit suicide at a time of weakness") and describes this object as being of "high importance".[83] The SCC however ultimately found it unnecessary to decide on *gross disproportionality* due to its finding on *overbreadth*.[84]

2.4 Section 1

2.4.1 Analytical Framework

Section 1 of the *Charter* states that:

> The *Canadian Charter of Rights and Freedoms* guarantees the rights and freedoms set out in it subject only to such reasonable limits prescribed by law as can be demonstrably justified in a free and democratic society.[85]

Section 1 is considered a saving section whereby the Crown has the opportunity (and burden) to demonstrate that an infringement of rights is justified. Specifically, legislation can be saved if the limit can be "reasonably and demonstrably justified in

[79] For preliminary discussion questioning overbreadth as a principle of fundamental justice see, Fehr (2020).

[80] *Carter SCC*, Footnote 1 at para 89.

[81] *Carter SCC*, Footnote 1 at para 89. Emphasis added.

[82] *Carter SCC*, Footnote 1 at para 90.

[83] *Carter SCC*, Footnote 1 at para 90.

[84] *Carter SCC*, Footnote 1 at para 90.

[85] *Charter*, Footnote 4.

a free and democratic society". Thus, justification under Section 1 is "a process of demonstration, not intuition or automatic deference to the government's assertion of risk".[86] After determining that the limit at issue is indeed "prescribed by law", the Section 1 test as developed by the courts (otherwise known as the "Oakes test")[87] involves a proportionality assessment:

> In order to justify the infringement of the appellants' s. 7 rights under s. 1 of the *Charter*, Canada must show that the law has a **pressing and substantial object** and that the means chosen are proportional to **that object**. A law is proportionate if (1) the means adopted are rationally connected to **that objective**; (2) it is minimally impairing of **the right in question**; and (3) there is proportionality between the deleterious and salutary **effects** of the law...[88]

Although it is difficult to justify a Section 7 limit under Section 1,[89] and the SCC has *never* found a limit on Section 7 to be justified to date, it is apparently not impossible:

> ... in some situations the state may be able to show that the public good — a matter not considered under s. 7, which looks only at the impact on the rights claimants — justifies depriving an individual of life, liberty or security of the person under s. 1 of the *Charter*. More particularly, in cases such as this where the competing societal interests are themselves protected under the *Charter*, a restriction on s. 7 rights may in the end be found to be proportionate to its objective.[90]

2.4.2 *Analysis*

2.4.2.1 Prescribed by Law and Pressing and Substantial Objective

These steps (i.e. "prescribed by law" and "pressing and substantial objective") were met, with the SCC stating as follows:

> Here, the limit is prescribed by law, and the appellants concede that the law has a pressing and substantial objective. The question is whether the government has demonstrated that the prohibition is proportionate.[91]

[86] *Carter SCC*, Footnote 1 at para 119.

[87] See *Carter SCC*, Footnote 1 at para 94; See also *R v Oakes* (1986) [1986] 1 SCR 103 at paras 69–71.

[88] *Carter SCC*, Footnote 1 at para 94. Emphasis added.

[89] For an example of one case where Section 7 infringement was justified under Section 1 see *Michaud*, Footnote 35; See also Footnote 38.

[90] *Carter SCC*, Footnote 1 at para 95. Emphasis added.

[91] *Carter SCC*, Footnote 1 at para 96. Emphasis added.

Observations/Comment

Regardless of the Claimant(s) concession, it is indeed regrettable that the SCC did not specifically clarify, state or restate the "pressing and substantial objective" of the law at the beginning of its Section 1 analysis.[92]

2.4.2.2 Proportionality

Deference

Before turning to the proportionality assessment, the SCC raises a preliminary question of what level of deference is owed to the government. Although this is not generally considered a formal stage of the Section 1 analysis, it is frequently done to determine the space within which Parliament is entitled to try to realize its policy preferences. To help answer this question, the SCC set out some key considerations[93]:

- "Proportionality does not require perfection" i.e. some deference to Parliament is owed;
- Section 1 only requires that limits be "reasonable"; and
- a "complex regulatory response" to a "social ill" requires a high degree of deference.

In relating the trial judge's observation that PAD involves "complex issues of social policy and a number of competing societal values" the SCC further commented that:

> Parliament faces a difficult task in addressing this issue; it must weigh and balance the perspective of **those who might be at risk in a permissive regime** against that of **those who seek assistance** in dying.[94]

While the SCC found that Parliament was owed a high degree of deference with respect to its "decision to impose an absolute prohibition on assisted death", according to the Court, the absolute prohibition could not be described as a "complex regulatory response", and therefore the SCC held that a lesser degree of deference was owed to Parliament.[95]

Observations/Comments

Within this preliminary step of Section 1, note how the Court's description of the interests to be balanced is again revised and sharpened:

---

[92] Compare to the trial judge's evaluation of "pressing and substantial" at Part XII.E and in particular para 1190: "I conclude that the objective of the legislation is, by imposing criminal sanctions on persons who assist others with suicide, to protect vulnerable persons from being induced to commit suicide at a time of weakness. The underlying state interest which this purpose serves is the protection of life and maintenance of the *Charter* value that human life should not be taken ..." *Carter BCSC*, Footnote 3 at Part XII.E and para 1190. Emphasis added; Compare to Footnote 61 and evaluation by Sopinka J in *Rodriguez*, Footnote 28.

[93] *Carter SCC*, Footnote 1 at para 97. Emphasis added.

[94] *Carter SCC*, Footnote 1 at para 98. Emphasis added.

[95] *Carter SCC*, Footnote 1 at para 98. Emphasis added.

- not a balance between the *autonomy and dignity* of those *seeking PAD* and the *sanctity of life and protection of vulnerable from being induced to commit suicide* specifically[96];
- not a balance between the *not vulnerable* and the *vulnerable* generally[97];
- but a balance between *seekers of PAD* and *those who might be at risk in a permissive regime*[98]

Instead of the object of the law (i.e. the impugned conduct with its attendant broader concerns and values) being the focusing point for delineating what sits on either side of the balance, under Section 1, the "right" as identified through Section 7 is launched over to the "other side" whereby the law's target is re-cast in relation to the newly identified "right". As described in the trial judgment:

> … it is the absolute nature of the prohibition **against** assisted suicide that requires justification, not the prohibition overall. In other words, the real question is whether the defendants have **demonstrated justification for criminalizing the rendering of assistance** in suicide **to persons such as Gloria Taylor.**[99]

Thus, all the criteria, concerns, and values that might be broadly explored on the "opposite side" of the scale are subsumed into what is, for all intents and purposes, a judicially-constructed PAD scheme based on the Claimant's rights and framed solely by criteria attendant with the Court's construction of those rights and the infringement(s) of those rights, whatever those criteria might be. Note that because it is not appropriate for the court to create a prescriptive regulatory scheme,[100] there is a risk that criteria implicit to the judicially-constructed, rights-based PAD scheme will not survive the end result.

A plain reading of the first part of the Section 1 test (see Sect. 2.4.1 above), appears to require the court to utilize the law's original object as previously identified in the Section 7 PFJ analysis.[101] This interpretation seems to be supported by the SCC's finding that a lesser degree of deference was owed to Parliament because the prohibition at issue was not a "complex regulatory response" (see Sect. 2.4.2.2. above). If, in applying its deference test, the SCC was obligated to specifically state the "social ill" targeted by the regulatory response (which the SCC did not), reason and logic would reveal the "social ill" to be the *inducement* of vulnerable persons to end their lives in a moment of weakness. This of course, is not how the Section 1 analysis unfolds because as already described, it is the limit on the identified right that has become the *de facto* target of the means taken by the law.

So not only do the constricting mechanisms of Section 7 shift focus away from the law's targeted mischief (i.e. *inducement* of the vulnerable) and towards a generalized

---

[96] *Carter SCC*, Footnote 1 at para 2.

[97] See *Carter SCC*, Footnote 1 at para 86.

[98] *Carter SCC*, Footnote 1 at para 98. Emphasis added.

[99] *Carter BCSC*, Footnote 3 at para 1171. Emphasis added.

[100] See discussion in *Carter SCC*, Footnote 1 at para 125.

[101] For further discussion of continuity with respect to the object of the law as between Sections 7 and 1, see Haak, Footnote 38 at Section III.1; See also *Alberta v Hutterian Brethren of Wilson Colony*, 2009 SCC 37, [2009] 2 SCR. 567 at para 76.

From mischief (*inducement*) ► to benefit (*protection*) ► to risk (associated with the realization of the Section 7 PAD right)

Fig. 2.1 Transformation of the object of the prohibition

notion of the "public good" or "benefit" arising from the law (i.e. *protection* of the vulnerable) to be addressed in Section 1, the mechanisms of the Section 1 analysis begin to re-assemble the general "benefit" of the law (the *protection* of the vulnerable) as an area of "risk" potentially associated with the realization of the newly identified "right". In so doing, the Court has effectively immunized the Section 7 right(s) from the full scrutiny that it might otherwise receive from a complete exploration of the "competing" rights, values and interests said to be sitting on the "other side" of the balance (Fig. 2.1).

These analytical machinations appear to thwart the substantive scope or positive effects of the law at issue (including "competing societal interests" that may themselves be protected under the *Charter*)[102] from being fully realized or re-instated within the judicial analysis. In other words, the objective of the law is only ever made relevant in relation to the identified right. Accordingly, the ongoing analysis not only risks dropping criteria implicit to the judicially-constructed PAD framework (e.g. *seeking*) but also risks losing sight of the positive benefits stemming from the law itself (e.g. deterring or sanctioning actors who may *induce*). These particular analytical or methodological threats arguably become increasingly acute as the judicial criteria associated with the infringement, and the mischief and benefits associated with the prohibition, are again transmuted into the broad notions of PAD *eligibility* and *safeguards*.

Rational Connection

To establish a rational connection, "the government need only show that there is a causal connection between the infringement and the benefit sought 'on the basis of reason or logic'".[103] Furthermore, according to the Court, a prohibition is a rational method of curtailing "risks" associated with an "activity".[104] Here the Court readily finds a "rational connection". According to the SCC:

> It is clearly rational to conclude that a law that bars all persons from accessing assistance in suicide will protect the vulnerable from being induced to commit suicide at a time of weakness. The means here are logically connected with the objective.[105]

Observations/Comment

Here we can directly observe the Court replacing the *seeking* requirement (critical to establishing the Section 7 infringements) with the activity of *accessing* assisted suicide. Although it may be implicit in the word "accessing" that the person is also

[102] *Carter SCC*, Footnote 1 at para 95.

[103] *Carter SCC*, Footnote 1 at para 99. Emphasis added.

[104] *Carter SCC*, Footnote 1 at para 100.

[105] *Carter SCC*, Footnote 1 at para 101. Emphasis added.

From mischief (*inducement*) ▶ to benefit (*protection*) ▶ to risk (associated with PAD) ▶ to safeguard (associated with PAD)

Fig. 2.2 Ongoing transformation of the object of the prohibition

"seeking", the SCC does not provide this kind of clarification and is yet another example of where differing interpretations can be provoked.

Furthermore, although the Court returns to the language of *inducement*, again note how inducement is contextualized in relation to PAD. By juxtaposing the "benefit" of the law (i.e. preventing persons from being *induced*) against what the rights holder is *seeking* (i.e. assistance in suicide), the analysis further establishes the foundation upon which the "benefit" is transformed into a "risk" associated with the specific activity of PAD and sets the stage for the "risk" to be funnelled into the concept of "safeguards" (discussed further below) (Fig. 2.2).

Additionally, at this stage it is fairly easy to perceive how:

- the scope, impact or relevance of broader "competing" societal values have been largely reduced and/or sidelined within the operation of the overall *Charter* analysis; and
- Section 1 seems to re-assert or re-play a form of Section 7 analysis, i.e. the evaluation of the nature or scope of the infringement from the Claimant's/rights-holder's perspective.[106]

Minimal Impairment

As described by the SCC:

> At this stage of the analysis, the question is whether the limit on the right is reasonably tailored to the objective. The inquiry into minimal impairment asks "whether there are less harmful means of achieving the legislative goal". The burden is on the government to show the absence of less drastic means of achieving the objective "in a real and substantial manner" … . The analysis at this stage is meant to ensure that the deprivation of *Charter* rights is confined to **what is reasonably necessary to achieve the state's object**.[107]

Specific to the facts of the case, the SCC continues:

> The question in this case comes down to whether the absolute prohibition on physician assisted dying, with its heavy impact on the claimants' s. 7 rights to life, liberty and security of the person, is the **least drastic means** of achieving the **legislative objective**. It was the task of the trial judge to determine whether a regime less restrictive of life, liberty and security of the person could address the **risks associated with physician-assisted dying**, or whether Canada was right to say that the risks could not adequately be addressed through the use of safeguards.[108]

---

[106] Notwithstanding the author's comment, the Sections 7 and 1 concepts are said to be analytically distinct, though "rooted in similar concerns". See discussion in *Bedford*, Footnote 35 at paras 124–128; For additional discussion and critique see for example, Carter, Different Questions, Footnote 38.

[107] *Carter SCC*, Footnote 1 at para 102. Emphasis added.

[108] *Carter SCC*, Footnote 1 at para 103. Emphasis added.

In short, whether the prohibition minimally impairs the Claimant's rights turns on whether the risks associated with PAD could be adequately addressed through the use of "safeguards", which the SCC describes as "the question that lies at the heart of this case".[109]

At the trial level, evidence included evidence from scientists, medical practitioners, ethicists, professional bodies, and public committees as well as experiences and data related to euthanasia and assisted suicide from other permissive jurisdictions. Apart from that which the trial judge used to overcome *stare decisis*[110] and to establish the infringement,[111] this evidence (and associated arguments) was sifted through and predominantly organized from the perspective of *safeguards* and the identification of areas of *risk* in relation to PAD practice.[112]

Based on the evidence reviewed by the trial judge which again included evidence from assisted death permissive jurisdictions[113] as well as that related to physicians' ability to reliably assess *inter alia* competence, voluntariness, coercion, undue influence and ambivalence,[114] the SCC (in agreement with the trial judge) rejected the possibility that an absolute prohibition was necessary to protect vulnerable persons from abuse and error *within* a PAD regime.[115]

As stated by the SCC, "While there are risks, to be sure, a carefully designed and managed system is capable of adequately addressing them".[116] Importantly, it is through the *safeguard* lens that the Court goes on to identify the "informed consent standard" as a possible standard for physicians to use for "patients who seek assistance in dying".[117]

---

[109] *Carter SCC*, Footnote 1 at paras 103–104. Emphasis added; see also Fig. 2.2.

[110] "The common law doctrine of *stare decisis* requires that cases involving materially the same facts and invoking the same legal principles be decided the same way". *Carter BCSC*, Footnote 3 at para 899. The trial court in *Carter* determined that it was not bound by *Rodriguez*, Footnote 28, an earlier Supreme Court of Canada decision which upheld the blanket prohibition on assisted suicide. *Carter BCSC*, Footnote 2 at Part X. The trial decision was successfully appealed to the British Columbia Court of Appeal which determined on the basis of *stare decisis* that the trial judge was indeed bound to follow *Rodriguez*. *Carter BCCA*, Footnote 28. The case was then appealed to the Supreme Court of Canada, the decision of which is the subject of this chapter.

[111] For example, aspects of palliative care were perceived by the Claimant (and others) as an unacceptable alternative. See for example *Carter BCSC*, Footnote 3 at paras 54–56 and 315(h).

[112] See *Carter BCSC*, Footnote 3 at Part IX.

[113] *Carter BCSC*, Footnote 3 at Part VIII.

[114] *Carter BCSC*, Footnote 3 at Part IX.

[115] *Carter SCC*, Footnote 1 at paras 102–107.

[116] *Carter SCC*, Footnote 1 at para 105.

[117] *Carter SCC*, Footnote 1 at para 106. Emphasis added. When applying the informed consent standard "to patients who seek assistance in dying", the trial court further cautioned that "physicians should ensure that patients are properly informed of their diagnosis and prognosis and the range of available options for medical care, including palliative care interventions aimed at reducing pain and avoiding the loss of personal dignity". See *Carter SCC*, Footnote 1 at para 106.

Further agreeing with the trial judge that the adoption of a PAD regulatory regime would not descend down a "slippery slope" to homicide,[118] the SCC thus concluded that the absolute prohibition was not minimally impairing.

Deleterious Effects and Salutary Benefits

The last step in the proportionality analysis "weighs the impact of the law on protected rights against the beneficial effect of the law in terms of the greater public good".[119] Having determined that the prohibition was not minimally impairing the SCC did not proceed with this step.

Observations/Comment

In advancing the "informed consent" standard as a tool that physicians could apply to patients who "seek" PAD[120] in order to guard against particular risks, the Court's analysis is consistent with its construction of the Section 7 infringement. That is, within its judicially-constructed rights-based PAD framework, the Court positions *informed consent* as a procedural safeguard (the process to be taken to ensure the eligibility criteria are met) and *seeking* as an eligibility requirement (the criteria that qualify a person for PAD). This is consistent with the court's elucidation of the Section 7 infringement as a denial of choice to a *seeker* of "termination of life". When the SCC translates its constitutional analysis into its "declaration of invalidity" (the "declaration"), however, the eligibility criterion of *seeking* is expressly absent, seemingly substituted with the words "clearly consents":

> *Declaration of Invalidity*
>
> We have concluded that the laws prohibiting a physician's assistance in terminating life (*Criminal Code*, s. 241(*b*) and s. 14) infringe Ms. Taylor's s. 7 rights to life, liberty and security of the person in a manner that is not in accordance with the principles of fundamental justice, and that the infringement is not justified under s. 1 of the *Charter*. To the extent that the impugned laws deny the s. 7 rights of people **like Ms. Taylor** they are void by operation of s. 52 of the *Constitution* Act, 1982. It is for Parliament and the provincial legislatures to respond, should they so choose, by enacting legislation consistent with the constitutional parameters set out in these reasons.
>
> The appropriate remedy is therefore a declaration that s. 241(*b*) and s. 14 of the *Criminal Code* are void insofar as they prohibit physician-assisted death for a competent adult person who **(1) clearly consents to the termination of life; and (2) has a grievous and irremediable medical condition** (including an illness, disease or disability) that causes enduring suffering that is intolerable to the individual in the circumstances of his or her condition. "Irremediable", it should be added, does not require the patient to undertake treatments that are not acceptable to the individual. The scope of this declaration is intended to respond to the **factual circumstances in this case** . We make no pronouncement on other situations where physician-assisted dying may be **sought**.[121]

[118] *Carter SCC*, Footnote 1 at para 120.

[119] *Carter SCC*, Footnote 1 at para 122. Emphasis added.

[120] *Carter SCC*, Footnote 1 at para 106.

[121] *Carter SCC*, Footnote 1 at paras 126 and 127. Emphasis added.

An argument could be made that the eligibility requirement of "seeking" is implicit in the following language: the "rights of people like Ms. Taylor"; that the declaration is intended to respond to "the factual circumstances in this case"; and that no pronouncement was being made "on other situations where assisted dying may be sought". However, making such an argument and unequivocal guidance from the Court are entirely different things.

Furthermore, "clearly consents" seemingly positioned by the SCC in its declaration as an *eligibility requirement*[122] is not equivalent to the "informed consent" standard also suggested by the SCC as a potential *procedural safeguard* to be applied to patients who *seek* assistance in dying. That is, the informed consent standard is a process[123] by which "clear consent" to PAD can be obtained. Having said that, it is easy to see how the phrase "clearly consents" could be interpreted (or misinterpreted) as the "informed consent" process. If the consent *eligibility requirement* for PAD is collapsed or conflated with the *procedure* for obtaining that consent, the relevance of the distinction between persons *seeking* PAD and persons who are not is further obscured. This in turn arguably increases the vulnerability of persons whose illness, disease or disability is perceived by others (health care professionals or otherwise) as "grievous and irremediable".[124]

2.5 Final Thoughts

Interestingly, the Federal MAID law created in response to *Carter*[125] situated the giving of "informed consent" to receive MAID as well as the making of a "voluntary request" for MAID as discrete *eligibility* criteria.[126] However, it is becoming increasingly apparent that as MAID practice unfolds within the Canadian provincial healthcare framework, many healthcare professionals and regulators do not appear inclined to interpret the "voluntary request" criterion as requiring the MAID request (or inquiry) to originate from the patient before a MAID discussion is embarked

[122] The second eligibility requirement set out by the SCC being: "grievous and irremediable medical condition (including an illness, disease or disability) that causes enduring suffering that is intolerable to the individual in the circumstances of his or her condition". *Carter SCC*, Footnote 1 at para 127.

[123] Informed consent involves a physician providing necessary information to a patient in order for them to make an informed decision about medical treatment. This information should include *inter alia* information about the nature of the proposed treatment, the anticipated outcomes, risks and alternatives. For further discussion of the informed consent doctrine see Osborne (2020) at Chapter 3.C.

[124] Evidence of increased vulnerability due to ableist perceptions and/or conflation of eligibility and process can perhaps be observed in recent media reports, for example: Peters (2021), Favaro et al. (2018), Bartlett (2017), Brewster (2022).

[125] *Bill C-14*, Footnote 3.

[126] *Bill C-14*, Footnote 3 at Section 3; See also *Criminal Code*, Footnote 2 at Sections 241.2(1)(e) and 241.2(1)(d).

upon.[127] This stands in stark contrast to the construction of PAD as a Section 7 right, which, as discussed, is based on persons *seeking* physician-assisted dying. Perhaps even more intriguing is the Preamble of Bill C-14,[128] the first bill to amend the *Criminal Code* and decriminalize PAD in response to the *Carter SCC* decision. Bill C-14's Preamble utilizes the word "seek" no less than three (3) times to clearly convey that the law is aimed at allowing persons *seeking* medical assistance in dying access to MAID. Specifically, the Preamble states:

> Whereas the Parliament of Canada recognizes the autonomy of persons who have a grievous and irremediable medical condition that causes them enduring and intolerable suffering and **who wish to seek medical assistance in dying:**
>
> ...
>
> Whereas, in light of the above considerations, permitting access to medical assistance in dying for competent adults whose deaths are reasonably foreseeable strikes the most appropriate balance between **the autonomy of persons who seek medical assistance in dying**, on one hand, and the interests of vulnerable persons in need of protection and those of society, on the other;
>
> ...
>
> And whereas the Government of Canada has committed to develop non-legislative measures that would support the improvement of a full range of options for end-of-life care, respect the personal convictions of health care providers and explore other situations — each having unique implications — in which a person **may seek access to medical assistance in dying**, namely situations giving rise to requests by mature minors, advance requests and requests where mental illness is the sole underlying medical condition;
>
> Now, therefore, Her Majesty, by and with the advice and consent of the Senate and House of Commons of Canada, enacts as follows ...[129]

Notwithstanding the Preamble, the word *seek* is conspicuously absent from the specific provisions of the MAID law; the provisions do not expressly include *seeking*

[127] Evidence of this can be found in current debates concerning access to MAID as part of provincial healthcare whereby an argument is being advanced that the *informed consent* requires doctors to introduce the MAID option to a patient who "appears to be eligible"; For example: "There is no provision in the law that prohibits healthcare professionals from initiating a discussion about MAiD .... Only physicians and nurse practitioners ... involved in care planning and consent processes have a professional obligation to initiate a discussion about MAiD if a patient might be eligible for MAiD. The discussion should include all treatment options, including palliative care and the option of MAiD. The appropriate timing of initiating a discussion about MAiD is determined by the clinical context. Healthcare professionals must not discuss MAiD with a patient with the aim of inducing, persuading, or convincing the patient to request MAiD." CAMAP, Footnote 21. Emphasis added.

[128] Bill C-14, Footnote 3.

[129] Bill C-14, Footnote 3 at Preamble. Emphasis added. It might be noted that when the *Criminal Code* is later amended in March 2021 by Bill C-7, Footnote 13, *inter alia* to remove the MAID eligibility criteria that death be "reasonably foreseeable", there is no mention that the law is contemplating persons who are seeking MAID. On the other hand, the Explanatory note to Bill C-7 does describe the MAID law and potential amendments in relation to persons who "seek" MAID. See Explanatory Note to Bill C-7: An Act to Amend the Criminal Code (medical assistance in dying) tabled October 21, 2020. Online: https://justice.gc.ca/eng/csj-sjc/pl/charter-charte/c7.html.

as an eligibility criterion, nor do they qualify or define the "voluntary request" criterion in relation to *seeking* but rather describe that "in particular", the request must not be made "as a result of external pressure". Granted Parliament is not beholden to the strictures of the *Carter* PAD scheme (or various interpretations thereof) and can ultimately draft MAID permissive legislation in whichever way it sees fit.[130] Given however, that ongoing MAID reform arguments continue to invoke notions of the rights stemming from the *Carter* case and furthermore, that MAID practice involves the legal intersection of criminal law and health care, it would be desirable to reach some sort of consensus as to what the Supreme Court of Canada did or did not establish in *Carter* as well as achieve legal clarity in respect of MAID eligibility criteria pursuant to the criminal law and distinct from the procedural criteria to ensure those eligibility criteria have been met.

Going back to *Carter*, the author would suggest that "the law" stemming from the *Carter* case is not only that which appears or does not appear in the Court's final "declaration of invalidity". And although the author has for the purpose of this Chapter, focused in particular on the concepts of *seeking* and *inducing* as well as the aspects of the *Charter* analysis that appear to sideline "competing" societal interests, there is still much more in *Carter* that can and ought to be explored.

2.6 Recommendations

Building on the concepts of *seeking*, and *inducing* within the context of the "competing" values in connection with the foregoing analysis, the author humbly offers the following recommendations:

- that Parliament clarify the legal principles and foundation upon which MAID practice has been decriminalized in Canada pursuant to the *Criminal Code*;
- that Parliament address the "seeking" criterion expressly within the MAID eligibility provisions of the *Criminal Code*;
- that Parliament clarify that the MAID eligibility criteria are distinct from procedural safeguards and clearly express, within the *Criminal Code*, the difference between MAID and other medical practices that might be indicated through general applications or interpretations of the medical doctrine of informed consent;
- notwithstanding that there is no "duty to live" and that autonomy and self-determination (as also reflected in the *Charter* rights to liberty and security) protect an individual's right to make decisions that may lead to death, locate an avenue(s) to clearly establish or re-assert, the "right to life" as meaning the protection of life and in relation to acts or omissions of the State;
- guarantee fundamental rights that reduce vulnerability to suicidality and promote an individual's wish to live (these may be, but are not limited to, a right to palliative

130 For an interesting discussion in this regard see, Sigalet (2022).

care, home care, housing, food, clean water and a basic living allowance) and establish the systems, resources and/ or research to help support same;

- educate and re-enforce that notwithstanding that MAID is being delivered as part of the Canadian health care system, MAID is a specific practice subject to the *criminal law* whereby there are and remain other equally important societal values and interests such as preventing or protecting all persons from being *induced* to commit suicide and which are subject to protections under the *Charter of Rights and Freedoms* as well as other human rights instruments including the *Convention on the Rights of Persons with Disabilities*[131];
- seriously reflect on whether judicial interpretation and application of the *Charter* (as it is currently interpreted and applied as demonstrated in the *Carter* decision) remains appropriate for resolving questions involving fundamental and highly complex societal issues like termination of life, particularly when there are competing interests which are themselves protected under the *Charter of Rights and Freedoms* and other human rights instruments such as the *Convention on the Rights of Persons with Disabilities.*

Acknowledgements Many thanks are owed to the editors, Dr. Jaro Kotalik and David Shannon, to Ashley Benard-Legris, and to Dr. Gerard Kennedy and Dr. M. Michelle Gallant for their helpful comments and suggestions. Thanks are also owed to my research assistant Kali Faingold B.A. (J.D. candidate 2024).

References

Cases

Alberta v. Hutterian Brethren of Wilson Colony, [2009] 2 SCR 567.
Canada (Attorney General) v Bedford, [2013] 3 SCR 1101.
Carter v Canada (Attorney General), 2012 BCSC 886.
Carter v. Canada (Attorney General), 2013 BCCA 435.
Carter v. Canada (Attorney General), [2015] 1 SCR 331.
Carter v. Canada (Attorney General), [2016] 1 SCR 13.
Law Society of British Columbia v. Andrews, [1989] 1 SCR 143.
R v Oakes, [1986] 1 SCR 103.
R v. Michaud 127 O.R. (3d) 81, 2015 ONCA 585.
Re B.C. Motor Vehicle Act, [1985] 2 SCR 486.
Rodriguez v. British Columbia (Attorney General), [1993] 3 SCR 519.
Truchon v. Canada (AG), 2019 QCCS 3792.

[131] *The Convention on the Rights of Persons with Disabilities*, 13 December 2006, 2515 UNTS 3 (entered into force 3 May 2008). Online: https://www.ohchr.org/en/instruments-mechanisms/instruments/convention-rights-persons-disabilities.

Legislation and Conventions

Bill C-7. 2021. *An Act to amend the Criminal Code (medical assistance in dying)*, S.C. 2021, c. 2.
Bill C-14. 2016. *Act to amend the Criminal Code and to make related amendments to other acts (medical assistance in dying)*, S.C. 2016, c. 3.
Canadian Charter of Rights and Freedoms, Part I of the Constitution Act, 1982, being Schedule B to the Canada Act, 1982 (U.K.). 1982. c. 11.
Convention on the Rights of Persons with Disabilities. 13 December 2006, 2515 UNTS 3.
Criminal Code of Canada, R.S.C. 1985, c. C-46.

Articles, Book Chapters, Reports and Commentary

Alberga, Hannah. 2022a. Toronto woman facing financial loss of long COVID begins process for medically assisted death. *CTV*, July 11. https://www.cp24.com/news/toronto-woman-facing-financial-loss-of-long-covid-begins-process-for-medically-assisted-death-1.5982576?cache=juzexmjvhq. Accessed January 9, 2023.
Alberga, Hannah. 2022b. Toronto woman in final stages of MAiD application after nearly a decade-long search for housing. *CTV News*, November 15. https://toronto.ctvnews.ca/toronto-woman-in-final-stages-of-maid-application-after-nearly-a-decade-long-search-for-housing-1.6145487. Accessed January 9, 2023.
Bartlett, Geoff. 2017. Mother says doctor brought up assisted suicide option as sick daughter was within earshot. *CBC News*, July 24. https://www.cbc.ca/news/canada/newfoundland-labrador/doctor-suggested-assisted-suicide-daughter-mother-elson-1.4218669. Accessed January 9, 2023.
Brewster, Murray. 2022. Former Paralympian tells MPs veterans department offered her assisted death. *CBC News*, December 1. https://www.cbc.ca/news/politics/christine-gauthier-assisted-death-macaulay-1.6671721. Accessed January 9, 2023.
CAMAP. 2022. *Bringing up medical assistance in dying (MAiD) as a clinical care option*. https://camapcanada.ca/wp-content/uploads/2022/02/Bringing-up-MAiD.pdf. Accessed June 9, 2022.
Carter, Mark. 2015. *Carter v Canada*: 'Societal interests' under Sections 7 and 1. *Saskatchewan Law Review* 78: 209.
Carter, Mark. 2017. Sections 7 and 1 of the charter after *Bedford*, *Carter*, and *Smith*: Different questions, same answers. *Criminal Law Quarterly* 64 (2): 108.
Coyne, Andrew. 2022. On assisted suicide, the slope is proving every bit as slippery as feared. *Globe and Mail*, January 17. https://www.theglobeandmail.com/opinion/article-on-assisted-suicide-the-slope-is-proving-every-bit-as-slippery-as/. Accessed June 9, 2022.
Downie, Jocelyn. 2022. In a nutshell: The continuing saga of Canada's medical assistance in dying legislation. *Impact Ethics*, May 17. https://impactethics.ca/2022/05/17/in-a-nutshell-correcting-the-record-about-medical-assistance-in-dying/. Accessed June 9, 2022.
Dufraimont, Lisa. 2014. *Canada (Attorney General) v. Bedford* and the Limits on Substantive Criminal Law under Section 7. *Supreme Court Law Review* 67: 483–503.
Farquhar, Ruth. 2022. Farquhar: Are the disabled choosing medically assisted death to escape poverty? *The Sudbury Star*, November 21. https://www.thesudburystar.com/opinion/columnists/farquhar-are-the-disabled-choosing-medically-assisted-death-to-escape-poverty. Accessed January 9, 2023.
Favaro, Avis, and Elizabeth St Philip. 2018. Chronically ill man releases audio of hospital staff offering assisted death. *CTV News*, August 2. https://www.ctvnews.ca/mobile/health/chronically-ill-man-releases-audio-of-hospital-staff-offering-assisted-death-1.4038841. Accessed January 9, 2023.

Favaro, Avis. 2022. Woman with chemical sensitivities chose medically-assisted death after failed bid to get better housing. *CTV News*, August 24. https://www.ctvnews.ca/health/woman-with-chemical-sensitivities-chose-medically-assisted-death-after-failed-bid-to-get-better-housing-1.5860579. Accessed January 9, 2023.

Fehr, Colton. 2020. Re-thinking the instrumental rationality principles of fundamental justice. *Alberta Law Review* 58 (1): 133.

Government of Canada. 2020. *Explanatory note to bill C-7: An act to amend the criminal code (medical assistance in dying).* https://justice.gc.ca/eng/csj-sjc/pl/charter-charte/c7.html. Accessed January 9, 2023.

Government of Canada. 2021. *Legislative background: Bill C-7: Government of Canada's legislative response to the superior court of Québec Truchon decision.* https://justice.gc.ca/eng/csj-sjc/pl/ad-am/c7/c7-eng.pdf. Accessed June 9, 2022.

Haak, Debra. 2022. Revisiting the analytical distinction between Section 7 and Section 1 of the charter: Legislative objectives, policy goals, and public interests. *Supreme Court Law Review.* https://papers.ssrn.com/sol3/papers.cfm?abstract_id=4235425. Accessed January 9, 2023.

Ho, Anita, Joshua S. Norman, Soodabeh Joolaee, Kristie Serota, Louise Twells, and Leeroy William. 2021. How does medical assistance in dying affect end-of-life care planning discussions? Experiences of Canadian multidisciplinary palliative care providers. *Palliative Care and Social Practice* 15: 1–14.

Hughes, Patricia. 2012. The reconciliation of legal rights. In Shaheen Azmi, Lorne Foster and Lesley Jacobs, eds. Balancing Competing Human Rights Claims in a Diverse Society: Institutions, Policy, Principles. Toronto: Irwin Law.

Joyal, Glenn. 2017. *The charter and Canada's new political culture: Are we all ambassadors now?* January 6. Canadian Constitution Foundation 2017 Law and Freedom Conference. https://theccf.ca/2207-2/. Accessed June 9, 2022.

Kruger, Brooke. 2022. Regina resident applied for medically assisted death after consistent surgical delays. *Global News*, November 30. https://globalnews.ca/news/9315789/regina-resident-medical-assistance-in-dying-surgical-delay/. Accessed January 9, 2023.

Leffler, Brennan, and Marianne Dimain. 2022. How poverty, not pain, is driving Canadians with disabilities to consider medically-assisted death. *Global News*, October 8. https://globalnews.ca/news/9176485/poverty-canadians-disabilities-medically-assisted-death/. Accessed January 9, 2023.

LeGal, Janine. 2022. *Battle with ALS brought overwhelming struggle for supports.* Winnipeg Free Press, November 12. https://www.winnipegfreepress.com/breakingnews/2022/11/12/battle-with-als-brought-overwhelming-struggle-for-supports. Accessed January 9, 2023.

Lemmens, Trudo. 2022. Turning human rights upside down with advance requests for maid. *Impact Ethics.* June 8. https://impactethics.ca/2022/06/08/turning-human-rights-upside-down-with-advance-requests-for-maid/. Accessed June 9, 2022.

Mulligan, Cynthia, and Meredith Bond. 2022. Ontario man not considering medically-assisted death anymore after outpouring of support. *City News*, November 15. https://toronto.citynews.ca/2022/11/16/ontario-medically-assisted-death-support/. Accessed January 9, 2023.

Office of the Parliamentary Budget Officer. 2020. *Cost estimate for bill C-7 "medical assistance in dying".* October 20. https://www.pbo-dpb.gc.ca/web/default/files/Documents/Reports/RP-2021-025-M/RP-2021-025-M_en.pdf. Accessed June 9, 2022.

Osborne, Philip H. 2020. *The law of torts.* Toronto: Irwin Law.

Parliament of Canada. 2015. *Historical information, bill C-14.* https://www.parl.ca/LegisInfo/en/bill/42-1/C-14?view=details. Accessed June 9, 2022.

Phillips, Andrew. 2022. We're all implicated in Michael Fraser's decision to die. *Toronto Star*, November 18. https://www.thestar.com/opinion/star-columnists/2022/11/18/were-all-implicated-in-michael-frasers-decision-to-die.html. Accessed January 9, 2023.

Peters, Gabrielle. 2021. Taking MAID way too far. *Maclean's*, February 4. https://www.macleans.ca/opinion/taking-MAID-way-too-far/. Accessed January 9, 2023.

Sigalet, Geoffrey. 2022. The truck and the brakes: Understanding the charter's limitations and notwithstanding clauses symmetrically. *Supreme Court Law Review* (2d) 105: 194–227.

Sikkema, Dan. 2019. Opinion: Federal leaders still don't seem to understand the attorney general's role. *National Post*, October 10. https://nationalpost.com/news/politics/election-2019/opinion-federal-leaders-still-dont-seem-to-understand-the-attorney-generals-role. Accessed June 9, 2022.

Tiedemann, Marlisa. 2019. Assisted dying in Canada After Carter v Canada. November 29. Library of Parliament Background Paper. https://lop.parl.ca/staticfiles/PublicWebsite/Home/ResearchPublications/BackgroundPapers/PDF/2019-43-e.pdf. Accessed June 9, 2022.

Traachtenberg, Aaron J., and Braden Manns. 2017. Cost analysis of medical assistance of dying in Canada. *CMAJ* 189 (3): E101–E105. https://www.cmaj.ca/content/189/3/E101. Accessed January 9, 2023.

United Nations. 2021. *Disability is not a reason to sanction medically assisted dying*, January 25. https://www.ohchr.org/en/press-releases/2021/01/disability-not-reason-sanction-medically-assisted-dying-un-experts?LangID=E&NewsID=26687. Accessed June 9, 2022.

Professor Mary J. Shariff Ph.D. is an Associate Professor of Law at the Faculty of Law, University of Manitoba, Canada. Professor Shariff received her Ph.D. from Trinity College, Dublin. In addition to teaching the Law of Contracts in the first year of the J.D. program, she teaches the upper year courses Law and Bioethics, Law and Religion and Animals and the Law. She also acts in a supervisory capacity for graduate students. Thesis supervisions have included research related to bioethics, climate change and non-human rights. She is published in the areas of end of life, assisted death and palliative care and is frequently engaged as a speaker on these topics. As a former Associate Dean, J.D. Program and Associate Dean, Research and Graduate Studies within the Faculty of Law, Professor Shariff is presently focusing on her research in law and death and dying, palliative care, long term care, aging and the rights of vulnerable persons, including older persons and persons with disabilities.

Chapter 3
From a Court Judgement to Federal Law

Travis Dumsday

Abstract The goals of this chapter are to provide a concise and accessible overview of: the key events leading up to the legalization in Canada of medical assistance in dying (i.e., assisted suicide plus voluntary active euthanasia) in June 2016; the subsequent amendments to that legislation, passed in March 2021, and some of the parliamentary and public discussion surrounding it; and an analysis of where the country may be headed in the near future with respect to additional legislative changes and their fallout in different regions of the country.

Keywords Assisted suicide · Canada · Eligibility · Euthanasia · Jurisdiction · MAID · Mental illness · Notwithstanding clause · Psychiatry · Supreme court

3.1 Introduction

Whether one views it in a positive or negative light, there can be little doubt that the Supreme Court of Canada's 2015 ruling in *Carter v Canada* was the most momentous federal court decision of the new century so far. It deemed the legalization of assisted suicide and voluntary active euthanasia (practices now referred to collectively under the joint label of 'medical assistance in dying' or MAID) to be constitutionally required. Chan and Somerville (2016, p. 144) write that *Carter v Canada* "is likely to go down as one of the most significant decisions ever issued by the Supreme Court of Canada." That ruling was translated into law by Parliament via Bill C-14, which received royal assent in June 2016.

However, that legislation was controversial before, during, and after its passage, and soon encountered multiple challenges in the provincial courts. The most significant of these came down in the form of a September 2019 ruling by Quebec Superior Court Justice Christine Baudouin, who deemed that the scope of eligibility laid out in Bill C-14 was too narrow (specifically, that restricting access to the terminally ill was

T. Dumsday (✉)
Concordia University of Edmonton, 7128 Ada Blvd., Edmonton, AB T5B 4E4, Canada
e-mail: travis.dumsday@concordia.ab.ca

J. Kotalik and D. W. Shannon (eds.), *Medical Assistance in Dying (MAID) in Canada*, The International Library of Bioethics 104,
https://doi.org/10.1007/978-3-031-30002-8_3

unconstitutional). This prompted the federal government to make amendments via Bill C-7, which received royal assent in March 2021. Moreover that bill held out the promise of further changes to come in the near future, and serious consideration of still *more* such changes, all of which would involve further broadening the eligibility criteria for receiving MAID.

Canadians, whether laypeople or academics, can surely be forgiven for any difficulties in keeping up with these shifts and keeping track of the still-unfolding debates. The aim of this chapter is to provide some assistance in understanding these developments: where we've been, where things are at now (as of spring 2022), and where things may be headed in the near future.

In the next section I summarize key components of *Carter v Canada* and how they were incorporated into federal legislation via Bill C-14. I also cover the legal challenges that bill immediately faced, culminating in the aforementioned Quebec Superior Court ruling. Then in section three I lay out the resulting amendments passed via Bill C-7 and some of the parliamentary and public debates related thereto. In the fourth and final section I engage in some informed speculation concerning the direction of MAID legislation and the debates surrounding it over the near term, suggesting that projected additional extensions of eligibility criteria may lead some provinces to invoke the notwithstanding clause in order to facilitate sidestepping their implementation. I further consider how the federal government might go about preempting such moves by putting a hold on certain components of Bill C-7.

3.2 *Carter v Canada* and Bill C-14

The proximate origins of this Supreme Court of Canada case go back ten years, to the 2012 ruling by the Supreme Court of British Columbia, also titled *Carter v Canada*.[1] That provincial court had declared unconstitutional those sections of the *Criminal Code of Canada* prohibiting assisted suicide[2] and voluntary active euthanasia.[3] Specifically, the court declared that those sections of the *CCC* violated plaintiffs' *Charter* rights to life, liberty, security of the person, and equality[4]; crucially, it found that they did so in ways *not* consistent with the principles of fundamental justice and

---

[1] *Carter v Canada* (2012).

[2] The criminal ban on assisted suicide was found in what had been Section 241(b) of the *CCC*, which read as follows: "Every one who aids or abets a person to commit suicide, whether suicide ensues or not, is guilty of an indictable offence and liable to imprisonment for a term not exceeding fourteen years." *Criminal Code* (1985), c. C-46, s. 241(b).

[3] This was found in what had been Section 14 of the *CCC*, which read as follows: "No person is entitled to consent to have death inflicted on him, and such consent does not affect the criminal responsibility of any person by whom death may be inflicted on the person by whom consent is given." *Criminal Code*, RSC 1985, c. C-46, s. 14.

[4] That is, those portions of the *CCC* were deemed by the BC Supreme Court to be incompatible with Sections 7, 12, and 15 of the *Charter*, which sections read as follows: "Everyone has the right to life, liberty and security of the person and the right not to be deprived thereof except in accordance with the principles of fundamental justice." (s 7) "Everyone has the right not to be subjected to any

not in ways that constituted reasonable limits on rights in a free democracy, thereby disallowing any loophole legal justifications for criminalizing MAID based on the relevant wording in Sections 7 and 1 of the *Charter*[5] (both of which sections had been deployed in the legal arguments of the Supreme Court of Canada when it had previously upheld the criminal prohibition on MAID in its 1993 *Rodriguez* decision).[6] Effectively then, a provincial supreme court had overturned[7] a prior ruling of the Supreme Court of Canada. This can of course happen in our system, though such rulings are subject to appeal, which is what happened in this case.

Thus in response, the Canadian government appealed to the BC Court of Appeal, which in 2013 overturned the Supreme Court of British Columbia and upheld the 1993 *Rodriguez* decision.[8] Criminal prohibition of MAID was preserved, but not for long.

After that setback, the plaintiffs appealed to the Supreme Court of Canada, which heard the case and issued its own *Carter v Canada* ruling in 2015. The federal court essentially affirmed the judgement of the Supreme Court of British Columba and deemed unconstitutional those sections of the *CCC* prohibiting assisted suicide and voluntary active euthanasia. In doing so they also effectively overturned their own prior ruling on the issue (the 1993 *Rodriguez* decision) and required that the Canadian government amend the *CCC* accordingly, such as to legalize and regulate MAID.[9]

In its ruling, the Supreme Court of Canada also laid down a basic framework for what those amendments would have to include in order to guarantee constitutionality:

> The appropriate remedy is therefore a declaration that s. 241(*b*) and s. 14 of the *Criminal Code* are void insofar as they prohibit physician-assisted death for a competent adult person who (1) clearly consents to the termination of life; and (2) has a grievous and irremediable medical condition (including an illness, disease or disability) that causes enduring suffering that is intolerable to the individual in the circumstances of his or her condition. "Irremediable", it

---

cruel and unusual treatment or punishment." (s 12) "Every individual is equal before and under the law and has the right to the equal protection and equal benefit of the law without discrimination and, in particular, without discrimination based on race, national or ethnic origin, colour, religion, sex, age or mental or physical disability." (s 15). *Canadian Charter of Rights and Freedoms*, ss. 7, 12 and 15, Part 1 of the *Constitution Act*, 1982, being Schedule B to the *Canada Act 1982* (UK), 1982, c 11.

[5] Section 1 reads as follows: "The *Canadian Charter of Rights and Freedoms* guarantees the rights and freedoms set out in it subject only to such reasonable limits prescribed by law as can be demonstrably justified in a free and democratic society." *Canadian Charter of Rights and Freedoms*, s. 1, Part 1 of the *Constitution Act*, 1982, being Schedule B to the *Canada Act 1982* (UK), 1982, c 11.

[6] *Rodriguez v British Columbia* (1993).

[7] Note that I am passing over some complexities here regarding the best descriptors to apply (whether 'overturning' or 'overruling' or some other term), and likewise passing over the question of whether the BC Court's stated intention to adhere to the principle of *stare decisis* in its ruling was actually met. My point is simply that the practical impact of the BC Court's ruling was that of striking down *Rodriguez*.

[8] *Carter v Canada* (2013).

[9] I have here skipped over the details of the legal reasonings set out in the four relevant court rulings (from 1993, 2012, 2013, and 2015). Readers interested in those details can consult my (2021, Chap. 1) for a summary.

should be added, does not require the patient to undertake treatments that are not acceptable to the individual.[10]

The court stated that it was temporarily suspending (for one year) the application of its decision, in order to give the federal government time to develop and pass a new legislative framework for MAID.

In theory the federal government needn't have implemented the ruling at all. Prime Minister Stephen Harper could have opted to invoke the notwithstanding clause, thereby effectively negating the court's ruling. It is not clear whether this option was ever seriously contemplated; when his justice minister, Peter MacKay, was asked about this by a reporter shortly after the decision (which came down on February 6, 2015), he declined to respond.[11]

For those who may profit from a quick refresher: the notwithstanding clause is found in Section 33 of the *Charter*, and grants authority to the federal government, as well as any of the provinces or territories, to pass and enforce laws that conflict with certain sections of the *Charter* (specifically Sections 2 and 7–15). By extension, it allows for a federal, provincial, or territorial government to pass and enforce legislation contravening court rulings pertaining to those sections of the *Charter*. In order to employ the notwithstanding clause, the relevant parliamentary body simply has to declare that they are doing so, in an act of the legislature, and renew the declaration in five years' time (if it wishes to continue the use of the clause).

The federal government could have used the notwithstanding clause to negate *Carter v Canada* and retain the criminal code prohibitions on assisted suicide and voluntary active euthanasia. The provinces and territories could have opted to do so as well, though the situation is more complicated in their case precisely because MAID touches on the criminal code, which is an area of federal jurisdiction. Still, there are ways that the provincial and territorial parliaments could have employed the clause on this issue (and indeed still could—more on this in Sect. 3.4).

The public heard little from the government concerning its plans until July 17, 2015, when MacKay and health minister Rona Ambrose announced the formation of a three-person expert panel charged with the task of developing a framework for implementing *Carter v Canada*. Before it could issue its recommendations, the federal election of October 19, 2015 brought the Trudeau Liberal government to power. The new government passed a motion to form a multi-party parliamentary committee to examine the issue, but delays in forming that committee, and the Supreme Court's looming February 6, 2016 deadline to pass relevant legislation, resulted in the government requesting an extension from the court. It was granted an extension, but only for four months, thus making June 6, 2016 the new deadline. In the meantime, the court also granted Quebec a legal exemption to continue with the implementation of its own MAID legislation, Bill 52, which had taken effect there in June 2014.[12]

---

[10] *Carter v Canada* (2015) at 390.

[11] For relevant reporting on this, see Ivison (2015).

[12] Quebec had earlier argued that rules concerning MAID fell under provincial jurisdiction, given the constitutional responsibility of the provinces for health care. There had been a legal debate over whether that claim was legitimate (the criminal code after all is under federal jurisdiction, and issues

The federal multi-party committee concluded its work and issued its report to the government, which in April 2016 tabled the first draft of what would become Bill C-14. It was substantially more conservative than what the multi-party committee had recommended. (That committee's set of suggestions had included permitting MAID for the legally competent mentally ill, for mature minors, and for legally incompetent individuals who had signed an advance directive prior to becoming legally incompetent—for instance Alzheimer's suffers writing in the early stages of the disease). Most controversially, the government's draft bill restricted eligibility for MAID to those for whom death was 'reasonably foreseeable'—while it did not specify a particular timeline (e.g., there was no requirement that a physician certify that the patient had six months to live or less),[13] the import seemed clear: assisted suicide and voluntary active euthanasia in Canada would be legally available only to the terminally ill, even if the precise borderlines of what counted as terminal illness were left indeterminate.

The provision that death had to be reasonably foreseeable caused a good deal of resistance in the Senate, where some senators feared that its inclusion rendered the draft bill in conflict with the requirements the Supreme Court laid down in *Carter v Canada* (where no restriction to the terminally ill was mentioned). The ongoing debate in the Senate delayed passage of the bill, such that the government missed the court-imposed June 6, 2016 deadline to pass MAID legislation. For a brief period following the passage of that deadline then, when it came to MAID the provisions mentioned in *Carter v Canada* were effectively the law of the land. Finally though, the government was able to push the legislation through the senate, and the final version of Bill C-14 received royal assent on June 17, 2016.

Key provisions of the bill included: (1) physicians, nurse practitioners, pharmacists and anyone assisting them in an approved case of MAID would be indemnified from criminal prosecution; (2) those receiving MAID had to be legally competent adults eligible to receive health care services funded by a Canadian province or territory; (3) they must have made a formal request for MAID in writing, before two independent witnesses with no known financial interest in their death and in the absence of any coercion; (4) they must be suffering from a grievous and irremediable illness, disease or disability prompting physical or psychological suffering that is enduring and intolerable, and their death must be reasonably foreseeable (though without necessarily having a prognosis specifying the length of time they have remaining); (5) under all but exceptional circumstances, there must be a ten-day waiting period between the approval of the formal request and the carrying out of the procedure.[14]

---

of MAID seem relevant to both health care and criminal justice); however, the 2015 *Carter* ruling effectively rendered that dispute moot. For more on Quebec's Bill 52 and the events leading up to its passage, see Martin (2016, Chap. 9).

[13] When the state of Oregon legalized assisted suicide (but not active euthanasia) in 1997, that was the timeline it specified.

[14] Bill C-14 (2016).

The bill faced immediate legal challenges, with the focus predictably falling on the 'reasonably foreseeable death' criterion. The BC Civil Liberties Association filed a suit in the BC Superior Court just eleven days after the bill's passage, arguing that that provision of the bill was unconstitutional.[15] Other challenges followed, with the result that a provincial court eventually declared that aspect of the bill to be in violation of the *Charter*; this came down in the form of a ruling in September 2019 by Quebec Superior Court Justice Christine Baudouin.[16] Shortly thereafter the federal election campaign got underway, with MAID policy becoming a wedge issue. The Trudeau government vowed to accept the Quebec court ruling and amend the legislation accordingly, while the Conservative opposition party, led by Andrew Scheer, promised to appeal the ruling. The Liberals won (though now with a parliamentary minority), and proceeded to fashion Bill C-7.

3.3 Bill C-7 and Its Expansion of Eligibility for MAID

The main point of Bill C-7 (which received royal assent on March 17, 2021) was to amend Canada's MAID legislation to rid it of the 'reasonably foreseeable death' eligibility criterion. Those applying for MAID must still be suffering from a grievous and irremediable medical condition causing enduring and intolerable physical or psychological suffering, but that medical condition no longer need be terminal. Somewhat more stringent review and oversight requirements have been put in place for the non-terminally ill, however, such as a mandatory 90-day assessment period before approval of the procedure. Safeguards for the terminally ill have been lessened, such that the original 10-day waiting period post-approval has been removed. Moreover, in all cases the requestor no longer needs two independent witnesses, but only one (and that witness can now be a healthcare provider, so long as s/he isn't the one who will carry out the MAID procedure). Bill C-7 also committed the federal government to study proposals for extending MAID eligibility to those suffering solely from mental illnesses; furthermore, in response to pressure from within the Senate, the government adopted the express intention of extending eligibility to that class of individuals after a two-year period of study and consultation.[17] During that period

[15] For further details on that suit, see Carter and Rodgerson (2018, pp. 798–799).

[16] See for instance the reporting by Curtis (2019), Ha and Grant (2019).

[17] The wording used in the bill itself does not make this latter intention clear, but the accompanying press release issued by the Department of Justice bears out the fact that this is the government's plan. It states that "the new law…temporarily excludes eligibility for individuals suffering solely from mental illness for 24 months, and requires the Ministers of Justice and Health to initiate an expert review tasked with making recommendations within the next year on protocols, guidance and safeguards for MAID for persons suffering from mental illness." Posted March 17, 2021 and available at www.canada.ca/en/department-justice/news/2021/03/new-medical-assistance-in-dying-legislation-becomes-law.html (accessed January 31, 2022).

See also this portion of the Justice Department's explanatory site on Bill C-7: "Canadians whose only medical condition is a mental illness, and who otherwise meet all eligibility criteria, will not

it will also consider extending eligibility to mature minors and permitting the use of advance directives.[18]

Bill C-7 underwent extended debate in both the House of Commons and the Senate. Its passage through the latter was comparatively smooth (at least by comparison with the aforementioned difficulties involved in obtaining Senate approval for its predecessor Bill C-14). It passed by a vote of 60–25, with 5 senators abstaining. Among those most resolutely opposed to the legislation was Senator Denise Batters: “I’ve been a mental health advocate in Saskatchewan for more than a decade since my husband passed away due to suicide after battling severe anxiety and depression. ‘Irremediable’ means it will never get better and you just can’t say that about mental illness….Mental illness is treatable. It’s not incurable.”[19] Prominent on the opposite side of that aspect of the legislation was Senator Stan Kutcher, a psychiatrist and proponent of extending MAID access to psychiatric patients. During the period in which the Senate was debating that issue specifically, he made the following comments in the Senate: “Intolerable suffering is a subjective, personal experience. It cannot be negated or delegitimized by anyone else’s valuation of that suffering….Persons who have intolerable suffering from a mental disorder do not have a second-class type of suffering. Their suffering must be taken just as seriously as we take the suffering of those who request MAID for any other medical condition.”[20]

Various public interest organizations were active in discussions on Bill C-7, both leading up to its passage and in its aftermath.[21] For instance the Canadian Mental Health Association (CMHA) objected to the government’s intention to extend MAID eligibility to those suffering solely from mental illnesses, with Phyllis O’Connor (executive director of CMHA Saskatchewan) giving the following remarks on the day after the bill’s passage: “We are definitely not in favour of allowing mental illness to be the reason to request medically assisted suicide. Mental illness, while we’re not discounting how serious the suffering can be, it’s not a terminal illness. There’s always hope, with proper supports and access to services, for recovery.”[22] By contrast, Dying with Dignity Canada issued a statement hailing it as “a momentous day for end-of-life rights in Canada.”[23]

---

be eligible for MAID until March 17, 2023. This includes conditions that are primarily within the domain of psychiatry, such as depression and personality disorders. It does not include neurocognitive or neurodevelopmental disorders, or other conditions that may affect cognitive abilities. This temporary exclusion will provide the Government of Canada with more time to study how MAID on the basis of a mental illness can be safely provided and to ensure appropriate safeguards are in place to protect those persons.”.

www.justice.gc.ca/eng/cj-jp/ad-am/bk-di.html (accessed January 31, 2022).

[18] Bill C-7 (2021).

[19] Her comments are included in reporting by Davis (2021).

[20] As reported by Bryden (2021a, b).

[21] I have opted to spend more time here on public engagement with, and reaction to, Bill C-7 than on Bill C-14, insofar as a good deal has already been written on the latter (including my own coverage in the opening chapters of Dumsday (2021)).

[22] See again Davis (2021).

[23] As reported by Bryden (2021a).

Disability advocacy organizations were particularly engaged in the process, with more than 300 openly opposing Bill C-7's expanded eligibility criteria; some suggested that this expansion, even though done in response to a court ruling (from the Quebec superior court), would itself face constitutional challenges due to its undermining of the rights of the disabled.[24] Some of these organizations also expressed concerns that despite a public consultation period, the passage of Bill C-7 through the House and Senate was rushed, and even biased. The Council of Canadians with Disabilities joined with other organizations in issuing the following statement:

> The Canadian disability-rights community remains united in denouncing Bill C-7 as an assault on the Equality Rights of people with disabilities. The flaws — and, indeed, the overt pro-MAID bias — which characterized the 'consultation' process leading up to the tabling of Bill C-7 are well documented. In short, everything from the online questionnaire to the in-person consultations were geared towards a predetermined outcome, namely, the expansion of Medical Assistance in Dying as a legally and socially sanctioned substitute for assistance in living that we seen in Bill C-7.[25]

Others suggested that such concerns were exaggerated, and that the safeguards put in place to ensure that MAID remains voluntary and uncoerced constitute adequate protection for the disabled. Jocelyn Downie, professor of law and medicine at Dalhousie University and a prominent commentator on MAID, argued that Bill C-14 had "prevented them from accessing MAID and was grounded in quite paternalistic concerns, basically protecting people from themselves."[26] On this view, prohibiting the non-terminal disabled from obtaining MAID is an affront to their dignity and a violation of their equality rights.

There were also expressions of concern regarding how MAID eligibility expansion might impact other vulnerable segments of society. Dr. Naheed Dosani testified before the Senate and later wrote the following in an op-ed piece:

> I was asked to testify — to share my perspective as a palliative care physician who provides street-based care for people with serious illnesses who also experience structural vulnerabilities like homelessness, poverty, and systemic racism. Why is this perspective integral to the discussion about Bill C-7? I work in a world where it is possible successfully to arrange for MAID in two weeks in an organized and efficient fashion. Yet, it takes years to get the people I care for into housing, months to get them income supports, and weeks to get mental health and harm reduction treatment essential to a good quality of life. I find this morally distressing. If we are making it easier for people to get MAID, we also have a moral obligation to ensure that people don't pursue MAID because they want to escape a society that doesn't adequately support their needs....I have witnessed how treating the social determinants of health for the people I care for can make a tremendous difference as to whether or not someone pursues MAID. 'Mary,' a woman in her 30s, is an example. She was on the streets living with untreated HIV, dealing with a substance use disorder and no health care team to meet her needs. Mary wanted to die through MAID. However, after addressing her emotional and physical pain, supporting her with housing, income, harm reduction services

[24] See Harris and Bryden (2020).

[25] Www.ccdonline.ca/en/humanrights/ [accessed March 20, 2022].

[26] As reported by Duong (2021).

and providing the kind of psychosocial supports that all Canadians deserve, she changed her mind because she now has a better quality of life.[27]

Indigenous leaders likewise spoke out on Bill C-7 in the debate leading up to its passage. Fifteen First Nations representatives and front-line healthcare workers addressed an open letter to Senators and MPs titled *Indigenous Peoples Should Not be Compelled to Provide or Facilitate Medical Assistance in Dying*. Signatories included former Senator Nick Sibbeston and former Lieutenant Governor of New Brunswick Graydon Nicholas. There they argued that "Bill C-7 goes against many of our cultural values, belief systems, and sacred teachings. The view that MAID is a dignified end for the terminally ill or those living with disabilities should not be forced on our peoples."[28] The letter went on to express the concern that Indigenous Canadians are already especially vulnerable to discrimination within the healthcare system and would be particularly vulnerable to receiving unsolicited counsel favouring the option of MAID.

Legal and medical professional organizations also contributed to the discussion, with the Canadian Bar Association supporting the extension of MAID eligibility to the non-terminally ill (including to those suffering solely from psychiatric conditions) ever since *Carter v Canada*.[29] The Quebec Psychiatric Association (Association des medecins psychiatres du Quebec) and the Canadian Psychiatric Association had also come out in favour of these eligibility extensions.[30]

3.4 MAID in Canada and the Coming Invocation of the Notwithstanding Clause

The legislative trend in Canada is clearly towards expansion of the eligibility criteria for MAID. But the federal government is moving too quickly, and in a manner too heedless of majority public opinion. Polling has shown consistently that the

---

[27] Dosani (2021). In a related vein, psychiatrist Maher (2020) shared his experience working with a young person requesting MAID: "A few days ago, a 30-year old patient with very treatable mental illness asked me to end her life. Her distraught parents came to the appointment with her because they were afraid that I might support her request and that they would be helpless to do anything about it. It's horrific that they have to worry that by going to a psychiatrist, their daughter might be killed by that very psychiatrist. That same patient told me, 'a doctor killing me is not suicide, it's totally different.' The perverse veneer of moral acceptability that follows from medicalizing what is patently not a medical act is infuriating in its obfuscatory power....I have been specializing in treatment-resistant mental illness for 17 years. I want to scream from the political and judicial mountaintop that 'treatment-resistant' does not mean untreatable. It means clinical experience and sophistication are needed....[E]very single person with severe mental illness can experience dramatic improvement in their symptoms and concomitant reductions in their suffering....I have yet to see a patient with irremediable suffering unless left untreated by inadequate availability of services.".

[28] Quoted in the reporting by Ruck (2021).

[29] See the summary by Steger (2021).

[30] As noted by Gupta and Downie (2022).

majority of Canadians (like the majority of those polled in other sampled industrialized nations, with the exception of Belgium and the Netherlands) does not approve of making MAID available to those suffering solely from psychiatric conditions.[31] On an individual level (even if not at the level of their national professional bodies) Canadian psychiatrists also remain opposed to that extension, and by a wide margin.[32] Consequently, if the present Liberal government persists in its intention to extend MAID eligibility to the mentally ill, they may pay a price for it in the next election. Moreover it is entirely possible that a future government will reverse this policy regarding MAID and mental illness; and since neither *Carter v Canada* nor any other federal or provincial court ruling has yet deemed this specific extension constitutionally necessary, no invocation of the notwithstanding clause would be needed by a future government to reverse it.

Would it be necessary to invoke the notwithstanding clause to go even further, and restore MAID eligibility to the narrower criteria originally laid down in Bill C-14 (which essentially restricted access to the terminally ill)? Not necessarily. Parliament might instead choose to repeal Bill C-7 entirely and then appeal the 2019 Quebec Superior Court ruling that judged unconstitutional that aspect of Bill C-14. Only if an appeals court (and ultimately the Supreme Court of Canada) upheld the constitutional necessity of extending MAID eligibility to the non-terminally ill would the government then have to invoke the notwithstanding clause in order legally to restore a more narrowly circumscribed set of access conditions.

It is also entirely possible that the Trudeau Liberals remain in power and that no such actions are taken in the near future. Would that mean that the MAID eligibility landscape would have to remain as presently envisioned? No.

Consider the following hypothetical: MAID eligibility extension proceeds as planned. Nevertheless, public opposition in some regions of the country remains high, prompting certain provinces to take action on their own. The premiers of Alberta and Saskatchewan declare that they won't permit the mentally ill to be euthanized in their jurisdictions. Lacking control over the criminal code, they cannot directly alter criminal laws pertaining to MAID. However, they can seek to restrict access to MAID by imposing new regulatory measures on healthcare practitioners, for instance declaring that physicians involved in euthanizing psychiatric patients would see their right to practice medicine in that jurisdiction revoked. Just because a medical procedure is legally permitted in Canada, that does not legally obligate a province to fund it, or even to permit its healthcare workers to participate in it without suffering professional penalties. There would likely be widespread public support for such measures in certain regions, such that the prospect of such actions being taken is not a far-fetched one—perhaps not *likely,* but still within the realms of political possibility. Whether or not such provinces could withstand subsequent federal pressure (e.g., punitively limiting health care funding transfers), and over the long-term, is another matter; it would likely depend on the depth and breadth of public opposition to the expansion of MAID eligibility to psychiatric patients.

[31] See my review of polling data in Dumsday (2021, pp. 131–132).

[32] See Rousseau et al. (2017).

None of that would strictly require a province or territory to invoke the notwithstanding clause, but a jurisdiction opting for such a course of action would likely be in a stronger position vis a vis future conflict with the federal government if it first followed the same steps suggested above: appeal the 2019 Quebec Superior Court ruling (which had required that MAID be available to the *non-terminally* ill) to the Supreme Court of Canada, and then if needs be invoke the notwithstanding clause to nullify any constitutional requirement to permit MAID access by the non-terminally ill (which would entail that they would likewise not have to grant MAID access to those suffering solely from psychiatric conditions).

If the federal government does proceed with MAID expansion to psychiatric patients (and especially if it does so on the compressed timeline presently envisioned),[33] the provinces ought to engage in the sort of resistance measures just laid out. However, an even better option would be for our federal Parliament itself to repeal Bill C-7 and proceed to appeal the 2019 Quebec Superior Court ruling, if needs be all the way to the Supreme Court of Canada. This would by itself put a pause on implementing MAID eligibility expansion and provide an opportunity for greater and more widespread public consultation in a post-pandemic environment far more conducive to it. It would also give the federal Supreme Court a say in the matter, which seems appropriate; it is unusual (and perhaps objectionable) for a lower court ruling to have such an outsized impact on a deeply controversial matter of vital national interest. On such matters the federal court ought to be heard.

Acknowledgements I would like to extend my sincere thanks to Jonas-Sebastien Beaudry, Jaro Kotalik, and David Shannon for their many helpful comments on earlier drafts of this chapter. The final product is much improved on account of their valuable input.

[33] Even if this change in MAID eligibility must proceed, surely it oughtn't to proceed so quickly. The psychiatric community needs more time to study best practices for integrating MAID into the existing standards of care for treating the mentally ill, more time to train psychiatrists to implement those practices effectively, and more time to settle regulatory complexities. E.g., what evidence-based medical criteria are available for psychiatrists to employ when distinguishing remediable from irremediable mental illness? (The latter is, after all, not a generally recognized diagnostic category within current psychiatry—from a medical perspective, it is essentially a legal fiction.) What range of mental illnesses should qualify one for MAID access? Should psychiatrists be allowed to suggest and/or encourage MAID as an option for their patients, or should they be required always to wait for the patient to raise the issue themselves? If the latter, what penalties should be put in place for those who engage in the former activity? Etc. It is difficult to believe that these and a host of other pressing questions can be adequately addressed within the current projected timeline (i.e., within the next ten months). As this chapter goes to typesetting in April 2023, I can provide a short update: in response to widespread pressure (especially from within the Canadian psychiatric community), the federal government did in fact announce a delay in its implementation of this eligibility extension, pushing it back a year to March 2024.

References

Legal

Bill C-14. 2016. An act to amend the criminal code and to make related amendments to other acts (medical assistance in dying), 1st session, 42nd Parliament.

Bill C-7. 2021. An act to amend the criminal code (medical assistance in dying), 2nd session, 43rd Parliament.

Carter v Canada. 2012. BCSC 1587.

Criminal Code, RSC. 1985.

Carter v Canada. 2013. BCCA 435.

Carter v Canada. 2015. SCC 5.

Canadian Charter of Rights and Freedoms, ss. 7, 12 and 15, Part 1 of the Constitution Act, 1982, being Schedule B to the Canada Act 1982 (UK). 1982. c 11.

Rodriguez v British Columbia. 1993. 3 SCR 519.

General

Bryden, Joan. 2021a. *Senate approves amendment to assisted-dying bill*. Canadian Press, February 9

Bryden, Joan. 2021b. *Canadians not near death gain access to assisted dying as senate passes bill C-7*. Canadian Press, March 17

Carter, Rose, and Rodgerson, Brandyn. 2018. Medical assistance in dying: Journey to self-determination. *Alberta Law Review* 55: 777–803.

Chan, Benny, and Margaret Somerville. 2016. Converting the 'right to life' to the 'right to physician-assisted suicide and euthanasia': An analysis of *Carter v Canada (Attorney General)*, Supreme Court of Canada. *Medical Law Review* 24: 143–175.

Curtis, Christopher. 2019. Quebec court nullifies euthanasia laws as too restrictive. *National Post*, September 12.

Davis, Stefanie. 2021. 'There's always hope': CMHA against bill C-7 changes that allow medical assistance in dying for people with mental illness. *CTV News Regina*, March 18.

Dosani, Naheed. 2021. If medically assisted death becomes more accessible for Canadians, we have a moral obligation to make living well—Through housing, mental health supports—Accessible too. *The Star*, February 11.

Dumsday, Travis. 2021. *Assisted suicide in Canada: Moral, legal, and policy considerations*. Vancouver, BC: University of British Columbia Press.

Duong, Diana. 2021. How can Canada safeguard those marginalized by society as MAID expands? *Canadian Medical Association Journal News*, March 19.

Gupta, Mona, and Downie, Jocelyn. 2022. Quebec's commission on MAID misses the mark on mental disorders. *Policy Options*, March 2.

Ha, Tu Thanh, and Grant, Kelly. 2019. Quebec court strikes down restriction to medically assisted dying law, calls it unconstitutional. *Globe & Mail*, September 12.

Harris, Kathleen, and Bryden, Joel. 2020. Proposed changes to assisted-dying law make death a viable option for people with disabilities, advocate fears. *CBC News*, December 17.

Ivison, John. 2015. The public backs court on decision. *National Post*, February 7.

Maher, John. 2020. Why legalizing medically assisted dying for people with mental illness is misguided. *CBC News Opinion*, February 11.

Martin, Sandra. 2016. *A good death: Making the most of our final choices*. Toronto, ON: HarperCollins.

Rousseau, Skye, Sarah Turner, Harvey Chochinov, Murray Enns, and Jitender Sareen. 2017. A national survey of Canadian psychiatrists' attitudes toward medical assistance in death. *Canadian Journal of Psychiatry* 62: 787–794.

Ruck, Agnieszka. 2021. First nations leaders say bill c7 goes against their beliefs and values. *Canadian Catholic News*, February 12.

Steger, Monika. 2021. *Update on medical assistance in dying in Canada.* British Columbia Law Institute [bcli.org] September 7. Accessed March 20, 2022.

Travis Dumsday (Ph.D, University of Calgary, 2010) is Associate Professor of Philosophy at Concordia University of Edmonton. He publishes regularly in applied ethics, philosophy of science, mediaeval philosophy, and Eastern Orthodox theology. He held the Canada Research Chair in Theology and the Philosophy of Science (2015–2020) at CUE.

Chapter 4
"We Should Not Lightly Assume": A Review of Legislative, Regulatory and Jurisprudential Developments of MAID Safeguards

David W. Shannon

Abstract In the Supreme Court of Canada decision of *Carter v. Canada* 2015, the Court approved medical assistance in dying on a limited basis. Central to the limitation was what the Supreme Court termed "a well-regulated regime". As a response to this decision, the government of Canada amended the *Criminal Code* in 2016 with Bill C-14. It was further amended with Bill C-7, which received Royal assent in March 2021. Healthcare and the implementation of its operations falls within provincial and territorial jurisdiction. As a result, the federal government, as well as each province and territory has been required to provide a MAID regulatory regime. To a greater and lesser extent, this has begun. This paper will provide an overview of regulatory measures, guidelines, and safeguards arising from the Carter decision. It will also consider whether the objectives articulated by the Supreme Court of Canada are being met by the federal and provincial legislatures in Canada, and whether they reflect sound public policy.

Keywords Regulations · Legislation · Guidelines · Irremediable · Safeguards · Criminal code of Canada

4.1 Introduction

The creation of a regulatory regime holds unique opportunities within the nexus of policy and law as well as in its intersection between the public, politicians, and professionals serving the end consumers. It may be argued that this is acutely so when applied to the termination of a life through medical assistance in dying (MAID).

It would also appear that both the Supreme Court of Canada (SCC) and the Federal Government have placed great confidence in the ability of regulations to ensure that MAID is delivered without mistakes or abuse. The SCC noted in *Carter v Canada* (Attorney General), "We should not lightly assume that the regulatory regime will

---

D. W. Shannon (✉)
Shannon Law, Thunder Bay, ON, Canada
e-mail: shannonlaw@tbaytel.net

J. Kotalik and D. W. Shannon (eds.), *Medical Assistance in Dying (MAID) in Canada*, The International Library of Bioethics 104,
https://doi.org/10.1007/978-3-031-30002-8_4

function defectively, nor should we assume that other criminal sanctions against the taking of lives will prove impotent against abuse"[1] before concluding at paragraph 147 that:

> Section 241 (*b*) and s. 14 of the Criminal Code unjustifiably infringes S. 7 of the Charter and are of no force or effect to the extent that they prohibit physician-assisted death for a competent adult person who (1) clearly consents to the termination of life and (2) has a grievous and irremediable medical condition (including an illness, disease or disability) that causes enduring suffering that is intolerable to the individual in the circumstances of.[2]

The Federal Government affirmed both its confidence in regulations and the central role that Provincial and Territorial Governments play in MAID when stating in the preamble of Bill C-7, "recognizing the provinces' jurisdiction over various matters related to medical assistance in dying, including the delivery of health care services and the regulation of health care professionals…"[3] With approximately 10,000 in 2021, and 7000 Canadian deaths through MAID in 2020, and indications that these numbers may increase, the question of whether the professed SCC confidence in the Federal, Provincial and Territorial governments is well placed or profoundly naïve, should be debated on such an important topic as the termination of individual Canadian lives. To open this discussion, a review of the applicable Criminal Code amendments, provincial legislation, and regulations and guidelines is a necessary point of departure. Much of the specific information related to the law of MAID can also be found in other chapters, and due to the limitations of this paper as well as the evolving nature of MAID law and regulations, this review will purposefully be broad in scope. However, findings will be provided with discussion on strengths, gaps and areas for future research, policy, or jurisprudential development.

4.2 The Federal Approach to Safeguards, *Criminal Code of Canada,* and MAID Regulations

4.2.1 Bill C-14: Exclusion from Criminal Liability and Eligibility Requirements

As a response to the SCC's conclusion that a total ban on euthanasia in Canada was unconstitutional, Bill C-14, *An Act to Amend the Criminal Code and to make related amendments to other Acts (medical assistance in dying)*[4] was introduced in the House

---

[1] Carter v Canada (Attorney General) (2015), para 120.

[2] Carter v Canada para 147.

[3] Bill C-7 (2021) [hereinafter "Bill C-7"] which received Royal Assent March 17, 2021. Online https://www.parl.ca/DocumentViewer/en/43-2/bill/C-7/royal-assent.

[4] Bill C-14, *An Act to amend the Criminal Code and to make related amendments to other Acts (medical assistance in dying)*, S.C. 2016, c. 3 at Section 241 [hereinafter "Bill C-14"] which received Royal Assent June 17, 2016.
Online https://www.parl.ca/DocumentViewer/en/42-1/bill/C-14/royal-assent.

of Commons on April 14, 2016. The Bill set out the requirements for the provision of MAID, established exemptions to various *Criminal Code* offences for physicians, nurse practitioners (NP), pharmacists and certain other persons who provide or assist in the provision of MAID, and broadly pointed to regulatory requirements.[5]

The amended Section 241 of the *Criminal Code* excludes physicians and NPs or individuals who aid a physician or NP, from culpable homicide if they provide MAID and/or a noxious substance. Pharmacists who dispense a substance prescribed for MAID[6] and persons who aid a person, at their explicit request, to self-administer a substance prescribed as part of the provision of MAID are exempt from criminal culpability.[7] Also, if a person has a reasonable but mistaken belief about any fact that is an element of the exemption, they may be exempted from the offences of homicide or aiding a suicide.[8]

4.2.2 *Bill C-14: Conditions of Eligibility*

The *Criminal Code* created criteria for MAID comprising of five substantive and numerous procedural safeguards.

Substantive Criteria

At a substantive level, eligibility is comprised of ensuring that the person: is eligible for government-funded health services in Canada[9]; is at least 18 years old and capable of making decisions with respect to his or her health[10]; has a grievous and irremediable medical condition[11]; has made a voluntary request for MAID that, in particular, was not made as a result of external pressure; and has given informed consent to receive MAID after having been informed of the alternative means that are available to relieve their suffering, including palliative care.[12] Furthermore, the *Criminal Code* clarifies that the phrase "grievous and irremediable medical condition" is defined as requiring all the following criteria:

- the person has a serious and incurable illness, disease or disability;
- the person is in an advanced state of irreversible decline in capability;
- the illness, disease or disability or the state of decline causes enduring physical or psychological suffering that is intolerable and cannot be relieved under conditions that the person considers acceptable, and

[5] *The Criminal Code of Canada* (R.S.C. 1985) at Section 241 [hereinafter *"Criminal Code"*].

[6] *Criminal Code*, note 5 at S. 241 (4) and at S. 245 (2).

[7] *Criminal Code*, note 5 at S. 241 (5).

[8] *Criminal Code*, note 5 at S. 227 (3). at S. 245 (2).

[9] *Criminal Code*, note 5 at S. 241.2(1)(a).

[10] *Criminal Code*, note 5 at S. 241.2(1)(b).

[11] *Criminal Code*, note 5 at S. 241.2(1)(c).

[12] *Criminal Code*, note 5 at S. 241.2(1)(d).

- natural death has become reasonably foreseeable, taking into account all of the medical circumstances, although a prognosis as to the specific length of time remaining is not necessary.[13]

Procedural Safeguards

In addition to the substantive criteria noted above, Bill C-14 also included procedural safeguards in the MAID process. Before the physician or NP provides MAID, they must:

- believe the conditions of eligibility in Section 241.2(1) listed above are satisfied[14];
- ensure that the request was made in writing and signed by the patient after a physician or NP had told the patient that the patient had a grievous and irremediable medical condition[15];
- be satisfied that the written request was signed and dated by the patient, or another person authorized to do so before two witnesses[16];
- ensure that the person is informed that they may withdraw the request at any time and in any manner, including immediately before the provision of MAID[17];
- ensure that another physician or NP has provided a written assessment confirming that the person meets the criteria[18];
- be satisfied that both physicians or NPs are independent[19];
- ensure that **10** clear days elapse between the day the request is signed and the day when MAID is provided[20] (this criterion was later repealed under Bill C-7), and
- provide a reliable means of communication in situations where a patient has difficulty communicating.[21]

The physician or NP must provide MAID with reasonable knowledge, care, and skill and in accordance with applicable provincial laws, rules, and standards.[22] A physician or NP who prescribes or obtains a substance for use in MAID must also inform the pharmacist of the intended purpose before that pharmacist dispenses the substance.[23]

---

[13] *Criminal Code*, note 5 at Section 241.2(2).

[14] *Criminal Code*, note 5 at Section 241.2(3)(a).

[15] *Criminal Code*, note 5 at Section 241.2(3)(b).

[16] *Criminal Code*, note 5 at Sections 241.2(3)(c) and 241.2(5).

[17] *Criminal Code*, note 5 at Section 241 (3.1)(k).

[18] *Criminal Code*, note 5 at Section 241.2(3)(e).

[19] *Criminal Code*, note 5 at Sections 241.2(3)(f) and 241.2(6).

[20] *Criminal Code*, note 5 at Section 241.2(3)(g).

[21] *Criminal Code*, note 5 at Section 241.2(3)(i).

[22] *Criminal Code*, note 5 at Section 241.2(7).

[23] *Criminal Code*, note 5 at Section 241.2(8).

4.2.3 *New Criminal Offences and Obligations to File Information*

Bill C-14 introduced new criminal offences for failing to comply with the safeguards, forging or destroying documents related to MAID, failing to provide the required information and contravening the regulations.[24] A breach in the safeguards outlined in Sections 241.3 and 241.4 are hybrid offences that may be prosecuted by way of indictment with a maximum of five years' imprisonment or as summary conviction offences with a maximum of 18 months' imprisonment.[25]

4.2.4 *Bill C-7*

Bill C-14 was subsequently amended creating a two-track approach to MAID, one where individuals' death was imminent, and another for those whom natural death was not naturally foreseeable. In the instance where natural death is not reasonably foreseeable, the applicant must wait at least 90 days before MAID can be administered, but this period can be shortened if the person is about to lose the capacity to make health care decisions, provided both assessments have been completed,[26] then immediately before MAID is provided, the practitioner must give the person an opportunity to withdraw their request and ensure that they give express consent.[27] Should a person whose natural death is reasonably foreseeable be found eligible to receive MAID, not only is the 10 day waiting period removed, but where there is a reasonable risk of their losing capacity to consent directly before MAID is carried out, an advanced written agreement can be arranged.

For individuals subject to the 90-day wait period, requests must be made in writing and signed by one independent witness. Requests must be made after the person has been informed that they have a "grievous and irremediable medical condition". Two independent physicians or NPs must provide an assessment and confirm that all the eligibility requirements are met. If neither of the two practitioners assessing eligibility has expertise in the medical condition that is causing the person's suffering, they must consult with a practitioner who has such expertise.[28] This assessment of patient eligibility, however, does not adequately address the dissenting report recommendations made by the Special Joint Committee on Physician-Assisted Dying in which they noted that:

---

[24] Bill C-14, note 4 at Sections 241.2(3)(b), 241.2(3)(i) and 241.4.

[25] *Criminal Code*, note 5 at Sections 241.3 and 241.4.

[26] *Criminal Code*, note 5 at Section 241.1 (3.1)(i); *Criminal Code*, note 5 at Section 241.1(3.1)(d).

[27] *Criminal Code*, note 5 at Section 241.1(3.1)(d).

[28] *Criminal Code*, Note 5 at Section 241.1(3.1)(a–k).

- not requiring a psychiatric assessment by a psychiatric professional for a patient "diagnosed with an underlying mental health challenge" would put vulnerable persons at risk.
- palliative care should be "offered and available to any person contemplating [MAID]."[29]

The person must also be informed of available and appropriate means to relieve their suffering, including counselling services, mental health and disability support services, community services, and palliative care, and must be offered consultations with professionals who provide those services. The person and the practitioners must have discussed reasonable and available means to relieve the person's suffering and agree that the person has seriously considered those means.[30] This approach presupposes that all health care practitioners are experts in disability supports, independent living modalities, social work, psychology, and palliative care. It adopts a potentially inadequate standard of information provision which could result in less informed decision-making and reduced sense of control among patients. Furthermore, it diminishes the importance of a full relationship between patient and healthcare provider.[31]

4.2.5 MAID and Foreseeability: A Jurisprudential Perspective

Bill C-7 does not clearly delineate where the line between immanent death, which does not require a wait period and where death that is not naturally foreseeable commences. Thus, there remains ambiguity around when the 90-day wait period begins. The courts have also been unclear through their deference to all medical practitioners. For example, in the matter of A. B. v. Canada, Justice Perell stated that:

> …natural death need not be imminent and that what is a reasonably foreseeable death is a person specific medical question to be made without necessarily making, but not necessarily precluding, a prognosis of the remaining lifespan.[32]

Arguably, this approach creates some confusion. From this wording, one may be left to believe that foreseeable is not imminent, but it is unreasonable to presume that

[29] Senate and House of Commons, Special Joint Committee on Physician-Assisted Dying (2016), p. 53.

[30] *Criminal Code*, Note 5 at Section 241.2(3.1)(g).

[31] See also RAA v SM (2021) where the Health Professions Appeal and Review Board of Ontario stated "that when discussion of Medical Assistance in Dying (MAID) arises, physicians should be aware of all alternatives that are available and the process, and if asked to start the process, they should assist the patient personally or refer the patient to a group that helps patients with the process and otherwise and to take no further action".

[32] A.B. v Canada (Attorney General) (2017) at para 79 [hereinafter "A.B. v Canada"].

it cannot be imminent, while not being time specific. It is person specific and requires that the person must have a natural death. The lack of clarity continued in A.B. v Canada, when Justice Perell decided not to decide, and then left MAID eligibility determination to the medical profession when stating, "These are matters at the core if not the whole corpus of medical knowledge and better known to doctors than to judges."[33]

This ambiguity therefore makes it difficult to determine in what circumstances the 90-day wait period is triggered and what safeguards would then apply. Empirical data on psycho-social adjustment to significant disability and/or chronic illness indicates that a 90-day wait period is a dangerous rush to judgement.[34] For example, individuals experiencing a newly acquired disability and/or chronic illness when given the opportunity to adjust through grief, psychological, physical and social supports, reported that their quality of life was approximately equivalent to able-bodied persons.[35] This suggests that the 90-day wait period is arbitrary and disconnected from accepted empirical evidence. It is also deeply flawed by omitting to include the need for a psychiatric consultation to probe whether there is depression or suicidality for which its suffering can be treated. The amendment that people can access MAID where death is not reasonably foreseeable may also reinforce negative stereotypes about disability and can therefore be viewed as discriminatory.[36]

Notwithstanding the possible pitfalls surrounding the triggering mechanism for the 90-day wait period, as was demonstrated in Sorensen v Swinnemar, the courts have made it clear they will not become involved in the monitoring or oversight of MAID eligibility.[37] In Sorensen v Swinnemar, the Nova Scotia Court of Appeal (NSCA) decided that 'neither this Court, nor the court below, should undertake a review of the assessments that found Mr. Sorenson to be eligible for MAID'. The spouse in this matter also did not have standing to attempt to prevent or delay Mr. Sorenson's receipt of MAID. The NSCA rationale for refusing to review the question of MAID eligibility was that there was no justiciable issue and the courts would not have the institutional capacity for such reviews except in matters relating to "challenges to the constitutionality of the MAID provisions, or questions as to whether policies governing MAID within a province or health authority are in compliance with the *Criminal Code*".[38]

While courts thus far will not become involved in the eligibility review of MAID, the federal and provincial governments have also not established a comprehensive regulatory regime that can identify abuse or meaningfully review the accuracy of

---

[33] A.B. v Canada, note 33 at para 81.

[34] For a thorough discussion, and expansion of these points, please see the chapter herein entitled, Marini and Villarreal: The Psychosocial Aspects of Adapting to Traumatic Non-Life-Threatening Disability.

[35] Ibid.

[36] Other chapters expand on this point further, through thorough discussion surrounding psychosocial adjustment to chronic illness and disability, ableism and discrimination.

[37] Sorenson v Swinemar (2020) [hereinafter "Sorensen v Swinemar"], and A.B. v Canada, note 33.

[38] Sorenson v Swinemar, note 36 at para 64.

MAID assessments. However, assessment reviews are frequently done in other legal realms, such as capacity to consent to treatment assessments.[39] Also, it is inadequate to presume that all physicians are equal in talent, analytical skills and are free of biases that may impair their ability to assess individual eligibility for MAID.

Although the courts have indicated that they do not have institutional capacity, they have left the door open to oversight related to the Constitution and Criminal Code. They have also indicated that should a province legislate some form of review tribunal or commission, then that would not be barred by the courts. Considering the lack of existing checks to ensure the appropriate provision of MAID, an administrative body analogous to the Consent and Capacity Board (CCB) in Ontario may be a very helpful model to build on within the context of MAID. At the CCB, from the date of application until hearing, the Board is legislated to have a maximum of seven days for scheduling.[40] They have a maximum 24 h to provide a decision, and if written reasons are requested, they have an additional maximum of four business days allowed.[41] These timelines would surely be sufficient to meet the "urgency" standard set by the NSCA within the context of MAID, especially where natural death is not reasonably foreseeable. In the 2019/20 period, the CCB demonstrated broad institutional capacity when receiving over 8600 applications of which many proceeded with complex matters heard by a multi-disciplinary tribunal made up of a lawyer, psychiatrist, and member of the public.[42] Documents included within these time limitations are corroborating evidence such as medical records with powers of cross-examination at the hearing.[43] It would not be necessary to review every MAID assessment, however, given that the outcome will be the death of an individual, it bears urgency to provide an administrative review body with institutional capacity to provide timely and effective oversight.

4.2.6 Federal Government of Canada Regulations

Further to the perceived procedural safeguards discussed above, the federal Minister of Health has also implemented regulations considered necessary to the collection, use and disposal of information regarding requests and provisions of MAID in order to monitor the practice.[44]

---

39 Starson v Swayze (2003).

40 Section 75, *Health Care Consent Act* (1996). Online: https://www.ccboard.on.ca.

41 *Health Care Consent Act*, see note 38.

42 Consent and Capacity BoardAnnual Report 2019/2020, p. 18; Consent and Capacity Board, see note 41, p. 6.

43 Consent and Capacity Board Rules of Practice, June 19, 2020, Rules 9, 10, 14, 15, 18, 19, 20, 25, 29, 30, 34.

44 *Criminal Code*, note 5 at S. 241.31(3). These monitoring requirements are addressed comprehensively in other chapters.

In response to this requirement, the Government of Canada created regulations for MAID in November 2018. The data collected includes procedural documentation, details about the patient, medical practitioners and pharmacists involved, as well as details about the request for MAID, such as referrals or transfers of the patient. In addition, the data collected includes information about the dispensing of a substance to effect MAID and its administration.

Once the data is collected, the regulations require Canada's Minister of Health to publish an annual report to the public with some, but not all, of the collected data.[45] Following the passage of Bill C-7, Canada's federal data collection and reporting system was updated. Canada's Minister of Health was given the authority to create new regulations regarding the collection of information about unwritten MAID eligibility assessments. The Minister also required that pharmacists who provide MAID drugs report data.[46]

The amended regulations arising from Bill C-7 also encouraged the development of federal collection of data related to race, Indigenous identity, and disability. The goal of which is to identify the presence of any existing individual or systemic inequality.[47] It is interesting to note that investigating instances of non-compliance with the eligibility criteria and procedural safeguards set out in the *Criminal Code* falls outside of the scope of the federal monitoring regime.[48] Furthermore, the monitoring regime is aimed at gathering and analyzing data about MAID from a societal perspective. It is fundamentally distinct from a process that seeks to assess individual physician or NPs' compliance with the *Criminal Code* exemptions or enforce penalties in the instance of errors or omissions.

4.3 Provincial Law and Regulations: An Overview

As this chapter is being written, the provinces have been very slow to create comprehensive, quasi-judicial or expert multi-disciplinary bodies to regulate or monitor the application of MAID at either a procedural or substantive level. The provinces primarily reference guidelines and defer case-by-case complaints to their respective Colleges. Quebec remains unique in the creation of a Commission with Ministry powers to investigate institutions and individual MAID related events, as well as requiring institutions to establish end-of-life protocols.[49] They have also demonstrated reluctance to establish an administrative body or tribunal to have specific individual review.

---

[45] Regulations for the Monitoring of Medical Assistance in Dying: SOR/2018-166, Canada Gazette, Part II, Volume 152, at Section 13(1).

[46] *Criminal Code*, Note 5 at Section 241.31.

[47] *Criminal Code*, Note 5 at Section 241.31(3).

[48] Regulations for the Monitoring of Medical Assistance in Dying: SOR/2018-166, Canada Gazette, Part II, Volume 152.

[49] Act Respecting End-of-Life Care, COLR c. S.-32-0001.

The Atlantic provinces: Newfoundland, Nova Scotia, New Brunswick, and Prince Edward Island report, as required by Federal legislation. Saskatchewan, British Columbia, and the Territories have adopted similar models of reporting.[50] Generally, oversight has been left to professional associations on a complaint-by-complaint basis. This responsibility then falls on the professional regulatory bodies, which are the applicable Colleges of Physicians and Surgeons, Nurses, or Pharmacists.[51] While this may streamline the delivery, it allows little transparency and public oversight of the complaint process.

Ontario amended several pieces of legislation to bring the province into compliance with the MAID related amendments to the *Criminal Code of Canada*. With this, MAID reporting and oversight is primarily reviewed by individual coroners.[52] The Coroners Act limits the liability of physicians, nurses, or care providers involved in the provision MAID for, "any act done or omitted in good faith."[53] The exception is in cases where negligence is alleged.[54] The Solicitor General is required to establish a process to review the provisions of this section.[55] The Coroners Act also establishes further parameters for oversight stating:

> Where a person dies as a result of medical assistance in dying, the physician or nurse practitioner who provided the medical assistance in dying shall give notice of the death to a coroner and, if the coroner is of the opinion that the death ought to be investigated, the coroner shall investigate the circumstances of the death and if, as a result of the investigation, the coroner is of the opinion that an inquest ought to be held, the coroner shall hold an inquest upon the body.[56]

---

[50] Medical Assistance in Dying FAQs, November 10, 2021, https://www.cpsbc.ca/files/pdf/PSG-Medical-Assistance-in-Dying-FAQ.pdf and Provincial Medical Assistance In Dying (MAID) Program A Resource for Facilities and Caregivers, Saskatchewan Health Authority, June 16, 202, https://www.saskhealthauthority.ca/sites/default/files/2021-06/A%20Resource%20for%20Facilities%20and%20Caregivers.pdf, https://www.saskhealthauthority.ca/sites/default/files/2021-06/A%20Resource%20for%20Facilities%20and%20Caregivers.pdf.

[51] Section 64 Medical Act chapter 38 of the Acts of 2011 as amended by (2014).

Section 86 Medical Practitioners Regulations made under Section 11 of the medical act S. N. S. 2011, College of Physicians and Surgeons of Nova Scotia (2021); The College of the Physicians and Surgeons of Labrador, Standard of Practice-Medical Assistance in Dying, March 11, 2017; Sections 39–56, Medical Act (2011); The College of Physicians and Surgeons of Prince Edward Island Policy on Medical Assistance in Dying, revised and approved by council, November 4, 2019.

Sections 35–59 Regulated Health Professions Act Prince Edward Island April 29, 2021.

Sections 29.1–38.5 Medical Act (1988); An Act to Amend an Act Respecting the New Brunswick Medical Society and the College of Physicians and Surgeons of New Brunswick, SNB 2013.

c. 48, regulation #9 professional misconduct, and Medical Assistance in Dying-Government of New Brunswick guidelines, Ministry of Health.

[52] Medical Assistance in Dying Statute Law Amendment Act (2017) amends the Coroners Act; Excellent Care for All Act (2010); Freedom of Information and ***Privacy Protection Act;*** Municipal Freedom of Information and Privacy Protection Act; Vital Statistics Act, and Workplace Safety and Insurance Act (1997).

[53] *Coroners Act* R.S.O. (1990), at Section 13.8(1).

[54] *Coroners Act*, Note 50 at Section 13.8(2).

[55] Coroners Act, Note 50 at Section 10.1(4).

[56] *Coroners Act*, Note 50 at Section 10.1(1).

It notes that the physician or NP and any other person who has knowledge of the death or who provided MAID shall provide the coroner with any information about the facts and circumstances relating to the death that the coroner considers necessary to form an opinion about whether the death ought to be investigated.[57]

Manitoba has created the Provincial Medical Assistance in Dying Clinical Team. It is an interdisciplinary team comprised of physicians, nurses, social workers, pharmacists, speech language pathologists, administrative personnel, and ethics specialists who serve as the central consultative and practical resource on MAID for health care providers, patients, and families in Manitoba. The primary functions of the Team are:

- To provide medical assistance in dying.
- To assist and support other qualified health-care professionals in providing medical assistance in dying.
- To provide and contribute to professional development and education regarding medical assistance in dying, and
- To participate in the monitoring and evaluation of the provision of medical assistance in dying in Manitoba.[58]

It also serves to provide a complaint resolution process.[59] Nonetheless, it says little about oversight or penalty in the instance of error or omission nor the importance of improving palliative care and disability support options, and therefore, adds little to the pre-existing federal requirements.

The Alberta Health Services (AHS) provides comprehensive multi-disciplinary care and oversight of MAID, in addition to other end-of-life care services, and makes available a patient concern resolution process as needed.[60] The Ombudsman's jurisdiction includes investigations of the complaint-handling process of regional health authorities.[61]

---

57 Coroners Act, Note 50 at Section 10.1(2).

58 See also Appellant v Manitoba Regional Health Authority, Maid Program (2018) where the Manitoba Health Appeal Board found that there were limitations to seeking MAID services when stating at paragraph 26: No patient in our health care system has the right to unlimited access. The Appellant's application for MAID was considered, he did receive an assessment. That the Respondent established limits beyond which it deemed expending additional health care resources unnecessary is a reasonable exercise of its discretion in how it allocates health care resources.

59 Shared Health Executive Policy, Medical Assistance in Dying, 350.140.127, December 1, 2017, https://sharedhealthmb.ca/wp-content/uploads/maid-complaint-resolution-protocol.pdf; WRHA MAID Complaint ResolutionProtocol.

60 Section 7 of the MEDICAL ASSISTANCE IN DYING (March 26, 2021, HCS-165-01) Policy Document, and AHS *Patient Concerns Resolution Process* pursuant to Alberta Regulation 28/2016.

61 Patient Concerns Resolution Process Regulation (2006).

4.4 Quebec: Act Respecting End-of-Life Care

The introduction in the Quebec legislature of Bill 52, *An Act Respecting End-of-Life Care*, on 12 June 2013, outlined the rights of persons at the end of life, and the organization and framework of end-of-life care in both the instance of palliative care and MAID. This includes powers of the Minister of Health and Social Services and Commission on end-of-life care.[62]

In addition to eligibility requirements like the *Criminal Code*, the Act *Respecting End-of-Life Care* provides additional safeguards by requiring that a physician verify the persistence of suffering and that the repeatedly expressed wish to obtain medical aid in dying remains unchanged by talking with the patient at different times. These discussions must be held at reasonably spaced intervals given the progress of the patient's condition. It is also required that the patient's request for MAID be discussed with any members of the care team who are in regular contact with the patient, with the patient's close relations, if the patient so wishes, and any other person the patient wishes to inform.[63]

The Act requires that all institutions develop an end-of-life policy, as well as promote a multiple disciplinary approach that involves health and social services professionals.[64] The Executive Director must provide a report to the Board of Directors that includes the number of end-of-life patients who received palliative care, the number of times continuous palliative sedation was administered, the number of requests for medical aid in dying, the number of times such aid was administered as well as the number of times medical aid in dying was not administered, including the reasons it was not administered.[65] Also, every palliative care hospice must inform persons of the end-of-life care it offers before admitting them.[66] This appears to go further, set a higher standard than informed consent than other provinces.

Quebec has established a Commission on end-of-life care ("the Commission").[67] The mandate of the Commission is to examine any matter relating to end-of-life care. For this purpose, it must, among other things:

(1) advise the Minister on any matter put before it by the Minister.
(2) evaluate the implementation of legislation with regard to end-of-life care.
(3) refer to the Minister any matter relating to end-of-life care that needs the attention of or action by the Government and submit its recommendations to the Minister.

[62] Act Respecting End-of-Life Care, COLR c. S.-32-0001.
[63] Act Respecting End-of-Life Care, Note 58 at Section 28(a–e).
[64] Act Respecting End-of-Life Care, Note 58 at Section 7.
[65] Act Respecting End-of-Life Care, Note 58 at Section 8.
[66] Act Respecting End-of-Life Care, Note 58 at Section 13.
[67] Act Respecting End-of-Life Care, Note 58 at Section 38.

(4) submit a report to the Minister, every five years, on the status of end-of-life care in Québec; and
(5) carry out any other mandate given to it by the Minister.[68]

The Commission is to submit an annual activity report, not later than the 30th of September each year, to the Minister. Also, complaints regarding end-of-life care made by any person to a local or regional service quality complaints commissioner must be given priority treatment. The same applies to complaints regarding end-of-life care made to the syndic of the Collège des médecins du Québec.[69]

This review indicates that Quebec has provided significantly more intellectual rigour and policy consideration than any other province or territory in Canada as it relates to MAID. While a detailed analysis is impossible within the length limitations of this chapter, Quebec's Act appears to consider in a detailed fashion MAID from both a procedural and substantive level at the proactive and reactive phases.

4.5 Findings and Recommendations

Based on the above broad review, gaps and recommended improvements to existing regulations are as follows:

1. Future regulations, guidelines or policy regarding MAID must be comprehensive and provide higher priority funding in their commitment to independent living, home care, and palliative care as alternatives to MAID. This would acknowledge the inextricable linkage of both a right to life and right to equality, creating regulations that meet a standard incorporating vulnerability and social disenfranchisement as a risk to error and abuse of access to MAID.
2. Regulations that identify or differentiate access to MAID through the no waiting track or the 90-day waiting track must be revisited on two fronts. First, correct the ambiguity as to the definition of irremediability or suffering. Second, adjustment to a diagnosis of chronic illness and or traumatic injury must have a waiting period that incorporates disability or palliative supports such as to allow for a psychosocial adjustment. This must be a comprehensive and well-designed approach that negates the arbitrary nature of assigning a 90-day limit to psychosocial adjustment that may take greater than one year.
3. Regulations must incorporate the utilization of a multidisciplinary approach. The narrow and singular use of two physicians and/or NPs overlooks the distinct skills that social workers, psychologists, palliative care specialists and independent living experts can offer in supporting patients through their psychosocial adjustment to illness.

---

68 Act Respecting End-of-Life Care, Note 58 at Section 38.

69 Act Respecting End-of-Life Care, Note 58 at Section 48.

4. Each province, through regulation, should create an oversight body by way of commission and/or tribunal that can review applications for MAID to ensure that procedural safeguards and thorough assessments have been conducted. Time limitations for a review must be limited to seven days in order to maintain the integrity of the oversight process.
5. The regulated MAID oversight body must also provide annual reports to the minister and the public, and be comprised of experts from related professions and stakeholder communities that possess multidisciplinary expertise.

4.6 Conclusions

The Supreme Court of Canada in Carter made a profound finding when indicating that a ban on physician assisted suicide infringed on an individual's right to life. Given the gravity of any discussion on the purposeful death of a person within a legal context, it is critical to prevent mistakes or abuse by professionals, or coercion of vulnerable persons at either an individual or social level. Due to the length limitations of this chapter, a comprehensive review of regulation safeguarding against such manipulation and/or criminality, and the question of discrimination or social disenfranchisement has not been fully addressed. However, it points to key areas for future discussions and research. The current regulatory regimes at federal, provincial, and territorial levels are at best superficial, and have not yet warranted the faith that the Supreme Court put into governments across Canada. The lower courts, following the Carter decision, have been highly deferential to these superficial regulations. It is therefore the job of legislatures to respond to gaps in their creation. This chapter has provided recommendations that would point toward a model to address the gaps in the regulatory regimes across Canada. Beyond the proposed principles to providing a safe and balanced regulatory regime, it will be critical to continue funding programs that remove stigma, discrimination, and disenfranchisement for persons living with disabilities and chronic illness and to provide comprehensive and effective palliative care funding. Lastly, strong, and effective enforcement mechanisms are of paramount importance in preventing abuse that may lead to the avoidable death of vulnerable Canadians.

References

Cases

A.B. v Canada (Attorney General). 2017. ONSC 3759.
Appellant v Manitoba Regional Health Authority, MAiD Program. 2018. CanLII 104055 (MB HAB).
RAA v SM. 2021. CanLII 18468 (ON HPARB). https://canlii.ca/t/jdp6t.
Sorenson v. Swinemar. 2020. NSCA 62 (CanLII).

Starson v Swayze. 2003. SCC 32, [2003] 1 S.C.R. 722.

Legislation

Act Respecting End-of-Life Care, COLR c. S.-32-0001.
An Act to Amend an Act Respecting the New Brunswick Medical Society and the College of Physicians and Surgeons of New Brunswick, SNB. 1989. c 45. https://canlii.ca/t/546h1
Bill C-7. 2021. *An act to amend the criminal code (medical assistance in dying)*, SC 2021, c 2. https://canlii.ca/t/551nn
Bill C-14. 2016. *An act to amend the criminal code and to make related amendments to other acts (medical assistance in dying)*, 1st session, 42nd Parliament.
Carter v Canada, (Attorney General). 2015. SCC 5.
Coroners Act R.S.O. 1990. c. C.37.
Criminal Code, RSC. 1985. c. C-46.
Excellent Care for All Act. 2010. SO 2010, c 14. https://canlii.ca/t/54qt0, S. 13.8-14.
Freedom of Information and Protection of Privacy Act, R.S.O. 1990. c. F.31.
Health Care Consent Act. 1996. S.O., c.2, Sched. A.
Medical Act. 2011. SNL 2011 c. M-4.02.
Medical Act Chapter 38 of the Acts of 2011 amended. 2014. c. 32, ss. 143–145.
Medical Act, RSPEI. 1988. c M-5. https://canlii.ca/t/54400.
Medical Assistance in Dying Statute Law Amendment Act. 2017. S.O. 2017, c. 7—Bill 84.
Medical Practitioners Regulations made under Section 11 of the Medical Act S.N.S. 2011. c. 38.
Municipal Freedom of Information and Protection of Privacy Act, R.S.O. 1990. c. M.56.
Patient Concerns Resolution Process Regulation, Alta Reg 124/2006. https://canlii.ca/t/52pdc.
Regulations for the Monitoring of Medical Assistance in Dying: SOR/2018-166, Canada Gazette, Part II, Volume 152.
Regulated Health Professions Act General Regulations, PEI Reg EC829/13. https://canlii.ca/t/5343m.
Workplace Safety and Insurance Act. 1997. S.O. 1997, c. 16, Sched. A.
Vital Statistics Act, R.S.O. 1990. c. V.4.

Articles

Alberta Health Services. 2021. *Medical assistance in dying*. Policy HCS-165-01. https://extranet.ahsnet.ca/teams/policydocuments/1/clp-med-assist-in-death-hcs-165-01.pdf.
College of Physicians and Surgeons Nova Scotia. 2021. *Professional standard regarding medical assistance in dying (MAiD)*. April 26, 2021 (amended May 5, 2021).
College of Physicians and Surgeons of British Columbia. 2021. Medical assistance in dying FAQs. *CPSBC*. https://www.cpsbc.ca/files/pdf/PSG-Medical-Assistance-in-Dying-FAQ.pdf.
College of Physicians and Surgeons of British Columbia. 2022. Practice standard medical assistance in dying. Last revised: January 17, 2022.
College of Physicians and Surgeons of Newfoundland and Labrador. 2017. Standard of practice medical assistance in dying. imis.cpsnl. https://imis.cpsnl.ca/web/files/2017-Mar-11%20-%20MAID.pdf.
Senate and House of Commons, Special Joint Committee on Physician-Assisted Dying. 2016. *Medical assistance in dying: A patient-centred approach*, 1st Session, 42nd Parliament, February 2016.

Saskatchewan Health Authority. 2021. Provincial medical assistance in dying (MAID) program a resource for facilities and caregivers. saskhealthauthority. https://www.saskhealthauthority.ca/sites/default/files/2021-06/A%20Resource%20for%20Facilities%20and%20Caregivers.pdf.
Winnipeg Regional Health Authority. n.d. *WRHA MAID complaint resolution protocol*. sharedhealthmb. https://sharedhealthmb.ca/wp-content/uploads/maid-complaint-resolution-protocol.pdf.

David Shannon, LLM, Barrister-and-Solicitor, David is a member of the Law Society Ontario practicing health law in Thunder Bay, Ontario. He received a Masters of Law degree at the London School of Economics, and has continued in his law practice and non-government organization leadership since then. His Dave Shannon Cross-Canada Tour in 1997 gained national media attention as he became the first quadriplegic to trek across Canada in a motorized wheelchair, and in 2009 he set another first by going to the North Pole celebrating the message that seemingly insurmountable barriers can be overcome.Shannon has won numerous prestigious awards and honours, such as the Order of Ontario and the Order of Canada. His academic and professional body of work has been expansive within the realm of human rights and health law. David has also been a sessional lecturer at Lakehead University and Negahnewin College at both the undergraduate and graduate level as well keynote speaker at numerous conferences across North America.

Chapter 5
The "Reasonably Foreseeable Natural Death" Requirement in Bill C-7, and Expressivist Harm

Jonas-Sébastien Beaudry

Abstract This chapter focuses on one possible criterion for eligibility to access Medical Assistance in Dying (MAID): the "reasonably foreseeable natural death" requirement. This requirement was enacted by Bill C-14, which legalized MAID in June 2016, and later repealed by Bill C-7, which expanded eligibility for MAID in March 2021. The chapter explains the reasonings behind the original adoption and the later elimination of this criterion, and some of the problems that removing this criterion creates from a disability perspective.

Keywords Carter · Truchon · Bill C-14 · Bill C-7 · Medical assistance in dying · Natural death reasonably foreseeable · Terminal · Expressivist harm

5.1 The Emergence of the "Reasonably Foreseeable Natural Death" Requirement

In Carter v. Canada (Attorney General),[1] the Supreme Court of Canada (SCC) declared the *Criminal Code*[2] provisions that made it illegal to aid or abet a person in committing suicide invalid. The Court held that the trial judge[3] made no error in concluding that "the evidence showed that a system with properly designed and administered safeguards offered a less restrictive means of reaching the government's objective"[4]—described in Rodriguez v. B.C. (Attorney General)[5] as "the protection of the vulnerable who might be induced in moments of weakness to commit suicide."[6]

[1] 2015 SCC 5 [*Carter*].

[2] R.S.C. (1985).

[3] See *Carter v Canada (Attorney General)* (2012).

[4] *Carter*, *supra* note 1 at para 31.

[5] [1993] 3 SCR 519, 107 DLR (4th) 342 [cited to SCR].

[6] *Ibid.* at 595.

J.-S. Beaudry (✉)
McGill University, Montreal, QC, Canada
e-mail: jonas-sebastien.beaudry@mcgill.ca

J. Kotalik and D. W. Shannon (eds.), *Medical Assistance in Dying (MAID) in Canada*, The International Library of Bioethics 104,
https://doi.org/10.1007/978-3-031-30002-8_5

As a result, in Carter, the SCC concluded that the *Criminal Code* provisions unjustifiably violated the rights to life, liberty, and security of competent adults seeking assistance to end their intolerable suffering caused by a "grievous and irremediable medical condition."

The Court issued a suspended declaration of invalidity in February 2015, allowing Parliament one year to amend the *Criminal Code* accordingly. This remedy specified that the relevant portions of the *Criminal Code*[7] would be void insofar as they prohibited MAID to the following category of patients:

1. a "competent adult person;"
2. "who clearly consents to the termination of life;" and
3. "has a grievous and irremediable medical condition."[8]

The Court defined "irremediable medical condition" as including not only terminal illnesses and diseases, but also disabilities and ongoing medical conditions that do not threaten life. The Court also specified that the term "irremediable" should not imply that patients must undertake treatments that they find unacceptable.[9]

In response, the federal Ministers of Justice and of Health created an External Panel to carry out consultations with medical authorities, interveners in *Carter*, and the public. The External Panel included a warning to Parliament in its final report:

> Eligibility criteria necessarily include some and exclude others ... whatever law is in place, it will likely be subject to Charter challenges by excluded individuals. It will therefore be incumbent on legislators to consider carefully which eligibility criteria are implemented to reduce the uncertainty that could result from repeated litigation.[10]

Soon after, the Senate and the House of Commons appointed a Special Joint Committee on Physician-Assisted Dying, which presented its report, *Medical Assistance in Dying: A Patient-Centred Approach*, in February 2016. The committee made 21 recommendations, including making MAID "available to individuals with terminal and non-terminal grievous and irremediable medical conditions that cause enduring suffering that is intolerable to the individual in the circumstances of his or her condition."[11]

The eventual result was Bill C-14. To be eligible for MAID, the Bill required that patients give free and informed consent, and suffer from a "grievous and irremediable medical condition," among other criteria.[12] The Bill defined "grievous and irremediable medical condition" as a condition making someone's natural death "reasonably

---

[7] *Supra* note 2 at s. 241(b), s. 14.

[8] *Carter, supra* note 1 at para 127.

[9] See *ibid*.

[10] *External Panel on Options for a Legislative Response to* Carter v. Canada*: Final Report* (2015).

[11] Parliament of Canada (2016) (Joint Chairs: Kelvin Kenneth Ogilvie and Robert Oliphant) at 13, online (pdf): www.parl.ca/Content/Committee/421/PDAM/Reports/RP8120006/pdamrp01/pdamrp01-e.pdf.

[12] Bill C-14 (2016a).

foreseeable, taking into account all of their medical circumstances, without a prognosis necessarily having been made as to the specific length of time that they have remaining," in addition to the following criteria:

1. having a "serious and incurable illness, disease or disability;"
2. being in an "advanced state of irreversible decline in capability;"
3. experiencing, as a result of (1) or (2), an "enduring physical or psychological suffering" that is subjectively intolerable and that cannot be relieved under subjectively acceptable conditions."[13]

5.2 Initial Rationales for, Defences, and Critiques of the "Reasonably Foreseeable Natural Death" Requirement in Bill C-14

Following its extensive consultation with "thousands of individual Canadians, experts, and organizations,"[14] the government ultimately took the position that the "reasonably foreseeable death" standard was necessary to balance the autonomy of individuals and the need to protect vulnerable lives.[15] The Bill's preamble contains the most explicit, textual, indication of this rationale, stating that:

> ...permitting access to [MAID] for competent adults whose deaths are reasonably foreseeable strikes the most appropriate balance between the autonomy of persons who seek medical assistance in dying, on one hand, and the interests of vulnerable persons in need of protection and those of society, on the other.[16]

Like the Minister of Justice, who said the Bill intended to avoid "promoting" a certain conception of the value of life (that is, a value such that suffering becomes a justification for premature death),[17] the Preamble also states that it aims to "affirm the inherent and equal value of every person's life and to avoid encouraging negative perceptions of the quality of life of persons who are elderly, ill or disabled."[18]

Nonetheless, the "reasonably foreseeable natural death" requirement was one of the most controversial components of Bill C-14. Indeed, these embryonic justifications leave much open to interpretation. They broadly point to the claim that

---

[13] See *ibid.* at s. 2. "Natural death" refers to a death caused by a medical condition rather than by a patient's decision or intervention.

[14] Bill C-14 (2016b) at 1005 (Hon. Jody Wilson-Raybould). See also footnotes 9 and 10 referring to particular groups and scholars connected to the disability community: Department of Justice Canada (2016) [Bill C-14 Legislative Background].

[15] Bill C-14 Legislative Background, *supra* note 14. See also MP Rankin's statement during debates on Bill C-14: "At the House and Senate committee, and again at the justice committee, some argued we could not afford to expand [access to MAiD to people whose natural death is not reasonably foreseeable] because to do so would pose an unacceptable risk to vulnerable persons." See Bill C-14 (2016c) at 1545 (Hon. Murray Rankin).

[16] Bill C-14, *supra* note 12, Preamble.

[17] Bill C-14 (2016b) at 1010 (Hon. Jody Wilson-Raybould).

[18] Bill C-14, *supra* note 12, Preamble.

MAID is such a fundamental encroachment on medical culture and our societal values—especially the value in protecting human life generally, and vulnerable lives specifically—that it must be constrained. This position implicitly denies that free and informed consent alone suffices to alleviate concerns. These statements may also suggest a strategy for preventing a "slippery slope" into more ethically problematic kinds of euthanasia. Still, a crucial question appeared unanswered: how was the federal government's concern for the vulnerability of people whose natural death is not reasonably foreseeable *different* from the concern that it presented in *Carter*?

This question focused the debate surrounding Bill C-14 on whether the federal legislator did, or could, contradict the SCC, and contained relatively little substantive discussion of the requirement's rationale. Perhaps the requirement's rationale was disregarded in part because critics assumed that the issue had been settled by *Carter* or because defenders of the requirement did not have much to add to the arguments presented in *Carter*. Regardless, to situate the reader with respect to the political evolution of this controversial issue, I provide a brief overview of the debate over the "reasonably foreseeable natural death" requirement here.

The fact that *Carter* did not mention any "proximity of natural death" requirement was used to support both sides of this debate. For critics of the requirement, the SCC's silence implied an exclusion.[19] To them, Bill C-14's deviation from *Carter*'s narrower set of conditions of eligibility *contradicted* the SCC's holding, and the "reasonably foreseeable natural death" requirement would therefore be unconstitutional, for the same reasons a complete ban on MAID was found to be unconstitutional.[20]

For defenders of the requirement, this silence meant that one could not ascertain whether the SCC would find the government's own balance of the values of autonomy and protection of vulnerable people unconstitutional.[21] They contended that the SCC only "intended to respond to the factual circumstances in [the case before

---

[19] See e.g. the following judicial statement, made by a Court assessing whether a non-terminally ill patient could avail themselves of the *Carter* criterion before Bill C-14 was passed: "In summary, the declaration of invalidity in Carter 2015 does not require that the applicant be terminally ill to qualify for the authorization. The decision itself is clear. No words in it suggest otherwise. If the court had wanted it to be thus, they would have said so clearly and unequivocally. They did not. The interpretation urged on us by Canada is not sustainable having regard to the fundamental premise of Carter itself as expressed in its opening paragraph, and does not accord with the trial judgment, the breadth of the record at trial, and the recommended safeguards that were ultimately upheld by the Supreme Court of Canada." *Canada (Attorney General) v EF* (2016) at para 41.

[20] For instance, Joseph Arvay, the lead counsel for the appellants in *Carter*, stated that Bill C-14 would not be constitutional. He stated that, when he framed the case and used the terms "grievous and irremediable" to qualify the conditions of eligibility, he did not mean "terminal," "end of life," "incurable," or a state in which natural death would be "reasonably foreseeable" [see Standing Committee on Justice and Human Rights (2016) at 0845–0850 (Joseph Arvay)].

[21] Consider Senator Sinclair's statement during the Bill's second reading before the Senate: "If the Supreme Court had wanted to very clearly say that people had to be in a terminal condition or not be in a terminal condition at the time the application is made, they would have said so in Carter. They didn't. [...] They didn't have evidence before them, and they were not asked to rule on it; therefore, we should take those comments with more than a grain of salt. Until the matter is argued before them with evidence, we don't know what they're likely to say about that particular point." See Bill C-14 (2016d) at 1210 (Hon Murray Sinclair).

the Court and made] no pronouncement on other situations where physician-assisted dying may be sought,"[22] thus leaving it to the government to set up a legislative MAID framework that would respect the *Charter* considerations raised in *Carter*.[23] This reflects the theory of dialogue between courts and Parliament relied on by the government, which contended that "the Supreme Court did not intend, in declaring the total prohibition invalid, to constitutionalize any particular legislative model."[24] Indeed, while courts have the mandate to decree that certain public policies violate *Charter* rights, they have also shown deference to legislative authority and capability in designing constitutionally valid frameworks. This deference is heightened when complex social, economic, or ethical issues requiring specific expertise, data gathering, or public consultations are implicated.[25]

Despite the reasonableness of this perspective on the division of powers generally, critics could still fairly demand that the government explain their reasons for concluding that a partial ban would withstand *Charter* scrutiny more fully. How could the *prima facie* unconstitutional harm caused to some by Bill C-14 be justified when the harm caused by a full ban was not?

5.3 Bill C-7: Repositioning of the Government on the "Reasonably Foreseeable Natural Death" Requirement

The arguments made by the Attorney General before the Superior Court of Quebec, in *Truchon c. Procureur général du Canada,*[26] suggest that the government did not intend to propose more detailed answers. Bill C-14 received royal assent on June 17th, 2016, and was immediately challenged by people wishing to access MAID but prevented from doing so because their natural death was not reasonably foreseeable.[27] In *Truchon,* the government provided little more than a repetition of the arguments

[22] *Carter, supra* note 1 at para 127.

[23] Illustrations of such critiques and defences can be found in the senators' statements: "Bill C-14, An Act to amend the Criminal Code and to make related amendments to other Acts (medical assistance in dying)", 2nd reading, *Senate of Canada Debates,* 42-1, vol. 150 (2 June 2016) at 1450–2220.

[24] Bill C-14 Legislative Background, *supra* note 14.

[25] See e.g. *R v Mills* (1999): "[I]t does not follow from the fact that a law passed by Parliament differs from a regime envisaged by the Court in the absence of a statutory scheme, that Parliament's law is unconstitutional … To insist on slavish conformity would belie the mutual respect that underpins the relationship between the courts and legislature that is so essential to our constitutional democracy." This case was cited during the public debates. See "Bill C-14, An Act to amend the Criminal Code and to make related amendments to other Acts (medical assistance in dying)", 2nd reading, *Senate of Canada Debates,* 42-1, vol. 150 (2 June 2016) at 1530 (Senator Baker).

[26] 2019 QCCS 3792 [*Truchon*].

[27] See e.g. *Lamb v Canada*, Notice of Civil Claim, Supreme Court of British Columbia, Vancouver Registry, No. S-165851.

made in *Carter*,[28] which were easily dismissed on the basis that the SCC's conclusions in *Carter* held equally for people whose death was not reasonably foreseeable.[29]

The government responded to *Truchon* by introducing Bill C-7, *An Act to amend the Criminal Code (medical assistance in dying)*, on October 5, 2020.[30] The Bill notably repealed the "reasonably foreseeable natural death" as a definitional criteria of "grievous and irremediable medical condition." It created two sets of safeguards: one for people whose natural death is reasonably foreseeable and one for those whose death is not reasonably foreseeable. For the former, the previous safeguard of a 10-day reflection period was eliminated, and the previous requirement that the written request for MAID be signed by two independent witnesses was changed to a requirement for a single signature.[31] For the latter, a 90-day assessment period was added along with a requirement that medical practitioners or nurse practitioners confirming the eligibility of the patient have an expertise in the condition that is causing the person's suffering, or to consult with a practitioner with such an expertise.[32] The latter also requires individuals be informed of "the reasonable and available means to relieve [their] suffering" and the practitioner authorizing MAID must be satisfied that "the person has given serious consideration to those means."[33]

At the same time, the introduction of Bill C-7 introduced normative changes to the values expressed by the MAID regime. Bill C-7's preamble removed the statement contained in Bill C-14's preamble to the effect that the government is committed to "avoid encouraging negative perceptions of the quality of life of persons who are elderly, ill or disabled."[34] Instead, Bill C-7 affirmed the value of life more generally, and the gravity of suicide as a general public health concern. The specific value of life of persons who are elderly, ill or disabled; the social denigration thereof, and their specific suicidality, were eliminated from Bill C-7's preamble, and replaced by a general commitment to affirming the constitutional value of life and because it recognizes suicide as a public health issue in need of other state-endorsed responses than assistance in suicide. Bill C-14's strategy to limit the impact of legalizing MAID on the value of life in Canadian society was to limit access to MAID to people who are on a "path toward death" anyway. Bill C-7's strategy to expand access to MAID to people who are not nearing death is to make it available only to people who suffer from a "serious and incurable illness, disease or disability."

---

[28] *Truchon*, *supra* note 39 at paras 518, 529–531.

[29] See *ibid.* at paras 519–522, 532–535.

[30] Bill C-7, *An Act to amend the Criminal Code* (medical assistance in dying) 2nd Sess, 43rd Parl, 2021, Preamble: "Whereas the Government of Canada has committed to responding to the Superior Court of Québec decision in *Truchon v. Attorney General of Canada*..." (assented to 17 March 2021) [Bill C-7].

[31] *Ibid.*, changing s.241.2(3) of the *Criminal Code*.

[32] *Ibid.*, changing s.241.2(3.1)(d) and (e) of the *Criminal Code*.

[33] *Ibid.*, changing s.241.2(3.1)(g) and (h) of the *Criminal Code*.

[34] Bill C-14, *supra* note 12, Preamble.

5.4 Refined Critiques of the Removal of the "Reasonably Foreseeable Natural Death" Requirement from a Disability Perspective

In the remainder of this chapter, I will expound upon the concern from within the disability community that legalizing assistance in dying for people whose natural death is not reasonably foreseeable risks implying that some lives are not worth living and perpetuates a trend of finding individual medical solutions to social problems. We may call (at least an interpretation of) such arguments "expressivist," because they point to the expressive harmfulness of a legal regime that makes it legal to end the lives of only a historically oppressed segment of the population: people with severe disabilities. Despite its targeted impact on this particular segment of the population, the "expressivist harm" is not an idiosyncratic problem affecting only a minuscule population. Anyone may acquire a grave medical condition rendering them eligible for MAID—thereby ushering them into this segment of the population with which they may not have previously identified.

Readers familiar with the objections to expressivist concerns in relation to prenatal testing may wonder whether similar objections are likely to arise in the context of MAID. One may argue, for instance, that the expressive harmfulness is an unfortunate by-product of necessary individual choices in the face of intransigent personal circumstances. Briefly, the "expressivist harm" examined here is distinguishable by not being the result of individual choices. Individual choices would, themselves, express only that "our society fails people with disabilities and their families by providing such scant resources to make their lives go as well as they can"[35]—to quote Eva Kittay's comments on individual mothers choosing not to bring a disabled child into the world. The harmful "expression" considered here is institutionalized and systemic; phrased and buttressed through state-endorsed legal and medical mechanisms that simultaneously fail to provide enough support to individuals choosing MAID.

In an open letter, dozens of organizations within the disability community urged the Minister of Justice not to follow *Truchon*, on the grounds that Bill C-7 effectively entrenched the stereotype that people living with severe disabilities suffer a "fate worse than death."[36] Although Bill C-7's preamble paid lip service to Convention for the Rights of Persons with Disabilities, the government was not influenced by the opinion of United Nations' Special Rapporteurs on the rights of persons with disability, older adults, and on poverty, who agreed that Bill C-7 did not recognize the equal value of the lives of certain historically vulnerable populations, and was thereby expressively harmful.[37]

---

[35] Kittay (2019) at 19.

[36] Open Letter to Hon. David Lametti, MP, Attorney General (2019).

[37] Quinn et al. (2021): "[u]nder no circumstance should the law provide that it could be a well-reasoned decision for a person with a disabling condition who is not dying to terminate their life with the support of the State". See also Chapter 14 of this collection on "MAID and the Right's to Life and Equality at International Law".

By limiting MAID to patients whose death is reasonably foreseeable, MAID could be considered not as a kind of suicide, but as a modality of dying. Death being inevitable in the relatively short term, MAID patients would be perceived as choosing the modality of their death rather than as choosing to forego years of existence. The problem with a regime without such a limit is not so much that people cannot reasonably judge, at an individual level, that their lives are not worth living. The problem is a political and social one: Bill C-7 opens a normative space in which various social actors, including medical experts and the state itself, can discuss the topic of "lives not worth living," by relying on concepts of quality of life and health care costs, amongst other arguments. This medico-legal space facilitates the cultural emergence of categories of human beings whose lives can be legally and morally disposed of. This is a dangerous door to open, considering the limited treatment that systemic ableism, internalized oppression, and social supports received in *Carter* and *Truchon*. Many disabled people have suffered from common ableist beliefs that they would be "better off dead," and the law would now give them a legal path to end an existence that is often socially devalued, whereas their fellow "healthy" Canadians would benefit from suicide prevention measures.[38]

The role of doctors and nurses in confirming eligibility for a "reasonable" suicide adds a veneer of objectivity to complex ethical considerations of quality of life, right to life, equality, and systemic discrimination. Safeguards must go beyond ensuring that people freely consent to ending their lives. As data (from Canada and other jurisdictions) suggests, people most often choose to die because of existential suffering connected with the loss of autonomy and the difficulty in enjoying activities that they used to enjoy.[39] Through autonomy-enhancing measures including therapy, financial support, at-home care, and disability and social networks, they might find their situation more bearable, embrace new rewarding roles and activities, and not feel that their lives are burdensome. Such changes would enhance their autonomy more significantly than merely rendering them eligible for MAID and put the focus on social rights; on revealing how ableism and ageism seep into the agency of many people—"sick" and "healthy" alike, "productive" and "dependent" alike—and how our laws and legal frameworks contribute to this process.

No one denies that Bill C-7 opens up access to medically provided death in the name of respect and compassion, but we should remember that oppression can be inadvertently hidden in good faith attempts to provide a benefit to a marginalized group. Bill C-7 was in line with a long history of judicial and legislative endeavours protecting "negative" freedom, that is freedom from limitations (such as a criminal ban) from the state,[40] and freedom to pursue one's goals once left to one's own

[38] See e.g. Council of Canadians with Disabilities (2020); Inclusion Canada (2021), Institute for Research and Development on Inclusion and Society (2020). Consider the case of others who have suddenly acquired severe disabilities: Spinal Cord Injury Canada (2020).

[39] See e.g. Health Canada (2021), Oregon Health Authority (2022).

[40] One may object to this claim because MAID is treated more like a positive right than other decriminalized activities (e.g. prostitution may be decriminalized, but there is no consideration of making it a right).

devices.[41] Such legal changes are cast as a narrative of freedom and autonomy; however, Bill C-7 is also part of a different story, one that puts a price on human life,[42] and uses social markers such as disability to construct a body that belongs to a special normative sphere, in which medical expertise can trigger a lessening of the rights to life and equality.

There were compelling reasons for this, but what should concern us is the failure to explore alternative MAID frameworks that would simultaneously pay attention to theories of "internalized oppression," "systemic injustice," "oppression," and cultural studies of ableism. I cannot explore such critiques in this chapter.[43] I only suggest that these narratives have been largely absent from the debates that have occurred so far. I speculate this is because the expressivist harm and social neglect examined in this chapter do not cohere, inherently or contingently, with mainstream interpretations of constitutional rights, partly entrenched in a libertarian ethos that is resistant to recognizing robust social rights, and partly entrenched in a liberal ethos traditionally premised on the myth that autonomy can be protected in abstraction of social and psychological factors like systemic and internalized kinds of oppression.

Acknowledgements The author thanks the editors of this collection and the peer-reviewer for their constructive comments, as well as his research assistant, Ms. Katrina Bland, for her diligent editorial work.

References

Beaudry, Jonas-Sébastien. 2018. The way forward for medical aid in dying: Protecting deliberative autonomy is not enough. 85 SCLR 335.

Beaudry, Jonas-Sébastien. 2021. Somatic oppression and relational autonomy: Revisiting medical aid in dying through a feminist lens. *UBCL Review* 53 (2): 241.

Bill C-7. 2021. *An act to amend the criminal code (medical assistance in dying)*, 2nd Sess., 43rd Parl. (assented to 17 Mar 2021).

Bill C-14. 2016a. *An act to amend the criminal code and to make related amendments to other acts (medical assistance in dying)*, 1st Sess., 42nd Parl. (assented to 17 June 2016).

Bill C-14. 2016b. *An act to amend the criminal code and to make related amendments to other Acts (medical assistance in dying)*, 2nd reading, *House of Commons Debates*, 42-1, No. 45 (22 Apr 2016).

Bill C-14. 2016c. *An act to amend the criminal code and to make related amendments to other acts (medical assistance in dying)*, 2nd reading, *House of Commons Debates*, 42-1, No. 57 (17 May 2016).

I concede that, if MAID is treated as a (positive) right, the state's willingness to pay to assist certain populations to die, but not to assist them in other social and financial ways could also contribute to the harmful expression discussed in this chapter.

[41] Consider scholarly critiques of this trend: Young et al. (2008), Flood et al. (2005).

[42] Office of the Parliamentary Budget Officer (2020).

[43] I begin doing this work in Beaudry (2018, 2021).

Bill C-14. 2016d. *An act to amend the criminal code and to make related amendments to other acts (medical assistance in dying)*, 2nd reading, *Senate of Canada Debates*, 42-1, vol. 150, (17 June 2016).
Canada (Attorney General) v EF. 2016. ABCA 155.
Carter v Canada (Attorney General). 2015. SCC 5.
Carter v Canada (Attorney General). 2012. BCSC 886.
Council of Canadians with Disabilities. 2020. *Brief of the council of Canadians with disabilities to the standing committee on justice and human rights regarding Bill C-7, an act to amend the criminal code (medical assistance in Dying)* (31 Oct 2020).
Criminal Code, R.S.C. 1985. c. C-46.
Department of Justice Canada. 2016. *Legislative background: Medical assistance in dying (Bill C-14)—Addendum: Bill C-14's requirement that natural death has become reasonably foreseeable (s. 241.2(2)(d))* (7 July 2016). www.justice.gc.ca/eng/rp-pr/other-autre/addend/index.html.
External Panel on Options for a Legislative Response to Carter v. Canada: *Final Report* (15 Dec 2015) at 53, online (pdf): *Department of Justice* www.justice.gc.ca/eng/rp-pr/other-autre/pad-amm/pad.pdf.
Flood, Colleen M., Kent Roach, and Lorne Sossin, eds. 2005. *Access to care, access to justice: The legal debate over private health insurance in Canada.* Toronto: University of Toronto Press.
Health Canada. 2021. *Second annual report on medical assistance in dying in Canada.*
Inclusion Canada. 2021. *An amendment to ensure the rights of persons with disabilities: Bill C-7, brief submitted to the standing committee on justice and human rights* (22 Feb 2021).
Institute for Research and Development on Inclusion and Society. 2020. *Bill C-7: Through a disability lens & proposed amendments; submitted to the standing senate committee on legal and constitutional affairs* (November 2020).
Kittay, Eva Feder. 2019. *Learning from my daughter: The value and care of disabled minds.* New York: Oxford University Press.
Lamb v Canada. Notice of Civil Claim, Supreme Court of British Columbia, Vancouver Registry, No. S-165851.
Oregon Health Authority. 2022. *Oregon death with dignity act: 2021 data summary.*
Office of the Parliamentary Budget Officer. 2020. *Cost estimate for Bill C-7 "medical assistance in dying"* (20 Oct 2020).
Open Letter to Hon. David Lametti, MP, Attorney General. 4 Oct 2019. Signed by 72 members and supporters of the Canadian Disability Community. https://inclusioncanada.ca/2019/10/04/advocates-call-for-disability-rights-based-appeal-of-the-quebec-superior-courts-decision-in-truchon-gladu/.
Parliament of Canada. 2016. *Medical assistance in dying: A patient-centred approach: Report of the special joint committee on physician-assisted dying* (February 2016) (Joint Chairs: Kelvin Kenneth Ogilvie and Robert Oliphant) at 13, online (pdf): www.parl.ca/Content/Committee/421/PDAM/Reports/RP8120006/pdamrp01/pdamrp01-e.pdf.
Quinn, Gerard, Olivier De Schutter, and Claudia Mahler. 2021. *Disability is not a reason to sanction medically assisted dying—UN experts* (25 Jan 2021), online: United Nations Office of the High Commissioner for Human Rights www.ohchr.org/EN/NewsEvents/Pages/DisplayNews.aspx?NewsID=26687.
Rodriguez v BC (Attorney General). 1993. 3 SCR 519, 107 DLR (4th) 342.
R v Mills. 1999. 3 SCR 668, 180 DLR (4th) 1.
Standing Committee on Justice and Human Rights. 2016. *Evidence*, 42-1, No. 14 (5 May 2016).
Spinal Cord Injury Canada. 2020. *Written response regarding changes to Bill C-7; brief submitted to the standing committee on justice and human rights* (27 Nov 2020).
Truchon c Procureur général du Canada. 2019. QCCS 3792.
Young, Margot, Susan Boyd, Gwen Brodsky, and Shelagh Day, eds. 2008. *Poverty, rights, social citizenship, and legal activism.* Vancouver: UBC Press.

Jonas-Sébastien Beaudry is an Associate Professor, jointly appointed to the Faculty of Law and the Institute for Health and Social Policy at McGill University, and a member of the Quebec Bar. His research focuses on human rights, ethics, and disability theory, and his most recent book is The Disabled Contract (CUP 2021). He has worked at the Supreme Court of Canada, the International Court of Justice, and has acted as a consultant on public policies for the Department of Justice, the Senate, and disability organizations.

Part II
Operational Issues in the MAID Program

Chapter 6
Integrating Medical Assistance in Dying into Health Care: Processes and Challenges

Alireza Bagheri and Saumil Dholakia

The history of liberty has largely been the history of the observance of procedural safeguards.
—Felix Frankfurter, Associate justice of the Supreme Court of the United States (McNabb v. United States, 318 U.S. 332. U.S. Supreme Court case, supreme.justia.com. March 1, 1943)

Abstract Medical Assistance in Dying or MAiD poses procedurally complex challenges that can be a major source of distress to both, the providers as well as participating patients and their family. While the health care providers can find the process emotionally overwhelming and applying the eligibility criterion and safeguards in the absence of objective procedural standards particularly challenging on one end, patients on the other end, can experience decisional bottlenecks and frustrating delays in accessing this service. This chapter critically evaluates the challenges at each stage of implementation of MAiD in Canada with Ontario as an illustrative example.

Keywords Medical Assistance in Dying (MAiD) · Healthcare provider · Procedural safeguards · Assisted death care pathways · MAiD pathways in Canada

A. Bagheri (✉)
Research Affiliate Center for Healthcare Ethics, Lakehead University, Thunder Bay, ON, Canada
e-mail: bagheria@yahoo.com

S. Dholakia
Department of Mental Health, The Ottawa Hospital, Ottawa, Canada
e-mail: saumil.dholakia@mail.utoronto.ca

Department of Psychiatry, University of Ottawa, Ottawa, ON, Canada

J. Kotalik and D. W. Shannon (eds.), *Medical Assistance in Dying (MAID) in Canada*, The International Library of Bioethics 104,
https://doi.org/10.1007/978-3-031-30002-8_6

6.1 MAiD Legislation in Canada

The MAiD legislation in Canada has evolved since its inception in 2016. The legislation's journey from public to the supreme court, from the supreme court to the parliament and then finally back to the public has seen many changes and amendments. While Bill C-14 attempted to balance patient autonomy and protection of the vulnerable population, the recently amended law (Bill C-7) most notably removed 'natural death has become reasonably foreseeable' as an eligibility criterion for MAiD (see Table 6.1 for details). Both oral self-administration of MAiD (assisted suicide) and injection of the lethal drugs by a physician or nurse practitioner (euthanasia) are acceptable in Canada. Because 99% of deaths are the outcome of lethal injections, the following description will focus on this modality of MAiD.

MAiD is integrated into the health care system and health regions across Canada have formed multidisciplinary MAiD teams to support access. Despite this degree of healthcare integration, the issue of MAiD in Canadian society remains a deeply polarizing one (Wasylenko 2017). Judicial, health care and societal understandings of key concepts such as suffering, dignified death and sanctity of life remain distinctive and diverse. To respect this diversity of opinions in such a contentious landscape, the legislation guards the right of healthcare providers to conscientiously object to participation in MAiD.

The purpose of this chapter is to describe Canada's MAiD process, beginning with the enquiry about MAiD until the person is informed of being ineligible or MAiD is delivered. We critically evaluate the challenges at each stage of implementation and suggest possible steps for improvement. We use MAiD in the province of Ontario as a representative example.

Step 1: MAiD enquiry: Helping make an autonomous, voluntary, well-informed choice

Case: *Mr. A.B., 55 years old, Ph.D. Scholar and author, married with no children, is diagnosed with advanced prostate cancer in summer of 2020. Bone metastasis and secondary vertebral fracture in late 2021 contributed to severe pain. He had lost his father to cancer at 51 and was well aware that he was on 'borrowed time'. He was referred to palliative care for adequate pain control by his oncologist.*

He shared an excellent relationship with his health care team- "My pain was never questioned", although his pain stopped responding to medical treatment in December 2021. Prognostically, his oncologist told him that 2022 must be 'his year of maximum efforts'. He liked the way it was reframed.

He expresses his desire to know more of the MAiD process.

First point of contact: As the case describes, members of the health care team (physicians, pharmacists, nurses, and nurse practitioners) are often the first point of contact. While this is true for most provinces in Canada, others, like Alberta, Yukon, Nunavut and Northwest territories have care coordinator services that also act as first point of contact (Silvius et al. 2019).

Table 6.1 MAiD legislation: criminal code of Canada incorporate changes with a two-track system from March 2021

Bill	Eligibility	Safeguards
Bill C-14	(1) A serious and incurable, illness, disease or disability; (2) In an advanced state of irreversible decline; (3) Intolerable physical or psychological suffering that cannot be relieved under conditions that the patient deems acceptable; (4) Where natural death had become reasonably foreseeable, without a prognosis having been made about length of time remaining and, (5) Provide a valid, informed consent, after having been informed of the means that are available to relieve their suffering, including palliative care	(1) Requests had to be made in writing and signed by two independent witnesses; (2) 10-day waiting period that could only be waived if the applicant was at risk of losing capacity; (3) Persons had to provide final consent immediately before the injection that would cause death
Bill C-7 reasonably foreseeable death track	As above + Final consent could be waived (Audrey's Amendment) as long as (1) they met all of the eligibility criteria; (2) they entered into an agreement in writing that included a specific day for the MAID procedure; (3) they were informed of the risk of losing capacity; (4) in the written agreement they consented to having MAID on that day if they had lost capacity; and, (5) at the time of administration they do not demonstrate refusal or resistance to have the substance administered	Request signed by one independent witness. No 10-day waiting period

(continued)

Table 6.1 (continued)

Bill	Eligibility	Safeguards
Bill C-7 death not reasonably foreseeable track	(1) A serious and incurable, illness, disease or disability; (2) In an advanced state of irreversible decline; (3) Intolerable physical or psychological suffering that cannot be relieved under conditions that the patient deems acceptable (4) The patient provides a valid, informed consent: – after having been informed of other means available to them, including counselling, mental health supports, disability supports, community services and palliative care; and – after having been offered consultation with relevant professionals, as available and applicable; and – after having discussed these means with the medical or nurse practitioner and given serious consideration to these means	Request signed by one independent witness There are at least 90 clear days between the day on which the first assessment begins and the day on which medical assistance in dying is provided, or—if the assessments have been completed and both of the physicians or NPs are of the opinion that the loss of the person's capacity to provide consent to receive medical assistance in dying is imminent—any shorter period that the physician or NP who is to provide MAID considers appropriate in the circumstances

Source Adapted from:
Bill C-14 2016. https://lop.parl.ca/sites/PublicWebsite/default/en_CA/ResearchPublications/LegislativeSummaries/421C14E
Bill C-7 2021. https://lop.parl.ca/sites/PublicWebsite/default/en_CA/ResearchPublications/LegislativeSummaries/432C7E

Challenges at this stage: At this point, the clinician will discuss and explore the request with the patient and provide information on all available treatment care options. There are many factors, both individual and systemic, that may drive a particular inquiry to explore assisted death. Societal influences like emphasis on individual choice for health care practices enhance patient autonomy, however, 'Good care' in fact, likely depends on the collaborative and continued work of health care professionals based on a sound platform of a therapeutic relationship and may have little to do with patient choices (Mol 2008)Furthermore, 'I want assisted death' may be an idiomatic expression for 'I want help', or a request to be given a reason to live (Muskin 1998). Frequently, a genuine desire to die is absent in patients who express a wish to hasten death (Hudson et al. 2006; Monforte-Royo et al. 2011) The prevalent role of depression and related suicidality needs further exploration and should be objectively assessed at this stage, if clinically indicated (Ruijs et al. 2011; Ganzini et al. 2008). From a health care systems perspective, another challenge at this step is making sure that patients are aware of the available options such as palliative care for optimum pain relief or allaying suffering. A direct conversation on the differences between MAiD and palliative care will go a long way in making patients more aware and help them reflect on their choices.

Step 2: Making a MAiD request: Patient chooses to pursue MAiD

Case continued: *In a joint session along with his spouse, Mr. A.B. told his oncologist, "We are people who turn up at the exact same time when we have said we will. I have left an everlasting mark on the human society with my books, which will remain on the shelf after I am gone. We live on our terms and die on our terms. I have decided to pursue MAiD.*

Procedural steps at this stage: Once the patient has decided to pursue MAiD, If the clinician chooses not to participate in provision of MAiD due to conscientious objections the clinician could be duty bound in most jurisdictions to refer the patient to another clinician, institution or agency that is willing to provide this service, while providing standard of care to the patient.

At this stage, the patient needs to make a written request for MAiD to their clinician. A patient's written request may take any form including a form Clinician Aid A, a text message or an e-mail. It must, be more than an inquiry or a request for information about MAID and the request must be in the format required by the Criminal Code as a safeguard when MAiD is provided (i.e., duly signed, dated and witnessed).

Challenges at this stage: Most of the challenges at this stage stem from the patient and health care provider's moral environment and the nature of patient-clinician relationship that influence the process of MAiD. While endogenous factors like previous personal or professional experiences, religious beliefs and emotional labor contribute to conscientious objections (Brown et al. 2021a, b). factors like professional identity, commitment to patient autonomy and the patient-clinician relationship contribute to defining the moral dimensions of care at this stage (Patel et al. 2020).

Step 3: MAiD assessments

Case: *Mr. A.B. undergoes two independent assessments, one by an anesthesiologist who is a MAiD provider and the other by his palliative care physician. This patient justified his application by the desire to control his time of death, 'to die on his own term'. This is not an acceptable criterion, the law requires that the patient has "enduring physical or psychological suffering that is intolerable", so this patient could be assessed ineligible for MAID. However, the patient meets other criteria ('irreversible decline in capacity' and 'serious and incurable illness') and does experience constant pain that has stopped responding to medical treatment. The wording of his MAiD application can be understood as his own internal justification arising from his philosophy of life. Both assessors after discussion agreed that Mr. A.B is eligible. His death was judged to be reasonably foreseeable. He was judged to be suffering from a grievous and irremediable medical cancer (advanced metastatic prostate cancer). His request for MAiD was judged to voluntary, without undue* *coercion, signed in presence of one independent witness,* and was well-informed of other means to relief of their suffering, particularly palliative care, but still wished to have MAiD.

While there were no concerns with capacity at this point, Mr. A.B had developed extensive liver metastasis and was at risk of losing capacity due to liver failure in the future.

Procedural steps at this stage: The central challenge at this stage is to screen for eligibility—most patients requesting MAiD would like to know whether they are eligible. While this may be straightforward in some cases who easily meet the criteria, the following are the procedural challenges that need astute navigation:

(a) It is vital at this stage that the assessors ensure that the patient completely meets all MAiD criteria. The Canadian MAiD Law requires them to be

   a. at least 18 years of age and capable of making decisions with respect to their health,
   b. be eligible for health services funded by Government of Canada,
   c. suffer from a grievous and irremediable medical condition, defined as
      i. as a serious and incurable illness, disease or disability that causes enduring suffering that is intolerable to the individual and
      ii. experiencing enduring physical or psychological suffering, due to the illness, disease, disability or state of decline, that is intolerable to the person and cannot be relieved in a manner that they consider acceptable; and,
      iii. where the person's natural death has become reasonably foreseeable, considering all of their medical circumstances, without requiring a specific prognosis as to the length of time the person has left to live.
      iv. resulting into an advance state of irreversible decline in capability.
   d. have made a voluntary request without any external pressure.

The recent Bill C-7 established a two-track assessment thereby allowing patients whose death is not reasonably foreseeable, meets all other criteria, respecting special set of safeguards to access MAiD. As much as possible, a shared decision making of *the* health care provider who could provide MAiD and the patient should be encouraged to determine if there is other, better option to deal with the patient's suffering or if MAiD is the only way. While emphasis on patient autonomy and control over their own end-of-life decisions is the main driver behind the legislation, the clinical response to legally valid requests is often influenced by a variety of individual dynamic factors including personal values and beliefs, the patient-clinician relationship, clinician's ability to assess and understand the nature of suffering, and the clinician's emotional and psychological response (Nuhn et al. 2018)as well as systemic factors of care-coordination, evolving social contact of care (example communities where they live, visibility of participant choices) and a patient-centered approach to care (Oczkowski et al. 2021; Brown et al. 2021a, b).

Challenges at this stage: The assessment often boils down to a case-by-case evaluation of a patient for each individual MAiD eligibility criterion. Table 6.2 summarizes some of the procedural challenges that an assessor may encounter at this stage. The terms “Incurable”, “foreseeable”, “Irreversible”, “intolerable” are not well defined in law or regulations, leaving large space for clinicians’ subjective assessment. As a result, a patient may be assessed as ineligible for MAiD by one physician, but another physician who may find them eligible. There is no legal requirement that either assessor has to see the applicant in person. In fact, with emergence of COVID 19 it appears that virtual assessment became common in most jurisdictions. If patient finds that the assessment made them ineligible, they can submit a new application for MAiD to another clinician, with no limit set in law on the number of assessments that a patient can initiate.

In Ontario, Clinical Aid B and C are tools that can be used by the physician or nurse practitioner to execute the two independent assessments (Medical Assistance in dying. MOHLTC, available from https://www.health.gov.on.ca/en/pro/programs/maid/).

While it is necessary for patient to have intact capacity throughout the MAiD process and the procedure, patients like Mr. A.B who are at risk of losing capacity anytime between the assessments and actual MAiD date have the option of signing the waiver of final consent (Audrey’s amendment) after having gone through an informed consent process.

The written agreement becomes permanently invalidated if a person demonstrates, by words, sounds or gestures, refusal to have the substance administered or resistance to its administration.

Involuntary words, sounds or gestures made in response to contact do not constitute a demonstration of refusal or resistance.

While not mandated by the law, practitioners may wish to encourage the patient to share their intentions and involve their family or loved ones in discussions about receiving MAiD. This way, family and friends will be aware of the patient’s wishes and more likely to respect them.

These assessments usually conclude with a review of patient’s understanding of the process and procedure for MAID including the need for vascular access, and risks, including burning on administration of medication, need for signed informed consent for the procedure and the patient’s option of withdrawal of consent at any time in the process, including up to administration of medications. If the patient was not found eligible for MAiD, assessors have the responsibility to do immediately what is in their competence to make the person comfortable and assure that other physicians and health care team and social services help them manage the condition and make adequate referrals.

Step 4: MAiD preparation and MAiD day

Case: *Mr. A.B. meets all eligibility criteria for MAiD and chooses the date as well as the place (home) for the MAiD procedure, to be administered by a physician. Mr. A.B.’s physician sends the copy of the consent form and order sheet to the pharmacy. All arrangement for a privacy inside his apartment along with necessary equipment (suction, IV access) are made. The pharmacist prepares two identical kits containing the prescribed doses of medications. The physician picks up the 2 kits*

Table 6.2 Procedural challenges during MAiD assessment

MAiD criterion	Relevant condition	Assessment challenge
Advance state of irreversible decline in capability	Major neuro-cognitive disorder	– Loss can be sudden or gradual, ongoing or stabilized – Someone can be in an advanced state of irreversible decline in capability in general terms, while still having moments of slight improvement – Patients need to have intact capacity to consent despite irreversible decline in cognitive and functional capability – Recent scoping review of Canadian literature points towards an emerging opinion on expanding access of MAiD to patients in advanced stages of dementia who may be incompetent
Assessing reasonably foreseeable nature of death (RFND)	A valid MAiD request	– Although no longer an eligibility criterion, practitioners instead must use it to determine which assessment track and hence which set of safeguards to apply to someone requesting MAiD – Requires a temporal, but flexible, connection between the person's overall medical circumstances and their anticipated death – People may decline toward death at different rates that aren't always predictable – An agreement between two independent MAiD assessors is not required, Law requires them to discuss and choose an opinion of a third assessor or consult another practitioner familiar with the patient's condition

(continued)

Table 6.2 (continued)

MAiD criterion	Relevant condition	Assessment challenge
Grievous and irremediable suffering-death not reasonably foreseeable	Psychiatric disorders	– Irremediability as an entry point vs. recovery model of care in psychiatric disorders-how do we define 'irremediability' in psychiatric disorders? – Defining intolerable suffering in context of wide range of psychiatric disorders—wide range, closely linked with the nebulous concept of 'treatment resistance' – Capacity assessment for MAiD-level of scrutiny and threshold of competence is challenging to define (Nicolini et al. 2020; Gaind 2020; Sinyor and Schaffer 2020; Anderson et al. 2022)

Adapted from Medical Assistance in Dying: Implementing a framework, 2023. Available from https://www.canada.ca/en/health-canada/services/medical-assistance-dying/guidance-reporting-summary/implementing-framework.html#a3

from the pharmacy 1 h prior to the procedure. Mr. A.B. has lost capacity at the time of the procedure and waiver of final consent was utilized.

Procedural steps at this stage: The Canadian 2020 Annual Report on MAiD shows that primary settings for the administration of MAiD in Canada were private residences (47.6%) and hospitals (28.0%), followed by palliative care units (17.2%) and long-term care facilities (5.7%).

Challenges at this stage: A qualitative process model flowchart study amongst MAiD practitioners in primary and secondary care setting in Nova Scotia, Canada pointed towards clinicians and pharmacists spending significantly high time to participate, raising questions about MAiD's sustainability, care-provider burnout, and uncompensated costs (Crumley et al. 2021)Difficulty obtaining intravenous access, longer than expected time to death, pain on injection and need for a second MAiD medication kit owing to technical failure of the first kit are common complications at this stage (Zworth et al. 2020). In a retrospective, hospital-based cohort study of sample adult patients who received MAiD in Ontario between 2016 and 2018 and those who received MAiD in 1 of 3 Canadian academic hospitals in 2019–2020 showed that death occurred within 3–15 min in most cases, and complications described above were rare (Stukalin et al. 2022). There is no clear definition of 'clinically-significant' prolongation of a MAiD procedure. In situations in which a very short time from procedure start until death is important (e.g., organ donation after MAiD) or for patients who are expected to have a prolonged death owing to physiologic robustness, it seems sensible to choose a medication protocol that is least likely to result in prolonged time to death.

Appropriate logistics is another challenge, especially for families who would like to accompany their loved one at their last moment of life. Most hospitals or facilities have a designated room for MAiD to accommodate family's comfort and privacy.

It also noteworthy that because currently, religion-affiliated hospitals in Canada do not allow the provision of MAiD to be performed in their facilities, patients have to be transferred to another facility or be discharged home for the last step. This may cause hardship and inconvenience for the patients as well as their families.

Step 5: Post MAiD death and after care

Case: *The physician administered the medications sequentially. Mr. A.B. breaths his last and dies at his home, without any complications.*

Physician MAiD report:

MEDICATION ADMINISTRATION START TIME: 1022 h. Lidocaine: 60 mg, Propofol: 1000 mg, Rocuronium: 200 mg, Bupivicaine: 400 mg, TIME OF DEATH: 1028 h.

Coroner notified and death certificate completed. Support provided for the family. Body released .

Procedural steps at this stage: The Clinician contacts the Office of the Chief Coroner to report the death of a patient due to MAiD. At this stage, The Office of the Chief Coroner determine if additional investigation steps are required. The Clinician will complete and sign the medical certificate of death (MCOD) in all cases except for the occasions when the coroner completes additional investigation.

Challenges at this stage: Exploring and incorporating patient and family's death-related religious/spiritual practices would be important at this stage to ensure a patient-centered approach. Patients without any particular religious affiliations may have other rituals or practices that they follow. 'Choreographing' the assisted death can be challenging but accommodating family requests as far as feasible, ensuring that patient has control over their death and that there is dignity and independence through this control would be crucial in navigating this final step.

A full pathway for MAiD with possible delays and decisional bottlenecks (represented by red diamond) is summarized in the flow chart depicted below in Fig. 6.1.

6.2 Practice Suggestions

While the Canadian MAiD legislation provided a 'top-down' legal framework to embed MAiD in medicine, integrating it into health care systems has its procedural challenges. Many of these challenges may stem from the moral environment of the health care provider, systemic challenges of a young legislation (2016–2021-amended) which is not physician-centered, and constantly evolving nature of the eligibility criteria (Table 6.3).

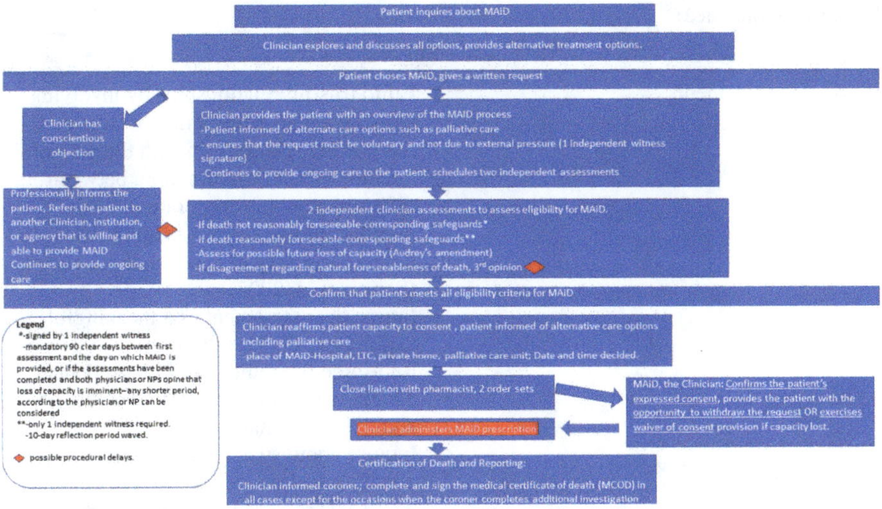

Fig. 6.1 MAiD pathway in Canada—flow chart

Table 6.3 Practice suggestions

Step	Suggestions
1. MAiD enquiry	– Be aware that patients may struggle to bring up MAiD, even if they are seriously considering it – Use direct language. Suggest that the patient involves spouse, children and trusted family in the decision-making while ensuring that the request remains voluntary without coercion
2. MAiD request	– Have clear institutional policy/protocol for referring patients and appoint a MAiD coordinator – Early and simultaneous involvement of palliative care (Cheng et al. 2022) – Policies should be supportive of conscientious objectors but mandate rapid referral/transfer of patients so as not impede timely response – A descriptive 'MAiD map' to patients can help keep track of next steps and hold the team accountable

(continued)

Table 6.3 (continued)

Step	Suggestions
3. MAiD assessments	– Build trustful relationships by involving clinicians who already have a strong therapeutic relationship with the patient. Establish Care co-ordination, between different teams, incorporate interdisciplinary care throughout the illness trajectory to ensure holistic care (Perron et al. 2022) – Frame the assessment as a meaningful, exploratory conversation – Reinforce flexibility and independence: patient can change their mind at any time and that is normal – Inquire about social and economic conditions of the person, to rule out that his MAiD request is driven by lack of suitable accommodation, experience of abuse, severe poverty, involuntary isolation and/or other systemic vulnerabilities – Plan to navigate scenarios where ambiguity in eligibility criteria is plausible—for example, planning and creating interdisciplinary teams who are comfortable with discussing, researching and, where justified, leading the implementation of MAiD in psychiatric disorders
4. MAiD preparation	– Identify patients at risk of losing capacity early – Prepare patients and families for the emotional nature of the waiting period and acknowledging that complexity of emotions they may be experiencing – Having a designated room for MAiD – At this stage, offer organ donation, if potentially eligible
5. Death and aftercare	– Take exceptional care to manage time as delays in provision can be distressing – Providing emotional support for the families – Training in cultural ethos of death and rituals around the dying process is crucial-engage spiritual care if available

(continued)

Table 6.3 (continued)

Step	Suggestions
6. Process Oversight	– A provincially empowered groups of physicians/nurse practitioners who are available and accessible to address patient, family and physician concerns and provide oversight in complex cases, and who can engage in providing standardized training for practitioners willing to provide MAiD – Accountability and transparency in care co-ordination (Simpson-Tirone et al., 2022)
7. Creating a care ecology and an ethically robust and sustainable organizational environment	– Creating an organizational ethics environment, establishing interdisciplinary groups that is value-focused, and cultivates open moral spaces that honors moral diversity of health care providers (Frolic et al. 2022)

6.3 Conclusion

MAiD traverses ethically overwhelming difficult territories and can be emotionally challenging for the patient, family if involved, and for the health care providers. A MAiD process that fully respect legal requirements of eligibility and safeguards and at the same time, encourages a shared decision making between the health care provider and the patient assures that if MAiD is carry out, it will meet both, regulatory oversight and standard of care expectations of patient and their family.

References

Anderson, B., S. Mihilli, M. Kumaresh, A. Kumaresh, R. Mirza, and C. Klinger. 2022. Advance care planning for seniors diagnosed with dementia: a scoping review of the canadian literature. *Canadian Journal on Aging/La Revue Canadienne Du Vieillissement* 41 (3): 377–403. https://doi.org/10.1017/S0714980821000283.

Bill C 14, *Act to amend the Criminal Code and to make related amendments to other Acts (medical assistance in dying)*, S.C. 2016, c.3.

Bill C 7, *An Act to amend the Criminal Code (medical assistance in dying)*, S.C. 2021, c.2.

Brown, J., D. Goodridge, L. Thorpe, and A. Crizzle. 2021a. "What Is right for me, is not necessarily right for you": The endogenous factors influencing nonparticipation in medical assistance in dying. *Quality Health Research* 31 (10): 1786–1800. https://doi.org/10.1177/10497323211008843.

Brown, J., D. Goodridge, L. Thorpe, and A. Crizzle. 2021b. "I am okay with it, but i am not going to do it": The exogenous factors influencing non-participation in medical assistance in dying. *Qualitative Health Research* 31 (12): 2274–2289. https://doi.org/10.1177/10497323211027130.

Cheng, E.Y., K. Mah, A. Al-Awamer, et al. 2022. Public interest in medical assistance in dying and palliative care. *BMJ Supportive and Palliative Care* 12: 448–456.

Crumley, E.T., S. Kelly, J. Young, et al. 2021. How is the medical assistance in dying (MAID) process carried out in Nova Scotia, Canada? A qualitative process model flowchart study. *BMJ Open* 11: e048698. https://doi.org/10.1136/bmjopen-2021-048698

Frolic, A., and P. Miller. 2022. Implementation of medical assistance in dying as organizational ethics challenge: A method of engagement for building trust, keeping peace and transforming practice. *HEC Forum* 34 (4): 371–390. https://doi.org/10.1007/s10730-022-09485-9. Epub 2022 Aug 24 PMID: 36001242.

Gaind, K.S. 2020. What does "irremediability" in mental illness mean? *The Canadian Journal of Psychiatry* 65 (9): 604–606. https://doi.org/10.1177/0706743720928656.

Ganzini, L., E.R. Goy, and S.K. Dobscha. 2008. Prevalence of depression and anxiety in patients requesting physicians' aid in dying: Cross sectional survey. *BMJ* 337: a1682. https://doi.org/10.1136/bmj.

Hudson, P.L., P. Schofield, B. Kelly, R. Hudson, A. Street, M. O'Çonnor, L.J. Kristjanson, M. Ashby, and S. Aranda. 2006. Responding to desire to die statements from patients with advanced disease: Recommendations for health professionals. *Palliative Medicine* 20: 703–710. https://doi.org/10.1177/0269216306071814.

Medical Assistance in Dying: Implementing the Framework. Government of Canada. 2023. Available from: https://www.canada.ca/en/health-canada/services/medical-assistance-dying/guidance-reporting-summary/implementing-framework.html#a3.

Medical Assistance in Dying. Ministry of Health and Long-term care, Ontario. Accessible from: https://www.health.gov.on.ca/en/pro/programs/maid/.

Mol A: The Logic of Care. 2008. *Health and the problem of patient choice*. Oxon: Routledge, 1.

Monforte-Royo, C., C. Villavicencio-Chavez, J. Tomas-Sabado, and A. Balaguer. 2011. The wish to hasten death: A review of clinical studies. *Psycho-Oncology* 20: 795–804. https://doi.org/10.1002/pon.1839.

Muskin, P.R. 1998. The patient-physician relationship. The request to die. Role for a psychodymamic perspective on physician-assisted suicide. *JAMA* 279: 323–328. https://doi.org/10.1001/jama.279.4.323.

Nicolini, M.E., Kim, S.Y.H., Churchill, M.E., and C. Gastmans. 2020. Should euthanasia and assisted suicide for psychiatric disorders be permitted? A systematic review of reasons. *Psychological Medicine* 50 (8): 1241–56 (Cambridge University Press).

Nuhn, A., S. Holmes, and M. Kelly. 2018. Experiences and perspectives of people who pursued medical assistance in dying: Qualitative study in Vancouver, BC. *Can Fam Clinician* 64: e380–e386.

Oczkowski, S.J.W., D.E. Crawshaw, and P. Austin, et al. 2021. How can we improve the experiences of patients and families who request medical assistance in dying? A multi-centre qualitative study. *BMC Palliative Care* 20: 185. https://doi.org/10.1186/s12904-021-00882-4.

Patel, T., Christy, K., and L. Grierson, et al. 2020. *Clinician responses to legal requests for hastened death: A systematic review and meta-synthesis of qualitative research.* BMJ Supportive & Palliative Care Published Online First: 29 June 2020. https://doi.org/10.1136/bmjspcare-2019-002018.

Perron, C., M.-E. Bouthillier, and E. Racine. 2022. Support structures for healthcare professionals involved in medical assistance in dying: Quebec, Canada, and the international landscape. *Healthcare Management Forum*. https://doi.org/10.1177/08404704221136847.

Ruijs, C.D.M., A.J.F.M. Kerkhof, G. van der Wal, and B.D. Onwuteaka-Philipsen. Depression and explicit requests for euthanasia in end-of-life cancer patients in primary care in the Netherlands: A longitudinal, prospective study. *Fam Practice* 28: 393–399. https://doi.org/10.1093/fampra/cmr006.

Silvius, J.L., A. Memon, and M. Arain. 2019. Medical assistance in dying: Alberta approach and policy analysis. *Canadian Journal on Aging* 38: 397–406.

Simpson-Tirone, M., S. Jansen, and M. Swinton. 2022. Medical Assistance in Dying (MAiD) care coordination: Navigating ethics and access in the emergence of a new health profession. *HEC Forum* 34: 457–481. https://doi.org/10.1007/s10730-022-09489-5.

Sinyor, M., and A. Schaffer. 2020. The lack of adequate scientific evidence regarding physician-assisted death for people with psychiatric disorders is a danger to patients. *Canadian Journal of Psychiatry*. Online first May 26, pp. 1–3. https://doi.org/10.1177/0706743720928658.

Stukalin, I., R. Olaiya, O.R. Oluwatobi, V. Naik, E. Wiebe, M. Kekewich, M. Kelly, L. Wilding, R. Halko, and S. Oczklowski. 2022. Medications and dosages used in medical assistance in dying: A cross-sectional study Jan 2022. 10 (1): E19–E26. https://doi.org/10.9778/cmajo.20200268.

The State of Knowledge on Medical Assistance in Dying Where a Mental Disorder Is the Sole Underlying Medical Condition. Ottawa (ON): The Expert Panel Working Group on MAID Where a Mental Disorder Is the Sole Underlying Medical Condition. Council of Canadian Academies, 2018. Available from: https://cca-reports.ca/reports/medical-assistance-in-dying/

Wasylenko, E. 2017. Becoming dead: Two solitudes? *Healthcare Manage Forum* 30 (5): 262–265. https://doi.org/10.1177/0840470417714886. PMID: 28929846.

Zworth, M., C. Saleh, I. Ball, et al. 2020. Provision of medical assistance in dying: A scoping review. *British Medical Journal Open* 10: e036054.

Alireza Bagheri, MD, Ph.D., is a physician bioethicist and is an Elected Fellow of the Hastings Center, USA. He is currently a clinical ethicist in Saskatchewan Health Authority. He is also a research affiliate in the Center for Health Care Ethics, Lakehead University, Canada. Dr. Bagheri served as a member of the UNESCO International Bioethics Committee (2010–2018), served as a member of the Board of Directors of the International Association of Bioethics (2008–2014). In addition to dozens of articles in academic journals and book chapters, his book publications include Abortion (2021); Indigenous Health Ethics (2021); Islamic Bioethics (2017); Global Bioethics (2015); Medical Futility (2013). In 2010, Dr. Bagheri received the National Razi (Rhazes) Medical Research Award for his work on Medical Futility and in 2018 he received Bioethics Leadership Award in recognition of his bioethics scholarship at the international level. Dr. Bagheri has studied and taught bioethics in Iran, Japan, Belgium, the United States and Canada, which strengthened his understanding of bioethics in a global context. As a palliative care physician, he brings extensive clinical experience to bioethical discussions as well as promoting close attention to the socio-cultural context especially in clinical decision making.

Saumil Dholakia, MBBS, MD, MHSc (Bioethics), (PG Diploma in Bioethics), is an assistant professor at the Department of Mental Health, The Ottawa Hospital and Department of Psychiatry, University of Ottawa. He is a CPSO-recognized specialist in psychiatry. Dr. Dholakia graduated as the inaugural clinical fellow in Geriatric Psychiatry and Clinical Pharmacology at the Department of Psychiatry and Behavioral Neurosciences, McMaster University and GeriMedRisk. Previously, Dr. Dholakia has worked as a consultant psychiatrist for 10 years with a faith-based, service driven Institution of Christian Medical College, Vellore, India. He is a Bioethicist with a Post-graduate Diploma from Bharathiar University Coimbatore and a Master's in Health Science degree from the Joint Center of Bioethics, University of Toronto. His capstone project was a systematic review on the emotional impact on health care professionals involved in Medical Assistance in Dying across the world under Dr. Alireza Bagheri and Dr. Alexander (Sandy) Simpson. His core area of work focuses on bridging the gap between science, evidence-based medicine, and health care: his work reflects the importance of a humanistic critical lens in medical practices, including exploring the intersection of Medical Assistance in Dying (MAiD), mental health care and its emotional impact on health care professionals.

Chapter 7
Monitoring of MAID: Deficits of Transparency and Accountability

Jaro Kotalik

Abstract Monitoring of MAID was recognised as an essential component of the MAID program. However, monitoring started late in the rollout of the program, and even at present is missing some essential elements that would be important for effective oversight and public accountability. Health Canada is using some of the data obtained from practitioners and pharmacists to produce Annual Reports on MAID, but these reports aim at providing only a "societal perspective," rather than an in-depth assessment of the program's performance. A major deficit is lack of monitoring and reporting on the compliance with eligibility criteria and procedural safeguards that should be coming from territorial and provincial governments, and that would provide assurance that MAID delivery complies with the Criminal Code. Monitoring is based entirely on self reporting and therefore it is recommended that Health Canada obtain supportive data by initiation of on-site retrospective reviews of randomly selected MAID deliveries across the country. In addition, it would be useful to enhance the quantity and quality of data collected from MAID assessors and providers. Organizations involved in MAID provisions (hospitals, regional health authorities and so on) should also report annually on MAID activities to their respective communities. Improvement of monitoring and reporting would address serious concerns among professionals, people with disabilities and advocates for vulnerable populations while providing for a higher level of transparency and accountability.

Keywords Medical assistance in dying · Monitoring of assisted dying · Public reporting on assisted dying · Accountability · Federal-provincial collaboration · Safeguards · Mandatory reporting · On-site reviews · Self-reporting on medical assistance in dying

---

J. Kotalik (✉)
Centre for Health Care Ethics Lakehead University and Northern Ontario School of Medicine University, 955 Oliver Road, Thunder Bay, Ontario P7B 5E1, Canada
e-mail: jkotalik@nosm.ca
URL: https://www.lakeheadu.ca/centre/chce

J. Kotalik and D. W. Shannon (eds.), *Medical Assistance in Dying (MAID) in Canada*, The International Library of Bioethics 104,
https://doi.org/10.1007/978-3-031-30002-8_7

7.1 Introduction

Ongoing monitoring and public reporting of euthanasia and suicide assistance has been considered essential elements of the assisted dying program in Canada from the very beginning as shown in court documents, in law and in regulations. When the Supreme Court of Canada delivered its benchmark *Carter* ruling that made physician-assisted dying, under certain conditions legal, the Court was aware that such a change of law may lead to misuse and harm, but it was persuaded that "a properly administered regulatory regime is capable of protecting the vulnerable from abuse or error." The Court concluded that "the risk associated with physician-assisted death can be limited through a carefully designed and monitored system of safeguards."

Directed by the Supreme Court, Parliament enacted Bill C-14 to amend the Criminal Code and created an exemption for "medical assistance in dying" (MAID) when provided by medical or nurse practitioners to persons who meet certain defined conditions. This Bill reflects very closely the concerns of *Carter*'s judgement with its multiple references to safeguards, monitoring and oversight. The Preamble of the bill which came into force in June 2016 stated that "robust safeguards, reflecting the irrevocable nature of ending life, are essential to prevent errors and abuse in the provision of medical assistance in dying." It also declares that "vulnerable persons must be protected from being induced, in moments of weakness, to end their lives."

The Bill charged the federal Minister of Health with the responsibility to make regulations for monitoring of medical assistance in dying, including analysis, interpretation, and publication of collected data. Failure of a practitioner to provide the required information was made an offence that could result in imprisonment for up to two years (Parliament of Canada 2016). In this chapter, we will be examining the evidence that in spite of the good intentions expressed by the Supreme Court and by the Parliament, the monitoring of MAID and its reporting of the results of public monitoring has not been a success: it started late in the rollout of the program, it is missing some essential elements that are important for effective oversight and accountability to the public and it reveals insufficient collaboration between federal and provincial/territorial authorities. To support these assertions, I will need to consider monitoring and reporting on the MAID program at both levels of governments.

7.2 Development of the MAID Monitoring and Reporting System

To understand the complexities of monitoring the medical assistance in dying (MAID) program implemented in Canada, it is necessary to be aware of the complexities of the Canadian federal system. In Canada, health care is a dual responsibility of the federal government, on one hand and of the ten provincial and three territorial governments on the other hand. Access to care, quality and cost of health care are to a great degree dependent on the good cooperation across these levels. Provincial and territorial governments are responsible for the actual delivery of health care to individuals and communities, while the federal government can influence or shape the health care delivery by passing laws valid for the whole country (like the law amending the Criminal Code that made MAID possible), by providing general and specific funds for health care programs, by coordinating, monitoring, creating statistics and reporting to the public. The enforcement of the Criminal Code and administration of justice within their own jurisdictions is largely the role of provincial, territorial, and local governments, with the federal government having an overarching responsibility.

To make a new federal law operate correctly requires the writing of regulations. Commonly, new Act passed by the Parliament only becomes effective at some future date as specified in the Act or on a date set by the Governor in Council. This allows time to create regulations and the necessary infrastructure. Yet, the amendment of a section of the *Criminal Code*, that became known as MAID Law became effective immediately, and MAID delivery was initiated in some locations across Canada within days of the law receiving the Royal Assent, in spite of the absence of monitoring and other regulations. As a result, for more than two years, MAID was being practiced without any federal monitoring whatsoever. The Canadian provinces and territories gradually provided their own temporary directives for reporting on MAID activities by MAID practitioners, but these directives were often only rudimentary and not harmonized with other jurisdictions. To fill the reporting gap, in April 2017, the federal Ministry of Health began to publish semi-annual Interim Reports on MAID that were based on voluntarily provided data by some provincial governments, noting that the information gathered was incomplete.

7.3 Federal Monitoring and Reporting

On November 1, 2018, about two years and four months after MAID delivery began, the long-awaited Regulations for Monitoring of MAID by the federal Ministry of Health went into effect (Government of Canada 2018a). The monitoring is based entirely on self-reporting. Physicians and nurse practitioners have a legal obligation to report on a dedicate internet portal various events associated with MAID, such as the receipt of patients' written request for MAID, provision of MAID by administering a substance, determination of ineligibility, withdrawal of request by the patient, death of patient from another cause and referral or transfer of care. The clinician will be asked about the application of each eligibility criteria and each procedural requirement. Negotiations that have taken place between provincial and federal authorities resulted in two streams of monitoring data. In about one half of the 13 Canadian provinces and territories, practitioners involved with MAID are expected to report to the federal government through a designated provincial or territorial body. Ontario instituted dual reporting responsibility to the Chief Coroner for competed MAID cases and to Health Canada for other reportable events. This mechanism then provides some governments with direct access to all data for their own investigations. In the other half of the provinces and territories, reports of practitioners go directly to the federal government, and from there, data can be transferred to provinces on quarterly basis. Data collected by this monitoring process were used by Health Canada to create Annual Reports, which are discussed in the next chapter of the book.

The Monitoring Regulations specify what information must be provided in the published Annual Reports by Health Canada, the federal authority. What is not listed among those items is the adherence by practitioners to eligibility criteria and to safeguards. This is most surprising and problematic because as explained in the earlier Chapters of this book, it is the adherence to eligibility criteria and to safeguards that protects medical and nursing practitioners against the charge of culpable homicide, prohibited by the Criminal Code of Canada. An explanation, for what to an external observer looks like a serious omission, can be found in the Regulatory Impacts Analysis Statement, a document that is not part of the Regulations but it follows Schedule 7 of the Regulations. It states about the federal monitoring that:

> (T)he monitoring regime is aimed at gathering and analyzing data about medical assistance in dying from a societal perspective. … Investigating instances of noncompliance with the eligibility criteria and procedural safeguards set out in the Criminal Code falls outside of the scope of the federal monitoring regime, and is under the purview of local law enforcement (Government of Canada 2018b).

The above description demonstrates that the monitoring in Canada is entirely different from the monitoring of similar regimes for assisted death in Benelux countries. In the Netherlands, for example, physicians must report all assisted death cases to one of five Regional Euthanasia Review Committees. A file is created for every case and the committee that consist of a physician, ethicist and lawyer and the chair determines if the practitioner has met all 'due care criteria'. If the criteria were not met, the case is referred to a Health Care Inspectorate of the Board of Procurator General. The committee's published annual reports and case materials are open to the public. The Canadian Medical Association (CMA) had recommended creating an independent national body that would receive and analyse all reports from across the country, but this recommendation was not adopted (CMA 2017).

7.4 Provincial and Territorial Monitoring and Reporting

What we have learned about the federal role suggests that at the provincial and territorial level monitoring, analysis and reporting is critical in order to assure that MAID is provided only under the conditions allowed by the Criminal Code. The problem is that up till now, no provincial or territorial government (with exception of Quebec) has released into the public domain a comprehensive report on the performance of its MAID program in their jurisdiction that would include this perspective. To their credit, Quebec and Ontario have provided partial data of this kind.

The province of Quebec operates their MAID program differently from other Canadian jurisdictions. The province established a commission on end-of-life care which reviews MAID reports and periodically releases its findings. The commission reported in April 2019 on 1354 cases and ruled that 13 cases (1%) did not comply with the law and in 41 cases (3%) there were missing data preventing the commission to reach a decision. Non-compliance was due to lack of eligibility in four cases and violation of safeguards in nine cases. The commission did not indicate what action was taken for non-compliance and failures to provide data, both being offenses in the *Criminal Code* (Commission sur les Soins 2020).

In Ontario, the Chief Coroner's office announced in 2018 that their team reviewed some 2000 cases of MAID (Office of the Chief Coroner 2018). There were some "compliance concerns with both the Criminal Code and the regulatory body policy expectations, some of which have recurred over time." The actual number, or a description of these cases was not provided. The team had engaged with practitioners on a case-by-case basis, sharing "learning opportunities". In addition, the office of the chief coroner of Ontario released a description of a ranking system indicating how this office will react to compliance deficiencies. For example, lack of compliance with the eligibility criterion of the requester ("capable of making decisions with respect to one's health") is classified as Level 5 and will lead to a report to the police. On other hand, lack of compliance with the requirement that a request for MAID is to be signed and dated in the presence of two independent witnesses is rated as Level 2 and will lead only to an "Educational Email" to the practitioner. The appropriateness of these responses to various types of non-compliance with MAID law and regulations could be debated, but it is reassuring that Ontario developed a systematic and rational approach to this issue.

Currently, the office of the Ontario Chief Coroner is issuing monthly statistics to interested parties. Tables provide the numbers of MAID cases, the average age of these persons, common underlying conditions, number of MAID providers and their specialty. However, there is no indication if these reports were examined for compliance with eligibility criteria and safeguards, nor if any action was taken for non-compliance. In fact, there is no evidence that the ranking system for assessing deficits, in reports from practitioners, which was announced in 2018, is currently being used. For other provinces, there are only basic and partial statistical data available on their websites or mass media, for example, from Nova Scotia (MacDonald 2021) and Manitoba (Shared Health Manitoba n.d.). These reports have a variety of formats, and the methodology of generating this information is unknown.

To sum up, publicly available documents provide no evidence that most provinces and territories monitor reports from the practitioners and pharmacists within their jurisdictions as it concerns the adherence to eligibility criteria and prescribed safeguards when providing MAID; and that, if they discover non-adherence, they investigate and take the actions as the law prescribes for transgression of the Criminal Code of Canada.

7.5 Current Monitoring Challenges

The utmost importance of monitoring is stressed several times in this document and the attached Impact Analysis Statement. We learned that "(T)he primary purposes of monitoring are to provide transparency and foster a public trust regarding the implementation of the new law." (Government of Canada 2018b).

If this is so, we have to ask: Is the method of monitoring, and quality and quantity of data collected and subsequent analysis of the data, such that it does fosters public trust? The answer is negative. There are a number reasons for that, most notable are:

1. Limited amount of data collected about MAID recipients

   Regulations for Monitoring of Medical Assistance in Dying of 2018 issued by Heath Canada defined objectives of the regulations very responsibly, as to support public accountability, support the protection of vulnerable individuals, identify trends, determine if the *Criminal Code* amendments of 2016 met their objectives and provide data for researchers. However, reading the Impact Analysis Statement attached to Regulartions reveals that the prevailing concern when developing these regulations was to achieve that monitoring is associated with a low administrative burden. After discussions with 'the medical community', that included representatives of colleges and Canadian Association of MAID Assessors and Providers, it was agreed by Health Canada to remove all information that might not be evident to the practitioners and required time to obtain it, such as a person's occupation. The description of a patient's "advanced state of irreversible decline in functional capability" was removed, even if this is one of the eligibility criteria that has to be fulfilled to make assistance in dying legal. All socio-economic questions were removed to be replaced with collecting the postal code. The requirement to estimate the amount of time by which MAID shortened a patient's life was initially included, but then removed on request of medical community because of "difficulties associated with making such assessment" (Supreme Court of Canada 2016, p. 19). This requirement is standard in other jurisdiction with assisted deaths, like Belgium and the Netherlands. The reporting timeline for MAID death was extended from 10 days to 30 days. As a result, if the practitioner's report would lead to some concerns, and examination of the cadaver would be desirable, it likely would be impossible. It was estimated that after all those adjustments of the reporting requirement, it will take a practitioner only 10 minutes to report on a death by MAID. The unfortunate result is that the amount of data, especially data about persons who received assisted death, is insufficient to fully assess how MAID operates. This becomes apparent upon examination of the federal Third Annual Reports on MAID, which is discussed in the next chapter of this book.
2. Lack of monitoring, analysing and reporting on adherence to eligibility criteria and procedural safeguards. Federal governments devolved to the provinces and territories all monitoring, enforcing and reporting on the adherence to those criteria that make MAID distinct from homicide. However, with a few exceptions,

those governments did not create any publicly visible organizational structures, processes or reports that would indicate that they assumed responsibility for this activity. The problem could be that such fundamentally new activity like MAID fell between the cracks of the two systems and did not receive attention from either side. Is this happening? Are there provisions in place that will ensure that the federally collected monitoring data from these jurisdictions will be, as the Regulation permits, regularly shared with local enforcement agencies? Will each province and territory develop their own way of analysing these reports, responding to noncompliance and to other concerns? Will Canadians be informed how the eligibility criteria and safeguards are respected in each province and territory, and what steps are being taken to deal with non-compliance? These are some of the serious questions that need to be addressed.

3. There appears to be no mention of MAID in the annual reports of hospitals, regional health authorities and other organizations that provide or coordinate MAID or credential MAID practitioners who offer home service. These annual or more frequent reports should not only provide relevant statistics but also information about quality control and quality assurance programs focused on MAID.

The consequences of the current lack of adequate communication about the performance of the MAID program are serious. This is a worrisome scenario to people living with chronic disabilities and to those who advocate for vulnerable groups of the population (Lemmens and Jacobs 2019). The *Canadian Disability Policy Alliance* (2020) is concerned about the term 'disability' among the eligibility criteria and is worried that this could be understood as an intention to permit MAID on the basis of disability alone. The *Coalition for Health CARE and Conscience* was gravely concerned when a woman with a disability presenting for hospital care was offered MAID (Hutchinson 2017). The Canadian Association for Community Living (2020) is concerned about the operation of safeguards in the MAID program and asserts that factors unrelated to patients' medical conditions can "make some people vulnerable to request an assisted death when what they really want and deserve is better treatment." The association developed, with the assistance of a large body of experts, the *Vulnerable Persons Standard* (VPS) and is advocating for its use in the MAID program (Canadian Association for Community Living 2020). Expressing similar concerns, a law professor and researcher called for more robust information about medical assistance in dying in Canada in order to help protect all vulnerable patients and to provide information on equality of access (VPS 2017). The *Council of Canadians with Disabilities* stated that "Canada's MAID regulations fall short", because among other deficiencies, they do not provide information on socio-economic factors of individuals requesting MAID, nor do they gather sufficient evidence on causes of 'intolerable suffering', one of the eligibility criteria for MAID. The association insists that these are critical data which, if extracted from the monitoring reports, could point to ways to improve end of life care. This 40,000 members' association concluded that "further safeguards are urgently needed to monitor the practice of MAID" (Council of Canadians 2018).

A legal scholar recently expressed concern about monitoring of MAID following growing external and internal criticism of adherence to eligibility criteria and safeguards in Belgium and Holland, where, like in Canada, physicians have broad discretionary powers and all information available about completed MAID cases is based on their self-reporting (Lemmens 2018). An academic palliative care physician wrote: "I remain concerned about the lack of provincial/territorial and national reporting standards for collection of data, reporting of actions and of experiences, and the robust monitoring practices related to medical assistance in dying…" (Wasylenko 2018). A team from the country's capital discovered that there is still inadequate provision of palliative care for those requesting MAID (Munro et al. 2020). Data collected by monitoring currently are insufficient to determine how widespread is this problem. It is not surprising, then, that an overview of MAID practice published in the World Medical Association Journal by a group of Canadian physicians reported "widening and loosening of already ambiguous eligibility criteria, the lack of adequate and appropriate safeguards … the failure of adequate oversight, review and prosecution for non-compliance with the legislation" (Herx et al. 2020).

These are some of the many voices across Canada expressing concerns or apprehension that can only be addressed effectively by a full and comprehensive reporting on the MAID program as was anticipated in the design of the program and as described above. It must be kept in mind that the enforcement of eligibility criteria and safeguard is beneficial to all involved, for obviously different reasons. Namely, the individual who does not want nor would consent to MAID; the individual requesting and consenting to MAID; the medical and nurse practitioner who wants to practice within the bounds of the law and could be pressured to provide MAID outside its bounds; to those who do not qualify; the consenting person interested in the options of postponement; finally, to all Canadians who would desire to trust the system and feel secure that their individual freedoms are respected. We can all feel secure in a community where law abiding citizens respect the boundaries of the law and the government upholds them. One may not agree that a particular "stop" sign on the street is necessary, but one nevertheless complies with the rule. Even MAID assessors and providers will benefit from regular, timely analysis and reporting of MAID data because it will assure them that they have followed the protocol correctly, that they are doing only what is legally permissible, and that there can be no retroactive action taken against them in the future. Accountability is only possible when there is transparency. These values are mentioned repeatedly in official MAID documents but further steps need to be taken to convert those values to reality in practice.

7.6 Recommendations

Considering these finding, the following three suggestions are made to address pressing issues in monitoring and reporting of MAID and enhance the transparency of the program.

1. *At the Local Levels*

The hospitals and other institutions where medical assistance in dying is taking place or which have credentialed MAID providers, would be wise to establish an internal review process of all instances of MAID, outcome of which would be reported to staff and public. It was suggested that such reviews could best be done by a multidisciplinary local committee that would include physicians, lawyers and bioethicists (Government of Canada 2015). Such a process would not require additional reports by MAID assessors and providers; rather it could use reports that are already required by federal or provincial authorities as well as originals of MAID requests and consents, consultations, nursing notes and records of delivery of MAID, including reports of adverse events. Such internal review process should be linked to the institution's quality assurance and quality control programs, so that MAID delivery would be treated as other programs where life of patients is at stake. Verification of the diagnosis of a 'grievous and irremediable condition,' the foremost eligibility criterion that supports a request for MAID, would appear to be a matter of due diligence and could best be done at this level. This internal review process would function best when under the oversight of boards of directors and its results published in the annual reports of the institutions.

2. *At the Provincial and Territorial Level*

The provincial and territorial governments in Canada need to provide evidence that they are analysing the data collected by the federal monitoring program in their territory and produce reports summarizing the past years of the program and then commence producing annual reports. These reports need to highlight the compliance with the MAID law and regulations regarding eligibility criteria, procedural safeguards and the administration of MAID. These reports would also indicate what actions were taken when non-compliance was detected. These provincial and territorial reports would complement the federal reports that are produced by Health Canada. It is understandable that at the initial phase of the MAID program, deficiencies found in reports would require education and the support of MAID providers and pharmacists, but now that the program has been in operation for several years, a different approach is needed.

3. At the Federal Level

It would be desirable to collect much more data from practitioners and pharmacists in order to better asses how medical and social needs of people could be taken care of, so that they do not have to shorten their life voluntarily by asking for MAID. For example, we should be able to better estimate current deficiencies in services, like palliative care and mental health. Better data would also be able to provide assurance that people with disabilities and vulnerable people in various categories are not affected disproportionately by MAID.

An inherent weakness of the current approach to monitoring and reporting of MAID is that it is based entirely on self-reporting of MAID assessors, providers and pharmacists. Therefore, MAID monitoring could be moved to another, higher level if the federal government would launch on-site reviews of a certain percentage of completed MAID cases, randomly selected across the country (Lemmens 2021). This could be set up similarly as inspections of clinical trials for human drugs that are currently conducted by Health Canada (Government of Canada 2015).

7.7 Conclusion

The Canadian Medical Association declared that the legalisation of medical assistance in dying is "the most profound change in medical practice in modern times" (CMA n.d.). Given the novelty of the practice, ongoing monitoring and public reporting of euthanasia and suicide assistance has been considered essential elements of assisted dying program in Canada from the very beginning in court documents, in law and in regulations. The monitoring system that was started according to federal regulations is based on mandatory reporting of MAID and MAID related events by physicians, nurse practitioners and pharmacists. Data required from each reporting party is prescribed by the Regulations. Health Canada set up an online dedicated portal. But given that health care delivery, into which MAID was embedded, is predominantly a provincial and territorial responsibility, each jurisdiction could choose a preferable flow of the data, either directly to Health Canada or through a designated provincial or territorial authority. Centrally collected data are being used to generate Annual Reports on MAID, the purpose of which is to provide 'societal perspectives'. However, monitoring, analysis and reporting of adherence to eligibility criteria and to procedural safeguards, that was from the start of the program recognized as critical in order to assure that MAID is provided only under the conditions allowed by the Criminal Code, is not being performed by Health Canada and is not included in Annual Reports. This important task was delegated to the federal and provincial authorities. Disturbingly, only two of those governments, Quebec, and to a limited degree, Ontario, are at present time reviewing and reporting on those monitoring data from their territory. It is possible that other governments are active in this area, but are not reporting their findings to the public? If so, they are ignoring the principles of transparency and accountability that are endorsed in many MAID documents. By the end of December 2021, over 30,000 Canadians died with medical assistance (Government of Canada 2020) but for a great proportion of MAID cases, we have no publicly accessible evidence that the eligibility criteria and safeguards prescribed by law were respected. Therefore, it is absolutely essential that all provincial or territorial governments and their agencies responsible for the oversight of MAID activities start collecting and analysing data from their territories, take appropriate actions and share their findings with the public, or where necessary, with law enforcement agencies. To assure the appropriate performance of the MAID program, it would be desirable for the federal government to undertake annual on-site reviews of a certain number of randomly selected MAID cases. Finally, to have health care institutions and regional health authorities take responsibility for quality control and quality assurance of their MAID programs and to keep their communities informed.

Monitoring, data analysis and public reporting are binding requirements of Canadian law but also can have an important educational value to health care professionals and users of health care; even more importantly, they demonstrate transparency and make the MAID program accountable to Canadians. At this time, the monitoring of MAID, which calls for effective collaboration of federal, provincial, territorial and local authorities, is a serious challenge that demands an urgent response.

References

Canadian Association for Community Living. 2020. *Safeguards in medical assistance in dying.* Vulnerable Persons Standard. https://inclusioncanada.ca/maid/.

Canadian Disability Policy Alliance. 2020. *MAID consultation statement of January 17, 2020.* http://www.disabilitypolicyalliance.ca/wp-content/uploads/2020/01/MAID-Consultation.pdf.

Canadian Medical Association (CMA). 2017. *CMA Policy: Medical Assistance in Dying.* https://policybase.cma.ca/viewer?file=%2Fmedia%2FPolicyPDF%2FPD17-03.pdf#page=1&search=&phrase=false.

Canadian Medical Association (CMA). n.d. *Medical assistance in dying. Are you prepared?* Accessed March 13, 2020. https://joulecma.ca/learn/medical-assistance-in-dying.

Commission sur les Soins de Fin de Vie. 2020. *Rapport Annuel d'Activités.* Gouvernement du Québec: Québec. https://www.recherchesoinspalliatifs.ca/wp-content/uploads/2020/10/RAA_2019-2020_Commission-soins-fin-de-vie-2020-10-20.pdf.

Council of Canadians with Disabilities. 2018. "Canada's Medical Assistance in Dying Regulations Fall Short." *Council of Canadians with Disabilities*, September 4th, 2018. http://www.ccdonline.ca/en/humanrights/endoflife/Media-Release-4Sept2018-MAiD-Regulations-Fall-Short.

Government of Canada. 2015. "About inspections of clinical trials for human drugs." https://www.canada.ca/en/health-canada/services/inspecting-monitoring-drug-health-products/drug-health-product-inspections/about-inspections-clinical-trials-human-drugs.html.

Government of Canada. 2018a. Regulations for the Monitoring of Medical Assistance in Dying: SOR/2018a-166. http://www.gazette.gc.ca/rp-pr/p2/2018a/2018a-08-08/html/sor-dors166-eng.html. Accessed March 13, 2022.

Government of Canada. 2018b. Regulations for the Monitoring of Medical Assistance in Dying: SOR/2018b-166. Regulatory Impact Analysis Statement. Background. http://www.gazette.gc.ca/rp-pr/p2/2018b/2018b-08-08/html/sor-dors166-eng.html. Accessed March 13, 2022.

Government of Canada. 2020. *An act to amend the criminal code (medical assistance in dying).* Technical Briefing, February 24, 2020. Ministry of Justice & Attorney General of Canada, Ministry of Employment, Workforce Development & Disability Inclusion, and Ministry of Health.

Herx, Leonie, Margaret Cottle, and John Scott. Open menu close menu the "normalization" of Euthanasia in Canada: The cautionary Tale continues. *World Medical Journal* 66 (2020).

Hutchinson, Carmela. 2017. Why are women with disabilities being given help to die, instead of help to live? *Coalition for Health Care and Conscience.* https://rabble.ca/health/why-are-women-disabilities-being-given-help-die-instead-help-live/.

Lemmens, Trudo. 2018. Charter scrutiny of Canada's medical assistance in dying law and the shifting landscape of Belgian and Dutch euthanasia practice Supreme Court Law Review Second Series 85459-544.

Lemmens Trudo. 2021. Dangers of a lax assisted death regime. *Impact ethics* (3 March 2016). https://impactethics.ca/2016/03/03/dangers-of-a-lax-assisted-death-regime/.

Lemmens, Trudo, and Laverne Jacobs. 2019. The latest medical assistance in dying decision needs to be appealed: Here's why. *The Conversation* (24 October 2019). https://theconversation.com/the-latest-medical-assistance-in-dying-decision-needs-to-be-appealed-heres-why-124955.

MacDonald, Michael. Nova Scotia's medically assisted dying program paused amid 'unprecedented' demand. 2021. *CBC News* (30 September). https://www.cbc.ca/news/canada/nova-scotia/nova-scotia-s-medically-assisted-dying-program-paused-amid-unprecedented-demand-1.6195412.

Munro, C., A. Romanova, C. Webber, M. Kekewich, R. Richard, and P. Tanuseputro. 2020. Involvement in palliative care in patients requesting medical assistance in dying. *Canadian Family Physician* 68(2):89.

Office of the Chief Coroner for Ontario. 2018. *Medical assistance in dying memorandum.* Official Memorandum. Government of Ontario. https://www.ontario.ca/page/medical-assistance-dying-memorandum#section-0.

Parliament of Canada. Bill C-14. Statues of Canada 2016. Chapter 3. Assented to June 17, 2016. https://www.parl.ca/DocumentViewer/en/42-1/bill/C-14/royal-assent. Accessed March 13, 2022.

Shared Health Manitoba. 2022. *Medical assistance in dying.* Accessed March 15, 2022. https://sharedhealthmb.ca/services/maid/.

Supreme Court of Canada. Carter v. Canada (Attorney General). 2015 February 06. Accessed March 13, 2022. https://scc-csc.lexum.com/scc-csc/scc-csc/en/item/14637/index.do.

Vulnerable Persons Standard (VPS). 2017. *VPS standard and compliance table* (September 28, 2017). https://static1.squarespace.com/static/56bb84cb01dbae77f988b71a/t/59ce974012abd96a0e1facfd/1506711361151/VPS+Standard+%26+Compliance+Table+-+Sept+2017.pdf.

Wasylenko, Eric. 2018. Reporting and Monitoring of Medical assistance in Dying. *VPS Secretariat.* http://www.vps-npv.ca/blog/2018/4/16/reporting-on-and-monitoring-of-medical-assistance-in-dying-maid

Jaro Kotalik, MD (Charles University, Prague), D.M.R.T. (University of Toronto), MA (Medical Ethics and Law, Kings College, University of London, UK), FRCPC, is a Professor, Division of Clinical Sciences, Northern Ontario School of Medicine; Adjunct Professor, Department of Philosophy; Chair, Executive Committee and Founding Director of the Centre for Health Care Ethics Lakehead University. Between 1984 and 2010, he held an appointment at the Faculty of Health Sciences, McMaster University, Hamilton, and retired as Clinical Professor. Between 1974 and 1994, he served as a radiation oncologist and later CEO of the Thunder Bay Regional Cancer Centre, and a Vice President of Cancer Care Ontario. After completing his studies in medical ethics and law, he dedicated most of his time to the field of bioethics. He served as an executive member of National Council on Ethics in Human Research and a member or chair of several research ethics boards and committees of the Royal College of Physicians and Surgeons. In his capacity as a consulting bioethicist, he has provided services to Health Canada, Public Health Agency of Canada, Swiss National Advisory Commission on Biomedical Ethics, Thunder Bay Regional Health Sciences Centre and other organizations. He published numerous research papers, 2 major consulting reports, and five book chapters. He is an Associate Member of the World Medical Association.

Chapter 8
Federal Annual Reports on MAID: Informative but Incomplete Picture

Jaro Kotalik

Abstract Annual Reports on Medical Assistance in Dying (MAID) issued by Health Canada on behalf of the Government of Canada strive to make essential information public, on this large national program that over five years has provided death to over 30,000 Canadians. The source of the information is self-reporting by physicians and nurse practitioners who offer to act as MAID assessors and providers, as well as self-reporting by cooperating pharmacists. Focus of this discussion is the latest available document in this series, the Third Annual Report on MAID published in July 2022 and providing data on 10,064 deaths by MAID in 2021. In addition to reviewing essential information about MAID recipients and MAID providers, the discussion is centered on the method of data collection, the rapid and unexpected annual increase of cases, the underlying medical conditions, the nature of suffering reported by MAID applicants, the use of palliative care, the decision-making about one's eligibility, the role of the reflection period, necessary requirement to obtain informed consent and the denial of assisted dying. In each case, the attention is drawn to missing or indeterminate elements, and essential but unavailable analysis of data, in order to identify further information that ought to be collected and used to generate future annual reports and to refine the delivery of MAID program and its oversight.

Keywords MAID · Assisted suicide · Euthanasia · Canada · Monitoring · Medically assisted dying · Public reporting · Nature of suffering · Denial of assisted dying · Palliative care · Euthanasia statistics

---

J. Kotalik (✉)
Centre for Health Care Ethics Lakehead University and Northern Ontario School of Medicine University, 955 Oliver Road, Thunder Bay, ON, Canada
e-mail: jkotalik@nosm.ca
URL: https://www.lakeheadu.ca/centre/chce

J. Kotalik and D. W. Shannon (eds.), *Medical Assistance in Dying (MAID) in Canada*, The International Library of Bioethics 104,
https://doi.org/10.1007/978-3-031-30002-8_8

8.1 Introduction

The revisions of the Canadian Criminal Code by Bills C-14 and C-7 (Statute of Canada 2016, 2021), later referred to as MAID Law, created an exemption in the Code for medical assistance in dying (MAID). This MAID Law charged the Minister of Health with the responsibility for regulation respecting the provision and collection, for the purpose of monitoring MAID, of information relating to requests for, and provision of, MAID. A previous chapter has already described the development and operation of the regulations and guidelines that formed the current monitoring system. This chapter will describe and discuss a recent federal report that was derived from this monitoring. The Third Annual Report (Health Canada 2022) provides the latest, publicly available information on MAID in Canada. When statistics are listed, I will round all percentages to the nearest full number and most other numbers to the nearest one hundredth.

8.2 Method of Data Collection

Before examining the content of the report, I will comment briefly on some aspects of data collection. Persons who are empowered to provide MAID in Canada are physicians and nurse practitioners. They have a legal obligation to report all written requests for MAID, the execution of MAID, and a number of other possible outcomes of MAID applications, as we will discuss later. Practitioners are asked to provide selected data on themselves, report on various characteristics of MAID applicants, and so on. Reporting is done directly to the federal Ministry of Health by an internet-based Data Collection Portal or, in some provinces, to a designated provincial or territorial authority which forwards the data received on a quarterly basis to the federal level (Government of Canada 2018).

Given that the adherence to the eligibility criteria and the safeguards are those conditions that differentiate the legal medical assistance in dying from a criminal act of culpable homicide, the method used to collect such data about MAID is critically important (Kotalik 2020). It is the only source of information about thousands of MAID deaths, which is available to the oversight authorities and the public. As yet, no rigorous analysis of the methodology was published.

We cannot be, at present, certain that the existing method of collecting data provides accurate information, especially its accuracy in accounting for each person who has received MAID. Unfortunately, the Death Certificate, which is another potential source for verifying the occurrence of MAID, remains unreliable. Bill C-14, which created the MAID program addressed the issue of Death Certificates. It charged the Minister of Health to establish guidelines for the information to be included on death certificates, which "may include the way in which to clearly identify medical assistance in dying as the manner of death" (Statute of Canada 2016, 241.31.3.31). It is the little word "may" that allowed authorities in some or even most provinces to order that no clue should be provided on the death certificate to indicate that a patient died from the administration of a toxic substance. In Ontario, for example, practitioners are instructed that "*(W)hen completing the death certificate, physicians : **must** list the illness, disease, or disability leading to the request for MAID as the cause of death; and **must not** make any reference to MAID or the drugs administered on the death certificate*" (Office of the Chief Coroner of Ontario 2021). No justification is provided for this unusual policy. One can speculate that it serves to hide the real cause of death from the next-of-kin and family that will, after the death of the person, obtain the certificate. The medico-legal process of MAID, from application, to assessment and administration of lethal injection follows the usual rules of privacy and confidentiality that compels healthcare staff to withhold the information that a patient is being assessed or scheduled for MAID. It is up to the individual to inform the next-of-kin. If the patient had withheld the information, and MAID was listed on death certificate, then survivors could be shocked.

This is ethically a highly problematic arrangement. Firstly, it makes it impossible to use a Certificate of Death file to verify the completeness of reporting on MAID. Secondly, this rule demands a physician to attach his signature to a lie and participate in deception. Such an act should never be requested of anyone, least of all of a health professional, especially by law and regulations. Asking a physician or a nurse practitioner, who has just injected a lethal dose of drugs into a patient's body, to state that the patient died of a pre-existing condition, for example, a cerebral infarction which he suffered months ago, is asking them lie, and thereby to act against the values of truthfulness, honesty and integrity. When a professional is forced to act against their professional moral code, and personal moral standards, the professional suffers a moral injury. An accumulation of these injuries will lead to burnout, indifference and detachment from morality, which is incompatible with good medical practice. Additionally, physicians who follow these instructions could possibly be charged of breaking section 380(1) of the Criminal Code that prohibits fraud.

A detailed analysis would likely show that a much more robust and thoughtful method of data collection is necessary in order to meet the Supreme Court of Canada expectations that, "the risk of physician-assisted death can be identified and very substantially minimized through a carefully designed system that imposes strict limits that are scrupulously monitored and enforced" (Supreme Court 2015, s.27).

8.3 Recipients of Medically Assisted Death

8.3.1 Number of Reported MAID Deaths

The Third Annual Report indicates that from June 2016 to December 2021 there were in Canada 31,664 deaths attributable to MAID. In 2021 only, there were about 10,000 death, which represents about 3.3% of all deaths. These figures are significantly higher than forecasted. Health Canada estimated that at steady rate, MAID will represents 2.05% of all deaths (Health Canada 2022). Also, the Supreme Court of Canada, in the judgment that made MAID possible, quoted an earlier Court of Appeal pronouncement, without questioning it, that Criminal Law is a "generally sound law" that has an extraordinary effect on a small number of individuals" (Supreme Court 2015, s.124), suggesting that only a small number of individuals would apply for assisted death. More recently, Lord Falconer, the former Lord Chancellor, advocating for assisted suicide in the United Kingdom, also grossly underestimated the likely impact of such a measure, stating that access to medically assisted suicide would "increase the dignity of the end of a quite a small number of people but to whom it would matter hugely" (Davenport 2019). Would the Supreme Court of Canada have issued the same judgment, if they had realized that it would not affect a "small number" but thousands of Canadians every year?

The number of MAID cases in Canada have been steadily increasing each year, 26.4% from 2018 to 2019, and 34.3% from 2019 to 2020, and 32.4% from 2020 to 2021, (see Fig. 8.1).

The Third Annual Report does not comment on the rapid increase in number of MAID procedures. The report indicated that deaths due to COVID-19 did not have any impact on MAID statistics. MAID deaths continue to represent a higher and higher percentage of total deaths in Canada. MAID deaths accounted for 2% of all deaths in 2019, and 2.5% in 2020, and 3.3% in 2021.

The Report does not comment on the accuracy of these statistics, particularly on the possibility that some deaths due to assisted dying are not counted, as has been a problem in other permissive jurisdictions. One way to verify the completeness of MAID data is to compare the information as reported by MAID providers and assessors with reports submitted to the same internet portal by pharmacists, who are required by law to report on the dispensing of lethal substances to practitioners. The Second Annual Report was the first report that offered a comparison of these

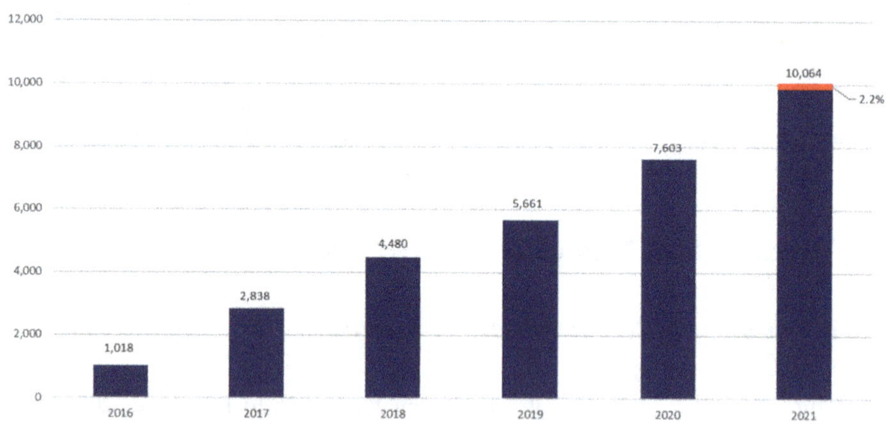

Fig. 8.1 Demonstrates the number of reported MAID deaths in Canada from 2016 to 2021. *Source* Health Canada (2022)

two sources of data. That Report stated that the program achieved 95% exact match, pointing out that the discrepancy could be due to an error in data entry. We also learned from the Report that, "Health Canada conducts follow-ups with jurisdictions to ensure completeness of the match in these cases" (Health Canada 2021). If we assume that the 5% discrepancy corresponds to the total official number of 21,538 cases and that this number was derived from reports of MAID providers, then it could mean that pharmacists reported issuing substances for an additional 1075 persons. In absolute terms, this is a large number, so it will be important that the next annual report provide a corrected number of MAID cases. Surprisingly, the Third Annual Report did not refer to data obtained from pharmacists and their match with data from MAID providers. This is a serious set-back in reporting.

8.3.2 Underlying Medical Condition of Persons Receiving MAID

Some of the statistics in the Annual Report are very predictable, and do not change much from year to year. For example, in 2021, the average age of patients who received MAID was about 76 years, comparing with 75 years in the two previous years. A slightly larger percentage of men (52%) than women, (48%) received MAID. Persons from urban and rural areas were equally represented. Also, not unexpected is the information that the most common underlying medical condition experienced by the applicants for MAID was cancer, which was recorded in 66% % of applicants. The next three medical conditions were cardiovascular disease, respiratory disease and neurological disease, that were reported in 19, 12 and 12% of MAID interventions (see Fig. 8.2).

Should we be surprised that such a high percentage of MAID deaths are related to cancer? For the past three decades, cancer has been the leading cause of death and responsible for about 28% of all deaths in Canada. In 2019, for example, there were 85,000 cancer deaths (Canadian Cancer Society 2021). Because in the same year, there were 5660 cases of MAID and of these, 70% reported cancer as the underlying medical condition, this tells us that only 4.6% of cancer deaths were associated with MAID.

However, the Annual Report does not provide enough details to assure us that this large cohort of cancer patients applying for MAID were able to obtain adequate medical care and support that would enable them to live out their life to a natural end. Also, the report does not provide any indication, whether all, or most of those cancer patients had an "incurable disease", which is one of the eligibility requirements. Across the board, about 35–40% of new cancer patients are considered to have incurable diseases. The Annual Report provided a breakdown of MAID cases by cancer sites, but did not separate localized and metastatic cases, did not provide stages of cancer, or indicate what was the interval between cancer diagnosis and MAID. We learned from the Report that only 1.3% of MAID practitioners were oncologists

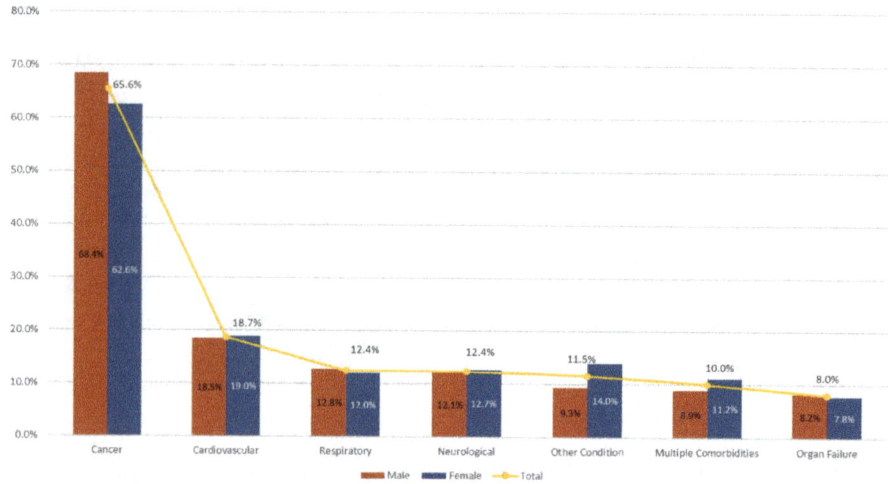

Fig. 8.2 Compares the profile of MAID recipients in 2019 and 2020. *Source* Health Canada (2022)

and that MAID practitioners obtained consultation with oncologists only in 12% of MAID cases, both suggesting that assessment of cancer patient for MAID eligibility likely relied on oncology consultation reports in patients' charts, if indeed these written consultations had been done. One recent Canadian study suggests that this is not always the case. A review that involved all cancer patients referred to a MAID program of a teaching hospital, about four and half years showed that of this group of 255 patients with cancer, who underwent MAID, 45 had a lung cancer and their records were examined in detail. The study revealed that out of 45 patients, 10 did not have a medical oncology consultation. Authors of the study stated, "Given the growing number of efficacious and well-tolerated treatment options in lung cancer, consultation with medical oncology may be reasonable to consider for all patients with lung cancer, who request MAID" (Moore et al. 2022). Even more disturbingly, 13% of patients in this study who received MAID with presumptive diagnosis of lung cancer, did not have a biopsy which is the gold standard of cancer diagnosis.

A family physician shared with me a disturbing situation in his practice. An older man complaining of cough was sent for chest-x-ray. When the radiology report indicated that an abnormality on the lung could be due to cancer, the patient immediately asked for MAID. It took considerable effort by the physician to convince the patient to go for an oncology consultation. At the cancer centre, the diagnosis of Hodgkin's disease was made. He completed therapy without complications and has now 90% chance of a long-term survival. This illustrates the fact that Canadians fear cancer more than any other disease and link cancer with dying (Ipsos 2012). People with newly diagnosed cancer often develop reactive depression and so it is important, as the Annual Reports insist, that people do not make the decision for MAID as the result of "a temporary period of despair" (Health Canada 2020, p.7). The Report does not provide us with intervals between diagnosis of cancer and MAID. We cannot be certain that all patients passed the possible temporary period of despair.

People with neurological conditions represented about 12% of people who receive MAID in 2021. The largest group of patients in this category (21%) were those diagnosed with amyotrophic lateral sclerosis (ALS). This is understandable, given that ALS is an incurable, progressive disease which almost always leads to death. But we must be concerned about accurate diagnosis. Other neurological diseases, like Lyme disease, or cervical myelopathy, which have much better prognosis can mimic ALS. There is no single definitive diagnostic test, and a false positive diagnosis is made in 10–20% of cases. The observation of progression of symptoms over a 6–10 months period is useful to confirm the diagnosis (Brooks et al. 2000). Hence, it would seem that if ALS is an underlying condition in an application for MAID, a consultation with a neurologist should take place, and MAID should be postponed, if possible, until about one year after the diagnosis. Again, the Annual Report does not provide the time interval between the diagnosis of a underlaying medical condition and the execution of MAID, because at present, this is not a reportable factor.

It is problematic that the Third Annual Report (as other annual reports) does not indicate the percentage of patients assessed for MAID, who previously had a mental disorder, were diagnosed with a depression, or were hospitalized for a mental health issue. Asking these questions would help the assessor to judge if the request for MAID could be a manifestation of suicidal ideation that was present before the underling illness, that could make the person MAID ineligible. If this is the case, the assessor could consider if the patient should be offered standard suicide prevention measures before exploring the possibility of MAID (Mishara and Weisstube 2018).

8.3.3 The Nature of Suffering as an Eligibility Criterion

The MAID Law states that to be eligible for MAID, the person must have besides a “serious and incurable illness, disease or disability” and “advanced state of irreversible decline in capability” also “intolerable and physical or psychological suffering that cannot be relieved in a manner that the patient finds to be acceptable”. By the time patients are seen by MAID assessors, patients are most likely well aware of the necessity to report some kind of suffering in order to qualify, and such knowledge may affect their responses. Yet, these particular messages from patients are still very important. The assessor is required to report how the patients describe their suffering on the federal MAID data collection portal. On this portal, the practitioner has a choice of 12 categories of suffering and can select, for the same patient, more than one category. Results of those reports are summarized in Fig. 8.3.

I will now comment on some of those categories of suffering and on problems associated with these categories.

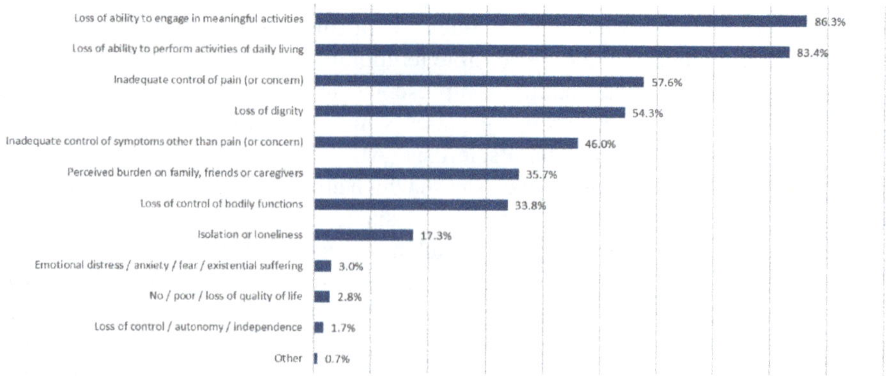

Fig. 8.3 Demonstrates the types of suffering experienced by patients who received MAID, as reported by their providers in 2021. *Source* Health Canada (2022)

(1) "loss of ability to engage in meaningful activities"—86%—Comment: lack of specifics make this item difficult to interpret and act on it; could these activities be reading a book or mountain climbing?
(2) "loss of ability to perform activities of daily living"—83%—Comment: lack of specifics; but if the activities of daily living are, for example, shopping, cooking meals and cleaning, would not the appropriate response be the provision of a homemaker rather than MAID?
(3) "inadequate control of pain (or concern)"—58%—Comment: coupling of pain and a concern or fear of pain make this category impossible to interpret; presence of intolerable pain should be immediately addressed by MAID assessor, or lead to urgent referral to standard palliative care; 'concern or fear of pain' calls for specialised psycho-social-spiritual counselling.

A major issue is that a great deal of suffering reported by MAID applicants is of a social and existential nature. There are two such separate but related categories: isolation and loneliness 17%, and emotional distress/ anxiety/ fear/ existential suffering 3%. If we apply these figures to the whole population, who received MAID just in 2021, that is 10,064, we have to conclude that about 1700 patients died mainly, largely or partially, due to loneliness. In another group of 300 persons, existential suffering, fear or anxiety were such a problem that they shared it with the clinician and they received MAID for relief. We have to wonder how MAID assessors are reacting to those patients who want to die because they are lonely. An Angus Reid Institute survey showed that 23% of Canadians suffer from extreme social isolation (Penning 2019) and so we know that social isolation and loneliness is a big social issue in Canada. Some speak about 'epidemics of loneliness" (Statistics Canada 2021), yet there are also simple, effective ways to relieve loneliness (Ahuja et al. 2021). Conclusions from this brief review of reported suffering of people applying for MAID is that the on-line questionnaire for assessors should be substantially refined, so it can more clearly pinpoint problems that are driving the request for MAID. This will make it possible to offer to MAID candidates a tailored assessment and therapeutic intervention and may eliminate the need for MAID. It would also reveal deficiencies in patients' services that may be driving the demand for MAID. Given the large number of psychological, social and existential problems that applicants experience, it would be prudent if all applicants for MAID be offered not only palliative care services, but also specialized psycho-social-spiritual assessment, and if the person wishes, an appropriate care. The two MAID physicians and nurses who act as MAID eligibility assessors ideally should identify a patient's problems, present alternatives and open avenues for support. But a multidisciplinary team may be required to fully explore social and community services that might allow a person interested in MAID to live. It would be tragic if a "life was ended as a result of failures of communication and assistance" (Kaufert and Koch 2003).

8.3.4 Palliative Care and Disability Support

Because MAID is made available only for people who experience "intolerable physical and psychological suffering" and because palliative care is the discipline that exists to provide relief of suffering, then it is expected that the Third Annual Report would contain substantial information about palliative care's availability, use, benefits, shortcomings, etc., in reference to persons who apply for MAID. Regretfully, only part of a page of the 47-page report is devoted to this topic.

It's reported that in 2021, 80% of MAID recipients received palliative care services but the duration of care was often limited, less than two weeks in 21%; two weeks to under one month in 17%; one month or more for 52%; and unknown in 8% of cases. MAID providers reported that their patients who did not receive palliative care had access to those services in 88% of cases. The Annual Report concluded "that palliative care continued to remain both available and accessible to individuals to who have received MAID." This is a highly questionable conclusion. Firstly, statistic in the Report suggest that palliative care was too short to be a reasonable trial of treatment for most of those patients. A recent international review concluded that, "evidence suggests that for a full benefit of palliative care to be realised, continuity by a multidisciplinary team is needed for at least three to four months" (Jordan et al. 2020).

Secondly, a recent Canadian study found that in patients presenting for MAID, only 27% had community palliative care physicians and about 60% of patients had some contact with palliative care. Writing from Canada's capital, which should have top notch health services, the authors concluded that, "There is still inadequate provision of palliative care for those requesting MAID" (Munro et al. 2020). Canadian research also showed that patients with advanced cancer benefitted from early palliative care intervention. Patients experienced decreased psychological distress, an improved quality of life, personal satisfaction and better relationship with health care providers (Mah et al. 2022), but the authors noted that this knowledge does not sufficiently affect the practice. Insufficient use of palliative care likely contributed to the high demand for MAID. It appears that the scarcity of palliative care data found in the Third Annual Report is not accidental, but rather a reflection of insufficient availability or use, of a service that could be of great benefit to patients with severe suffering.

The last revision of the Criminal Code in Bill C-7 definitely increased many Canadians' concerns about MAID's negative impact on the life and wellbeing of people with disabilities (CCD 2016). Ontario's former lieutenant governor stated that the Bill "opens up a catastrophic Pandora's box" (Gillmore 2021). So, it remains disappointing that this Annual Report devotes only two short paragraphs to disability issues. Firstly, we learned that 43% of MAID recipients reported that they required disability support services and 87% of those received such services. 'Disability' is not defined, and may be an aggregate of short term disability created by an acute illness prior to the MAID, or it may be a long term disability, which is the main concern of the disability community. Also the term "disability support services" is not defined and may range from specialized professional services in the home of a patient with quadriplegia and on respirator, to "meals on wheels" services, and this may make the current statistics almost useless. To better address this issue, the disability community suggested that a brief but mandatory "vulnerability assessment" of those interested in MAID be requested to determine if the lack of services may be possibly driving requests for MAID (CCD 2016). Clearly, there is a need to collect much more detailed data on disability during the assessment for MAID and to report the information in a meaningful fashion.

8.3.5 *Period of Reflection After Person Submitted Request for MAID*

A diagnosis of serious or incurable disease is frequently accompanied by mental depression, hopelessness and anxiety, and although these reactive symptoms are often temporary, they may be accompanied by suicidal ideation. Research involving 200 patients, who were terminally ill with cancer showed that 45% had transient wishes to die, but only 8.5% had sustained wishes to die. In order for the consent for MAID to be informed, the patient with reactive depression needs to be informed that it is temporary, and be offered help such as social support, pharmacotherapy and psychological or spiritual counselling. These measures may be effective even in patients with advanced illness (Chochinov et al. 1995). If depressive symptoms do lift and the patient is still interested in MAID, then doubts about a patient's capacity to consent can be removed.

The awareness of this situation is one reason why the original MAID Law (C14) mandated a 10 days period for reflection after the person signed the application and before MAID could be administered. The Law permitted this safeguard to be waived if both MAID assessors agreed that the person's death was imminent or that the person was at risk of losing capacity to give their final consent. Both the First and

Second Annual Reports indicate that in practice this reflection period was shortened in about 34% of cases.

This reflection period, even if frequently reduced, was effective. The Report indicates that 3.8% and 2.5% of applicants for MAID during 2019 and 2020 respectively withdrew their requests because they changed their mind, or because palliative measures were sufficient. This safeguard resulted in 521 patients avoiding having their life shortened. The revision of the MAID Law in 2021 led to the elimination of this critical reflection period for people whose death was reasonably foreseeable. It was suggested that some patients during this 10 days period may refuse pain medication, in order to remain capable to give the final consent for MAID just before the procedure. A better response to this situation, rather than removing the reflection period would have been to assure that these patients could receive a well balanced pain control. Now, with no refection period prescribed, a person may sign the request for MAID and die by MAID in the same day, as long as two clinicians are able to complete their assessment (in person or on line) and find the person eligible for MAID. Such an approach is clearly in conflict with the MAID Law preamble that states, "Vulnerable persons must be protected from being induced, in moments of weakness, to end their life" (Statute of Canada 2016, c 3, preamble).

8.3.6 Denial of MAID to Applicant

The number of persons whose request for assisted death was denied was indicated to be 602 in 2019, and 566 in 2020, representing 8% and 6% of written requests for that year. In 2021, the number of persons found ineligible dropped further to 487, which represents 4% of written requests. Two main reasons for denial were "not capable of making decisions with respect to health" and "natural death not reasonably foreseeable". The Third Annual Report, does not provide a breakdown of those numbers by provinces or territories. These numbers are surprisingly low in comparison with other jurisdictions. In the Netherlands, for example, physicians refuse 2 out of 3 requests and this is not considering request-inquiries of which only one out of ten result in assisted death (Kimsma 2010). The Annual Report explains that many assessments of ineligibility in Canada were made verbally without a written request, and therefore were not reported and counted in the official numbers.

There are other possible explanations. Nurses and nurse practitioners involved in MAID noted that it was more intensely stressful and time consuming to determine someone's ineligibility than approving an application. These nurses also noted that, "particularly resourceful persons were able to achieve their goal of having MAID even if they had been deemed ineligible" (Pesut et al. 2021, p.24). One other possible reason for a low rate of application rejections in Canada may be the particular nature of decision-making expected of practitioners in Canada. Their chief role is to determine if the person asking for MAID meets the formal medico-legal criteria of eligibility as defined by the Criminal Code. A Canadian MAID assessor is expected to accept and record the patient's description of their suffering and cannot raise doubts about it or make any judgment (Government of Canada 2022).

In contrast, in the Netherlands "physician and patient together have come to a conclusion that, for euthanasia to be allowed, there is no other reasonable solution available" (Kimsma 2010). Also, in Canada, as will be discussed later, the MAID provider is most often a stranger to the patient, whereas in the Netherlands, most often it is the patient's family physician.

The Report does not indicate how many persons applied more than once for MAID, even if this fact should be known to Health Canada given that every application is reportable. Keeping track of requests for MAID that were not approved is important. It was suggested that the database of all requests should be continuously updated and open to MAID assessors and providers. "The risk to guard against is a sort of 'physician-nurse practitioner shopping' in which a suicidal patient simply goes from one health-care provider to another until one is found willing to accede to the procedure" (Dumsday 2021, p.145). Knowing that a person made one or more previous applications for MAID should not disqualify the person from being assessed again for eligibility, but it would be helpful in making a the new assessment.

8.4 Providers of Medically Assisted Death

The Third Annual Report indicates that 10,064 MAID interventions reported in 2020 were carried out by 1577 unique practitioners, 94% of them were physicians and the remaining were nurse practitioners. Nurse practitioners are active in all jurisdictions but can provide MAID only in 7 out of 13 provinces and territories. Because there are in Canada about 86,000 physicians, we can calculate that 1.8% of Canadian physicians in 2021 provided MAID. Of those who did get involved, 36% provided MAID to only one patient. Another 48% provided two to nine MAID procedures. The remaining 16% of practitioners delivered 10 or more intervention in 2020. Some providers within this group made MAiD the primary focus of their practice. One of them, Dr. Stefanie Green, estimated providing MAID to hundreds of patients and wrote a book about it (Green 2022).

Most MAID providers (68%), like Dr. Green, are family physicians, 10% work in medical specialties, most often in palliative care and anesthesiology.

We need to note that when MAID providers were asked how they determined that the patient's request was voluntary, only 14% of providers reported that they relied on the knowledge of the patient from a time prior to the MAID assessment. This suggests that in 86% of cases, the MAID assessor was a stranger to the patient. Yet, it is the patient's family physician who, for years, knows and treats the patient and possibly other members of the family, and therefore would be in the best position to judge if the patient's request for MAID is authentic and consistent with the patient's values and behaviour, or if it could be the result of undue external influences. A clinician who has never treated the patient and is called only to deliver the service may not be in a position to detect any form of external pressure on the patient. On other hand, in the present system, it seems that the majority of physicians who do not wish to be involved in MAID delivery can always call on a colleague.

But it is of concern that in 59% of cases, MAID assessors proceed in their decision making without consulting other professionals, who would know the patient better, or have a wider knowledge of treatment options relevant to the patient's condition. If a consultation did take place, it was with nurses in 41% of cases, with primary care providers 34% of cases, palliative care specialists 25% of cases and social workers in 23% of cases. Other physicians and oncologists were asked for input in 12%, and psychiatrists in 5%, of assessments. In order to obtain informed consent in Ontario, the MAID assessor has to "discuss all treatment options with patients including associated risks and side effect" (CPSO 2021). This must be a difficult condition to fulfil for physicians generalists unless specialists are involved, given the range of treatment options now available for cancer and a other diseases.

8.5 Summary

1. The Third Annual Report on Medical Assistance in Dying (MAID) in Canada is derived from mandatory reports of physicians, nurse practitioners and pharmacists involved in the MAID program into an internet-based data portals set up by Health Canada and provincial authorities. These reports are the only source of information that is now, and ever will be available to oversight authorities and the public concerning the death of thousands of Canadians.
2. There has been 10,064 deaths attributable to MAID in 2021, representing 3.3% of all deaths in Canada. Since the launch of the MAID program in June 2016 until December 2021, there have been 31,664 deaths by MAID. There has been a large incremental increase in MAID cases every year, such as 32% from 2020 to 2021. Only about 4% of those persons who applied for MAID were deemed ineligible.
3. In 2021, there have been about 1600 self-selected MAID assessors and providers. 94% of those were physicians, which correspond to about 1.8% of all Canadian medical practitioners. The Report gives the impression that MAID assessors often proceed to judge the eligibility for MAID alone, without consulting other professionals who would know the patient better or have a wider knowledge of treatment options relevant to the patient's condition. The most common underlying medical condition of persons who received MAID was cancer (70%), followed by cardiovascular, respiratory and neurological diseases. The Report provides very limited information on these conditions.
4. Data were collected on the nature of suffering for those who received MAID, but the Report often conflates existing suffering with a fear of future suffering, and such data cannot provide guidance for a desirable course of action. There is a suggestion that uptake of MAID interventions is driven at least to a certain degree by the limited use or availability of palliative care.

5. About one of five patients who received MAID complained of loneliness, isolation, emotional distress, fear, anxiety or existential suffering. Because several sources of suffering could have been reported for every MAID applicant, it is not clear from the Report how many people may have received MAID for no other indication than those of psychological and social states. Also, there is no information on what, if any, intervention was offered to those patients. Very little information was collected about disability and no data were collected about the economic situation of MAID recipients.
6. The Third Annual Report does contribute to informed dialogue as Canada's MAID regime continues to evolve but also points to the urgent need to enhance the quantity and quality of data collection and improve the data analysis and reporting, in order to achieve the level of accountability and transparency that a program of this nature and scope demands.

8.6 Recommendations

1. Advance the standard of MAID assessment for patients asking for MAID by developing guidelines for a thorough, multidisciplinary evaluation of patient's medical, social, economic and psycho-spiritual issues that are driving the patient's request for MAID, and be prepared to offer a trial of assistance tailored to each patient.
2. Improve the collection and reporting of better quality and quantity of data concerning palliative care which was provided to MAID requestors.
3. Collect from MAID assessors richer data on the underlying medical conditions, such as method of diagnosis and stage of cancer and the time intervals between the diagnosis of the underlying medical condition and MAID administration.
4. Collect from all practitioners information if the assessment for MAID eligibility by each practitioner was done in person or was virtual.
5. Undertake rigorous analysis and refinement of the on-line questionnaires used for data collection from MAID assessors and providers, so that they could provide better, more detailed information on palliative care, disability and the nature of suffering of the MAID requester.
6. Re-institute the 10 days reflection period for persons whose death is reasonably foreseeable in order to avoid a hasty choice of death in moment of weakness, while assuring optimal pain and symptom control during this period.
7. Create a database of all written requests for MAID and make it accessible to physicians and nurse practitioners in order that they can determine if a person asking for MAID requested the intervention before.

References

Ahuja, M. K. Lewis, L. Kim, S. Owais, et al. 2021. Evaluation of a student-older adult telephone befriending program to reduce social isolation during the COVID-19 pandemic: A pilot Evaluation. *Canadian Geriatric Journal* 24(3): 258.

Brooks, B.R., R.G. Miller, M. Swash, and T.L. Munsat. 2000. El Escorial revisited: Revised criteria for the diagnosis of amyotrophic lateral sclerosis. *Amyotrophic Lateral Sclerosis and Other Motor Neuron Disorder* 1 (5): 293–299. https://doi.org/10.1080/146608200300079536.

Canadian Cancer Society. 2021. Cancer statistics at a glance. https://cancer.ca/en/research/cancer-statistics/cancer-statistics-at-a-glance

Chochinov, Harvey Max, Keith G. Wilson, Murray Enns, Neil Mowchun, Sheila Lander, Martin Levitt, and Jennifer J. Clinch. 1995. Desire for death in the terminally ill. *The American Journal of Psychiatry* 152 (8): 1185–1189.

CCD. Council of Canadians with Disabilities. 2016. *CCD Submission to Special Joint Committee on Physician Assisted Dying*. January 28. http://www.ccdonline.ca/en/humanrights/endoflife/SJCPAD-28jan2016.

CPSO. 2021. Medical assistance in dying. Policy. College of Physicians and Surgeon of Ontario. April 2021. https://www.cpso.on.ca/Physicians/Policies-Guidance/Policies/Medical-Assistance-in-Dying.

Davenport, Liam. 2019. Cancer patients deserve choice of medically assisted dying. *Medscape*, 18 November, 2019. https://www.medscape.co.uk/viewarticle/cancer-patients-deserve-choice-medically-assisted-dying-2020a1000yxm.

Dumsday, Travis. 2021. *Assisted Suicide in Canada. Moral, Legal and Policy Considerations*. Vancouver: UBC Press.

Gillmore, Meagan. 2021. *'Catastrophic Pandora's Box': Disabled Ontarians Speak Out Against Proposed MAID Law*. TVO.org. March 03. https://www.tvo.org/article/catastrophic-pandoras-box-disabled-ontarians-speak-out-against-proposed-maid-law.

Government of Canada. 2018. Regulations for the monitoring of medical assistance in dying: SOR/2018-166. *Canada Gazette, Part II* 152 (16). http://www.gazette.gc.ca/rp-pr/p2/2018/2018-08-08/html/sor-dors166-eng.html.

Government of Canada. 2022. *Guidance for Reporting on Medical Assistance in Dying*. https://www.canada.ca/en/health-canada/services/medical-assistance-dying/guidance-reporting-summary.html.

Green, Stefanie. 2022. *This is Assisted Dying*. Simon & Schuster.

Health Canada. 2020. *First Annual report on medical assistance in dying in Canada*. https://www.canada.ca/en/health-canada/services/medical-assistance-dying-annual-report-2019.html

Health Canada. 2021 *Second Annual Report on Medical Assistance in Dying in Canada in 2020*. Health Canada. https://www.canada.ca/en/health-canada/services/medical-assistance-dying/annual-report-2020.html

Health Canada. 2022. *Third Annual Report on Medical Assistance in Dying in Canada 2021*. https://www.canada.ca/en/health-canada/services/medical-assistance-dying/annual-report-2021.html

Ipsos. 2012. *Ontarians Fear Cancer*. Poll conducted on behalf of Canadian Cancer Society. https://www.ipsos.com/en-ca/ontarians-fear-cancer-70-most-ahead-heart-disease-59-alzheimers-57-diabetes-41-multiple-sclerosis

Jordan, R.I., M.J. Allsop, Y. El Mokhallalati, and C.E. Jackson, et al. 2020. Duration of palliative care before death in international routine practice: A systematic review and meta-analysis. *BMC Medicine* 18 (368), 26 November. https://doi.org/10.1186/s12916-020-01829-x.

Kaufert, Joseph, and Thomas Koch. 2003. Disability or end-of-life? Competing narratives in bioethics. *Theoretical Medicine* 24: 459–469.

Kimsma, G.K. 2010. Death by request in The Netherlands: Facts, the legal context and effects on physicians, patients and families. *Medicine, Health Care and Philosophy* 13 (4): 355–361. https://doi.org/10.1007/s11019-010-9265-0.

Kotalik, Jaro. 2020. Medical assistance in dying: Challenges of monitoring the Canadian program. *Canadian Journal of Bioethics* 3 (3): 202–209.

Mah, Kenneth, Nadia Swami, Brenda O'Connor, Breffni Hannon, Gary Rodin, and Camilla Zimmermann. 2022. Early palliative intervention: Effects on patient care satisfaction in advanced cancer. *BMJ Supportive and Palliative Care* 12 (2): 218–225. https://doi.org/10.1136/bmjspcare-2020-002710.

Mishara, B.L., and D.N. Weisstube. 2018. Is suicide prevention absolute? *Crisis* 39 (5): 313–317.

Munro, Camille, Anna Romanova, Colleen Webber, Michael Kekewich, Ryelle Richard, and Peter Tanaseputro. 2020. Involvement of palliative care in patients requesting medical assistance in dying. *Canadian Family Physician* 66 (11): 833–842.

Moore, Sara, Chloé Thabet, and Paul Wheatley-Price. 2022. Brief report: Medical assistance in dying in patients with lung cancer. *JTO Clinical Research Reports* 3 (2): 1000283.

Office of the Chief Coroner. 2021. *Process overview and checklist: Reporting a MAID death to the office*. https://health.gov.on.ca/en/pro/programs/maid/docs/occ_checklist_en.pdf.

Penning, R. 2019. *Extreme Social Isolation and Loneliness Affect Almost One Quarter of Canadians*. Cardus. Ottawa. https://www.cardus.ca/news/news-releases/extreme-social-isolation-and-loneliness-affect-almost-one-quarter-of-canadians/.

Pesut, Barbara, Sally Thorne, David Kenneth Wright, et al. 2021. Navigating medical assistance in dying from Bill C-14 to Bill C-7: A qualitative study. *BMC Health Services Research* 21: 1195.

Statistics Canada. 2021. *Canadian Social Survey: Loneliness in Canada*, released 2021–11–2. https://www150.statcan.gc.ca/n1/daily-quotidien/211124/dq211124e-eng.htm.

Statute of Canada. 2016. Chapter 3. *An act to amend the criminal code and to make related amendments to other Acts (medical assistance in dying)*. Assented to June 17, 2016. Bill C-14. https://www.parl.ca/DocumentViewer/en/42-1/bill/C-14/royal-assent.

Statute of Canada. 2021. Chapter 2. *An act to mmend the criminal code (medical assistance in dying)*. Assented to March 17, 2021. Bill C-7. https://parl.ca/DocumentViewer/en/43-2/bill/C-7/royal-assent.

Supreme Court of Canada. 2015. Carter v. Canada (Attorney General). 2015–02–06. https://scc-csc.lexum.com/scc-csc/scc-csc/en/item/14637/index.do.

Jaro Kotalik, MD (Charles University, Prague), D.M.R.T. (University of Toronto), MA (Medical Ethics and Law, Kings College, University of London, UK), FRCPC, is a Professor, Division of Clinical Sciences, Northern Ontario School of Medicine; Adjunct Professor, Department of Philosophy; Chair, Executive Committee and Founding Director of the Centre for Health Care Ethics Lakehead University. Between 1984 and 2010, he held an appointment at the Faculty of Health Sciences, McMaster University, Hamilton, and retired as Clinical Professor. Between 1974 and 1994, he served as a radiation oncologist and later CEO of the Thunder Bay Regional Cancer Centre, and a Vice President of Cancer Care Ontario. After completing his studies in medical ethics and law, he dedicated most of his time to the field of bioethics. He served as an executive member of National Council on Ethics in Human Research and a member or chair of several research ethics boards and committees of the Royal College of Physicians and Surgeons. In his capacity as a consulting bioethicist, he has provided services to Health Canada, Public Health Agency of Canada, Swiss National Advisory Commission on Biomedical Ethics, Thunder Bay Regional Health Sciences Centre and other organizations. He published numerous research papers, 2 major consulting reports, and five book chapters. He is an Associate Member of the World Medical Association.

Chapter 9
Organ Donation After Medical Assistance in Dying

Ryan Tonkens

Abstract Here I consider some of the ethical and philosophical issues at the intersection of medical assistance in dying (MAiD) and deceased organ donation (DOD). Three possible objections about inherent aspects of the practice of DOD after MAiD are considered, and rejected. The bulk of the chapter examines recent calls to keep decisions about DOD and MAiD separate, and to clarifying the nature of the ethical concerns underlying effort to protect patients from undue pressure and coercion. Several insights are revealed as a result of this analysis, including that there is good reason to think that sometimes it may be morally justifiable to allow an eligible patient to pursue MAiD because they wish to donate their organs after death. Noting that more research is needed, I close with several recommendations for policy and practice.

Keywords Medical assistance in dying · Deceased organ donation

9.1 Introduction

The current situation in Canada is a rare one. On one hand, deceased organ donation (DOD) is legal. It happens about 550–700 times per year (Canadian Institute of Health Research 2020), and the requisite resources, equipment and facilities for organ transplantation are available, at least in major urban centres. On the other hand, as of 2016, medical assistance in dying (MAiD) is also legal in Canada, in the form of self-administered medical assistance in dying (i.e. physician assisted suicide) and clinician-administered medical assistance in dying (i.e. voluntary euthanasia). Because of this, by extension, the practice of DOD after MAiD is also legally

R. Tonkens (✉)
Centre for Health Care Ethics, Lakehead University, Thunder Bay, Canada
e-mail: rtonkens@kakeheadu.ca

Northern Ontario School of Medicine University, Thunder Bay, ON, Canada

J. Kotalik and D. W. Shannon (eds.), *Medical Assistance in Dying (MAID) in Canada*, The International Library of Bioethics 104,
https://doi.org/10.1007/978-3-031-30002-8_9

permitted, and is already being practiced in some provinces and territories.[1] With this combination of life-ending and life-saving practices, though, comes important bioethical issues that require attention.

The purpose of this chapter is to present a critical analysis of some of the central ethical issues at the intersection of DOD and MAiD, and to generate a set of recommendations for policy and practice. In the next section I offer a brief description of the current context in Ontario, Canada, before examining specific ethical issues in detail in subsequent sections, including whether DOD after MAiD should be permitted *at all* and, if so, how DOD after MAiD ought to be orchestrated in practice so as to protect donors from undue pressure or coercion.

9.2 Current Practice in Ontario

In Ontario, organ recruitment, allocation and transplantation is orchestrated under the auspice of the *Gift of Life Act*, R.S.O. 1990, c. H.20. The Trillium Gift of Life Network (TGLN) oversees organ recovery and transplantation in Ontario hospitals, many of which have a TGLN coordinator working on site.

According to the Office of the Chief Coroner/Ontario Forensic Pathology Service MAiD Data, as of December 2021 there has been 77 cases of DOD after MAiD in Ontario since 2016 (the year MAiD became legal in Canada). Many types of organs and tissues may be recovered and transplanted after MAiD, including those most in demand, such as kidneys, heart valves, lungs and livers.

The following is a description of the current process for DOD after MAiD in Ontario, as recommended by TGLN:

The TGLN coordinator approaches all patients who have made a formal written request for MAiD, after *at least one* qualified medical professional has confirmed that the patient meets the eligibility criteria for receiving MAiD.[2] As outlined in TGLN's "Organ and Tissue Donation Following Medical Assistance in Dying: Program Development Toolkit" (2018, p.11), the 'TGLN coordinator will conduct the "donation discussion" with the MAiD candidate' in *all* cases where the patient is eligible for organ donation (i.e. has an Ontario Health Card and no medical contraindication).

In the current context of DOD after MAiD in Ontario, the patient is not the one initiating the conversation about organ donation. The TGLN coordinator will have a donation discussion with the patient. If consent is provided by the patient after

[1] At the time of writing, confirmed cases of DOD after MAiD have been reported from at least Ontario, Quebec, British Columbia and New Brunswick. Other provinces are advertising the possibility of DOD after MAiD, e.g. Nova Scotia: https://www.nshealth.ca/legacy-life/donation-information-0.

[2] This means that a second independent medical professional would still need to confirm the patient's eligibility for MAiD. Interestingly, in their report for Canadian Blood Services, Downar et al. (2019) hold the view that proceeding after only 1 assessment is not enough: "Discussions concerning donation should happen only after [...] patients have been found eligible for MAiD by 2 independent assessments" (E607).

the donation discussion, then DOD after MAiD can proceed. After consent has been obtained, "the TGLN Coordinator and the most responsible physician will facilitate a call with the Office of the Chief Coroner to seek permission from the MAiD nurse investigator to proceed with donation" (ibid., p.11). There is then ongoing donor testing and screening for assessment of the patient's medical suitability for donation. "Following the MAID provision and the pronouncement of death by two physicians, the patient is transferred to the operating room where organ recovery begins". Because the donor is alive at the time when they are having discussions with the TGLN coordinator about their wishes for organ donation, the issue of 'family veto' of the deceased person's wishes to donate (Toews and Caulfield 2016) does not emerge in the context of DOD after MAiD.[3]

Both of the options for receiving MAiD in Canada (i.e. self-administered and clinician-administered) could be performed in hospital, and thus the prospect of DOD after MAiD does not rest on the manner in which MAiD is carried out. Indeed, recent reviews have reported that DOD after MAiD is being practiced even in some situations where the MAiD patient (donor) chooses to die at home (e.g. Mulder et al. 2022a, b). For example, Mulder et al. (2022a, b) describe a case of MAiD and deceased donor lung transplantation, wherein the beginning of the MAiD protocol is initiated at the patient's home as per their request, and then the unconscious patient is transferred to an awaiting ambulance. Once the patient's death has been confirmed, "a non-perfused, in situ lung preservation technique is applied" (Mulder et al. 2022a), and the deceased patient is transported to hospital. Once at the hospital, transplantation of the lungs occurs.

9.3 Is Deceased Organ Donation After MAiD Morally Permissible?

Some have argued that there is something *inherently* unethical about MAiD. For example, the view that doctors should never kill their patients (e.g. Kass 1989), or that MAiD presents people with a "problematic choice", by which they are made worse off relative to before, when they did not have this choice to make (e.g. Dworkin 1982; Velleman 1992). However, I must leave these and other such concerns to the side, as they are not new problems that emerge at the intersection of DOD after MAiD, but rather are attached to the seemingly undecidable question about the ethics of assisted dying more generally (Cholbi and Varelius 2023).

The first ethical concern to consider in the context of DOD and MAiD is the most fundamental, to do with whether DOD after MAiD ought to be allowed *at all*, and, if so, under what circumstances. Keeping in mind that we are assuming a Canadian context where both DOD and MAiD are legally permissible, the question is: Is deceased organ donation after MAiD morally acceptable? In one of the first philosophical treatments of this topic, Shaw (2015) asks: What would be morally

[3] Thank you to Kiley Perrier for bringing my attention to this important nuance.

troublesome about harvesting the organs of a consenting person who has requested to die, given that harvesting organs from consenting people *who did not want to die* is legally permitted, and widely (though not unanimously) heralded as a morally justifiable enterprise? In short, having a personal or professional objection to the practice of MAiD does not by itself produce an argument against allowing people who die from MAiD opportunity to donate their organs.

Wilkinson and Savulescu (2012) argued that a competent patient who has provided informed consent can have life-sustaining measures withdrawn in hospital and have their organs recovered for transplantation into others in need. Their view applies to DOD after MAiD as well. According to Wilkinson and Savulescu, allowing such practices "would be a rational improvement over current practice [in 2012 in the United Kingdom] regarding withdrawal of life support. It would give individuals the greatest chance of being able to help others with their organs after death. It would increase patient autonomy" (41). DOD after MAiD would make it more likely that the donor's organs are successfully transplanted, because much about the procedure can be planned for ahead of time, and the person would be dying in a context equipped for this purpose. It is less clear what Wilkinson and Savulescu have in mind about 'increasing patient autonomy'. On my reading, this is about increasing the legal options available for end of life care/assisted dying, rather than increased options about organ donation; if there are any rights that attach to the donor side of organ donation, they do not extend so far as to entitle anyone to actually donate their organs, especially if implicating others is necessary for exercising this putative right.

Downar et al. (2019) also answer the above question in the affirmative. In a report commissioned by Canadian Blood Services, Downar et al. Recommendation #1 states: "Medically suitable, conscious and competent patients who provide first-person consent to end-of-life procedures should be given the opportunity to donate organs and tissues. Patients who seek MAiD [...] should not be prohibited from donating organs and tissues" (E605).

Conversely, Somerville argues against allowing DOD after MAiD: "a critical question in a context of legalized euthanasia is what does maintaining as much respect for death and life as possible in this context require that we not do? I propose that one such requirement should be that euthanasia is not linked *in any way* with organ transplantation" (emphasis added, 365).

Somerville hints that one *inherent* ethical problem with DOD after MAiD is a Kantian one, to do with the donor being treated merely as a means (an object, thing, prop) and not also as a rational and autonomous being with ends of their own: "In euthanasia by donation, the person is overtly employed for a further purpose, even more so than in organ donation after euthanasia" (603).[4] However, this concern is not unique to DOD *after MAiD*; if it applies at all, it applies to all transplantation of human organs (Wilkinson and Savulescu 2012, 37). Moreover, there is nothing about DOD *after MAiD* that makes it more likely that the donor will be treated

[4] It is not clear that this concern is more pronounced in the context of death *by* donation, and so I will take Somerville's concern to apply equally to deceased organ donation *after* medical assistance in dying.

merely as a means, compared to standard deceased organ donation. And, a plausible application of Kantian moral philosophy to modern Western bioethics suggests that this version of the categorical imperative does not rule out people *willingly and knowingly consenting to* 'being used' in this way (i.e. agreeing to allow their organs to be given to others in need after their death)—this is one way that their being an "end in themselves" can be acknowledged and respected.

Perhaps it may be replied that, in the context of DOD after MAiD, the donor is being exploited (unfairly taken advantage of) or coerced (threatened with harm for non-compliance), e.g. by those in the business of recruiting organs for transplantation. This concern may be amplified to the extent that donation rates in Canada continue to be insufficient for meeting ever-rejuvenating demand. In the context of DOD after MAiD, it may be worried that e.g. TGLN coordinators or people in need of organs and their families *may* come to see such patients as "organ suppliers", and thus compromise proper respect for the donors (i.e. as persons with dignity).

I think that we ought to be vigilant in monitoring for exploitation and coercion in the context of DOD and MAiD (more on which below). However, in response to this concern, Downar et al. (2019) and others are calling for rigorous consent processes—which are already in place in Ontario for both organ donation and for MAiD—as well as limiting involvement of the TGLN coordinator until *after* MAiD is requested and eligibility is (at least partially) confirmed. Thus, the possible reply here is important, but does not undermine the prospect of DOD after MAiD. Furthermore, to bolster this protection, it may be prudent to follow Downar et al., rather than TGLN, in requiring that eligibility for MAiD be confirmed by *two* qualified medical professionals (rather than "at least one"), before which the donation discussion should not occur.

Another possible, inherent ethical concern with the coupling of DOD and MAiD, according to Somerville, is that this practice "so obviously detracts from the claim that euthanasia is used solely for the benefit of the person euthanized" (363). However, standard definitions of "euthanasia" (e.g. "a good death") are not incompatible with the idea that the death could also pose some (direct or residual) good for others as well, e.g. for surviving family members who are saved from incurring further debt. Moreover, it is important to keep in mind that, in the standard case of DOD after MAiD, the person's death was going to be happening anyhow, because they have decided to pursue MAiD for reasons of their own. That someone's good death is later on revealed to also be a good thing for others does not render the concept of "euthanasia" meaningless or misleading in this application of the term to the context of DOD after MAiD.

I want to raise a third concern, by way of acknowledging systemic racism and the potential for this to influence practice at the intersection of DOD and MAiD. For example, it is now common knowledge that e.g. Indigenous patients in Ontario and Canada more generally often experience inequity in medical care, and are impacted by systemic racism on the front lines. Compared to other groups in Canada, they have historically been given lower priority in access to transplantable organs when they are in need. Many elements of these populations make them vulnerable, and special attention is needed to protect all people who may be vulnerable to harm in the context of DOD and MAiD, including Indigenous people. Therefore, an important ethical concern at the intersection of MAiD and DOD is to ensure that no one is unjustly discriminated against within this context, and that practices are fair and inclusive. Monitoring at the intersection of DOD and MAiD needs to continue in order to make sure that these practices are not discriminatory, racist, or exclusionary.

If there are no inherent moral problems with the practice of DOD after MAiD, then the ethical question shifts: Can DOD after MAiD be practiced in a manner that upholds appropriate ethical standards?

9.4 Ethical Concerns with DOD After MAiD

9.4.1 Conversations at the End of Life

Downar et al. (2019) Recommendation #4 states that "the decision to proceed with MAiD must be separate from, and must precede, the decision to donate" (E609). According to these authors, organ donation must not be the reason for pursuing MAiD; requests for MAiD should come prior to discussion and decision about organ donation.

The deep-rooted concern with making sure that organ donation is not the cause of death stems from concern that doctors would gain a reputation for 'killing patients for organs', and thereby lose public trust. Professional codes of ethics and the ethos of the medical profession have strong emphasis on doing no harm, in Western bioethics often articulated in the principle of non-maleficence. Thus, perhaps the insistence on keeping DOD and MAiD decisions separate, and the idea that the latter should always come prior to the former (for people who are eligible for MAiD) is because, even though MAiD is legal, there is pronounced residual concern that harvesting organs after MAiD would give the medical profession a bad reputation, even though the patient/donor is willingly and knowingly making a request for both MAiD and DOD.

One possible response[5] to this concern is to say that the law in Canada does not currently reflect the belief that it is never morally acceptable for doctors to kill their patients. Even prior to the legalization of MAiD in 2016, physicians were legally permitted to provide terminal palliative sedation, and also were legally required to accommodate sound requests for the removal of life sustaining treatment. Secondly, under current circumstances where MAiD is legal, appeal to the principle of non-maleficence as the basis for requiring that MAiD decisions come prior to donation decisions would need to attach to *something specific about the combination of DOD and MAiD* that was harmful to the donor, or some other relevant party involved. I discuss concerns about harm due to undue pressure below. If there are no other significant harm-based concerns, then the claim that the request for DOD must come

[5] Another way to respond to this concern is to claim a morally relevant difference between acts and omissions.

after the request for MAiD in order to prevent undue harm to the patient is not compelling.

Importantly, Downar et al. recommendation seems to preclude having the DOD discussion prior to the MAiD discussion in *all* cases. But this rules out too much. For example, some people that come to request MAiD have already registered to be organ donors. If someone is a registered organ donor (e.g. as part of driver's licensing), would this be sufficient for disqualifying them from pursuing MAiD later on, because their decision about organ donation has come prior to the decision about MAiD? Presumably not.[6] Furthermore, in places like the Canadian province of Nova Scotia, there is now a presumed consent ('opt-out') organ donation scheme in place. I contend that people who meet the eligibility criteria for MAiD and who live in Nova Scotia should not be disqualified from MAiD because their decision to be a donor—i.e. to not opt-out—has come prior to the decision to pursue MAiD.[7]

I have not seen any arguments in the literature in favour of allowing the donation decision to come prior to the MAiD decision, and it is hard to think of any compelling reasons to allow this practice in cases where the person requesting MAiD is not already a registered organ donor—even those who may want to pursue MAiD in order to donate their organs (discussed below) could do so without the donation decision coming prior to the MAiD decision. Thus, it seems wise to follow Wilkinson and Savulescu's (2012) recommendation in the context of DOD and MAiD as well: "Whichever transplantation policy is adopted, we should ensure that decisions about withdrawal of life-sustaining treatment are separated from decisions about organ donation" (47). At the same time, there is good reason to modify Downar et al.'s recommendation:

In cases where the person pursuing MAiD does not already have a registered organ donation decision, the decision about whether or not to pursue MAiD should precede the decision to donate. However, having a previous registered preference to be a deceased organ donor should not be the reason why any eligible person is denied access to MAiD.

[6] There may be a morally relevant difference between this case and a patient, who is eligible for MAiD, having conversations about organ donation prior to making a decision about whether to request MAiD. The difference will depend on who the conversation involves (e.g. friends, family physician, TGLN coordinator, etc.) and the degree to which the decision is informal or formal (e.g. telling a friend of one's decision to donate is less formal than telling the TGLN coordinator that you would like to do so).

[7] There is an interesting empirical question about whether it is even possible to separate the donation question from the MAiD decision. Buturovic (2021) argues that doing so is practically unrealistic. They conclude that: "Psychologically unrealistic assumptions about a sequential process that allows for a clear separation between the two allows advocates of organ donation in MAiD patients to proceed unchallenged" (706).

9.4.2 Pressure to Donate and the Reason for Requesting MAiD

Much more concerning is the prospect of some people being influenced to make decisions about MAiD and/or about DOD because of the coupling of those practices. This is a more compelling basis for the claim that decisions about MAiD ought to come prior to decisions about DOD.

Downar et al. (2019) recommend that the decision about MAiD come before the decision about organ donation not just because of concerns about doctors being seen as killers or 'organ stealers'. The main ethical concern behind wanting to maintain this chronology is to uphold purity in the decision of the patient about whether to choose to request MAiD, independent of desire to be a deceased organ donor. Indeed, Downar et al. (2019) suggest that the "most important ethical concern that arises" from the combination of MAiD and organ donation is 'the possibility that the decision to proceed with MAiD could be driven by the desire to donate organs' (E609). Unfortunately, these authors do not offer explanation for what the problem is that they see with MAiD being driven by the desire to donate.

I think this concern can be divided into three distinct concerns, each requiring attention:

(1) Desire to pursue MAiD because of a desire to donate after MAiD: Someone could freely and without undue pressure desire to pursue MAiD for the purpose of being a deceased organ donor (and for relieving their own intolerable suffering);
(2) Pressure coming from others to follow through with MAiD even after one has changed one's mind: The person who is eligible for MAiD and has previously made a request for MAiD is pressured to not change their mind, because they have agreed to be a donor; and
(3) Pressure to request MAiD in order to be a deceased organ donor: The person who is eligible for MAiD was not going to request MAiD, but has been pressured to do so, in order to make their organs available for transplantation.

I consider each of these concerns, beginning with #3.

It is important to consider the possibility that some people could be pressured by others to request MAiD *for the purpose of making their organs available for transplantation*. This is because *this* reason for requesting MAiD—and for being pressured to request MAiD—would not be a live one if DOD after MAiD were banned.

Pressure to Die to Donate: Imagine a person, G, who meets all of the eligibility criteria for access to MAiD, but does not desire to request MAiD. After a recent meeting with a friend/family member/local mobster, J, who is a known tissue match with G, G changes their mind—perhaps because of a financial incentive offered by J. G has their mind changed by J, and will now request MAiD. G is being influenced by J to pursue MAiD, in order to make their vital organs available to J after death.

Legal provision of MAiD in Canada explicitly requires both informed consent and that the person "make a voluntary request for MAID that is not the result of outside pressure or influence" (Canada MAiD website). The scenario described in *Pressure to Die to Donate* is such that the request for MAiD is the result of pronounced "outside pressure or influence". Thus, this would not be legally permitted. Of course, an important practical question is whether the safeguards in place are sufficient for catching such cases, and also whether there is anything about DOD after MAiD that makes such cases more likely and/or more difficult to detect, compared to undue pressure to pursue MAiD more generally.

There is opportunity for the person (potential MAiD patient) to avoid such circumstances, and there might be sufficient safeguards available. In the first instance, a person who is eligible for MAiD but does not want to pursue it could choose to keep their health information private, especially from people they do not trust. The existence of explicit, established eligibility criteria means that a very narrow population would be the target of such pressure. There are also several formal opportunities for the person to express their concerns or withdraw their request for MAiD. It does seem plausible to suggest that appropriate safeguards could be implemented to make *Pressure to Die to Donate* unlikely.

Of course, we can imagine incentives large enough to convince G to pursue MAiD in order to donate their organs, and it is possible that this pressure could go undetected. Is there enough here to justify a ban on DOD after MAiD? A Utilitarian may be quick to point out the possible benefits that would be forfeited by those waiting for the transplantable organs. Nevertheless, the prospect of some people who meet the eligibility criteria and end up receiving MAiD even though they do not actually want to die is unsettling, and safeguards should be explored to address this concern in practice. One way to prevent cases like *Pressure to Die to Donate* is to place a moratorium on *directed* donation of organs recovered after MAiD. This alone may be sufficient for squashing this kind of pressure, since it would be futile for people known to the patient to pressure them to die in order to make their organs available for donation, for their organs will be given to someone else instead.

9.4.3 *Desire to Die to Donate*

Consider now the prospect of a person who meets the eligibility criteria for MAiD and who wants to die, and for the purpose of being a deceased organ donor, as well as relieving their intolerable suffering (#1 above). Perhaps it is unlikely that their organs will remain viable for transplantation if the patient puts off requesting MAiD. Or perhaps a close family member is now in need.[8] Downar et al. (2019) suggest that the desire to donate one's organs ought not to be the motivation for requesting

[8] Shaw (2015) goes so far as to suggest that "any jurisdiction where assisted dying is legal could solve its organ scarcity problem by investing in facilitation of organ donation after assisted suicide and/or euthanasia" (187). While I do not share Shaw's optimism about the degree to which organ

MAiD. Unfortunately, there has been very little explanation given for *why* this ought to be the case.

Current (as of April 2023) MAiD eligibility requirements in Canada include: The person must "(1) have a serious illness, disease or disability, (2) be in an advanced state of decline that cannot be reversed, and (3) experience unbearable physical or mental suffering from your illness, disease, disability or state of decline that cannot be relieved under conditions that you consider acceptable". Also, as noted above, they must not be unduly pressured or influenced in their decision (Canada MAiD website). In cases where a person meets these criteria and decides to request MAiD *because they want to be an organ donor*, what, if anything, has gone morally wrong? Mulder et al. (2022a, b) suggest that the only explanation for an eligible patient desiring to pursue MAiD *in order to be a deceased organ donor* is because they 'believe they are more valuable dead than alive'.[9] Unfortunately, these authors do not offer any explanation or justification for this claim. As stated, this claim is overly simplified, if not untrue; some people who meet all of the MAiD eligibility criteria may be doing so out of a desire to help others in need, and their motivation may be grounded in a sense of solidarity and altruism, rather than a sense of loss of value. Moreover, the important provision that MAiD not be the result of undue pressure does not by itself require that all MAiD patients have exclusively self-regarding reasons for pursuing MAiD. For example, a person who meets all of the eligibility criteria for MAiD in Canada and who also hopes that their death brings financial or psychological relief for their loved-ones is not automatically disqualified from having access to MAiD thereby. The same ought to be the case for people who meet the eligibility criteria for MAiD and who also want to be a deceased organ donor.

The concern about making sure that no one is killed who does not desire to die does not hold here, for the person *does* desire to die. It is their reason for having this preference that some find problematic. Yet, as long as they meet the objective criteria for eligibility for MAiD outlined above, and are not being coerced into choosing MAiD (e.g. the decision is informed and voluntary), then there is no obvious justification for probing the subjective reasoning behind the person's decision to pursue MAiD. Indeed, there is no reason why the person would need to disclose this information about desiring to donate, until after the MAiD decision has been made an confirmed.

Consider the following scenario

Loving grandparent: Imagine a grandparent, B, who meets all of the eligibility requirements for MAiD, and is not being pressured in any way to pursue MAiD. B has decided to request MAiD and is motivated primarily by the reason that they

---

donation after MAiD could solve the issue of scarcity, they are correct to highlight the *potential* increased supply.

[9] Mulder et al. (2022a, b) write: "A donation as a primary reason, that is, MAiD patients thinking that they are more valuable dead than alive, would be unacceptable and harm the MAiD cause" (2774).

would like to make sure that their grandchild receives the vital organ(s) that they need.

Are there any compelling reasons to prevent *Loving Grandparent* from pursuing MAiD for this reason?

Perhaps it is thought that it would be difficult to distinguish between *Loving Grandparent*, who is deeply devoted to their grandchild, and other cases where e.g. the parents of the grandchild are deviously influencing the grandparent to make this decision to pursue MAiD, when, without such undue influence, B would not be requesting MAiD. However, *this* concern is no longer about *the genuine desire to die to donate*, but rather has transformed into another variant of *Pressure to Die to Donate*, and so the responses provided above apply. Additionally, we do feel qualified to make such distinctions between *Loving Grandparent* and *Devious Parent* in the context of living donation of non-vital organs, where risk of death is often in play for the donor.

While more discussion is warranted on this topic, I have provided a plausible criticism of Downar et al.'s assumption that *all cases* of seeking MAiD for the purpose of being an organ donor are obviously morally unjustified. At least, in order to convincingly maintain a blanket moratorium on pursuing MAiD in order to be an organ donor, Downar et al. need to explain why people such as *Loving Grandparent* ought to be stopped from pursuing MAiD even though they are free from undue pressure and meet all of the objective eligibility criteria for receiving MAiD.

9.4.4 Changing One's Mind About MAiD or DOD

Consider #2 above, regarding pressure from others to follow through with MAiD after one has changed one's mind. According to Somerville, 'a link between organ donation and euthanasia could be a barrier to the person's changing their mind about wanting euthanasia' (362). Somerville is not alone in suggesting that a link between MAiD and DOD could be 'a barrier to someone changing their mind' about MAiD. Is this worry well-placed?

Change of Heart: Imagine a patient who qualifies to receive MAiD and who requests MAiD, and subsequently decides to donate their vital organs after death. Imagine that, days before the provision of MAiD, this person changes their mind, and no longer desires to receive MAiD. However, because of a feeling they have, or perhaps because of reactions or comments from others, they continue with the MAiD request, even though they have changed their mind and no longer desire MAiD.

This concern is significant, as it is grounded on the possibility that people could be killed when they do not actually want to die—which clearly differentiates *Change of Heart* from *Loving Grandparent*.

Might some such people *not* express their change of heart *because* they feel pressure to continue with the previous decision to pursue DOD after MAiD? Perhaps. How likely is this possibility to come to fruition in practice? On my reflection, I suggest "not very likely". *If* MAiD and organ donation are allowed to happen in the same patient *at all*, then this concern will persist. However, protections could help avoid this, including more detailed and targeted information available to patients, as well as a special final consent process for MAiD patients who want to be organ donors. For example, perhaps no waiver of final consent should be permitted in these circumstances. Moreover, more explicit, targeted conversations about this possible scenario could occur in the early stages of conversations with the TGLN coordinator; 'One of the foremost facts you need to be clear about is that you are not letting anyone down or a bad person for making a decision to not donate your organs after MAiD. You are free to change your mind at any time and for any reason, about organ donation and about MAiD, and no one will judge you for your decisions no matter what'. *If* there are any legitimate concerns about coercion coming from organ donation coordinators themselves, perhaps because of a perception that they are rewarded based on how many organs they recruit, then some independent party could be the one having such conversations instead—or else standard practice could be changed such that benefits for organ donation coordinators are not in any manner connected to rates of organ recruitment.

9.5 Conclusion

Nothing inherently unethical about the combination of DOD and MAiD was identified here. Arguments were developed that (i) motivate deeper critical engagement with the hitherto unexamined assumption that donation decisions should in all cases come after decisions about MAiD, and (ii) bring into question widespread concerns about people pursuing MAiD in order to donate their organs (and relieve their suffering), especially being unduly pressured to do so, in case the combination of these practices is allowed.

While further discussion is needed, I propose the following recommendations:

Recommendation 1: Modify Downar et al.'s recommendation about the sequencing of the DOD and MAiD decisions, in order to properly recognize that no eligible person should be denied MAiD because they are already a registered organ donor.

Recommendation 2: Consider precluding directed donation in the context of DOD after MAiD, in circumstances where there are legitimate doubts about whether the intended recipient is unduly pressuring the patient.

Recommendation 3: Include a question in the MAiD assessment and consent process about "Are you being pressured to pursue MAiD by someone who wants you to donate your organs?" This *could* include the organ donation coordinator explicitly.

Recommendation 4: Include targeted conversations about *Change of Heart*-type scenarios in the early stages of the donation discussion. For example: 'You are free to change your mind at any time and for any reason'.

Recommendation 5: Update Canadian Blood Services recommendations about DOD and MAiD to reflect changes to MAiD law which came in 2021. For example,

Downar et al. (2019) Recommendation #15 states that: "If a conscious and competent patient provides first-person consent to donate after [withdrawal of life-sustaining measures] but subsequently loses decisional capacity, there is a strong case for proceeding with donation after WLSM because the patient was adequately informed about the decision by a trained donation expert and gave consent in the context of their illness and an anticipated imminent death. *However, if a patient loses capacity prior to the MAiD procedure, then MAiD procedures cannot be carried out*" (emphasis mine, E606). In April 2021, MAiD law in Canada changed, such that under some circumstances patients not able to provide final consent immediately prior to receiving MAiD can still receive MAiD, e.g. where 'the eligible patient has made a written arrangement with their practitioner in which they consent in advance to receive MAID on a chosen date if they no longer have capacity to consent on that date'. In cases where final consent to MAiD is waived, it is reasonable to presume that previous organ donation decisions hold.

References

Buturovic, Z. 2021. Procedural safeguards cannot disentangle MAiD from organ donation decisions. *Journal of Medical Ethics* 47 (10): 706–708.

Canadian Institute of Health Research. 2020. Organ replacement in Canada: CORR annual statistics, 2020 https://www.cihi.ca/en/organ-replacement-in-canada-corr-annual-statistics-2020#:~:text=For%20all%20solid%20organs%2C%20the,the%20same%20as%20in%202010.

Cholbi, M., and J. Varelius (eds.). 2023. *New directions in the ethics of assisted suicide and Euthanasia*. 2nd edition. Springer.

Downar, J., S.D. Shemie, C. Gillrie, M.C. Fortin, A. Appleby, D.Z. Buchman, and M.D. Sharpe, et al. 2019. Deceased organ and tissue donation after medical assistance in dying and other conscious and competent donors: guidance for policy. *Canadian Medical Association Journal* 191 (22): E604–E613.

Dworkin, G. 1982. Is more choice better than less? *Midwest Studies in Philosophy* 7: 47–61.

Government of Ontario. *Gif of Life Act*. https://www.ontario.ca/laws/statute/90h20

Kass, L. 1989. Neither for love nor money: Why doctors must not kill. *The Public Interest* 94: 25.

Mulder, J., H. Sonneveld, D. Van Raemdonck, J. Downar, K. Wiebe, B. Domínguez-Gil, A. Healey, B. Desschans, A. Neyrinck, A.P. Blanco, and I. van Dusseldorp. 2022a. Practice and challenges for organ donation after medical assistance in dying: A scoping review including the results of the first international roundtable in 2021. *American Journal of Transplantation*.

Mulder, J., H. Sonneveld, A. Healey, and D. Van Raemdonck. 2022b. The first international roundtable on "organ donation after circulatory death by medical assistance in dying" demonstrates increasing incidence of successful patient-driven procedure. *American Journal of Transplantation* 22 (3): 999–1000.

Shaw, D.M. 2015. Saving lives with assisted suicide and euthanasia: organ donation after assisted dying. In *New Directions in the Ethics of Assisted Suicide and Euthanasia*, 185–192. Springer.

Somerville, M. 2019. Does it matter how we die? Ethical and legal issues raised by combining euthanasia and organ transplantation. *The Linacre Quarterly* 86 (4): 359–365.

Toews, M., and T. Caulfield. 2016. Evaluating the "family veto" of consent for organ donation. *Canadian Medical Association Journal* 188 (17–18): E436–E437.

Velleman, J.D. 1992. Against the right to die. *Journal of Medicine and Philosophy* 17 (6): 665–681.

Wilkinson, D., and J. Savulescu. 2012. Should we allow organ donation euthanasia? Alternatives for maximizing the number and quality of organs for transplantation. *Bioethics* 26 (1): 32–48.

Ryan Tonkens, earned a Ph.D. in Philosophy from York University (Toronto) in 2012. Dr. Tonkens is a bioethicist and researcher at the Lakehead University *Centre for Health Care Ethics*, Associate Professor in Bioethics and Philosophy at the Northern Ontario School of Medicine, and Adjunct Professor in the Department of Philosophy at Lakehead University. The scope of Dr. Tonkens' research in theoretical ethics, applied ethics and bioethics is wide, including topics in reproductive ethics, ethics of AI and robotics, ethics of human gene editing (e.g. "Vulnerable groups an the hollow promise of benefit from human gene editing", *Bioethics* 2021), and ethical issues in the context of organ recruitment and transplantation (e.g. "The most good you can do with your kidneys: Effective altruism and the organ-shortage problem", *Journal of Medicine & Philosophy* 2021). Dr. Tonkens is an award-winning teacher.

Chapter 10
Medical Practice in Presence of MAID

Amy Hendricks

Abstract In an interview with Dr. Amy Hendricks of Antigonish, Nova Scotia she discusses the challenges of providing end-of-life care in a rural, remote practice. She articulates her experience as a practicing physician that has provided insights and sensitivity to the importance of a person's contribution at all stages of life.

Keywords Indigenous rights · Physician conscience · Organ donation · Mental health · Rural health care · Effective referral · Autonomy · Agency

1. *Dr. Hendricks, the first 14 years of your life as a physician was spent among First Nations and Inuit patients. Did this experience influence your approach to death and dying, and your opinions about MAID?*

I was frequently humbled to be at the bedside of a dying elder during my years in the north. Over the years I spent there, I found myself questioning the "standard" approach to bioethics in Canada, which places autonomy above most other values when we are offering patients and families choices at the end of life. My own white, North American culture places less value on older people since we tend to focus very much on our ability to do things, whether in the professional or leisure realms. Many people see their lives as less valuable—to them and to others—when they can no longer act as agents due to frailty, disability, dementia or life circumstances. These two values—autonomy and agency—seem to be very important to people in my own culture, and they are both principles that focus on a person as an individual, instead of as a member of a community.

What I witnessed in the north was tremendous sense of connection—that a person did not belong simply to themselves, but to a whole community. A family meeting to discuss goals of care often involved a large group of people including a community leader, a spokesperson, a translator, and more distant relatives. It wasn't unusual to

A. Hendricks (✉)
Dalhousie University, Halifax, NS, Canada
e-mail: DrAmy.Hendricks@nshealth.ca

J. Kotalik and D. W. Shannon (eds.), *Medical Assistance in Dying (MAID) in Canada*, The International Library of Bioethics 104,
https://doi.org/10.1007/978-3-031-30002-8_10

have a nursing baby and a few teenagers included. In this situation, asking "What do you think Mabel would want in these circumstances?" didn't always seem to be the best approach. Mabel was connected to them, and each person needed time with her. On one occasion, an elderly man spoke through a translator to tell me that, while the family appreciated my efforts to save his sister, it was the Creator—not I—who would decide how long she would live. The conversation changed and I learned about my patient—who she was, how she had cared for her family, how many children she raised, how hard she worked at their camp every summer. She was the only remaining person in his generation, besides the elder who had spoken. Eventually we got around to code status, but it was a footnote to the conversation that ultimately showed me who my patient was to a large network of people.

On another occasion, I was informed by a family spokesperson that every member of a small community would need time at the bedside of an intubated patient, "so that she can bless them". The concept of blessing passing from one generation to another—despite all the medical gear surrounding my critically ill patient—tore apart some of my pre-conceived notions about what it means to be in relationship with others. The concept that someone's presence and spirit—not their ability to directly communicate or *do something*—could be a gift to another person, was mind-blowing to me as a western-trained physician. And yet, in this concept was a richness that I had not encountered in my own culture, whether in medicine or in my day-to-day life.

I can't say exactly how these experiences should inform my approach to death and dying. But I have developed the sense that placing the highest value on autonomy (in deciding what choices should be available) and agency (in defining what aspects of our life have meaning) makes us less human, less connected to each other, than what I witnessed in indigenous communities. If we truly considered our elders to be the keepers of wisdom, and defined our personhood partly through our community connections, we might be having very different conversations. I believe that First Nations and Inuit communities should be listened to—not only to understand their own history and aspirations, but also to show us another way of thinking about what it means to be human. My own culture has a lot to learn.

2. *Did MAID ever come up when you were working in the north?*

Because of the timing of my departure, deliberate action by a physician to hasten a patient's death was not legal during the time that I was practicing medicine in a predominantly indigenous setting. However, there was some public debate at the time, including before the CMA general council. I didn't have the sense that physician leaders in the Northwest Territories were necessarily seeking out their patients' and community members' perspectives and concerns regarding euthanasia/MAID. It seemed to be something that was arriving in the north through a non-indigenous legal and medical model, not because local communities were asking for it. For me, this raised concerns that the rights of indigenous communities to live in ways that honoured their traditional teachings might not be considered as changes came very quickly, first through federal courts and then through legislation—and then into medical practice.

I was concerned that the change in legislation would bring unique challenges to First Nations and Inuit communities. Suicide, particularly among indigenous youth, is a major problem and affects all generations in tightly-knit communities. The terrible history of colonization, as well as poor housing, employment and education opportunities for young people, can give rise to feelings of hopelessness. New developments that will expand MAID eligibility to people with mental health disorders could easily prompt young people to seek medically assisted death as a "solution" to difficult life circumstances and hopelessness. When there are already far too few mental health resources available to indigenous people, who will decide which suicides are to be prevented and which ones are to be facilitated?

I don't have the sense that these concerns, similar to those articulated by Dr. Alika Lafontaine and several aboriginal leaders in an open letter in February 2021,[1] have been addressed by legislators, regulatory authorities or health care providers. The rush to bring MAID to everyone who wants it, regardless of the harms that may ensue, have left vulnerable people and communities exposed to some frightening possibilities. We have been told that age, geography, and a psychiatric diagnosis shouldn't hold people back from being able to access MAID. Where does this leave a remote community that is struggling to support its youth through difficult times, and helping to develop their sense of identity as a person of First Nations or Inuit heritage, in a world that can be so hostile? Is MAID, as a treatment option for depression, really something that should be on the table? And what are we telling our young people about the value of their lives, if we offer to facilitate their suicides?

3. *As a community-based internist, intensive care is part of your daily work. Some people don't want a lot of intrusive medical care at the end of their lives, and seek to preserve their dignity by accessing MAID at a time and in a place that they can designate. Do you see this as an important option for Canadians?*

I've had the privilege of caring for many patients across the spectrum of medical acuity—during a heart attack or while they are on a ventilator, in a general medical ward, in the office when their medical issues are more stable, and sometimes with home visits towards the end of life. At no time have I felt that deliberately hastening death should be a part of my practice. But across the spectrum of care there are many choices that patients and families have about what is done, where, and how. Treating people with dignity means that we don't oblige them to submit to treatments and investigations, and there are many decision points when we should question what is being done, and why.

A few years ago, I cared for a friend in the hospital—a very articulate man who had lived well at home into his 90s. When some tests demonstrated a more sinister process at work than my initial diagnosis, we had a conversation about moving from an aggressive to a palliative approach. He immediately asked about MAID, which surprised me; until I reflected on who he was as a person. He was one to face a problem and figure out what to do, then set to work.

[1] White et al. (2021).

As we stripped away the various interventions that we had previously hoped would help him, I considered how intrusive medical care can be. No scans. No blood tests—they hurt. No blood pressures; the automatic cuff would only interrupt his sleep. Let's get his own pajamas, blankets and pillows. Play the music that he loves, and turn on a familiar movie to watch with his family, as they brought in their favorite meal. An ICU nurse encouraged my patient's wife to join him for a snuggle, and he died with peace and intimacy. (I think we need some queen size beds in our hospital.) I think his real desire was not for MAID, but for the right kind of care, which would dignify his last days by letting him die as he had lived. Once we created a home-like space for him and his family—in the ICU room, since he died within a couple of days and no palliative bed was available—he did not bring up MAID again, and he gave his family the last few memories that they will always cherish.

I believe that we need to do better at providing the setting and circumstances that will bring a dying patient peace and comfort. At the hospital where I work, that includes a beautiful palliative care ward, music therapists, and a strong spiritual care service. I've seen an ICU nurse take a patient on a ventilator out to the hospital garden on a warm day, to feel the breeze and see the flowers; but that sort of approach is not common, and perhaps it should be—even if a patient is not actively dying.

In response to the question above, I would ask another question: If a person's death is foreseeable within days, how much benefit is there in controlling the exact date and time of death? What price will be paid for this? Will a physician be obliged to violate her conscience? Will a nurse be forced to administer a lethal injection? Will a pharmacist be fired if he refuses to provide a lethal substance? And will the individuals involved in providing compassionate care to thousands of patients—like the team mentioned above—be considered acceptable collateral damage if they are forced out of their professions?

4. *Are you really concerned you could lose your job over MAID?*

When I entered medical school, aiding someone to commit suicide—including in the medical setting—was a criminal offence. The situation has changed so quickly in the last 6 years that a physician is now obliged, under current legislation, to provide an "effective referral" or transfer of care for MAID if a patient asks for it. My sense of what medical care involves, and what I role I play in patients' care at the end of life, has not changed; and I will not refer a patient to be killed.

I therefore represent a problematic group of physicians for legislators and regulatory authorities. I work 80 h per week administering care that very few people have the inclination and skills to provide. I cannot be replaced by a nurse practitioner, a family practitioner, or an academic physician who has not maintained a broad and procedurally based practice. I serve a rural, underserviced population of patients, and my department currently has bare-bones staffing; the loss of a single physician could lead to closure of our regional hospital ICU and related surgical services. Yet I cannot practice without a license, and my license could be rescinded if it is determined that I have not met requirements of the law in my medical practice.

There is an alternative to the "effective referral" requirement, which has not been pursued in Canada for reasons that are not clear to me. A separate system for euthanasia/MAID provision, that does not require involvement by physicians who are conscientious objectors to the practice, would allow patients to self-refer. This would allow me to offer a patient all of the care options that I *could* provide for them, without forcing me to choose between unemployment and participation in a process that I find to be ethically unacceptable.

5. *Aren't you constricting a patient's freedom if you don't mention MAID as an option for severe illness?*

I guess that's a possibility, depending on how you look at the doctor-patient relationship, and whether you take freedom of conscience seriously. But there is another possibility that is worth discussion. When patients seek out MAID, they may not be provided with other options that could really make a difference in their lives, because the assessment is MAID-focused instead of being truly patient-focused.

I encountered a patient with severe heart failure a couple of years ago. I hadn't met him previously and he was not followed by our heart failure clinic, which is led by a highly trained nurse practitioner under my guidance. However the patient had been referred for MAID (on the basis of heart failure) to an academic physician in Halifax—three hours from his home. She assessed him and documented that he would be appropriate for MAID, and that she would be happy to facilitate this at any time. However the patient was not referred to our heart function clinic, which would have supported the management of his disease—and his quality of life—by offering better medical therapy than he was taking at the time. He was not aware of a specialist service closer to home that could help him—including providing phone outreach—and he simply could not afford travel to the academic centre for medication adjustments. He had some personality traits that made it harder than usual to provide care, but the NP and I took him on, stuck with him, and really worked to control his heart failure symptoms.

Patients who live in rural or remote circumstances, with social isolation and financial challenges, have a right to high quality care regardless of their disease process. If they ask for MAID as a cry for help or an expression of suffering, they should be first offered the treatments (including, perhaps most importantly, the therapeutic relationships) that can help to alleviate their suffering so that they can live well. At the very least, no one should offer to kill the patient instead of providing an effective referral to high quality medical services close to their home.

6. *Have you seen a relationship between MAID requests and poverty in other situations?*

Not yet in my own practice, since MAID is not something I bring up with my patients. But I am concerned by a recent high-profile case, of a woman with multiple chemical sensitivities who could not access affordable housing without exposure to smoke and other chemicals. She requested, and received, MAID on this basis. This situation is deeply disturbing; I don't think any of us want Canada to be a country where housing options are so bleak for people of low income, that they are driven to suicide—which is then provided as a service of the medical system.

If MAID is seen as a solution to suffering that is actually caused by poverty and poor housing, the better option is obviously to provide better services and support for the vulnerable. The burden of poverty disproportionately affects people living with disabilities, previous trauma, addictions, and mental health disorders, and these individuals need care—not elimination by request. I was alarmed that the physicians involved in the MAID case mentioned above did not step back from completing the process, despite knowing that the cause of her suffering was completely treatable—with decent housing options. I hope that the option of MAID does not, in the future, give our government an "easy out" when services do not align with the complex needs of a patient. The concept of MAID being an expression of choice and freedom begins to erode, when people are not offered the basic resources they need to live with dignity. And if a person's care needs cannot be met by their family members and are not funded by the public system, it may not be long before the "right to die" becomes the "duty to die".

7. *Are there any other groups of patients that you think might be particularly vulnerable under current MAID legislation?*

I have been concerned by the possibility of organ donation being coupled with MAID, particularly in the context of MAID for psychiatric disease. As the mother of two teenaged boys—one of whom is treated for ADD, with excellent control of his symptoms currently—I often hear of aspirations and disappointments over dinner. One of the most basic aspirations of young people is to make a difference for good in someone else's life, and this desire can be amplified in the days of social media when almost anyone can be a hero (at least for a moment) in the eyes of millions.

The most shattering experiences of my career have involved providing ICU care to young victims of suicide. Some have survived—at times with disability—and some have died. All these situations have been tragic on multiple levels—the agony of grieving parents, the regret of friends who wish they had done or said something differently, the depth of a person's unshared suffering, and the young life cut short. In some cases, my patients might have sought MAID, coupled with organ donation, if they could have accessed it as an alternative to the route of suicide they chose. And in some cases, MAID would have provided certain death when the patient actually survived the suicide attempt—with treatment options available, and possibly many good years ahead. Despair does not need to be permanent.

I am deeply concerned, as a mother and as a citizen, that offering the option of MAID with organ donation could incentivize suicide, particularly if a young person experiencing depression has a deep desire for their life—and death—to be meaningful to others. I would hope that medical professionals involved in transplant would recoil at the thought of using organs harvested from a young donor whose cause of death was MAID for psychiatric reasons. But the temptation to use the organs of a physically healthy young person—whose death could be timed perfectly with several transplant recipients' operations—might be too strong, and people may justify this act on the basis of a greater common good. I would be horrified, as a mother, to think that if one of my sons developed a depth of depression that led him to consider death, he could be taken apart organ by organ by various surgeons, rather than receiving the best possible psychiatric care and support.

I am not aware of any safeguards in proposed legislation that will prevent this situation. Criteria for MAID are only expanding, and I fear that our culture may be shifting away from the belief that a depressed or despairing person's life truly matters, and their suffering must be addressed—that the prospect of suicide must stimulate efforts for prevention, not facilitation. This sounds dystopian, but given the rapid change in MAID criteria over the last five years, it may not be far away. Again, the people and communities most at risk (of social-media influence to seek MAID with organ donation) would be those who are already marginalized and under-resourced for mental health resources and social services.

8. *What do you propose as an alternative to the dystopian possibilities you mention above?*

We all need compassion and understanding. I hope that I can be understood as a professional, whose personal moral compass—while inconvenient to the current MAID processes—has motivated me to serve rural and remote Canadians for twenty years. I hope that persons experiencing distress or despair, related to life circumstances and poverty, will be met with understanding and a system that is more flexible and easier to navigate, so that they are not driven to look to suicide. I hope that all people will be treated as precious and unique, especially during the times when they feel that they have no value or no future. I hope we can recognize that we are connected deeply to each other; and that autonomy is not necessarily a higher value than community. I hope we will take time to reflect on the experience of the past five years, and consider carefully whether the course we are charting is truly a path towards greater humanity and compassion.

Reference

White, Tyler, Graydon Nicholas, Nick Sibbeston, Alika Lafontaine, Lana R. Potts, Rosella Kinoshameg, Bob Sutherland, et al. 2021. *Indigenous peoples should not be compelled to provide or facilitate medical assistance in dying.*

Dr. Amy Hendricks is a Canadian-trained general internal medicine specialist who trained at McGill University from 1994 to 2002. She spent the first fourteen years of her career in the Northwest Territories and Nunavut, being based in Yellowknife but traveling widely to small communities. She has spent the last six years in rural Nova Scotia, and has served in a number of medical leadership positions. Her scope of practice includes intensive care, outpatient cardiology and internal medicine. She continues to travel to rural Cape Breton on a regular basis, bringing internal medicine care closer to home for thousands of patients.

Part III
Palliative Care and MAID

Chapter 11
Palliative Care: Captive and Casualty of *Carter v Canada*?

Mary J. Shariff

Abstract When the *Carter v Canada* case found the prohibition against euthanasia and physician-assisted suicide unconstitutional, it simultaneously launched an existential threat to the discipline and vocation of palliative care—the view that there is no distinction between palliative care and physician-assisted death. This confronting narrative is often supported by reference to the *Carter* case, but what did the court actually say? The chapter examines the *Carter* case to present the court's findings and observations with respect to palliative care. The chapter also explores subsequent Federal legislative initiatives establishing the practice of "medical assistance in dying" (MAID) as well as initiatives taken to improve palliative care. The reader will perhaps find it interesting to know that the trial court's findings are *consistent* with the definition, practice and purposeful vision of palliative care—to address suffering with *life-affirming* care while not hastening or postponing death. Legal and clinical clarity *could be* achieved if Canadian law and policy-makers *uniformly* adopted and *cohesively* advanced this definition. Yet such clarity does not appear to be forthcoming. It is hoped that this detailed review provides groundwork that can help increase transparency and in turn offer assistance towards securing palliative care as a right of *all* Canadians.

Keywords Palliative care · Palliative sedation · End of life · End-of-life · Euthanasia · Assisted suicide · Assisted death · Physician-assisted suicide · Physician-assisted dying · Medical assistance in dying · MAID · MAiD · *Carter v. Canada* · Termination of life · Right to die · Right to life · Right to liberty · Right to security · Equality · Human rights

---

M. J. Shariff (✉)
Faculty of Law, University of Manitoba, Winnipeg, Manitoba, Canada
e-mail: mary.shariff@umanitoba.ca

Centre on Aging, University of Manitoba, Winnipeg, Manitoba, Canada

J. Kotalik and D. W. Shannon (eds.), *Medical Assistance in Dying (MAID) in Canada*, The International Library of Bioethics 104,
https://doi.org/10.1007/978-3-031-30002-8_11

11.1 Introduction

Universal access to palliative care across Canada has still to be achieved. Yet when Canada decriminalized termination of life practice or "medical assistance in dying" (MAID)[1] in 2016, "palliative care" as a means for relieving suffering, was (and still is) expressly included for achieving informed consent to MAID.[2]

While the role of palliative care practice in Canada obviously goes beyond its reference in MAID law, over the past several years beginning with the case *Carter v Canada*,[3] palliative care has become entangled with and impressed upon by MAID law reform. A review of MAID dialogue including court cases, parliamentary processes, government and commissioned reports, academic and other articles reveals several points of interaction, argument and commentary on palliative care.[4] From the perspective of palliative care which "seeks to identify and alleviate suffering across the trajectory of a life-threatening illness, including care at the end of life, and supports living well while dying",[5] this dialogue can appear perplexing for what seems to be a growing indifference towards its definition, the ongoing dissemination of inaccuracy in terms of its care goals and practices, as well as hostility towards what should have been a fairly non-controversial, common-sense argument: that universal access to palliative care in Canada should be established concurrent with if not prior to the legalization and delivery of MAID. Gaining a deeper understanding of the Canadian MAID-palliative care dialogue is particularly pressing given a persistent argument that palliative care and MAID should be on some form of "end of life" continuum and given that this remains a health care regulatory puzzle yet to be solved.

While not attributing any intention to subvert or subsume palliative care practice into MAID on the part of different actors at various stages in the MAID law reform process, the ongoing confrontation for palliative care posed by MAID does become much less confounding when palliative care is explored not as a *goal* in itself but as a *means* for achieving MAID decriminalization—the subject of this chapter. *Carter*

---

[1] In Canada, "medical assistance in dying" or "MAID" refers to a medical practice which includes a form of assisted suicide (whereby a medical or nurse practitioner prescribes or provides a substance to a person, at their request, for self-administration in order to intentionally cause their own death) as well as euthanasia (whereby a medical or nurse practitioner administers a substance to a person, at their request, to intentionally cause their death). *Criminal Code of Canada*, R.S.C. (1985), c. C-46 at Section 241.1 [hereinafter "*Criminal Code*"].

[2] Bill C-14, (2016), c. 3 at Section 241.2(1)(e) [hereinafter "Bill C-14"]. Bill C-14 received Royal Assent June 17, 2016. Online: https://www.parl.ca/DocumentViewer/en/42-1/bill/C-14/royal-assent.

[3] This discussion examines both the trial decision, *Carter v Canada (Attorney General)* (2012) BCSC 886 [hereinafter "*Carter BCSC*"] as well as the Supreme Court of Canada decision, *Carter v. Canada (Attorney General)* (2015) SCC 5, [2015] 1 S.C.R. 331 [hereinafter "*Carter SCC*"]. Together these cases will be referred to as "*Carter*" or "*Carter v Canada*".

[4] For full discussion in Canadian context see, Shariff and Gingerich (2018). Online: https://papers.ssrn.com/sol3/papers.cfm?abstract_id=3191962; For a recent scoping review of the relationship between palliative care and assisted death see Gerson et al. (2020).

[5] Herx (2015). Online: https://www.ncbi.nlm.nih.gov/pmc/articles/PMC4399615/.

v Canada,[6] the case which held the *Criminal Code* prohibition of "euthanasia" and "physician-assisted suicide"[7] to be unconstitutional, arguably provides an apt starting point for such an investigation due to the trial court's extensive review of palliative care evidence within a constitutional claim for the decriminalization of MAID.

In light of the forgoing, following a brief synopsis of the 2015 Supreme Court of Canada decision (*Carter SCC*)[8] in Part 2, the examination of the MAID-palliative care dynamic begins in Part 3. Part 3 attempts to set out as accurately and objectively as possible, the specific evidentiary findings and observations on palliative care from the 2012 trial decision (*Carter BCSC*).[9] As will be described, the trial court's findings and observations are *consistent* with the definition and vision of palliative care noted above. These findings and observations, however, are quickly obscured and muddied as law and evidence are drawn into the court's examination of whether the prohibition against "physician-assisted dying" was a breach of *Charter* rights—the subject of Part 4. Of particular focus here is the trial court's inclusion of an "ethical debate" and the acceptance of an argument asserting an ethical basis upon which assisted death practice *could* be established, seemingly distinct from the overall legal analysis. Specific judicial statements associated with the "ethical debate" are included in the subsequent *Carter SCC* decision[10] and are critical to understanding the ways in which palliative care has been made vulnerable or subject to ongoing MAID reform and regulatory dialogue. Part 5 continues with a summary of the palliative care narratives that track from the *Carter* case through to subsequent legislative initiatives including: Bill C-14,[11] the 2016 Federal legislative amendment to the *Criminal Code of Canada*[12] to permit MAID; Bill C-7,[13] the 2021 Federal amendment that expanded MAID eligibility for persons whose deaths are "not reasonably foreseeable"; and

---

[6] *Carter*, Footnote 3.

[7] The terms "Physician-assisted death" and "physician-assisted dying" are used interchangeably throughout the *Carter BCSC* decision and were defined by the court to include both the acts of: "physician-assisted suicide" (whereby the patient intentionally kills themselves with assistance of a physician or someone acting under a physician's direction); and "voluntary euthanasia" (whereby the patient's life is intentionally terminated by a physician or someone acting under a physician's direction). See Carter *BCSC* at paras 23 and 37–39; In Canada, these terms are superseded by the term "medical assistance in dying" or "MAID" when the *Criminal Code of Canada* was amended to allow these practices in 2016 through the passing of Bill C-14. See *Criminal Code*, Footnote 1.

[8] *Carter SCC*, Footnote 3.

[9] *Carter BCSC*, Footnote 3.

[10] Note that the trial decision was successfully appealed to the BC Court of Appeal on the ground that the trial judge was bound to follow *Rodriguez v. British Columbia (Attorney General)* (1993), an earlier Supreme Court of Canada decision which upheld the blanket prohibition on assisted suicide. *Carter v. Canada (Attorney General)* (2013). The case was then appealed to the Supreme Court of Canada. *Carter SCC*, Footnote 3.

[11] Bill C-14, Footnote 2.

[12] *Criminal Code*, Footnote 1.

[13] See Bill C-7 2021), c. 2 [hereinafter "Bill C-7"] which received Royal Assent March 17, 2021. Online: https://www.parl.ca/DocumentViewer/en/43-2/bill/C-7/royal-assent.

Bill C-277, a 2017 Federal initiative specifically taken towards improving palliative care.[14]

The discussion concludes in Part 6 with the observation that in the result, the *Carter v Canada* judgment operates as legal witness and deposit for *what is palliative care* but also continues to be a source for arguments that exacerbate and perpetuate distortions of palliative care. It is hoped that this detailed review of *Carter v Canada* through the lens of palliative care can help to "correct the record" so to speak, and in turn offer some assistance towards palliative care's establishment as a right of *all* Canadians,[15] responsive to yet free of the constraints and idiosyncrasies imposed by MAID advocacy and law reform.

11.2 Summary of *Carter SCC*

11.2.1 The Plaintiffs' Claim

The main plaintiffs in the *Carter* case were Gloria Taylor (an individual with amyotrophic lateral sclerosis or "ALS", a fatal neurodegenerative disease), the daughter and son-in-law of Kay Carter (who suffered from spinal stenosis and for whom they had helped arrange for assisted death in Switzerland) as well as a physician willing to participate in assisted death if it was not prohibited.[16] Challenging several provisions in the *Criminal Code*,[17] the plaintiffs argued that the prohibition of "physician-assisted dying" was an unjustifiable infringement of their rights under the *Canadian Charter of Rights and Freedoms* (the *Charter*).[18]

It should be noted here that while the trial court found Ms. Taylor's Section 15 right to equality was engaged as well as the Section 7 rights to "life, liberty and security" of Ms. Taylor *and* those who would assist,[19] the Supreme Court of Canada (SCC) confined its decision to the Section 7 rights of those "who seek assistance in dying",[20] agreeing with the trial judge that:

---

[14] Bill C-277 (2017), c. 28, [hereinafter "Bill C-277"] which received Royal Assent December 12, 2017. Online: https://parl.ca/DocumentViewer/en/42-1/bill/C-277/royal-assent.

[15] Because this chapter is specifically examining the Federal MAID legislation and subsequent Federal initiatives in relation to palliative care, this chapter does not include examination of the right to palliative care as established through provincial Quebec medical aid in dying legislation which was implemented prior to the Federal MAID law. See Quebec legislation, *An Act respecting end-of-life care*, CQLR, c. S-32.0001.

[16] *Carter BCSC*, Footnote 3 at para 44.

[17] *Carter BCSC*, Footnote 3 at para 100.

[18] *Canadian Charter of Rights and Freedoms* 1982), c. 11 [hereinafter "*Charter*"].

[19] *Carter BCSC*, Footnote 3 at para 1 [sic]. Note that the claimed infringement of s. 7 liberty right of Lee Carter and Hollis Johnson was based on the risk of incarceration for helping a "loved one" obtain an assisted death. *Carter BCSC*, Footnote 3 at para 17.

[20] *Carter SCC*, Footnote 3 at para 69.

- The right to life was engaged "in the sense of a right *not* to die".[21] The prohibition on physician-assisted suicide had the effect of "forcing" some individuals to make an earlier decision to end their life while still physically able to do so out of fear that they would incapable of doing so later.[22]
- The rights to liberty and security were engaged by interfering with fundamentally important and personal medical decision-making and personal autonomy involving control over one's bodily integrity.[23]

11.2.2 Absolute Prohibition a Violation of Section 7 Rights

Also in agreement with the trial judge, the SCC found that the *absolute* prohibition of physician-assisted dying pursuant to Section 241 (b) of the *Criminal Code* violated the Section 7 life, liberty and security rights "of competent adults who are suffering intolerably as a result of a grievous and irremediable medical condition" and this infringement was not justified under Section 1 of the *Charter*.[24]

11.2.3 The Declaration of Invalidity

In the result, the SCC declared that *Criminal Code* Section 241(b) and Section 14 (the provision that does not permit a person to consent to their own death) were[25]:

> of no force or effect to the extent that they prohibit physician-assisted death for a competent adult person who (1) clearly consents to the termination of life and (2) has a grievous and irremediable medical condition (including an illness, disease or disability) that causes enduring suffering that is intolerable to the individual in the circumstances of his or her condition.

This "declaration of invalidity" was suspended for a total of 16 months to allow time for Parliament to respond.[26]

---

[21] *Carter BCSC*, Footnote 3 at para 1322.

[22] *Carter SCC*, Footnote 3 at paras 57–58; See also *Carter BCSC*, Footnote 3 at paras 1322, 17.

[23] *Carter SCC*, Footnote 3 at paras 65 and 66; See also *Carter BCSC*, Footnote 3 at paras 1302–1304.

[24] *Carter SCC*, Footnote 3 at paras 3 and 126.

[25] *Carter SCC*, Footnote 3 at para 147.

[26] Note that the Supreme Court initially suspended the declaration of invalidity for 12 months and subsequently granted an additional 4-month extension. See *Carter SCC*, Footnote 3 at para 147 and *Carter v. Canada (Attorney General)*, 2016 SCC 4, (2016) 1 S.C.R. 13.

11.3 Palliative Care Evidence in *Carter BCSC*

11.3.1 Definition of Palliative Care

The definition of palliative care as set out by the trial court for the purposes of its reasons (and used for this discussion) is as follows[27]:

> "Palliative care" means care provided to people of all ages who have a life-limiting illness, with little or no prospect of cure, and for whom the primary treatment goal is quality of life. The treatment is aimed at alleviating suffering – physical, emotional, psychological, or spiritual – rather than curing. It aims neither to hasten nor to postpone death, but affirms life and regards dying as a normal process. It recognizes the special needs of patients and families at the end of life, and offers a support system to help them cope.

The court also observed[28]:

> [W]hile the weight of the evidence is that proper use of opioids in palliative care does not cause death, opioids can have that effect if used inappropriately.

Although both palliative care and physician-assisted death are *responses* to suffering,[29] it is clear from the above that *Carter* applied a bright-line *legal* distinction[30] between palliative care and assisted death in that palliative care focuses on *quality of life*, does *not* aim to hasten death, and, with respect to pain medications administered pursuant to palliative care, does *not* cause a patient's death when *properly* administered.[31] On the other hand, intentionally terminating the life of the patient is the *sine qua non* if not the *raison d'être* of assisted death.[32]

---

[27] *Carter BCSC*, Footnote 3 at para 41. Emphasis added; Note that this definition is consistent with the widely accepted World Health Organization (WHO) definition of palliative care. See for example: WHO (2018) at page 5. Online: https://apps.who.int/iris/bitstream/handle/10665/274559/9789241514477-eng.pdf?ua=1); See also: the International Association for Hospice and Palliative Care (see for example, *Palliative Care Definition*. Online: https://hospicecare.com/what-we-do/projects/consensus-based-definition-of-palliative-care/definition/) and the Canadian Hospice Palliative Care Association and the Canadian Society of Palliative Care Physicians (see for example, *CHPCA and CSPCP - Joint Call to Action*. Online: https://www.cspcp.ca/wp-content/uploads/2019/11/CHPCA-and-CSPCP-Statement-on-HPC-and-MAiD-Final.pdf). At the time of writing, the WHO definition was currently under review.

[28] *Carter BCSC*, Footnote 3 at para 199. Emphasis added; See also *Carter BCSC*, Footnote 3 at para 195.

[29] See for example, *Carter BCSC*, Footnote 3 at paras 38, 39 and 342.

[30] The judicial finding in *Carter BCSC* that there is no "bright-line ethical distinction" between other end-of-life practices and MAID is discussed further below.

[31] *Carter BCSC*, Footnote 3 at para 195; See also discussion *Carter BCSC*, Footnote 3 at paras 196–199.

[32] See *Carter BCSC*, Footnote 3 at paras 37–39.

11.3.2 Clinical Limits of Palliative Care

Palliative care can reduce requests for euthanasia or lead to their retraction, however, it is not considered a "panacea" for all suffering.[33] As put by the court:

> … even the very best palliative care cannot alleviate all suffering, except possibly through sedation to the point of persistent unconsciousness (palliative sedation).[34]
>
> … despite the best possible palliative care, some patients suffer pain that cannot be alleviated, according to expert witnesses called by both the plaintiffs and the defendants. As well, symptoms can cause suffering other than pain (such as nausea, vomiting, and shortness of breath) that cannot be alleviated even by the best palliative care. Some patients experience what is called "existential suffering", meaning, for example, a profound sense of loss of dignity.[35]

11.3.3 Definition of Palliative Sedation

As defined by the court[36]:

> "Palliative sedation" means the intentional administration of sedative medication to reduce a patient's level of consciousness, with the intent to alleviate suffering at the end of life. It includes both intermittent and continuous sedation, as well as both superficial and deep sedation. It may be accompanied by the withdrawal of artificial hydration and nutrition. The phrase "terminal sedation" is occasionally used synonymously with palliative sedation.

The court did *not* conclude the administration of palliative sedation within palliative care hastened or caused death. Rather, the court observed[37]:

> In the context of palliative care, it is fairly widely accepted that when a patient is close to the end of life, and is experiencing symptoms that are severe and refractory (that is, resistant to treatment), it is ethical practice for her physician to sedate her and maintain her in a state of deep, continuous unconsciousness to the time of death, with or without providing artificial hydration or nutrition ("terminal sedation" or "palliative sedation").
>
> …
>
> Palliative sedation remains somewhat controversial and is typically made available only when the patient is thought to be near death (normally, within one week), although it is not always possible to be accurate in such assessments.
>
> …
>
> However, it cannot be assumed that palliative sedation, even accompanied by withdrawal of hydration, necessarily hastens death when it is administered to patients who are already in the final stages of dying.

---

33 *Carter BCSC*, Footnote 3 at paras 188–191.

34 *Carter BCSC*, Footnote 3 at para 4. Emphasis added.

35 *Carter BCSC*, Footnote 3 at para 190. Emphasis added; See also *Carter BCSC*, Footnote 3 at paras 309 and 823.

36 *Carter BCSC*, Footnote 3 at para 42. Emphasis added.

37 *Carter BCSC*, Footnote 3 at paras 200–202. Emphasis added.

The trial court, however, suggested that *if* palliative sedation *did* indeed result in hastened death, the *legality* of such an outcome could attach to law related to:

- informed consent (including the patient's right to refuse or discontinue treatment)[38];
- the withdrawal of life-sustaining treatment based on the "best interests" of an incompetent patient whose treatment preferences are not known[39]; or
- the permissibility of potentially life-shortening symptom relief where the physician's *intention* is to ease pain at end of life.[40]

Note again that the last point above is an observation about *the law* and physician acts that are not subject to criminal sanction because of *intention,*[41] and is not a judicial finding of fact that palliative care or palliative sedation hastens death. It might, however, be noted here that the SCC summary of the trial judge's findings with respect to the effects of palliative sedation does not appear entirely accurate[42]:

> She [the trial judge] found that current unregulated end-of-life practices in Canada — such as the administration of palliative sedation and the withholding or withdrawal of lifesaving or life- sustaining medical treatment — can have the effect of hastening death...

11.3.4 The Legal Limits of Palliative Care and Palliative Sedation Treatment

The court described the limits of palliative care and palliative sedation as follows[43]:

> ... while palliative care including palliative sedation may relieve the suffering of many, for some persons it may be unavailable (due to the nature of their illness) or unacceptable to them (because they value maintaining consciousness and the ability to communicate, feel that death while under palliative sedation will be difficult for their families to observe, worry that they will in fact maintain consciousness, or for other reasons). Thus, they may be required to continue to undergo physical pain or psychological suffering or both, possibly exacerbated by terrible fear about what is yet to come.

Apart from the court's description as to *why* palliative sedation in particular might be *unacceptable* to a given patient, the *unacceptability* of palliative care or palliative sedation ultimately links to a patient's overarching right to refuse or discontinue treatment and is thus an operative legal limit as it is with all medical treatment decision-making by competent patients i.e. that individuals have a "veto" right with respect to impingement on their bodies.[44]

---

[38] *Carter BCSC*, Footnote 3 at paras 226.

[39] *Carter BCSC*, Footnote 3 at para 223. Note that the court recognized that this is a controversial area of the law. See *Carter BCSC*, Footnote 3 at paras 22, 312 and 227–230.

[40] *Carter BCSC*, Footnote 3 at para 226.

[41] *Carter BCSC*, Footnote 3 at para 225; See also *Carter BCSC*, Footnote 3 at para 231(c).

[42] Carter SCC, Footnote 3 at para 23. Emphasis added.

[43] *Carter BCSC*, Footnote 3 at para 1328. Emphasis added.

[44] *Carter BCSC*, Footnote 3 at para 341.

11.3.5 Lack of Universal Availability of Palliative Care Across Canada

The court described and recognized at different points in its judgment that palliative care was "far from universally available in Canada".[45] Additionally, when referencing Senate and Parliamentary reports describing the "patchwork" of palliative care services across Canada in terms of quality and accessibility, the court stated, "Many parts of Canada have no palliative care services at all."[46]

11.3.6 Impact or Risk to the Development of Palliative Care

Distinct, yet connected to the lack of palliative care availability, was evidence from ethicists, palliative care specialists and professional bodies such as the Canadian Medical Association that raised a concern that reforming Canadian criminal law to permit physician assisted death should *not* be pursued *until* adequate palliative care services were made available to all Canadians.[47] The court largely organized this and related arguments into a question of the *impact* or potential risk to palliative care[48] that could arise from a *change* in existing law. In exploring this question (as will be discussed further below), the court looked at the experiences of certain assisted death permissive jurisdictions to conclude that the effect on the "palliative care system" by a change in the law to permit physician assisted death "would not necessarily be negative".[49]

11.4 Tracking Palliative Care Evidence Through *Carter*

11.4.1 Limits of Palliative Sedation from the Plaintiff Perspective

After setting out the Plaintiffs' claim (described in 2.1 above) and definitions along with some history behind the *Criminal Code* prohibition against aiding suicide,[50] the trial court related the intense suffering as described by the key plaintiff, Gloria Taylor. This included fear of progressive decline and being reliant on others, "stripped

[45] *Carter BCSC*, Footnote 3 at para 4; See also para 309.

[46] *Carter BCSC*, Footnote 3 at para 708.

[47] *Carter BCSC*, Footnote 3 at paras 253, 270, 274.

[48] *Carter BCSC*, Footnote 3 at para 314(l).

[49] *Carter BCSC*, Footnote 3 at para 736; See also *Carter BCSC*, Footnote 3 at para 9.

[50] *Carter BCSC*, Footnote 3 at Parts II–IV.

of dignity and independence".[51] With no "pre-set, trigger moment",[52] Ms. Taylor described that there will be some point "where life is not worthwhile" and expressed her desire for a "dignified goodbye" and "final closure".[53] Ms. Taylor found palliative sedation, "repugnant" and not a viable *alternative* to physician-assisted death.[54]

11.4.2 The "Ethical Debate"

The trial court then turned to expert opinion evidence,[55] particularly that of medical practitioners and ethicists. The court encircled this evidence by essentially asking if physician-assisted death is "right or wrong"[56]—the "ethical debate". It is within this "ethical debate"[57] that palliative care evidence is predominantly presented, summarized and organized. Evidence revealing specifics and distinctions between palliative care and physician-assisted dying (e.g. does not hasten death versus causes death) is ultimately not of particular interest to the court due to the court's focus on the individual claim that constitutional rights were being violated by the prohibition of physician-assisted death. That said, the "ethical debate" is itself curious, given that the results of this debate do not appear to be substantively applied to the *Charter* analysis.[58]

In any event, to answer the question of "right or wrong", the trial court poses two additional questions[59]:

(1) If the law permitted it, would it ever be ethical, <u>in an individual case</u>, for a physician to assist a competent and informed patient who <u>requests</u> hastened death?
(2) Even if it were ethical to provide such assistance <u>in an individual case considered in isolation</u>, would a change in the existing law or policy pose a threat to other persons (particularly those who are vulnerable due to age or disability) or to <u>broader societal values (such as general respect for human life, building a</u>

---

[51] *Carter BCSC*, Footnote 3 at para 52.

[52] *Carter BCSC*, Footnote 3 at para 54.

[53] *Carter BCSC*, Footnote 3 at para 54.

[54] *Carter BCSC*, Footnote 3 at para 56; See also *Carter BCSC*, Footnote 3 at para 315(h).

[55] *Carter BCSC*, Footnote 3 at Part VI.

[56] *Carter BCSC*, Footnote 3 at para 164.

[57] *Carter BCSC*, Footnote 3 at Part VII; See also *Carter BCSC*, Footnote 3 at paras 173 and 178. According to the court, ethical principles were relevant because they "shape law and medical practice" and because "legal and constitutional issues are derived from and shaped by societal values" *Carter BCSC*, Footnote 3 at paras 165 and 317.

[58] See *Carter BCSC*, Footnote 3 at paras 944, 1336 and 1369. For additional analysis on whether the "ethical debate" bears any relation to the analysis on the constitutionality of the law see Sean Murphy (2017), pp. 361–362.

[59] *Carter BCSC*, Footnote 3 at para 161. Emphasis added; See also *Carter BCSC*, Footnote 3 at para 164.

strong palliative care system and maintaining appropriate relationships between physicians and patients)?

On the first question, the trial court found that it could be ethical to provide physician-assisted death by: (i) observing that certain reputable physicians "would find it consistent with their ethical principles to hasten death if legal to do so"[60]; and (ii) accepting a logic argument (or "ethical framework")[61] that combined existing law and medical practices where legal sanctions have not been imposed even though the outcome is death or "highly likely to be death".[62] This law included:

- *Informed Consent and Omissions*: "End-of-life" practices such as withholding or withdrawal of life-sustaining treatment on the basis of informed consent and a patient's right to accept or refuse medical treatment are lawful *even if* foreseeably resulting in a patient's death or hastened death.[63]
  - Values/ goals reflected: acting in accordance with a patient's wishes and upholding the principles of autonomy and self-determination (also consistent with "best interests").
- *Intention*: "Life-shortening symptom relief" (for example, the administration of medications designed for symptom relief and pain control) is legally permissible where the physician's *intention* is to ease pain, *even if* a physician knows the medication *may* hasten death.[64]
  - Values/ goals reflected: "best interests" and "relief of suffering".

The evidence and details of palliative care practice (for example that the proper use of opioids does *not* actually cause death[65]) were not relevant to the logic argument which *extracts and synthesizes* mid-level principles or "values" reflected in the laws that do not sanction certain physician acts even if a patient's death results (e.g. patient's wishes and best interests, autonomy, self-determination and relief of suffering) and *de-emphasizes* the discrete legal underpinnings.[66] Accordingly, with respect to the ethical arguments presented, the trial court first observes (repeated in the SCC judgment[67]) that[68]:

---

60 *Carter BCSC*, Footnote 3 at paras 319 and 343; See also Carter SCC, Footnote 3 at para 23.

61 *Carter BCSC*, Footnote 3 at para 335; See also *Carter BCSC*, Footnote 3 at para 236.

62 *Carter BCSC*, Footnote 3 at para 335; See also discussion at *Carter BCSC*, Footnote 3 at paras 331–333.

63 *Carter BCSC*, Footnote 3 at paras 207, 220 and 222; See also *Carter BCSC*, Footnote 3 at para 231(a) and (b).

64 *Carter BCSC*, Footnote 3 at para 225; See also discussion *Carter BCSC*, Footnote 3 at para 231(c) and paras 332–334; Sometimes also described as the "double effect doctrine". *Carter BCSC*, Footnote 3 at paras 198 and 314(c).

65 *Carter BCSC*, Footnote 3 at para 199.

66 See discussion in *Carter BCSC*, Footnote 3 at paras 325–333.

67 *Carter SCC*, Footnote 3 at para 23.

68 *Carter BCSC*, Footnote 3 at para 335. Emphasis added.

> The preponderance of the evidence from ethicists is that there is no ethical distinction between physician-assisted death and other end-of-life practices whose outcome is highly likely to be death.

The trial court then finds these ethical arguments "persuasive" and agrees that in an individual case, a "bright-line *ethical* distinction" between physician-assisted death and other "end-of life" practices where there is a high probability of hastened death is "elusive".[69] Note how these statements risk creating an impression that palliative care practice de facto incorporates practices that hasten death, when all the court has done is agree with certain ethicists that when one looks at the *values* reflected in the grouping of laws that do not sanction physicians for acts that cause or risk hastened death, it is difficult to find an "ethical" distinction between those acts and physician-assisted death.

The trial court thus concludes with the observation that the *intentional* taking of a life could be considered *ethical* in *limited* and *exceptional* circumstances based on the patient's *wishes* (requires competency and a fully informed, voluntary decision) and *best interests* (consistent with respecting patient's autonomy) in order to *relieve suffering* (grievously ill and suffering that cannot be alleviated from the patient's perspective).[70] The Supreme Court summarized this as follows[71]:

> Based on these findings, the trial judge concluded that, while there is no clear societal consensus on physician-assisted dying, there is a strong consensus that it would only be ethical with respect to voluntary adults who are competent, informed, grievously and irremediably ill, and where the assistance is "clearly consistent with the patient's wishes and best interests, and [provided] in order to relieve suffering …."

11.4.3 Effects on Palliative Care System and the Evidence from Permissive Jurisdictions

Evidence and arguments related to the lack of palliative care in Canada including that access to palliative care should be secured *prior to* legalizing physician-assisted death[72] are marshaled into the trial court's second line of inquiry regarding *impact* on palliative care and whether, amongst other things, a change in law or policy could pose a *threat* to "broader societal values" including "building a strong palliative care system".[73]

---

[69] *Carter BCSC*, Footnote 3 at paras 335 and 338; See also *Carter BCSC*, Footnote 3 at paras 1336 and 1369.

[70] *Carter BCSC*, Footnote 3 at paras 342 and 358; Referenced in Carter *SCC*, Footnote 3 at paras 23 and 24.

[71] Carter *SCC*, Footnote 3 at para 24. Emphasis added.

[72] *Carter BCSC*, Footnote 3 at para 314(l).

[73] *Carter BCSC*, Footnote 3 at para 161.

In examining potential impact on Canada's palliative care system, the court looked at evidence related to the assisted death schemes in permissive jurisdictions.[74] Following a detailed review, the trial court stated[75]:

> My review of the evidence regarding Oregon, the Netherlands and Belgium suggests that in those jurisdictions, legalization of assisted death has not undermined palliative care; on the contrary, palliative care provision has improved since legalization by some measures.
>
> Few conclusions, however, can be reached about the possible impact on palliative care from a change in Canadian law regarding physician-assisted death.
>
> ...
>
> ... while a change in the law to permit physician-assisted death could affect the palliative care system, predictions as to how would be speculative. I find that the evidence establishes that the effects would not necessarily be negative.

These findings are similarly relayed in the SCC judgment.[76]

While little can be said about the trial court's conclusion in this regard, a couple of points might still be made. First, at the time the Belgium *Act on Euthanasia* was passed in 2002,[77] Belgium simultaneously passed legislation establishing a right to palliative care.[78] This kind of initiative might have been flagged as an important consideration in developing Canadian safeguards, but it is lost in judicial translation. That is, while safeguard evidence from permissive jurisdictions was explored, it was only utilized by the court to demonstrate that safeguards *can be* used rather than as evidence of the types of safeguards that *could* or *should be* used[79] (keeping in mind, of course, the proper role of parliament/legislatures in ultimately developing regulatory regimes).[80] Second, as far as the immediate judicial inquiry was concerned, it is implicit to the *Carter* case that the lack of universal access to palliative care in Canada was treated as a "not my problem" problem. Indeed, the trial court not only later points the problem of palliative care availability in Canada right back at the defendants (the Attorneys General of British Columbia and Canada) but also gives a nod to the advocacy argument of MAID legalization proponents, that postponing decriminalization of assisted death until palliative care is established is a "form of hostage-taking" in that the suffering of grievously-ill individuals who wish to die serves as "leverage for improving the provision of adequate palliative care".[81]

---

[74] Jurisdictions reviewed by the court: Oregon and Washington, Netherlands, Belgium, Switzerland, Luxembourg, Montana, Colombia. *Carter BCSC*, Footnote 3 at Part VIII.

[75] *Carter BCSC*, Footnote 3 at paras 731–732, 736. Emphasis added.

[76] *Carter SCC*, Footnote 3 at paras 26 and 107.

[77] *The Belgian 28 May 2002 Act on Euthanasia*, B.S. 22 June (2002).

[78] *Carter BCSC*, Footnote 3 at para 507.

[79] *Carter BCSC*, Footnote 3 at para 646 et seq.

[80] *Carter BCSC*, Footnote 3 at para 1386.

[81] *Carter BCSC*, Footnote 3 at para 1274; For a similar argument see *Provincial -Territorial Expert Advisory Group on Physician-Assisted Dying* (2015) at page 20 [hereinafter "Provincial-Territorial Report"]. Online: https://www.health.gov.on.ca/en/news/bulletin/2015/docs/eagreport_20151214_en.pdf.

11.4.4 Palliative Care as Part of Informed Consent for Physician-Assisted Dying

When the court goes on to specifically consider the "feasibility" of safeguards in the Canadian context, it proceeds by first identifying what are the *risks* for which safeguards would be required. This is not a consideration of *all* risks, but those which might be identified based on the evidence before the court. To this end, the trial judge organizes the evidence into five areas: competence, voluntariness, informed consent, ambivalence and socially vulnerable individuals.[82] Palliative care evidence is considered throughout this discussion and includes evidence related to the *limits* of palliative care in addressing suffering.[83] As the discussion unfolds, palliative care surfaces for its potential to help mitigate risk. When the court advances "informed consent" as a key criteria or safeguard for persons seeking assisted death,[84] palliative care emerges as part of the requirement to advise patients of treatment options[85]:

> The evidence as to informed consent permits me to conclude that, just as physicians routinely assess the requirements for informed consent in patients seeking or refusing medical treatment, it would be feasible to *require* informed consent for physician-assisted death. In that context, particular care would be required to ensure a patient is properly informed of her diagnosis and prognosis, given the seriousness of the decision. As well, the range of treatment options described would have to encompass all reasonable palliative care interventions, including those aimed at loss of personal dignity.

The SCC decision includes the trial judge's findings and observations with respect to using the informed consent standard (with palliative care positioned as a treatment option) for physician-assisted death.[86] Note how the application of the *informed consent standard* to physician-assisted death as a medical treatment causing death appears to be a conceptual extension of the "no ethical distinction" argument and the values associated with decisions to refuse or withdraw from life-sustaining treatment.[87] Because of this, the court may be taken as positioning informed consent more like an eligibility criterion (or qualifier) for physician-assisted death rather than a procedural safeguard (to ensure the criteria are met).

In bringing its examination of risk to a close, the trial court concluded[88]:

> My review of the evidence… leads me to conclude that the risks inherent in permitting physician-assisted death can be identified and very substantially minimized through

[82] *Carter BCSC*, Footnote 3 at para 761.

[83] *Carter BCSC*, Footnote 3 at para 823.

[84] See *Carter BCSC*, Footnote 3 at paras 795–98, 815, 837, and 843; See also *Carter SCC*, Footnote 3 at para 106.

[85] *Carter BCSC*, Footnote 3 at para 831. Emphasis added; See also discussion in *Carter BCSC*, Footnote 3 at paras 816–821.

[86] *Carter SCC*, Footnote 3, at paras 104–106, 115.

[87] See *Carter BCSC*, Footnote 3 at para 1151.

[88] *Carter BCSC*, Footnote 3 at para 883; See also *Carter SCC*, Footnote 3 at para 27 and paras 105–106.

a carefully-designed system imposing stringent limits that are scrupulously monitored and enforced.

The SCC agreed, stating[89]:

> We agree with the trial judge that the risks associated with physician-assisted death can be limited through a carefully designed and monitored system of safeguards.

Notwithstanding the notion of stringent limits, in advancing "informed consent" as a key operative criteria for assisted death, the court (unintentionally or not) is not only giving future legal teeth to the "no ethical distinction" argument, it is unequivocally positioning assisted death as one medical treatment among many (thus immediately challenging the notion of "last resort" option)[90] and establishing a basis for one of, if not the broadest formulations for euthanasia practice in the world to date.[91]

11.4.5 The Legal Analysis

Leaving aside the ethical and safeguard discussions, the trial court enters into the legal analysis which it principally organizes around Section 241(b),[92] the provision in the *Criminal Code* that prohibits aiding suicide the purpose of which, as described by the court, is to protect "*vulnerable persons* from being *induced* to commit suicide at times of weakness".[93] When the trial court examines the infringement of the Section 7 rights to life, liberty and security against the narrow purpose of 241(b), the rest of the judgment very much reads like a *fait d'accompli* because it was conceded that "not every person who wishes to commit suicide is vulnerable" and there are persons, such as Ms. Taylor, "who have a considered, rational and persistent wish to end their own lives".[94]

The trial court, with the SCC in agreement, thus found the prohibition to be *overbroad*[95] and not in accordance with the *principles of fundamental justice* because the "blanket prohibition sweeps conduct into its ambit that is unrelated to the law's objective".[96] In other words, the absolute prohibition was overbroad because it cast a

[89] *Carter SCC*, Footnote 3, at para 117.

[90] For argument and evidence in this regard see *Carter BCSC*, Footnote 3 at paras 315(o), 381 and 533.

[91] See discussion commencing at Footnote 125 and associated text.

[92] See *Carter SCC*, Footnote 3 at para 20.

[93] *Carter BCSC*, Footnote 3 at para 1362. Emphasis added. According to the Court, the purpose, however, was grounded in the respect for and the desire to protect human life; See *Carter BCSC*, Footnote 3 para 1190; See also *Carter BCSC*, Footnote 3 at para 86.

[94] See discussion in *Carter BCSC*, Footnote 3 at paras 1136 and 1122; See also *Carter BCSC*, Footnote 3 at para 86.

[95] Note that the trial court found the infringement to be both overbroad and grossly disproportionate. See *Carter BCSC*, Footnote 3 at paras 18, 1371 and 1378.

[96] *Carter SCC*, Footnote 3 at para 86 and *Carter BCSC*, Footnote 3 at paras 1362–1371.

net wider than necessary to protect vulnerable persons from being induced to commit suicide at a moment of weakness.

The absolute prohibition could also not be saved by Section 1 of the *Charter* because it was not *minimally impairing*. That is, a less harmful or drastic means could achieve the legislative goal.[97] As described by the trial court[98]:

> a less drastic means of achieving the legislative purpose would be to keep an almost-absolute prohibition in place with a stringently limited, carefully monitored system of exceptions allowing persons in Ms. Taylor's situation – grievously and irremediably ill adult persons who are competent, fully-informed, non-ambivalent and free from coercion or duress – to access physician-assisted death.

In agreeing with the trial judge's minimal impairment evaluation[99] and that a "permissive regime with properly designed and administered safeguards was capable of protecting vulnerable people from abuse and error,"[100] the SCC *inter alia* re-stated the trial judge's conclusion that it would be possible for physicians to apply the "informed consent standard" to requests for assisted death, which would include patients being "informed of their diagnosis and prognosis and the range of available options for medical care, including palliative care interventions aimed at reducing pain and avoiding the loss of personal dignity.[101]

In the result, consistent with the trial judgment, the SCC declared Sections 241(b) and 14 to be[102]:

> of no force or effect to the extent that they prohibit physician-assisted death for a competent adult person who (1) clearly consents to the termination of life and (2) has a grievous and irremediable medical condition (including an illness, disease or disability) that causes enduring suffering that is intolerable to the individual in the circumstances of his or her condition.

11.4.6 Observations

What can be readily observed with respect to the construction and operationalization of palliative care evidence in the *Carter* case is that given that it was an individual rights' claim, palliative care was of most interest to the court not for *what it is* but rather for *what it was not* to the individual actively seeking physician-assisted death. Accordingly, the factual limits of palliative care in being able to address all suffering in combination with the legal right to refuse medical treatment helped to demonstrate in a general way, a factual "need" for physician-assisted dying. At the

---

[97] *Carter SCC*, Footnote 3 at para 102.

[98] *Carter BCSC*, Footnote 3 at paras 16 and 1243. Emphasis added; See also *Carter SCC*, Footnote 3 at para 121.

[99] *Carter SCC*, Footnote 3 at para 121.

[100] *Carter SCC*, Footnote 3 at para 105.

[101] *Carter SCC*, Footnote 3 at para 106.

[102] *Carter SCC*, Footnote 3 at para 147; See also *Carter BCSC*, Footnote 3 at para 1367.

same time, because palliative care does address much suffering, the court leveraged it to help achieve informed consent for physician-assisted dying. Note how this leveraging is confined to individual requests for physician-assisted death and falls short of endorsing palliative care as a systems-level safeguard or alternative to assisted death.[103]

The court does *not* equate palliative care with physician-assisted dying. However, as discussed above, the "ethical debate" does risk leaving that impression. It is regrettable that the trial court did not circle back to its definition of palliative care and link its findings with respect to palliative care clinical goals and practices. Without this type of clarification, it is challenging not to conflate or mistake the ethics argument that would support an extension to physician-assisted death in certain circumstances as also providing an interpretation of *what palliative care is* or the impression that *palliative care hastens or causes death.*

Very little of the trial judgment findings with respect to palliative care are included in the SCC decision. When the statement the "preponderance of the evidence from ethicists is that there is *no ethical distinction* between physician-assisted death and *other end-of-life practices whose outcome is highly likely to be death*"[104] is repeated by the SCC as part of a detailing of the evidence and arguments that led to the trial judge's conclusion that physician-assisted dying could be ethical in certain circumstances,[105] this statement does not have the benefit or nuances of the trial analysis and is presented without clear explanation as to its overall relevance to the constitutional analysis.

Yet this statement has become somewhat of a sound bite[106] which going forward, has become problematic for those advocating the importance of maintaining the distinction between the practices of palliative care and MAID—regardless of whether one agrees with MAID practice or not. This appears to be particularly so for a variety of reasons including: imprecision with respect to the judicial findings regarding palliative sedation[107]; mistaking or asserting the court's acceptance of the ethical argument as a factual conclusion rather than the court finding the argument to be "persuasive"[108]; and because an ethical argument (judicially-advanced) carries with it a force of logic able to unmoor legal foundations and chip away at professionally held clinical, legal and moral distinctions.

Furthermore, because the SCC decision is the highest and final court decision in Canada, its statements carry significant societal impact. Notwithstanding that

---

[103] Compare to *Carter BCSC*, Footnote 3 at para 315(o).

[104] *Carter BCSC*, Footnote 3 at para 335.

[105] *Carter SCC*, Footnote 3 at paras 23 and 24.

[106] See for example: Stuart Chambers (2016a). Online: https://impactethics.ca/2016/04/19/incremental-medical-assistance-in-dying/; Lett (2017). Online: https://www.winnipegfreepress.com/opinion/analysis/the-torturous-ethics-of-assisted-death-430492323.html.

[107] See Footnote 42 and associated text.

[108] *Carter BCSC*, Footnote 3 at para 335; For further discussion see analysis in McMorrow (2018) at footnote 172; For examples see: Stuart Chambers (2016b) Ottawa Citizen. Online: https://ottawacitizen.com/opinion/columnists/chambers-why-suffering-children-should-have-the-right-to-assisted-dying/

the court was not directly addressing the issue of palliative care, a hazy picture of palliative care nonetheless emerges from its judgment, and it is this hazy picture of palliative care that has disseminated into broader socio-political narratives.

11.5 Legislative Responses

11.5.1 Bill C-14

In 2016, in response to Carter, the Federal government passed Bill C-14[109] to amend the *Criminal Code* to allow medical and nurse practitioners[110] to provide "MAID" in specific circumstances. Consistent with Carter, information about palliative care as a means to relieve suffering is included within informed consent, one of the MAID *eligibility* criteria.[111] Bill C-14 also appears to pick up the *Carte*r thread regarding potential impacts on the development of palliative care by including a provision committing the government to conduct a parliamentary review of the MAID law as well as "the state of palliative care in Canada" within five years of the law coming into effect.[112]

Presumably, providing information about palliative care to individuals requesting MAID would also carry with it an actual offer of palliative care. Even if it did, however, this alone does not presuppose the scope of palliative care actually available or accessible. While this in turn raises questions related to voluntariness and meaningful choice,[113] challenging the validity of informed consent on this basis would likely transform into having to defend against the "hostage-taking" argument described in *Carter*.[114] The Bill C-14 amendments do not establish a right to palliative care, although the Federal government did describe a commitment to "working with provinces, territories and civil society to facilitate access to palliative and end-of-life care" in the bill's preamble.[115]

In terms of MAID eligibility, under the Bill C-14 amendments, in addition to giving informed consent, a person must also: be eligible for government health services; be at least 18 years of age and capable of making health decisions; have

[109] Bill C-14, Footnote 2.

[110] See Bill C-14, Footnote 2 at Sections 241.1 and 241.2(3). Note that Canada is the only jurisdiction in the world to permit the medical practice of euthanasia by non-physicians. Compare to *Carter BCSC*, Footnote 3 at para 1389.

[111] Bill C-14, Footnote 2 at Section 241.2(1)(e).

[112] Bill C-14, Footnote 2 at Section 10.

[113] See for example discussion in Provincial-Territorial Report, Footnote 81 at page 20: "The Advisory Group understands that the validity of patient consent lies at the heart of the assessment of eligibility for physician-assisted dying and that lack of access to quality palliative care might, in some specific cases, threaten various elements of a valid consent".

[114] See Footnote 81 and associated text.

[115] Bill C-14, Footnote 2 at Preamble.

a "grievous and irremediable medical condition"; and make a voluntary request.[116] The criteria "grievous and irremediable medical condition" further required that four additional criteria be met[117]:

(a) "serious and incurable illness, disease or disability";
(b) "advanced state of irreversible decline in capability";
(c) "enduring physical or psychological suffering" (caused by that illness, disease or disability or state of decline) that is "intolerable to them and that cannot be relieved under conditions that they consider acceptable"; and
(d) "natural death has become reasonably foreseeable".

Procedural safeguards pursuant to Bill C-14 included: practitioner to be of opinion all eligibility criteria are met; a witnessed, signed request; person is informed and given opportunity to withdraw request; written confirmation by independent practitioner that criteria met; and a 10 day waiting period between the signed request and the provision of MAID (or shorter period if death or loss of capacity is imminent).[118]

11.5.2 Bill C-7

In March 2021, in apparent response to a 2019 decision of the Superior Court of Québec,[119] the MAID law was amended by Bill C-7 to extend MAID access to persons whose natural deaths are *not* reasonably foreseeable.[120] In short, Bill C-7 eliminated the foreseeability of natural death as an eligibility criterion (by removing it from the definition of "grievous and irremediable medical condition") and created two discrete procedural tracks for MAID based on whether natural death is ("track one")[121] or is *not* ("track two") foreseeable.[122] Compared to track one, the procedural safeguards introduced by Bill C-7 for track two are said to be more stringent and require, for example, practitioners to either have "expertise" in the condition causing the person's suffering or consult with another practitioner with that expertise.[123]

---

[116] Bill C-14, Footnote 2 at Section 241.2(1).

[117] Bill C-14 Footnote 2 at Section 241.1(2)(a)–(d). Emphasis added.

[118] Bill C-14, Footnote 2 at Section 241.2 (3).

[119] In *Truchon v Canada (AG)* (2019) the Superior Court of Québec which *inter alia* found the reasonably foreseeable death requirement in Bill C-14 to be unconstitutional. Note that the Superior Court of Québec also struck down a similar end of life criterion in Quebec's provincial end-of-life legislation, SQ (2014), c 2. This decision was not appealed by the Attorney General of Canada or Quebec.

[120] Bill C-7, Footnote 13.

[121] *Criminal Code*, Footnote 1 at Section 241.2(3).

[122] *Criminal Code*, Footnote 1 at Section 241.2(3.1).

[123] *Criminal Code*, Footnote 1 at Section 241.2(3)(e.1); Note that Bill C-7 also amended certain procedural safeguards for track one (i.e. where death reasonably foreseeable) including removal of the 10 day waiting period.

With respect to palliative care, it is retained as part of the information required to meet the *informed consent* criterion for MAID along with the other eligibility criteria: 18 years of age and capable; grievous and irremediable medical condition; and voluntary request.[124] It is worth noting here how informed consent (and palliative care along with it) is indeed positioned within the MAID law as an *eligibility criteria* (i.e. what qualifies a person for MAID) rather than a *procedural safeguard* (i.e. to ensure the eligibility criteria are met).[125] With the Bill C-7 amendments, however, palliative care is also added as a *procedural safeguard* for track two whereby the practitioner must ensure that the person has not only been *informed* of the means available to relieve their suffering, including, where appropriate, "counselling services, mental health and disability support services, community services and *palliative care*" but has also been "*offered* consultations with relevant professionals who provide those services or that care".[126] Furthermore, the practitioner must agree with the person that reasonable and available means to alleviate the person's suffering "have been discussed" and given "serious consideration"[127] (keeping in mind that the person always retains the right to determine what treatment is acceptable to them and can refuse offered treatment[128]).

While the above-described *safeguards* (i.e. "informed" and "offered") have been described as a "clarification of informed consent"[129] in respect of persons whose deaths are *not* reasonably foreseeable, it is difficult to not also view this as a benefit under the law in terms of, for example, increased access to an increased scope of palliative care—a benefit denied to suffering persons under track one whose deaths *are* reasonably foreseeable. This arguably makes little sense from the perspective of palliative care nor from the perspective of equality. Furthermore, if the *doctrine of informed consent* requires doctors to introduce the MAID option to a patient who "appears to be eligible", even though the patient themselves has not requested or inquired about MAID (a controversial argument currently being advanced by some MAID proponents and providers[130]), then it would seem that palliative care must also be offered and made available. Lack of access or availability of palliative care (or other needed supports for that matter) in such circumstances would be a clear violation of rights to life, liberty and security, if not also equality, discriminating

---

[124] *Criminal Code*, Footnote 1 at Section 241.2(1) "Eligibility for medical assistance in dying".

[125] *Criminal Code*, Footnote 1 at Section 241.2(1)(e).

[126] *Criminal Code*, Footnote 1 at Section 241.2(3)(g). Emphasis added.

[127] *Criminal Code*, Footnote 1 at Section 241.2(3)(g).

[128] *Criminal Code*, Footnote 1 at Section 241.2(2)(c). There have been unofficial reports that refusal of treatment for certain medical conditions can render track two patients (i.e. death not reasonably foreseeable) eligible for track one (i.e. death becomes reasonably foreseeable if for example, the patient refuses required medications).

[129] Government of Canada, Legislative Background: Bill C-7: Government of Canada's Legislative Response to the Superior Court of Québec *Truchon* Decision (2021-09-01) at pages 11–12. Online: https://justice.gc.ca/eng/csj-sjc/pl/ad-am/c7/c7-eng.pdf.

[130] CAMAP. *Bringing up Medical Assistance in Dying (MAiD) as a clinical care option.* Online: https://camapcanada.ca/wp-content/uploads/2022/02/Bringing-up-MAiD.pdf.

against all persons with life-limiting illnesses and disability who are perceived to meet MAID criteria.

Within Bill C-7, the Federal government re-committed itself *inter alia* to review the MAID law and "the state of palliative care in Canada" within 30 days of Bill C-7 coming into force, with a report to Parliament to follow within one year after the start of the review. Although this report was not available at the time of writing, it should be noted that there is again nothing in the amended MAID law that obligates the government to make palliative care universally available to all Canadians who could benefit from it, except perhaps the extent to which such a commitment might arise in connection with MAID requests. This in turn raises the question of the extent to which palliative care resources (including funding) are being directed or re-directed towards MAID practice compared to increasing access to palliative care generally to improve the quality of life for *all* Canadians suffering from life limiting illnesses.

While it is outside the scope of this particular discussion to consider the *impact* of MAID legalization on access to palliative care, according to the Second Annual Report on MAID 2020 (based on data from written requests and 7,595 MAID cases across Canada for 2020 and prior to the Bill C-7 amendments)[131]:

> During 2020, the majority of MAID recipients (82.8%) received palliative care. Of the MAID patients who did not receive palliative care during 2020, 88.5% were reported to have had access to these services had they wished to avail themselves of palliative care.

Of the 82.8% (6,115) MAID recipients who received palliative care, 19.0% (1,163) received palliative care services for a duration of two weeks to less than a month, and 18.1% (1,104) received services for less than 2 weeks.[132] Current reliable data with respect to availability and access to palliative care separate from MAID requests is not readily available.[133]

11.5.3 Bill C-277

On May 30, 2016 before Bill C-14 came into effect, private members Bill C-277, *An Act providing for the development of a framework on palliative care in Canada* (Framework) was introduced into Parliament.[134] The bill came into force December

[131] Health Canada (2021, June) at pages 5, 20 and 21 [hereinafter "Health Canada 2020 Report"]. Online: https://www.canada.ca/content/dam/hc-sc/documents/services/medical-assistance-dying/annual-report-2020/annual-report-2020-eng.pdf.

[132] Health Canada 2020 Report, Footnote 131 at pages 20 and 21. For a critical analysis of how MAID data points to a failure of the Canadian health care system see Gallagher (2020, October 19) Policy Options. Online: https://policyoptions.irpp.org/magazines/october-2020/lack-of-palliative-care-is-a-failure-in-too-many-maid-requests/

[133] For discussion and examples of data that is available see, Canadian Institute for Health Information (2018). Online: https://www.cihi.ca/sites/default/files/document/access-palliative-care-2018-en-web.pdf.

[134] Bill C-277, Footnote 14.

12, 2017 and is a direct response to concerns arising out of proceeding with assisted death without also ensuring the option and availability of high-quality palliative care to all Canadians.[135] Bill C-277, however, does not establish or commit to a *right* to palliative care. Rather the goal was to create a Framework that could help *improve access* to palliative care via hospitals, home care, long-term care facilities and residential hospices.[136]

The Minister of Health tabled the Framework on December 4, 2018.[137] In addition to providing guiding principles and identifying best practices, the Framework sets out goals and a range of priorities to improve palliative care within four priority areas, namely[138]:

- education and training;
- support measures for palliative care providers;
- research and data collection; and
- measures to facilitate equitable access across Canada (with focus on underserviced populations).

Though responsive to MAID,[139] the Framework is broader, stating that its purpose is to provide "a structure and an impetus for collective action to address gaps in access and quality of palliative care across Canada" under a collective vision, "that *all Canadians with life-limiting illnesses live well until the end of life*".[140] The Framework also articulates the bright-line distinction between the act of MAID and palliative care[141]:

> MAID is a medical intervention which intentionally ends a person's life at their request, while palliative care intends neither to hasten nor postpone death, and should be available over the continuum of care of a life-limiting illness.

Following the Framework, in 2019 the Federal Government issued its *Action Plan on Palliative Care* (Action Plan).[142] The Action Plan sets out initiatives "that fall under federal responsibility" with implementation to be overseen and coordinated by Health Canada[143] and collaboration from partners including provinces and territories

[135] Bill C-277, Footnote 14 at Preamble.

[136] Bill C-277, Footnote 14 at Section 2(1).

[137] Canada, Health Canada (2018) (Ottawa: Government of Canada, December 4) at Appendix A [hereinafter "Framework"]. Online: https://www.canada.ca/content/dam/hc-sc/documents/services/health-care-system/reports-publications/palliative-care/framework-palliative-care-canada/framework-palliative-care-canada.pdf.

[138] Framework, Footnote 137 at pages 3 and 17.

[139] MAID is identified as one of the "drivers of change". See Framework, Footnote 137 at page 5.

[140] Framework, Footnote 137 at page 3. Emphasis added; See also Framework, Footnote 137 at page 14.

[141] Framework, Footnote 137 at footnote vii. Emphasis added.

[142] Canada, Health Canada (2019) (Ottawa: Government of Canada, August) [hereinafter "Action Plan"]. Online: https://www.canada.ca/content/dam/hc-sc/documents/services/health-care-system/reports-publications/palliative-care/action-plan-palliative-care/action-plan-palliative-care-eng.pdf.

[143] Action Plan, Footnote 142 at page 1.

and other stakeholders.[144] In brief, the Action Plan sets out goals consistent with the priority areas identified in the Framework[145]:

> The Action Plan lays out Health Canada's five-year plan to tackle issues uncovered through the development of the Framework. It aims to improve quality of life for people living with life-limiting illness, families and caregivers, and enhance access, quality of care and health care system performance. It complements current financial support to provinces and territories under the Common Statement of Principles on Shared Health Priorities.

The Action Plan (like the Framework and Bill C-277 above it) does not come anywhere close to articulating or establishing a right to palliative care, focusing instead on improving palliative care through improvements to overall system performance. While there are no doubt challenges for the Federal government to advance palliative care given that health care is predominantly provincial jurisdiction,[146] without establishing a right to palliative care, improvements to system performance can arguably only go so far and will likely continue to fall short of that which has been achieved in respect of access to MAID.[147] Nevertheless, it is worth mentioning that within the Action Plan, the Government of Canada describes that it has "made palliative care a priority"[148] and does so without any explicit reference to MAID.

11.6 Concluding Thoughts

In terms of the *Carter* case and how palliative care was organized and marshaled through the judicial analysis, it must be kept in mind that the court was operating within a very narrow analytical bandwidth—the constitutionality of the *Criminal Code* provision(s) prohibiting seeking persons from accessing "physician-assisted death". This rendered palliative care relevant only to the extent of that analysis with the vast majority of the concerns raised in relation to access to palliative care sidelined to be presumably picked up in a different legal forum. Thus within *Carter*, palliative care was not directly approached on the basis of its substantive goals, role or positioning within health care nor on the basis of it potentially being a human right in itself.[149] This is important to contemplate because although some of the *Carter* dialogue is clearly a jumping off point for subsequent Federal palliative care initiatives, the possibility of an increased commitment to palliative care arising out of *Carter* was remote from the outset.

---

[144] Action Plan, Footnote 142 at page 6.

[145] Action Plan, Footnote 142 at page 2. Emphasis added.

[146] Action Plan, Footnote 142 at page 2.

[147] For a recent study examining stakeholder perspectives on whether progress has been made pursuant to Canada's palliative care Framework, Footnote 137 and Action Plan, Footnote 142, see Pesut et al. (2022). Online: https://bmcpalliatcare.biomedcentral.com/counter/pdf/10.1186/s12904-022-01074-4.pdf.

[148] Action Plan, Footnote 142 at page 1.

[149] For recent analysis of palliative care as a human right, see Brennan and Gwyther (2021).

At the same time, given the voluminous evidence before the court in *Carter*, not least of which was evidence from other permissive jurisdictions which established access or rights to palliative care prior to or concurrently with rights to assisted death, along with the finding that palliative care was "far from universally available in Canada,"[150] there was opportunity for the court to clearly articulate a perspective on the goals and significance of palliative care distinct from assisted death. Such a clarification would have increased the likelihood of the court's findings with respect to palliative care being more accurately disseminated to the broader public which in turn would have also helped shield palliative care from the court's "ethical debate" dicta, the impact of which continues to render palliative care particularly vulnerable to the ever present "no ethical distinction" sword. With the utmost respect, the Supreme Court of Canada bears some responsibility as well in terms of ensuring its analysis reflected an accurate presentation of the trial court's findings with respect to palliative care and by providing at least some context for the portions of the "ethical debate" it chose to include in its decision.

The realization of palliative care as a form of gatekeeper in the subsequent MAID legislation could be viewed as a preliminary step towards establishing a right to palliative care, particularly as the amendment extending MAID access to persons whose natural deaths are *not* reasonably foreseeable requires not just information but an actual offer of palliative care consultation. This step, although crafted as a MAID safeguard, arguably moves palliative care closer to a human right and also demonstrates one way in which the Federal government can locate jurisdiction. Taken a step further, with respect to MAID, once such a benefit (i.e. an offer of palliative care consultation to a person who is suffering from a life-limiting illness) has been extended by government, the right to equality would arguably require that it be extended to all similarly suffering persons. Additionally, *if* the doctrine of *informed consent* permits or requires Canadian clinicians to introduce MAID to patients who simply "appear to be eligible", this arguably would also create a concomitant right to palliative care as previously discussed.[151]

All of this begs the question: What purpose could it possibly serve—other than to work a grave injustice—to continue to delay recognition of a right to palliative care having established within Canadian health care, death treatment as a means to end suffering and embedding palliative care as an alternative; meaningful choice and informed consent require it, and as described above, equality would demand it. It would seem that the issue can no longer be a question of whether there is a right to palliative care, but rather a question of implementation. This begs the next question, what exactly is Canada waiting for?

---

[150] *Carter BCSC*, Footnote 3 at para 4.

[151] See Footnote 130 and associated text.

11.7 Recommendations

In light of the forgoing, the following recommendations are humbly asserted:
Federal, provincial and territorial governments should work together to:

- adopt and promote through legislation, a uniform definition of palliative care that expressly distinguishes palliative care practice from the life terminating act of MAID;
- develop and implement a framework to allow palliative care practice to continue to develop unfettered by, yet responsive to, MAID and its ongoing reform; and
- expressly recognize a right to palliative care for all Canadians living with life-limiting illnesses.

Acknowledgements Thanks are owed to the editors Jaro Kotalik and David Shannon, research assistants Kyle Wilfer and Michael Siklosi and to the Legal Research Institute at the Faculty of Law, University of Manitoba.

References

Cases

Carter v Canada (Attorney General), 2012 BCSC 886.
Carter v. Canada (Attorney General), 2013 BCCA 435.
Carter v. Canada (Attorney General), 2015 SCC 5.
Carter v. Canada (Attorney General), 2016 SCC 4.
Rodriguez v. British Columbia (Attorney General), [1993] 3 SCR 519.
Truchon v. Canada (AG), 2019 QCCS 3792.

Legislation

An Act respecting end-of-life care, SQ 2014, c 2.
An Act respecting end-of-life care, CQLR, c. S-32.0001.
Bill C-7. 2021. *An Act to amend the Criminal Code (medical assistance in dying)*, S.C. 2021, c. 2.
Bill C-14. 2016. *Act to amend the Criminal Code and to make related amendments to other Acts (medical assistance in dying)*, S.C. 2016, c. 3.
Bill C-277. 2017. *An Act providing for the development of a framework on palliative care in Canada*, S.C. 2017, c. 28.
Canadian Charter of Rights and Freedoms, Part I of the *Constitution Act, 1982*, being Schedule B to the *Canada Act, 1982* (U.K.), 1982, c. 11.
Criminal Code of Canada, R.S.C., 1985, c. C-46
The Belgian Act on Euthanasia of 28 May 2002, B.S. 22 June 2002.

Articles

Brennan, Frank, and Liz Gwyther. 2021. Human Rights Issues. In *Oxford Textbook of Palliative Medicine*, 6th ed., ed. Nathan I. Cherny, Marie T. Fallon, Stein Kaasa, Russell K. Portenoy, and David C. Currow. Oxford: Oxford University Press.

CAMAP. 2022. Bringing up Medical Assistance in Dying (MAiD) as a clinical care option. https://camapcanada.ca/wp-content/uploads/2022/02/Bringing-up-MAiD.pdf. Accessed 9 June 2022.

Canadian Hospice Palliative Care Association and Canadian Society of Palliative Care Physicians. 2019. Joint call to action. https://www.cspcp.ca/wp-content/uploads/2019/11/CHPCA-and-CSPCP-Statement-on-HPC-and-MAiD-Final.pdf. Accessed 9 June 2022.

Canadian Institute for Health Information. 2018. Access to palliative care in Canada. https://www.cihi.ca/sites/default/files/document/access-palliative-care-2018-en-web.pdf. Accessed 9 June 2022.

Chambers, Stuart. 2016a. Incremental medical assistance in dying. *Impact Ethics*. https://impactethics.ca/2016a/04/19/incremental-medical-assistance-in-dying/. Accessed 9 June 2022.

Chambers, Stuart. 2016b. Chambers: Why suffering children should have the right to assisted dying. *Ottawa Citizen*, March 3. https://ottawacitizen.com/opinion/columnists/chambers-why-suffering-children-should-have-the-right-to-assisted-dying/. Accessed 9 June 2022.

Gallagher, Romayne. 2020. Lack of palliative care is a failure in too many MAID requests. *Policy Options*, October 19. https://policyoptions.irpp.org/magazines/october-2020/lack-of-palliative-care-is-a-failure-in-too-many-maid-requests/. Accessed 9 June 2022.

Gerson, Sheri M., Gitte H. Koksvik, Naomi Richards, Lars Johan Materstvedt, and David Clark. 2020. The relationship of palliative care where assisted dying is lawful: A systematic scoping review of the literature. *Journal of Pain and Symptom Management*. 59 (6): 1287–1303.

Government of Canada. 2021. Legislative background: Bill C-7: Government of Canada's legislative response to the Superior Court of Québec Truchon Decision. https://justice.gc.ca/eng/csj-sjc/pl/ad-am/c7/c7-eng.pdf. Accessed 9 June 2022.

Health Canada. 2018. Framework on palliative care in Canada, Appendix A. https://www.canada.ca/en/health-canada/services/health-care-system/reports-publications/palliative-care/framework-palliative-care-canada.html#appa. Accessed 9 June 2022.

Health Canada. 2019. Action plan on palliative care: Building on the framework on palliative care in Canada. https://www.canada.ca/content/dam/hc-sc/documents/services/health-care-system/reports-publications/palliative-care/action-plan-palliative-care/action-plan-palliative-care-eng.pdf. Accessed 9 June 2022.

Health Canada. 2021. Second Annual Report on MAID 2020. https://www.canada.ca/content/dam/hc-sc/documents/services/medical-assistance-dying/annual-report-2020/annual-report-2020-eng.pdf. Accessed 9 June 2022.

Herx, Leonie. 2015. Physician-assisted death is not palliative care. *Current Oncology* 22 (2): 82–83. Online: https://www.ncbi.nlm.nih.gov/pmc/articles/PMC4399615/. Accessed 9 June 2022.

International Association for Hospice and Palliative Care. 2019. Palliative care definition. https://hospicecare.com/what-we-do/projects/consensus-based-definition-of-palliative-care/definition/. Accessed 9 June 2022.

Lett, Dan. 2017. The tortuous ethics of assisted death. *Winnipeg Free Press*, June 24. https://www.winnipegfreepress.com/opinion/analysis/the-torturous-ethics-of-assisted-death-430492323.html. Accessed 9 June 2022.

McMorrow, Thomas. 2018. MAID in Canada? Debating the constitutionality of Canada's new medical assistance in dying law. *Queen's Law Journal* 44 (1): 69–120.

Murphy, Sean. 2017. Legalization of assisted suicide and euthanasia: Foundational issues and implications. *BYU Journal of Public Law* 31 (2): 333–394.

Pesut, Barbara, Sally Thorne, Anne Huisken, David Kenneth Wright, Kenneth Chambaere, Carol Tishelman, and Sunita Ghosh. 2022. Is progress being made on Canada's palliative care framework and action plan? A survey of stakeholder perspectives *BMC Palliative Care* 21(1): 1–16.

Provincial-Territorial Expert Advisory Group on Physician-Assisted Dying. 2015. Provincial-Territorial Expert Advisory Group on Physician-Assisted Dying: Final Report.

Shariff, Mary J., and Mark Gingerich. 2018. Endgame: Philosophical, clinical and legal distinctions between palliative care and termination of life. *Second Series Supreme Court Law Review* 85: 225.

World Health Organization. 2018. Integrating palliative care and symptom relief into primary health care: A WHO guide for planners, implementers and managers. https://apps.who.int/iris/bitstream/handle/10665/274559/9789241514477-eng.pdf?ua=1. Accessed 9 June 2022.

Mary J. Shariff, Ph.D. is an Associate Professor of Law at the Faculty of Law, University of Manitoba, Canada. Professor Shariff received her Ph.D. from Trinity College, Dublin. In addition to teaching the Law of Contracts in the first year of the J.D. program, she teaches the upper year courses Law and Bioethics, Law and Religion and Animals and the Law. She also acts in a supervisory capacity for graduate students. Thesis supervisions have included research related to bioethics, climate change and non-human rights. She is published in the areas of end of life, assisted death and palliative care and is frequently engaged as a speaker on these topics. As a past Associate Dean, J.D. Program and Associate Dean, Research & Graduate Studies within the Faculty of Law, Professor Shariff is presently focusing on her research in law and bioethics, death and dying, palliative care and the rights of vulnerable persons, including older persons and persons with disabilities.

Chapter 12
Palliative Care and Medical Assistance in Dying

Leonie Herx and Ebru Kaya

Abstract Palliative care differs fundamentally from MAiD in its philosophy, intention and approach. While access to MAiD has been guaranteed as a right in Canada, access to palliative care and other supports for living, including home and disability services, has not. There has been expansion in MAiD services but no similar investment in palliative care necessary to improve its access, and in some areas, MAiD expansion has come at the cost of reduced access to palliative care resources. Very few Canadians have access to specialist palliative care which can negatively impact patient outcomes. Without access to high-quality palliative care, some patients may feel that MAiD is their only option because their suffering has been inadequately addressed or they perceive that their families or social supports must carry an excessive burden. This lack of access to palliative care should be considered a failure of the health care system and medical error. Palliative care must remain distinct from MAiD to ensure clarity and to avoid the potential for people to refuse palliative care services because they fear it might hasten death. Canada needs to prioritize investment in and enhancement of palliative care services.

Keywords Palliative care · MAiD · MAiL · Suffering · Serious illness · Life-threatening · End-of-life · Symptoms · Safeguards

12.1 Distinctions Between Palliative Care and Medical Assistance in Dying

Palliative care is a holistic approach to care that focuses on enhancing the quality of life for persons living with life-threatening illnesses and their families by means of

L. Herx (✉)
Queen's University, Kingston, ON, Canada
e-mail: Leonie.Herx@kingstonhsc.ca

E. Kaya
University of Toronto, Toronto, ON, Canada
e-mail: Ebru.Kaya@uhn.ca

J. Kotalik and D. W. Shannon (eds.), *Medical Assistance in Dying (MAID) in Canada*, The International Library of Bioethics 104,
https://doi.org/10.1007/978-3-031-30002-8_12

early identification and impeccable assessment and treatment of symptoms, including physical, psychosocial and spiritual concerns, and supports a person in living well until they die; it does not hasten death and sees dying as a normal part of life. In contrast, medical assistance in dying (MAiD) intends to cause the immediate death of a person at their request, through the administration of a lethal dose of drugs (Health Canada 2018).

Because palliative care fundamentally differs from MAiD in its philosophy, intention and approach, national and international palliative care associations have all released documents declaring that palliative care is a practice distinct from MAiD, including the Canadian Society of Palliative Care Physicians (2019), Canadian Hospice Palliative Care Association (2019), American Academy of Hospice and Palliative Medicine (2016), Australia New Zealand Society of Palliative Medicine (2021), European Association for Palliative Care (Radbruch et al. 2016), and International Association for Hospice and Palliative Care (De Lima et al. 2017). The Canadian Medical Association has also recognized that palliative care and MAiD are distinct (2016).

In addition to the philosophical and clinical distinctions between MAiD and palliative care, Canadian legislation has established an inherent legal distinction between the two (Shariff and Gingerich 2018). Canadian MAiD legislation states that requests for MAiD must be 'voluntary' and that 'informed consent' requires a person to have been "informed of the means that are available to relieve their suffering, *including palliative care*". Shariff and Ginsberg explain how palliative care is thereby positioned as a key element for consent in MAiD and serves as a general safeguard, and that this legal distinction must be maintained for the benefit of the public interest. A conflation of these two distinct practices would therefore result in the loss of palliative care as an inherent safeguard for MAiD.

12.2 Ongoing Conflation of MAiD and Palliative Care

The intentional use of the term 'Medical Assistance in Dying' (or 'Medical Aid in Dying', in Quebec) by the federal government to refer to euthanasia and assisted suicide has contributed to the confusion of assisted death with palliative care. Canadian palliative care organizations, the Canadian Hospice Palliative Care Association (CHPCA) and Canadian Society of Palliative Care Physicians (CSPCP) advocated against the use of this euphemistic language, arguing that palliative care aids those who are dying in order to help them live as fully as possible until their natural death, but does not intentionally end life (CSPCP 2019; CHPCA and CSPCP 2020). They also expressed concern that the linking of assisted death and palliative care perpetuates the myth that palliative care hastens death and may prevent patients from seeking early palliative care interventions.

When asked for his thoughts on MAiD, Dr. Balfour Mount, the 'father of palliative care in Canada', said, "The first thing I would say is that the very name of that intervention, 'medical aid in dying', is misleading rhetoric. Medical aid in dying is what I have been doing [*in palliative care*] for 50 years. This bill is not talking about medical aid in dying really, it is not talking about ending the suffering, but instead, ending the sufferer" (Mount n.d.).

In spite of ongoing national advocacy regarding the terminology for MAiD and education on the distinct natures of palliative care and MAiD (CSPCP 2015a; CSPCP 2016), conflation of palliative care and MAiD continues to be perpetuated in both public and healthcare spheres, including in the general media and peer-reviewed medical journals. For example, in a 2019 Canadian Guideline for Parkinson Disease, published in the Canadian Medical Association Journal (Grimes et al. 2019), not only was MAiD listed directly as a palliative care support, but it was the only specific support stated, implying that MAiD is the only intervention palliative care has to offer patients with Parkinson's Disease. When CSPCP asked for a correction to where MAiD was placed (Herx 2019) no correction was made by the authors or the journal.

Despite the clear distinction between palliative care and MAiD, some MAiD proponents are trying to push for MAiD be seen as part of the 'continuum' of palliative care and refer to MAiD as an "extension" of palliative care (Buchman 2019) or as another tool in the "palliative care basket" (Kutcher 2018). This misrepresentation is been further perpetuated by federal politicians who do not understand palliative care (Kaya and Herx 2022).

In their brief to the federal joint parliamentary committee undertaking the Statutory Review of the Medical Assistance in Dying Legislation, the CSPCP stated, "Conflating palliative care with MAiD ensures palliative care remains associated with 'end of life' and this has created unnecessary and unintended barriers to the integrated and timely access to palliative care services that are needed early on in the disease trajectory. Patients not contemplating death but who may significantly benefit from palliative care, may be denied palliative care services by their clinicians, due to its association with MAiD. Patients may also refuse palliative care services out of fear of it hastening their death (CSPCP 2022).

To better differentiate MAiD and palliative care, we have recommended using the term "Medical Assistance in Living" (MAiL) in reference to palliative care (Kaya and Herx 2022; Herx 2022).

12.3 How Palliative Care Addresses Suffering and a Desire to Die

Over the past 40 years, palliative care has evolved as a discipline with a growing evidence base that shows early integration of palliative care improves quality of life, reduces anxiety and depression, reduces caregiver distress, and may even help people to live longer. An integrated palliative approach to care should be provided as part of the management of all serious illness—starting soon after diagnosis—but due to the ongoing stigmatization and misunderstanding of palliative care as "end of life care", it is usually only considered in the last weeks of life which has limited its early integration and uptake into many settings (CSPCP 2016, 2017; Quality End-of-Life Care Coalition and CHPCA 2015).

In the setting of a life-limiting or life-threatening illness, desire for hastened death and talking about wanting to die are common expressions of grief and loss as a person tries to come to terms with the possibility of their own death. Research by renowned Canadian palliative psychiatrist, Dr. Harvey Chochinov, has shown that in this population, desire to die is most often transient and dissipates within 2 weeks (Chochinov et al. 1995, 1999).

These expressions of distress are normalized and supported through skilled palliative care interventions which seek to understand the nature of the suffering and how to address it. Holistic dignity-conserving palliative care interventions such as Dignity-Conserving Therapy, developed by Dr. Chochinov (2002), help to restore purpose, meaning, and hope in the face of losses that accompany life-threatening illnesses and help a person to focus on living.

With this type of palliative care support, healing is possible even in the face of many losses, but it takes time to build the therapeutic relationship necessary to accompany a person on their journey so they do not feel abandoned or alone in their suffering. "Palliative care is not a singular event, but rather a process of caring applied over time" (Chochinov and Mount 2014).

Sensitively exploring desire to die statements, in order to understand the nature of the person's suffering and to determine what supports can be offered to address it, to accompany the person in their suffering, and to let them know that they matter and are not alone (Mount 2020), is a fundamental component of palliative care.

Even though we know that many requests for hastened death are indications of suffering, that can be ameliorated by palliative care, rather than being a specific request for a termination of life procedure (CSPCP 2019), far too often these expressions of suffering are now met with an automatic consult to the MAiD team without palliative care being involved (Herx et al. 2020). As stated by the CSPCP, "No patient should choose MAiD because of lack of access to palliative care" (CSPCP 2019).

A recent study exploring how the Canadian MAiD legislation affects end-of-life care discussions between patients and multidisciplinary specialist palliative care providers determined that "It remains essential to address psychosocial and existential suffering, but also to provide timely palliative care to ensure suffering is addressed before it is deemed irremediable. Palliative care provided late in the disease trajectory, is like a person having pneumonia for days and coming to the hospital being unable to breathe. Even though they are put on life support systems, they often die because the infection has spread to so many other key organs. Hence, clarification is required regarding assisted dying as an intervention of last resort. Furthermore, professional and institutional guidelines need to better support palliative care providers in maintaining their holistic standard of care, as they navigate the legislative and cultural shifts" (Ho et al 2021).

12.4 Access to Palliative Care and Informed Consent for MAiD

Only about 30–50% of Canadians who need it have access to some kind of palliative care of unknown quality, and very few have access to specialist palliative care to address more complex issues despite over 90% of Canadians not wanting an assisted death with MAiD. There are significant disparities in the ability to access palliative care across the country, especially in rural and remote regions where there may be no palliative care at all. Whether you will get the palliative care you need, depends on where you live, how old you are, what you are dying from, and how close you are to death. You are also less likely to receive palliative care if you are poor, Indigenous, homeless, or incarcerated (CSPCP 2016, 2017; Quality End-of-Life Care Coalition of Canada and CHPCA 2015).

Lack of access to necessary care can have a negative impact on a patient's sense of autonomy and dignity, which are common reasons for MAiD requests (Wiebe et al. 2018). This begs the question then as to whether a request for MAiD can be considered "informed" and "voluntary" if there is limited or no access to palliative care—a reality faced by many Canadians.

Patients have told us that they were scared they would not have a peaceful death without MAID. The reality is that patients don't have access to enough information, or sometimes even correct information, about palliative care. Many are not aware how palliative care can help relieve their pain or other symptoms to help them live well until they die. Or that natural dying and death are most often comfortable, and any complex symptoms can be well-managed by specialists in palliative care. Well-meaning public figures and clinicians who have a strong bias toward MAiD, but don't understand palliative care, cause confusion by perpetuating myths that dying necessarily involves intense pain and suffering. As a result, too many Canadians believe that a natural death is uncomfortable and undignified (Kaya and Herx 2022).

In the Final Report of the External Panel on Options for a Legislative Response to *Carter v. Canada* (2015), the panel stated that "a request for physician-assisted death cannot be truly voluntary if the option of proper palliative care is not available to alleviate a person's suffering" Shariff and Gingerich (2018) similarly asked, "If there is not a genuine commitment (resource or otherwise) behind an offer of palliative care upon which a patient can rely, can it truthfully be said from a regulatory perspective that a patient has meaningful choice?".

The International Association for Hospice Palliative Care (IAHPC) has stated that, "…no country or state should consider the legalization of euthanasia or PAS (*physician assisted suicide*) until it ensures universal access to palliative care services" (De Lima et al. 2017).

During their Carter deliberations in 2015, the Supreme Court of Canada (SCC) recognized the important role palliative care plays in alleviating suffering that may lead to requests for assisted death and discussed what possible impacts implementing a law for assisted death prior to securing universal access to palliative care might have on the development of a strong palliative care system. They ultimately suggested that legislation for assisted death could be used to leverage improvements to palliative care (*Carter v. Canada* 2015).

The importance of palliative care was recognized by parliament when Bill C-277 was unanimously passed in 2018, establishing the National Framework for Palliative Care in Canada (Health Canada 2018), but this did not come with the required financial supports and federal infrastructure for implementation. Stakeholders made it clear that the framework for palliative care had already been developed through years of work by the national palliative care associations in The Way Forward National Framework (2015), and what was really needed was support for its implementation such as via a well-funded secretariat (CSPCP 2017).

As noted recently by MP Marilyn Gladu, the original sponsor of Bill C-277, at the parliamentary Special Joint Committee on Medical Assistance in Dying, "The government pledged $6 billion over 10 years in the 2018 budget for palliative care. Sadly, what's happened is that they've spent only $200,000 of that, and it was bucketed together with MAID, which was never the intent. The World Health Organization has been clear that palliative care does not hasten death; that's a totally different thing from MAID" (Gladu 2022).

Gladu asked witnesses presenting to the committee the following question, "We know that the framework was also looking at addressing the 60,000 missing resources in terms of palliative care doctors, nurses and other providers. Would you agree that the government should meet its commitment of $6 billion over 10 years for palliative care and address the gap of missing resources? Dr. Buchman, I'll start with you". Dr. Buchman replied, "I agree one hundred per cent. It's a shame that the funding did not follow the framework. That's been the consistent place in Canada. We have wonderful plans, but they end up sitting on a shelf because the funding never comes" (Gladu 2022).

In 2020, the federal government had another opportunity to enshrine access to palliative care as part of the Canada Health Act, when MAiD legislation was being amended under Bill C-7. National palliative care organizations, CHPCA and CSPCP, advocated for palliative care to be deemed a fundamental healthcare right under the Canada Health Act (CHPCA and CSPCP 2019), which would require provinces to adequately fund and make palliative care accessible as they had done for MAiD, but the government once again failed to take this opportunity. Despite palliative care continuing to be a safeguard for MAiD, there is nothing in the current MAiD legislation that requires palliative care to be made accessible.

As stated by Dr. Romayne Gallagher in her expert witness testimony to the Special Joint Committee on MAiD in May 2022, "Parliament has voted three times in favour of funding a palliative care strategy. It's 2022, and it's yet to be implemented. Canadians will suffer with life-limiting illnesses whether they access MAiD or die naturally" (Gallagher 2022).

As a person approaches the end of life, the inevitable physical decline is viewed in Western European culture as undignified, full of suffering and a burden on caregivers. This belief is not shared by other cultures. There is a specific demographic that has been leading the advocacy for legalization and expansion of MAiD in Canada—white Canadians, who are often well-off and well educated. But it is not this demographic who will choose MAiD because of a lack of access to care and supports needed to live. By funding MAiD only, the Canadian government is providing a significant amount of public resources to a group of Canadians who are already privileged while continuing to underserve the rest of our society who do not share these same values or privileges.

12.5 Involvement of Palliative Care in Patients Requesting MAiD Since Legalization

Health Canada Reports on MAiD (2019, 2020, 2021) have concluded that a lack of access to palliative care is not driving MAiD requests, and that in 2020, the "vast majority" of persons receiving MAiD had palliative care. However, this is not possible to conclude from data that relies on self-reporting by MAiD providers. There is no information regarding the quality or quantity of palliative care, how it was delivered, or who was providing it and their level of expertise, all of which are essential to understand the situation (Herx 2022; CSPCP 2022).

As per the CSPCP, "Although Health Canada asserts that over 80% of individuals receiving MAiD in 2019 and 2020 reportedly received palliative care services, the adequacy of the services offered was not determined, nor was it articulated how palliative care services were defined, what criteria were used to determine the involvement of palliative care, and if these involvements were adequate. Clear data regarding availability, access, quality and types of palliative care (e.g. specialist palliative care, primary palliative care or palliative approach) are essential, not only for those requesting or receiving MAiD, but in general, in order to better inform areas for improvement. Currently we do not have the effective data in Canada about palliative care quality of accessibility to draw any meaningful conclusions about palliative care for those receiving MAiD" (CSPCP 2022).

Even in the context of the limitations and potential biases of the Health Canada reports, we can see that a significant number of people receiving MAiD had no palliative care at all (15%) or palliative care much too late to address their severe distress and "irremediable suffering", including at the time of MAiD request (18%) or within short weeks of the request (19%). Another 2.9% are unknown. Taken together, this means that over 35% of Canadians who received MAiD had almost no palliative care and almost half had little to no palliative care. This is consistent with the results of a Canadian study by Munro and colleagues (2020) which showed that 40% of patients had no palliative care involvement prior to requesting MAiD.

Dr. Romayne Gallagher describes how this constitutes a failure of the healthcare system and should be seen as medical error. The root cause of this medical error is not MAiD itself, but the lack of quality palliative care in the previous weeks, months and years of the disease trajectory—a known therapy that the system fails to provide (Gallagher 2020; Gallagher et al. 2020).

MAiD was initially established as an "exception" and was said to be intended for very extenuating circumstances, not as a routine option for ending life as a solution to any suffering that could be addressed through readily available health and social supports that help people live a dignified life, such as palliative care. Since Bill C-7, we now see countless cases in the media—and in our own medical practices—of people requesting and receiving MAiD due to a lack of basic supports needed for living, including lack of palliative care, homecare, and disability care (Herx 2022; Herx et al. 2020; Kaya and Herx 2022).

12.6 Replacement and Erosion of Palliative Care with MAiD

In the deliberations of *Carter v. Canada* (2015), the Supreme Court of Canada determined that "few conclusions" could be reached about the impact of MAiD legalization on palliative care and suggested that MAiD legislation could be used to leverage improved access to palliative care. We now have six years of lived experience with legalized MAiD and its implementation in Canada, and the reality is that there have not been significant investments in palliative care nor improvements in the accessibility of palliative care.

Moreover, we are seeing significant deleterious effects on palliative care, including diminished resources, replacement of palliative care with MAiD, and increasing distress experienced by palliative care clinicians (Herx et al. 2020; Ho et al. 2021; Mathews et al. 2021; Herx 2022; CSPCP 2022), including:

- Lumping of MAiD into "Palliative and End of Life Care" programs in some hospitals and community services which has resulted in the utilization of existing scarce palliative care resources to provide MAiD. For example, in Ontario, some palliative care nurses are tasked with administration and coordination of MAiD which has been taking up an increasing proportion of their roles—to the point that nurses have left their jobs because they were not able to provide palliative care. In Alberta, British Columbia and Ontario, some Hospice Palliative Care Nurse Practitioners and physicians use their time in paid palliative care roles to provide MAiD, at the direction of health system administrators.
- In Ontario, palliative care billing codes are used by doctors to capture and bill for their MAiD work at the direction of the Ministry of Health, despite strong advocacy against this from palliative care medical associations. The use of these palliative care codes means that Ontario MAiD providers are eligible for special financial premiums intended as an incentive to provide palliative care.
- Some health authorities require hospices and palliative care units to admit patients for the sole purposes of administering MAiD which decreases access to already limited specialized palliative care beds.

- Patients unwilling to access palliative care services because they associate it with MAiD and are afraid palliative care will hasten their death, or that MAiD will be provided without their consent.
- Patients requesting palliative care and the MAiD team sent instead.
- Patients accessing MAiD due to lack of palliative care availability.
- Increasing moral distress in palliative care clinicians due to forced participation in MAiD through health authorities or provincial governments mandating MAiD be provided on site at hospices and palliative care units when it is counter to the philosophy of care and vocation of the staff. This is resulting in retention difficulties and early retirements from palliative care, exacerbating and accelerating the already critical shortages in specialist and generalist palliative care physicians and other team members.
- Increasing moral distress in palliative care clinicians when patients refuse palliative care interventions out of fear it will make them ineligible for MAiD.
- Palliative care clinicians seen as obstructing access to MAiD. When concerns about safeguards for MAiD have been raised by palliative care clinicians, e.g., questions about patient eligibility for MAiD such as lack of capacity, these concerns are sometimes dismissed, ignored, or scrutinized as "obstructing access" to MAiD, and even investigated in some regions. This has also happened to some palliative care clinicians when discussing palliative care options to persons requesting MAiD. A similar situation has been reported in Belgium where raising palliative care as an option to those requesting euthanasia is now seen as an obstruction to accessing euthanasia (Devos 2021).

12.7 Impacts of Changes to MAiD Safeguards on Palliative Care Practice

The SCC stated in their *Carter v. Canada* ruling (2015) that a carefully designed system of safeguards would be needed to minimize harm to vulnerable persons and the C-14 legislation (2016) established certain safeguards for persons with a reasonably foreseeable death that were put in place way to ensure that there were basic mechanisms in place to reduce the risk of coercion to MAiD and wrongful death. Bill C-7 subsequently removed several of the safeguards including the requirement for two independent witnesses, capacity at the time of MAiD, and removal of the 10-day reflection period between written request for MAiD and administration the lethal drugs (2021).

The removal of the 10-day reflection period in patients with a reasonably death can now result in people requesting and receiving MAiD potentially on the same day—effectively making a person's worst day their last day, due to despair or existential distress that may have improved with time (CSPCP 2021; Chochinov 2021). CSPCP stated that, "Assertions that requests for MAiD only occur after careful consideration do not reflect our clinical experience. Desire for death and requests for MAiD are often expressions of grief, anger, loss and despair as a response to a devastating diagnosis or change in condition. This "lament"—a cry for help—now often results in a MAiD consult even before palliative care has been involved. Dr. Harvey Chochinov's published work in psychiatry and palliative care demonstrates that desire to die in the terminally ill fluctuates and often dissipates within two weeks. The 10-day reflection period strikes an important balance, allowing MAiD to be requested while protecting people who might change their minds and go on to live many weeks, months or even years."

CSPCP (2020) also recommended that it was important to maintain the requirement for capacity to consent at the time of MAiD provision, stating that a person should be able to change his/her decision up until the time of the procedure. And such an opportunity is lost if a person loses capacity after previously consenting and a substitute decision maker is thereby deciding for a person when MAiD should be provided, even though the person may no longer view things the same way.

12.8 Bringing up MAiD to All Potentially Eligible Patients as Part of Informed Consent

The Canadian Association of MAiD Assessors and Providers (CAMAP) has proposed that clinicians must bring up MAiD to all potentially eligible patients, as part of informed consent for palliative care (2022).

When a trusted doctor brings up an option for care, because of the trust and power differential inherent in the physician–patient relationship, this is often seen as a recommended course of action. As Finlay and Carlile (2011) stated, "…the agreement of a trusted professional to consider assisted dying as a therapeutic option could all too easily signal to the patient, however unwittingly, that a hastened death was the appropriate course of action to contemplate in that patient's clinical condition."

To safeguard against any possibility of subtle or overt pressure on patients, the CSPCP recommended that physicians should not initiate a discussion about MAiD or suggest the option of MAiD unless brought up by a patient (CSPCP 2020).

In other jurisdictions where physician assisted death is legalized, this risk of coercion due to the power imbalance and differential of expertise present in the physician–patient relationship is addressed directly in the law or policy regulating assisted death by requiring discussions about MAiD to be patient-initiated (Victoria State Government 2017).

12.9 Recommendations for Improving Palliative Care in the Context of Canada's MAiD Regime

Currently Canadians have a right to medical assistance in dying but not a right to medical assistance in living. Since palliative care is positioned as a safeguard that can alleviate suffering leading to some MAiD requests, it cannot be a true safeguard if it is not accessible and integrated early into the course of serious illness before suffering becomes irremediable. Having accessible high quality palliative care needs to be a universal health care right in Canada.

We recommend the following actions for the federal government to take in order to improve palliative care and give Canadians meaningful choices in the context of Canada's current MAiD regime.

1. Canadians need medical assistance in living (MAiL) and the way to provide this is to have consistent universal access to high quality palliative care that is accessible where and when they need it, regardless of diagnosis, postal code, or income level. This could be accomplished by deeming palliative care an essential service under the Canada Health Act.
2. Federal funding is required to provide the necessary infrastructure and resources to implement the key priorities outlined in the Framework for Palliative Care in Canada (Health Canada 2018) and Action Plan for Palliative Care in Canada (Health Canada 2019).
   (i) Guarantee sustained investment in a competent and interdisciplinary workforce for palliative care. Integrate the core competencies for primary and specialist level palliative care outlined in the Canadian Interdisciplinary Competency Framework (Canadian Partnership Against Cancer and Health Canada 2021) into training curricula for all health care professionals providing palliative care, linked to national reporting requirements.
   (ii) Require palliative care programs across the country to collect and report standardized patient-reported outcomes and quality improvement measures through accreditation standards.
   (iii) Fund a National Centre for Palliative Care Education and Research (completely separate from MAiD, unlike the current Health Canada portfolio) with representation from the national palliative care organizations, to ensure standards and accessibility for quality palliative care are advancing.
   (iv) Fund a national public education campaign to destigmatize palliative care, debunk myths that palliative care hastens death, and improve awareness that ordinary dying is most often peaceful and expert palliative care supports can readily address more complex symptoms.
3. MAiD programs must be administered and resourced separately from palliative care programs to protect limited palliative care resources, and to ensure these two different services (palliative care and MAiD) are not confused by patients and families (CSPCP 2019; CSPCP 2020).

4. There must be safe institutional spaces for Canadians who do not want MAiD and are afraid MAiD will be provided without their consent. Long term care homes, hospices, and palliative care units must be allowed to offer safe spaces free from MAiD. Like for other medical procedures, accessibility does not mean that a procedure is provided in every setting of care, but that patients are routinely moved between sites or have to travel to access specialized services.
5. There must be meaningful conscience protection for palliative care professionals with respect to MAiD. Physicians who do not wish to participate directly or indirectly in MAiD should have their integrity and fundamental freedoms, including freedom of conscience, protected.
6. MAiD must be a patient-initiated discussion. Like other jurisdictions, Canada must build into legislation a safeguard to mitigate potential pressure from clinicians raising unsolicited MAiD to patients by requiring MAiD discussions to be patient-initiated.
7. More oversight for MAiD. Adherence to safeguards, assessment, access, provision of MAiD, and professional disagreements about eligibility for MAiD must be more thoroughly scrutinized in order to ensure public confidence in the MAiD system, as recommended by Dr. Kotalik (2020).

References

American Academy of Hospice and Palliative Medicine. 2016. Statement on physician-assisted dying. http://aahpm.org/positions/pad.

Australia New Zealand Society of Palliative Medicine. 2021. Australian and New Zealand society of palliative medicine position statement: The practice of Euthanasia and physician-assisted suicide. https://www.anzspm.org.au/c/anzspm?a=sendfile&ft=p&fid=1635983800&sid.

Bill C-7. 2021. An Act to amend the Criminal Code (medical assistance in dying), S.C. 2021, c. 2 which received Royal Assent March 21. https://parl.ca/DocumentViewer/en/43-2/bill/C-7/royal-assent.

Buchman, S. 2019. Bringing compassion to medicine and to the CMA. *Canadian Medical Association*. https://www.cma.ca/dr-sandy-buchman

Canadian Association of MAiD Assessors and Providers. 2022. Bringing up medical assistance in dying as a clinical care option. https://camapcanada.ca/wp-content/uploads/2022/02/Bringing-up-MAiD.pdf.

Canadian Hospice Palliative Care Association. 2019. Policy on hospice palliative care and medical assistance in dying (MAiD).

Canadian Hospice Palliative Care Association and Canadian Society of Palliative Care Physicians Joint Call to Action. 2019. https://www.cspcp.ca/joint-statement-from-chpca-and-cspcp-regarding-palliative-care-and-maid/.

Canadian Medical Association. 2016. CMA policy: Palliative care. https://www.cma.ca/sites/default/files/2018-11/cma-policy-palliative-care-pd16-01-e.pdf.

Canadian Partnership Against Cancer & Health Canada. 2021a. *The Canadian interdisciplinary palliative care competency framework*. Toronto, ON.

Canadian Society of Palliative Care Physicians. 2015. Key messages: Physician-hastened death. October. http://www.cspcp.ca/wp-content/uploads/2015/10/CSPCP-Key-Messages-FINAL.pdf.

Canadian Society of Palliative Care Physicians. 2016. How to improve palliative care in Canada—a call to action for federal, provincial, territorial, regional and local decision-makers. http://www.cspcp.ca/wp-content/uploads/2016/11/Full-Report-How-to-Improve-Palliative-Care-in-Canada-FINAL-Nov-2016.pdf.

Canadian Society of Palliative Care Physicians. 2017. Palliative care: A vital service with clear economic, health and social benefits. http://www.cspcp.ca/wp-content/uploads/2017/02/Economics-of-Palliative-Care-Final-EN.pdf.

Canadian Society of Palliative Care Physicians. 2019. Key messages: Palliative care and medical assistance in dying (MAID). https://www.cspcp.ca/wp-content/uploads/2019/05/CSPCP-Key-Messages-PC-and-MAiD-May-2019-FINAL.pdf.

Canadian Society of Palliative Care Physicians. 2020. Submission to the standing committee on justice and human rights on Bill C-7: An act to amend the criminal code (medical assistance in dying), October 27. https://www.ourcommons.ca/Content/Committee/432/JUST/Brief/BR10927610/br-external/CanadianSocietyOfPalliativeCarePhysicians-e.pdf.

Canadian Society of Palliative Care Physicians. 2022. CSPCP submission to special joint committee on medical assistance in dying April 2022. https://www.cspcp.ca/wp-content/uploads/2022/05/CSPCP-submission-to-Special-Joint-parliamentary-committee-on-expanding-MAiD-Apr-26_.pdf

Carter v Canada (Attorney General). 2015. SCC 5 [2015] 1 SCR 331 [Carter]. https://scc-csc.lexum.com/scc-csc/scc-csc/en/item/14637/index.do

Chochinov, H.M. 2002. Dignity-conserving care—A new model for palliative care: Helping the patient feel valued. *JAMA* 287 (17): 2253–2260.

Chochinov, H.M. 2021. Briefing to the Senate of Canada: Standing committee on legal and constitutional affairs regarding Bill C-7. https://sencanada.ca/content/sen/committee/432/LCJC/Briefs/HarveyMaxChochinov_e.pdf.

Chochinov, H.M., and B. Mount. 2014. Physician hastened death: Awaiting a verdict. *CMAJ Blog*. October 14. https://cmajblogs.com/physician-hastened-death-awaiting-a-verdict/.

Chochinov, H.M., D. Tataryn, J.J. Clinch, and D. Dudgeon. 1999. Will to live in the terminally ill. *Lancet* 354 (9181): 816–819.

Chochinov, H.M., K.G. Wilson, M. Enns, N. Mowchun, S. Lander, M. Levitt, and J.J. Clinch. 1995. Desire for death in the terminally ill. *American Journal of Psychiatry* 152 (8),1185–1191.

De Lima, L., R. Woodruff, K. Pettus, J. Downing, R. Buitrago, E. Munyoro, C. Venkateswaran, S. Bhatnagar, and L. Radbruch. 2017. International Association for Hospice and palliative care position statement: Euthanasia and physician-assisted suicide. *Journal of Palliative Medicine* 20 (1): 8–14.

Devos, T. (ed.). 2021. *Euthanasia: Searching for the Full Story*. Switzerland: Springer Nature. https://doi.org/10.1007/978-3-030-56795-8#about.

External Panel on Options for a Legislative Response to *Carter v. Canada* (2015). Final report [Internet]. https://www.justice.gc.ca/eng/rp-pr/other-autre/pad-amm/pad.pdf.

Finlay, I., and A. Carlile. 2011. A question of safety. *Living and Dying Well*. https://www.livingand dyingwell.org.uk/sites/default/files/LDW%20-%20Report%20-%20Safeguards.pdf.

Gallagher, R. 2020. Lack of palliative care is a failure in too many MAiD requests. Policy Options October 19. https://policyoptions.irpp.org/magazines/october-2020/lack-of-palliative-care-is-a-failure-in-too-many-maid-requests/.

Gallagher, R. 2022. Joint special committee on medical assistance in dying. Meeting 6 Evidence. https://parl.ca/DocumentViewer/en/44-1/AMAD/meeting-6/evidence.

Gallagher, R., M.J. Passmore, and C. Baldwin. 2020. Hastened death due to disease burden and distress that has not received timely, quality palliative care is a medical error. *Medical Hypotheses* 142, 109727.

Gladu, 2022. Joint special committee on medical assistance in dying. Meeting 4 Evidence. https://parl.ca/DocumentViewer/en/44-1/AMAD/meeting-4/evidence.

Grimes, D., M. Fitzpatrick, J. Gordon, et al. 2019. Canadian guideline for Parkinson disease. *Canadian Medical Association Journal* 191 (36), E989–E1004. https://www.cmaj.ca/content/191/36/E989.long.

Health Canada. 2018. Framework on palliative care in Canada. https://www.canada.ca/en/health-canada/services/health-care-system/reports-publications/palliative-care/framework-palliative-care-canada.html.

Health Canada. 2019. Action plan on palliative care—Building on the framework on palliative care in Canada. https://www.canada.ca/en/health-canada/services/health-care-system/reports-publications/palliative-care/action-plan-palliative-care.html.

Health Canada. 2020. First annual report on medical assistance in dying in Canada 2019. https://www.canada.ca/en/health-canada/services/medical-assistance-dying-annual-report-2019.html.

Health Canada. 2021. Second annual report on medical assistance in dying in Canada. https://www.canada.ca/en/health-canada/services/medical-assistance-dying/annual-report-2020.html.

Herx, L. 2019. Response to Canadian guideline for Parkinson disease—Clarification on palliative care & MAiD. *Canadian Medical Association Journal* 191 (36). https://www.cmaj.ca/content/191/36/E989/tab-e-letters#re-canadian-guideline-for-parkinson-disease---clarification-on-palliative-care--maid.

Herx, L. 2022. Brief to the Special Joint Committee on medical assistance in dying, April 24. https://www.ourcommons.ca/Content/Committee/441/AMAD/Brief/BR11740612/br-external/HerxLeonie-e.pdf

Herx, L., M. Cottle, and J. Scott. 2020. The "normalization" of euthanasia in Canada" the cautionary tale continues. *World Medical Journal* 66: 28–39.

Ho, A., J.S. Norman, S. Joolaee, K. Serota, L. Twells, and L. William. 2021. How does medical assistance in dying affect end-of-life care planning discussions? Experiences of Canadian multidisciplinary palliative care providers. *Palliative Care & Social Practice* 15: 1–14.

Kaya, E., and L. Herx. 2022. Assisted dying must not be confused for palliative care. *Globe and Mail*. https://www.theglobeandmail.com/opinion/article-assisted-dying-must-not-be-confused-for-palliative-care/.

Kotalik, J. 2020. Medical assistance in dying: challenges of monitoring the canadian program. *Canadian Journal of Bioethics* 3 (3): 202–209.

Kutcher, M. 2018. Navigating MAiD on PEI. *Canadian Medical Association*. https://www.cma.ca/dr-matt-kutcher.

Mathews, J.J., D. Hausner, J. Avery, B. Hannon, C. Zimmermann, and A. al-Awamer. 2021. Impact of medical assistance in dying on palliative care: A qualitative study. *Palliative Medicine* 35 (2), 447–454.

Mount, B. n.d. The father of palliative care in Canada, physician Balfour Mount on the legacy of Cicely Saunders, the start of palliative care, and the true meaning of medical aid in dying. https://www.mcgill.ca/palliativecare/portraits-0/balfour-mount.

Mount, B. 2020. *Ten Thousand Crossroads: The Path As I Remember It*, p234. Montreal, QC: McGill-Queen's University Press.

Munro, C., A. Romanova, C. Webber, M. Kekewich, R. Richard, and P. Tanuseputro. 2020. Involvement of palliative care in patients requesting medical assistance in dying. *Canadian Family Physician* 66 (11): 833–842.

Quality End-of-Life Care Coalition of Canada and Canadian Hospice Palliative Care Association. 2015. The way forward national framework; a roadmap for an integrated palliative approach to care. http://www.hpcintegration.ca/media/60044/TWF-framework-doc-Eng-2015-final-April1.pdf.

Radbruch, L., C. Leget, P. Bahr, and C. Müller-Busch, J. Ellershaw, F. de Conno, P.V. Berghe, and Board Members of EAPC. 2016. Euthanasia and physician-assisted suicide: A white paper from the European Association for Palliative Care. *Palliative Medicine* 30 (2), 104–116.

Shariff, M., and M. Gingerich. 2018. Endgame: Philosophical, clinical and legal distinctions between palliative care and termination of life. *Second Series Supreme Court Law Review* 85: 225–293.

Victoria State Government. 2017. Voluntary Assisted Dying Act. https://content.legislation.vic.gov.au/sites/default/files/2020-06/17-61aa004%20authorised.pdf.

Wiebe, E., J. Shaw, S. Green, K. Trouton, and M. Kelly. 2018. Reasons for requesting medical assistance in dying. *Canadian Family Physician* 64 (9): 674–679.

Dr. Leonie Herx, MD, Ph.D., CHE, CCFP (PC), FCFP is Chair of the Division of Palliative Medicine and Associate Professor in the Department of Medicine, Queen's University, a former President of the Canadian Society of Palliative Care Physicians, and Chair of the Royal College Specialty Committee in Palliative Medicine. In her clinical work she is a palliative medicine consultant, Medical Director of Palliative Care at Kingston Health Sciences Centre and Clinical Director of Palliative Care at Providence Care Hospital. She was an expert witness before the Canadian House of Commons and Senate committees examining the expansion of MAID in 2020-2021 through Bill C-7 and to the Joint Parliamentary Committee on MAID in 2022. She is a founding member of Physicians Together with Vulnerable Canadians. Leonie is passionate about advocating for the integrity of holistic palliative care and building capacity for palliative care across the healthcare system to ensure that all Canadians have access to high quality palliative care, where and when they need it.

Ebru Kaya, MB, BS, MRCP graduated from Guy's and St. Thomas's Hospitals Medical Schools, London, UK in 1998 and after her residency in internal medicine, went onto specialize in palliative medicine. Ebru is an Associate Professor of Medicine and Division Director at the University of Toronto in the Division of Palliative Medicine and the President of the Canadian Society of Palliative Care Physicians. She is also the site lead for the Toronto General Hospital palliative care team and Program Director for the subspecialty program in palliative medicine. Her passion is non-cancer palliative care, and her vision is to ensure that all Canadians have access to excellent palliative care, regardless of their disease, age, cultural, religious or socioeconomic background. She currently divides her time between provision of clinical services, teaching, and in her leadership roles for University Health Network, the Division of Palliative Medicine and the Canadian Society of Palliative Care Physicians.

Chapter 13
Suffering as a Criterion for Medical Assistance in Dying

John F. Scott and Mary M. Scott

Abstract Canada has followed the pattern of Benelux nations by legislating suffering as the pivotal eligibility criterion for euthanasia/assisted death without requiring terminal prognosis as is needed in most permissive jurisdictions. This chapter will explore the relationship between suffering and Medical Assistance in Dying (MAID) and the ways in which suffering is understood in the Supreme Court of Canada, the federal Criminal Code legislation and by health care assessors. Based on this analysis, we will argue that the resulting suffering eligibility criterion leaves the law open to unintended forms of interpretation, thus instituting perhaps the most unbounded and risk-prone form of assisted death in the world. A review of the literature on suffering, the wish to die, and reasons for MAID requests found suffering of severe illness to be frequently associated with unstable mental health, social, and existential variables that are often amenable to therapy or remediation. While suffering remains a powerful motivating force to seek opportunities to relieve distress, its use as an eligibility criterion for MAID has significant ambiguity in its' characterization of 'enduring' and 'irremediable' and thus fails to protect against abuse for the vulnerable.

Keywords Suffering criterion · Medical assistance in dying · Wish to die · Autonomy · Total pain

---

J. F. Scott (✉)
University of Ottawa, Ottawa, ON, Canada
e-mail: jscott2@uottawa.ca

M. M. Scott
Ottawa Hospital Research Institute, ON Ottawa, Canada
e-mail: mscot070@uottawa.ca

J. Kotalik and D. W. Shannon (eds.), *Medical Assistance in Dying (MAID) in Canada*, The International Library of Bioethics 104,
https://doi.org/10.1007/978-3-031-30002-8_13

13.1 Suffering and MAID in Canada's Courts, Legislature and Health Care System

The origins of Canada's use and interpretation of suffering as a criterion for Medical Assistance in Dying (MAID) can be found in the legal proceedings that culminated in the Carter v Canada (2015) decision of the Supreme Court of Canada (SCC) which declared that the Criminal Code's prohibition of assisted suicide and homicide was contrary to the Charter of Rights and Freedoms when applied to those suffering as a result of "grievous and irremediable medical condition, including illness, disease or disability" ("Carter v. Canada" 2015). The ruling specified the amendments required to remedy this violation of rights; its vocabulary and proposed suffering criterion became the framework for subsequent legislation (Canada 2016).

The judgement highlighted the story of one of the appellants, Gloria Taylor, diagnosed in 2009 with amyotrophic lateral sclerosis (ALS), a terminal progressive neurodegenerative disease. The court described Ms. Taylor's rationale for seeking assisted death:

> By 2010, Ms. Taylor's condition had deteriorated to the point that she required a wheelchair to go more than a short distance and was suffering pain from muscle deterioration. She required home support …described as an assault on her privacy, dignity, and self-esteem. She continued to pursue an independent life but… steadily losing the ability to participate fully in that life. 'I do not want my mode of death to be traumatic for my family members. I want the legal right to die peacefully, at the time of my own choosing. There will come a point when I will know that enough is enough… this is the point where life is just not worthwhile.' ("Carter v. Canada" 2015)

Gloria Taylor's 2011 affidavit focused primarily on her fearful anticipation of death and future suffering from deterioration. "I live in apprehension that my death will be slow, difficult… painful, undignified…stripped of dignity and independence…an ugly death" ("Carter v. Canada" 2015).

The theme of autonomy is prominent in her request. A good death for her is pain-free, dignified, and fast but also "at a time of my own choosing." This loss of perceived autonomy over the timing and manner of death is portrayed by the SCC as a form of suffering, seemingly the pivotal one in the mind of the jurists. In August 2012, the Superior Court of BC gave Ms. Taylor permission to access assisted death. However, she did not proceed, dying 4 months later, not of ALS or MAID, but of a sudden illness caused by colon perforation. After her death, her son is quoted as saying "It was all about having a choice; if it got too bad, then she would have the ability to do something on her own terms" (Friscolanti 2012). Taylor's action not to proceed with MAID implies that 'intolerable suffering' was not present at the time of her exemption, but rather unbearable suffering was anticipated and feared. The failure of the court to define suffering or to analyze the application of the suffering criterion in Taylor's specific case (particularly concerning present versus future suffering) played a critical role in creating Canadian legislation with a more ambiguous and unbounded suffering criterion than other jurisdictions.

Table 13.1 Criterion of suffering for MAiD

Grievous and irremediable medical condition
(1) A person has a *grievous* and irremediable medical condition only if they meet all of the following criteria: **(a)** they have a serious and incurable illness, disease or disability; **(b)** they are in an advanced state of irreversible decline in capability; **(c) that illness, disease or disability or that state of decline causes them enduring physical or psychological suffering that is intolerable to them and that cannot be relieved under conditions that they consider acceptable**; and **(d)** their natural death has become reasonably foreseeable, taking into account all of their medical circumstances, without a prognosis necessarily having been made as to the specific length of time that they have remaining

In response to SCC Carter v Canada, the Parliament of Canada in June 2016 passed legislation setting out criteria for assisted death eligibility (Canada 2016) (Table 13.1).

In 2021, Canada's Parliament further amended this section of the Criminal Code to permit those whose natural death is not in the foreseeable future to qualify for MAID but did not further change or clarify the suffering criterion ("QCCS 3792 (CanLII)" 2019).

The amendment employs a cascade of superlative adjectives to describe the medical condition and the suffering that must exist to qualify for MAID: grievous, irremediable, serious, incurable, advanced state of decline, irreversible, enduring, and intolerable. The initial impression is a law designed to be a remedy for the last resort cases—the rare exceptions where severe pain and other forms of distress persist, despite all attempts to achieve relief. However, suffering and these qualifying adjectives are not further clarified or defined by the legislation. In the debate stage, the legislators specified "suffering that is intolerable to them and that cannot be relieved under conditions that they consider acceptable." The lack of both definition and objective professional parameters for assessing suffering leaves the suffering criterion open to interpretation. Furthermore, the only requirement in the law to ensure that other treatment options have been offered and explained is that the assessor must ask if the patient is aware of palliative care. Asking a critically ill person if they are aware of the palliative approach without prior knowledge of the dying process or the clinical options to alleviate suffering may compromise free and informed consent. Some palliative care physicians have warned that patients and family are vulnerable to misinformation and may perceive the offer of assisted death as a recommendation from professional experts.

While at least 11 countries have some form of legalized assisted suicide or euthanasia/assisted death, only the three Benelux nations have a comparable suffering criterion without prognostic boundaries. Yet, Netherlands, Belgium, and Luxembourg still require physician judgment to confirm the presence of severe suffering and absence of reasonable alternatives (Rietjens et al. 2009) while physicians in Canada view their role more passively as accepting the patient's request for MAID as evidence of suffering (Mishara and Kerkhof, 2018). Federal and provincial governments, courts and physician regulatory bodies have interpreted this criminal code amendment as providing all Canadians with the 'right' to access MAID as part of a universal health care system with some calling for a duty for physicians to inform all seriously ill or disabled persons of their right to this option. While a MAID assessor must acknowledge the presence of suffering, their professional assessment of suffering plays no role in determining eligibility (H. Canada 2020). If the patient with a qualifying illness or disability says they have suffering and states they are aware of palliative care and other potential supports, they will be deemed eligible.

There is a complex tension in the dynamics of the bill: legal remedy for unrelieved suffering and/or affirmation of individual autonomy when one no longer wishes to live. While the language of the legislation might appear to limit eligibility to severe irremediable suffering, the open-ended 'intolerability' clause allows interpretation to focus entirely on individual autonomy. The centrality of suffering in the legislation may be pivotal in securing public support for what is a radical change in the law of suicide and homicide. Yet in its bedside application,' suffering' appears to have become a code-word for thwarted and restrained autonomy or threatened loss of autonomy in the future. Some argue for abandoning this 'welfare' argument (i.e., assisted death is justified for unrelieved suffering) and instead adopting an exclusively autonomy-centered view of assisted death (Davis and Mathison 2020) although public support may yet be still too grounded in the suffering rationale to tolerate this shift.

Does this concept of suffering put at risk increasingly broader expansions in eligibility? In March 2021 legislation expanded to include any serious illness or disability even if death was not in the foreseeable future. In March 2023 eligibility is scheduled to expand to patients with psychiatric diagnoses, although calls for re-evaluating this timeframe may delay its application. Based on political signals, we can expect expansion to include mature minors and those who wish to prepare advanced directives because of early dementia and other progressive illness that could impact capacity. Using the same equality under the Charter argument as Truchon vs Quebec ("QCCS 3792 (CanLII)" 2019), other forms of suffering might qualify, even beyond those associated with 'illness, disease and disability.' Is the suffering of serious illness and disability different or more severe than other forms of suffering? For example, if someone wishes to die because of their deep suffering surrounding the death of a child, and are judged to be capable, can it not be argued that such a person has the right to access MAID using the suffering criterion?

In conclusion, this passive and mercurial interpretation of suffering as a key criterion for MAID entitles Canada with having the most unbounded form of assisted death in the world.

13.1.1 The Nature of Suffering and the Wish to Die

The political rationale justifying MAID legislation relies heavily on the concept of irremediable suffering that has failed to respond to treatment and support. Furthermore, it assumes that suffering in people with illness or disability can be a stable, enduring, and irreversible reality on which a request for MAID can be based. Yet the literature on the nature and characteristics of suffering are complex and ambiguous without a consistent definition of unbearable suffering (Dees et al. 2010), casting considerable doubt on its capacity to act as a criterion for MAID eligibility. Qualitative research on the phenomenon of suffering associated with severe illness and assisted death present a set of dominant themes. These emerged when patients described specific thoughts and feelings associated with their suffering (Abraham et al. 2006; Bruce et al. 2011; Cassell 1998; Krikorian and Limonero 2012; Rodríguez-Prat et al. 2017; Sacks 2013; Yalom 1980). Particularly relevant is the research that developed themes from the subset of patients who experience their illness as suffering and distinguish them from those who do not describe their illness as suffering (Mount et al. 2007). Notably, a significant proportion of terminally ill patients do not describe their illness experience as suffering even when their symptoms or disease characteristics are similar to those who do report 'suffering.' The major themes emerging from all these studies on those with life-limiting illness are:

- Helplessness (loss of control/ autonomy, powerless, voiceless, trapped)
- Hopelessness (nothing to look forward to, even in immediate future)
- Dependence (loss of physical independence, need for care by others)
- Meaninglessness (purposeless, lack of intellectual or shared cultural framework to make sense of illness or disability.)
- Disconnection (cut-off, abandoned, alone, lonely, no sense of connection to shared community or shared meaning)
- Disintegration (falling apart, scattered, loss of personal integrity or wholeness)
- Worthlessness (not worth living, loss of self-esteem, disgust)
- Victimization (unjust/ unfair, shouldn't be happening, undeserved)
- Sense of being a burden (burden on family/friends, drain on resources)
- Happening too fast (unexpected, no warning, too soon, too young)
- Loss of Dignity
- Inability to live in the present: either stuck in the past (guilt, regrets, resentments, blame, old trauma) or stuck in future (fearful anticipation, dread, diffuse anxiety).

It is crucial in our analysis to note that most of these themes are also common to literature on depression and suicide (Chochinov 2009; Macleod 2011). Furthermore, comparable themes and vocabulary also emerge from studies of trauma and loss including bereavement, rape, incest, torture, military invasion or family breakdown (Bahnson 1985). Research to date does not allow us to make measurable associations between themes described as suffering in patients with life-threatening illness and those associated with other types of traumatic or distressing experiences. However, all the literature suggests that suffering is less about the characteristics of the external threat and more closely tied to the intrapersonal dynamic and self-perception of the meaning of the threat.

For several decades before euthanasia and assisted death were legalized, the wish to die and the desire for hastened death were being documented in some terminally-ill patient populations (Belar et al. 2021a, b; Chochinov et al. 1998; Erdmann et al. 2021; Kremeike et al. 2020; Ohnsorge et al. 2019). Since the introduction of assisted death legislation, there has been additional publication of prospective qualitative data from patients (Dees et al. 2011; Ganzini et al. 2009; Nuhn et al. 2018), as well as retrospective data from family members (Ganzini et al. 2009; Georges et al. 2007), and from providers in the form of compulsory government reports, volunteer physician surveys or analysis of physicians' notes post-mortem (Authority 2021; H. Canada 2021; Selby and Bean 2019; Simmons et al. 2022). We summarize the literature on the nature of suffering associated with assisted death by using Saunders' foundational concept of 'total pain' with its four domains of suffering: physical, psychological, social, and spiritual (Saunders 1978).

13.1.1.1 Physical

Most studies, particularly those derived directly from patients, confirm that physical symptoms and medical treatment issues are seldom the predominant reason for the wish to die or the request for assisted death (Ganzini et al. 2009; Nuhn et al. 2018; Wiebe et al. 2018). Nonetheless, these physical issues, both present and anticipated, may form the central challenge or threat to personal integrity (Dees et al. 2011). The important issue may be the perceived meaning of the cancer or the pain and the fearful anticipation of potential future threats to this integrity. The differences in MAID rates across diagnoses may be important in understanding suffering. There were 10,064 MAID deaths in Canada in 2021 representing 3.3% of all deaths, but this rate varies significantly by diagnosis (H. Canada 2021). For example, in 2020 MAID was performed in 6.3% of all cancer deaths, but only 1.6% of all cardiovascular deaths (H. Canada 2020). Similar patterns are reported in other jurisdictions (Maessen et al. 2010). The variability of MAID deaths across diagnostic categories may represent true differences in symptom, psychosocial or sociodemographic profiles between diseases (e.g., speed of disease or symptom progression may change suffering experience), but the literature suggests that cultural fears and misinformation may play an even more important role. For example, symptom burden, incapacity levels and prognosis of heart failure is equivalent in scope and severity to cancer (Bekelman et al. 2009). Yet research has demonstrated that heart failure is viewed culturally as a chronic disease with which one lives while cancer and ALS evoke deep anxiety and fear of a horrible death. This may point to the importance of public education in reducing false fears of suffering.

13.1.1.2 Psychological

Most studies document the central importance of psychological, emotional, and mental health components in the patient's rationale for assisted death. The formal diagnosis of depression and anxiety is challenging in the presence of serious physical illness since somatic symptoms (e.g., loss of appetite) and certain thought patterns (e.g., preoccupation with death, feeling a burden) are less discriminating in this setting (Rayner et al. 2009, 2011; Wilson et al. 2000a). Nonetheless, several studies have found significant levels of depression in terminally ill patients and these levels are substantially higher among those who express a desire for hastened death (Ellwood 2007; Rosenstein 2011; Wilson et al. 2000a). A substantial body of literature exists on the association between depression and the desire to die or request for a hastened death (Block and Panel 2005; Chochinov et al. 1994, 1998; Rabkin et al. 2015; Stutzki et al. 2012, 2014; Verschueren et al. 2019; Wilson et al. 2000b). Furthermore, a wish to die often co-existed with a wish to live (Bryan et al. 2016; Chochinov et al. 1999; Verschueren et al. 2019), suggesting patients simultaneously grapple with the complexity of living and dying interchangeably, a pattern common to those contemplating suicide (Heisel et al. 2016).

It is important to note the presence of literature reporting little to no depression symptomatology in those receiving MAID, yet it must also be considered that much of this is derived from retrospective MAID provider sources in contrast to the patient-derived literature (Maessen et al. 2010; Pearlman et al. 2005). Furthermore, many of the common reasons listed by patients seeking assisted death have strong links to the depression and suicide literature—sense of loneliness, feeling disconnected, sense of being a burden, loss of dignity, loss of self, hopelessness, purposelessness, worthlessness (Abbey et al. 2006). Some researchers and clinicians have employed less well-defined terms including 'demoralization' (Kissane et al. 2001, 2004) or 'existential distress' (LeMay and Wilson 2008; Watt et al. 2022) to describe the shifts in mood, thought and coping that develop in a portion of patients facing grave illness. Palliative care and psychosocial oncology professionals (Canadian Association of Psychosocial Oncology 2017) report that such thoughts and feelings occur at some point in many patients but persist only in a minority. These depression-like symptoms have been shown to fluctuate over the course of illness (Price et al. 2011), aligned with the theory that this suffering may be comparable to a grief process over both present and anticipated losses. While Canada has not yet accepted psychiatric diagnosis as the sole qualifying condition, co-existing mental health diagnoses, especially those with significant suicide risk (e.g., depression, borderline personality disorder or psychosis) may well already influence MAID requests (Brodeur et al. 2021; Nicolini et al. 2020). Further research is urgently required to examine suffering related to psychiatric disease, determining irremediability in mental health disorders and mapping ways in which co-existing mental health diagnoses impact MAID decisions in non-psychiatric qualifying illness.

13.1.1.3 Social

Social and environmental factors are elicited in several studies. Inflection points that are associated with the wish to die include increases in dependency, relapses or episodes of deterioration, and moving into a different care plan or facility (Erdmann et al. 2021). Several reasons for requesting MAID align with the social suffering literature (e.g., loneliness, isolation, disconnection, fear of losing independence or inability to live at home, feeling a burden or financial drain on family and loss of social roles (Noyes and Clancy 1977)). There is conflicting evidence as to whether socio-economic factors change the frequency of MAID requests (Verhofstadt et al. 2017). Evidence from the first three years of legalized MAID found those living in neighborhoods with lower income were less likely to receive MAID (Redelmeier et al. 2021), although descriptive results of an analysis with linked census income data found a disproportionate amount of MAID requests were from those with low socioeconomic status (Tran et al. 2021). One Canadian survey indicated those with lower socioeconomic status face multiple barriers to healthcare, were more likely to have poorer knowledge of end-of-life care practices and to have positive attitudes to MAID (Bérubé et al., 2022). Furthermore, higher rates of assisted death requests are found among those living alone, as well as those with dismissive attachment style, distrust, and avoidance of others (Oldham et al. 2011; Selby et al. 2020; Smith et al. 2015). Disabled rights organizations worldwide report that social factors make them disproportionately vulnerable to making a request for MAID because of inadequate housing, home care, disability supports in combination with a growing societal devaluing of their lives which they believe has been aggravated by efforts to promote MAID (Behuniak 2011). This same combination of poverty, inadequate alternative services and social devaluing may also place racialized, frail elderly and Indigenous peoples as well as underhoused and other vulnerable, socially disadvantaged populations at higher risk of being offered MAID and of their accepting MAID because of their perception of having no other good option. In recent months, the North American press has reported on multiple such cases (Raikin 2022) with media reports noting concerns that MAID legislation and practice needed to ensure protection of the vulnerable (Brassolotto et al. 2022). Research is urgently required to explore these concerns more rigorously.

13.1.1.4 Spiritual/Existential

Many of the themes elicited in qualitative studies of patients include items that are either centered on or have significant overlap with this spiritual or existential domain—meaning, hope, dignity, guilt, forgiveness, injustice, wholeness, peace, connection with the divine or higher power (Edwards et al. 2010). The distinction between psychological and spiritual suffering is almost impossible to draw. However, given the significance and frequency of spiritual, philosophical, or religious themes, further recognition that interventions should include spiritual care professionals and cultural communities is warranted. Risks of using assisted death to resolve a spiritual or existential crisis have been identified by many commentators. The work on Dignity therapy (Fitchett et al. 2015), Meaning therapy (Breitbart et al. 2004) and Hope therapy (Cheavens and Guter 2018) all aim at influencing a combination of existential, psychological and social factors in those facing life-threatening illness. There are many examples of the link between spirituality and death, including the 2022 report “Beginning the Journey into the Spirit World” which illustrates a culturally sensitive approach to providing palliative care to Indigenous peoples that honours and incorporates spirituality. Considering spiritual care as a treatment of suffering does not have clear guidelines; however, evidence on suffering and death suggest spirituality is a tool people use to find meaning and solace.

13.2 Suffering and Autonomy

Autonomy stands out as a dominant theme in the literature on suffering and the wish to die, transecting the four domains above. The literature suggests that the cry for autonomy at both individual and societal levels is but one of many deeper intertwined forms of suffering that demand our recognition and attention.

It has been proposed that a sharp distinction can be made between ‘wish to die or hasten death’ which arises from transient, remediable suffering and a request for MAID that arises from stable irremediable suffering (Canadian Association of Psychosocial Oncology 2017). In this theory the MAID request is a non-deviating cerebral decision based on a long-held individualistic, self-actualizing philosophy of life that prizes autonomy. In one Canadian study, almost half of patients receiving MAID described a long-standing philosophical belief in assisted death that pre-dated diagnosis of severe illness or disability (Davis and Mathison 2020; Li et al. 2017) Notably, this study was conducted in downtown Toronto during the first year of Canadian legalization when early adopters may have been more likely to have pre-existing views on assisted death (Selby et al. 2020).

On the other hand, this theory proposes that a ‘wish to die or wish to hasten death’ arises, often transiently, from psychological and existential suffering but frequently responding to psychosocial interventions (Canadian Association of Psychosocial Oncology 2017). The literature on the experience of those nearing death does not support such a sharp distinction between transient/ remediable and stable/irremediable suffering. In fact, the literature suggests that a significant proportion of persons receiving MAID may have transient and remediable forms of suffering. Research to date provides little insight on how to address autonomy-based suffering in isolation from the emotional turmoil and masked forms of hidden grief and death anxiety that are common in those at the end of life (Bachner et al. 2011; Dong et al. 2014, 2016). Attempts to distinguish personality traits most associated with specific reasons for assisted death requests may prove to be an important avenue of research (Selby et al. 2020). In the 2020 Canada MAID report, providers checked “loss of control/ autonomy/ independence” as a reason for MAID request in only

3.1% of cases (H. Canada 2021). Yet the overwhelming conclusion of the patient-derived literature is that desire for control, maintaining mental capacity, physical independence and most especially decisional choice are all frequently stated reasons to seek hastened death.

It is notable that jurisdictions permitting only assisted suicide report much lower rates of assisted death and much slower escalation profiles than nations or states that permit physician administered death in which the vast majority choose a physician lethal injection—all but 7 MAID deaths in 2021 in Canada (Borasio et al. 2019). While multiple factors may be at play, particularly trust in physicians' competence to carry out the task, this discrepancy does lead us to nuance the desire for autonomy. We need more research to unravel if some patients feel less comfortable taking responsibility for suicide. By deferring to professionals, MAID may be understood as a 'treatment' that physicians offer or recommend. Given physicians' psychological profiles including control characteristics, fear of death and sense of failure when confronted with suffering (Brooks et al. 2011; Noble et al. 2008), there may be concern that the rapid escalation of physician administered death could represent physicians' need for control and not simply the patients' desire for autonomy (Libbey 2021).

13.3 Conclusion

The spectre of severe irremediable suffering associated with serious illness or disability has provided the central rationale and justification for MAID, conjuring up the evocative image of writhing in pain and emotional despair. However, the evidence from the international literature portrays a different reality. Those patients who describe their illness experience as one of 'suffering' are not referring, predominantly, to physical symptoms but to psycho-social-existential variables that fluctuate widely over time and have been shown to be responsive to therapy and support. The research evidence does not allow a professional to assess with any confidence that such distress is irremediable, and there is evidence that patients' self-assessment changes over time. With no objective parameters to define suffering, its ambiguous, unbounded, intrapersonal nature makes it ill-suited to the task of legally distinguishing or defining the group of persons who are eligible for assisted death.

Furthermore, the suffering criterion for MAID eligibility appears to have been reinterpreted by Canadian jurists, legislators, and physicians to mean thwarted autonomy or threatened loss of autonomy in the future. When a patient requests MAID and the checklist is subsequently completed by two assessors, the legislation permits this to be designated irremediable suffering. Given the literature on the nature of suffering as transient, intrapersonal, and meaning-based, the present legislation and medical assessment regime fail to provide an effective guide for eligibility or a safeguard against abuse. In fact, the centrality of the suffering criterion in Canadian legislation and its failure to require professional confirmation of its irremediability leaves the law vulnerable to expansions based on the equality provision of the Charter of Rights and Freedoms, perhaps even beyond the boundaries of "serious and incurable illness, disease and disability".

Nonetheless, suffering continues to act as a powerful motivating force for action. It was this word, far more than autonomy, that motivated public opinion and legislators around the world to turn over centuries-old law on suicide and homicide. Focusing on the concept of suffering can continue to provide the dynamic energy required to mobilize our individual and collective compassion. Despite vigorous disagreement on vision and tactics, both MAID enthusiasts and skeptics seek compassionate responses to suffering. Research reveals the enormous potential to relieve suffering and the huge gaps in our efforts to do so. Hopefully, common ground can be found so that, together, we can lessen the burden of remediable suffering.

References

Abbey, J.G., B. Rosenfeld, H. Pessin, and W. Breitbart. 2006. Hopelessness at the end of life: The utility of the hopelessness scale with terminally ill cancer patients. *British Journal of Health Psychology* 11 (2): 173–183.

Abraham, A., J.S. Kutner, and B. Beaty. 2006. Suffering at the end of life in the setting of low physical symptom distress. *Journal of Palliative Medicine* 9 (3): 658–665.

Act to amend the Criminal Code and to make related amendments to other Acts (Medical assistance in dying, C-14), C-14 § 14 (2016).

Authority, O.H. 2021. Death with dignity act annual reports. Retrieved from Oregon, USA. https://www.oregon.gov/oha/ph/providerpartnerresources/evaluationresearch/deathwithdignityact/pages/ar-index.aspx.

Bachner, Y.G., N. O'Rourke, and S. Carmel. 2011. Fear of death, mortality communication, and psychological distress among secular and religiously observant family caregivers of terminal cancer patients. *Death studies* 35 (2), 163–187. Retrieved from http://ovidsp.ovid.com/ovidweb.cgi?T=JS&CSC=Y&NEWS=N&PAGE=fulltext&D=med5&AN=24501834.

Bahnson, C.B. 1985. The phenomenology of responses to external and internal threats of annihilation in war and disease. In *Psychiatry the State of the Art*, pp. 389–395. Springer.

Behuniak, S.M. 2011. Death with "dignity": The wedge that divides the disability rights movement from the right to die movement. *Politics and the Life Sciences* 30 (1): 17–32.

Bekelman, D.B., J.S. Rumsfeld, E.P. Havranek, T.E. Yamashita, E. Hutt, S.H. Gottlieb, J.S. Kutner, et al. 2009. Symptom burden, depression, and spiritual well-being: A comparison of heart failure and advanced cancer patients. *Journal of General Internal Medicine* 24 (5): 592–598.

Belar, A., M. Arantzamendi, Y. Santesteban, J. López-Fidalgo, M. Martinez, M. Lama, C. Centeno, et al. 2021a. Cross-sectional survey of the wish to die among palliative patients in Spain: One phenomenon, different experiences. *BMJ Supportive & Palliative Care* 11 (2): 156–162.

Belar, A., M. Martinez, C. Centeno, J. López-Fidalgo, Y. Santesteban, M. Lama, and M. Arantzamendi 2021b. Wish to die and hasten death in palliative care: a cross-sectional study factor analysis. *BMJ Supportive & Palliative Care.*

Bérubé, A., D. Tapp, S. Dupéré, A. Plaisance, G. Bravo, J. Downar, and V. Couture. 2022. Do socioeconomic factors influence knowledge, attitudes, and representations of end-of-life practices? A cross-sectional study. *Journal of Palliative Care*, 08258597221131658.

Block, S.D., and A.C. Panel. 2005. Assessing and managing depression in the terminally ill patient. *Focus* 132(2), 209–319.

Borasio, G.D., R.J. Jox, and C. Gamondi. 2019. Regulation of assisted suicide limits the number of assisted deaths. *The Lancet* 393 (10175): 982–983.

Brassolotto, J., A. Manduca-Barone, and P. Zurbrigg. 2022. Medical assistance in dying: A review of related Canadian news media texts. *Journal of Medical Humanities*, 1–20.

Breitbart, W., C. Gibson, S.R. Poppito, and A. Berg. 2004. Psychotherapeutic interventions at the end of life: A focus on meaning and spirituality. *The Canadian Journal of Psychiatry* 49 (6): 366–372.

Brodeur, J., P.S. Links, P.-E. Boursiquot, and N. Snelgrove. 2021. Medical assistance in dying for patients with borderline personality disorder: Considerations and concerns. *The Canadian Journal of Psychiatry*, 0706743721993645.

Brooks, S.K., C. Gerada, and T. Chalder. 2011. Review of literature on the mental health of doctors: Are specialist services needed? *Journal of Mental Health* 20 (2): 146–156.

Bruce, A., R. Schreiber, O. Petrovskaya, and P. Boston. 2011. Longing for ground in a ground (less) world: A qualitative inquiry of existential suffering. *BMC Nursing* 10 (1): 1–9.

Bryan, C.J., M.D. Rudd, A.L. Peterson, S. Young-McCaughan, and E.G. Wertenberger. 2016. The ebb and flow of the wish to live and the wish to die among suicidal military personnel. *Journal of Affective Disorders* 202: 58–66.

Canada, H. 2020. First annual report on medical assistance in dying in Canada, 2019. In Government of Canada Ottawa.

Canada, H. 2021. Second annual report on medical assistance in dying in Canada 2020.

Canadian Association of Psychosocial Oncology. 2017. *Role of Psychosocial Oncology in Medical Assistance in Dying (MAID): Position Statement*, pp. 1–17. Toronto, ON: CAPO.

Carter v. Canada. 2015. (Attorney General), 35591 (Supreme Court of Canada 2015).

Cassell, E.J. 1998. The nature of suffering and the goals of medicine. *Loss, Grief & Care* 8 (1–2): 129–142.

Cheavens, J.S., and M.M. Guter. 2018. Hope therapy. In *The Oxford Handbook of Hope*, 133–142.

Chochinov, H.M., ed. 2009. *Handbook of Psychiatry in Palliative Medicine*. Oxford¸USA: Oxford University Press.

Chochinov, H.M., K.G. Wilson, M. Enns, and S. Lander. 1994. Prevalence of depression in the terminally ill: Effects of diagnostic criteria and symptom threshold judgments. *American Journal of Psychiatry* 151 (4): 537–540.

Chochinov, H.M., K.G. Wilson, M. Enns, and S. Lander. 1998. Depression, hopelessness, and suicidal ideation in the terminally ill. *Psychosomatics* 39 (4): 366–370.

Chochinov, H.M., D. Tataryn, J.J. Clinch, and D. Dudgeon. 1999. Will to live in the terminally ill. *The Lancet* 354 (9181): 816–819.

Davis, J., and E. Mathison. 2020. The case for an autonomy-centred view of physician-assisted death. *Journal of Bioethical Inquiry* 17 (3): 345–356.

Dees, M., M. Vernooij-Dassen, W. Dekkers, and C. Van Weel. 2010. Unbearable suffering of patients with a request for euthanasia or physician-assisted suicide: An integrative review. *Psycho-Oncology* 19 (4): 339–352.

Dees, M.K., M.J. Vernooij-Dassen, W.J. Dekkers, K.C. Vissers, and C. Van Weel. 2011. 'Unbearable suffering': A qualitative study on the perspectives of patients who request assistance in dying. *Journal of Medical Ethics* 37 (12): 727–734.

Dong, S., P. Butow, A. Tong, M. Lovell, and M. Agar. 2014. "Longing for an end to the emotional rollercoaster": A qualitative study on patients' attitudes and experiences of multiple symptoms in advanced cancer. *Psycho-Oncology* 23: 221–222. https://doi.org/10.1111/j.1099-1611.2014.3695.

Dong, S.T., P.N. Butow, A. Tong, M. Agar, F. Boyle, B.C. Forster, M.R. Lovell, et al. 2016. Patients' experiences and perspectives of multiple concurrent symptoms in advanced cancer: A semi-structured interview study. *Supportive Care in Cancer* 24 (3): 1373–1386. https://doi.org/10.1007/s00520-015-2913-4.

Edwards, A., N. Pang, V. Shiu, and C. Chan. 2010. The understanding of spirituality and the potential role of spiritual care in end-of-life and palliative care: A meta-study of qualitative research. *Palliative Medicine* 24 (8): 753–770.

Ellwood, A.-L. 2007. The desire for hastened death among cancer patients near the end of life: Are hopelessness, depression, and quality of life mediated by the valuation of life construct? *Dissertation Abstracts International: Section B: The Sciences and Engineering* 67(11-B), 6734. Retrieved from http://ovidsp.ovid.com/ovidweb.cgi?T=JS&CSC=Y&NEWS=N&PAGE=fulltext&D=psyc5&AN=2007-99010-088.

Erdmann, A., C. Spoden, I. Hirschberg, and G. Neitzke. 2021. The wish to die and hastening death in amyotrophic lateral sclerosis: A scoping review. *BMJ Supportive & Palliative Care* 11 (3): 271–287.

Fitchett, G., L. Emanuel, G. Handzo, L. Boyken, and D.J. Wilkie. 2015. Care of the human spirit and the role of dignity therapy: A systematic review of dignity therapy research. *BMC Palliative Care* 14: 8. https://doi.org/10.1186/s12904-015-0007-1.

Friscolanti, M. 2012. Gloria Jean Taylor, 1948–2012. *MacLean's*, October 25. Retrieved from https://www.macleans.ca/society/gloria-jean-taylor-1948-2012/.

Ganzini, L., E.R. Goy, and S.K. Dobscha. 2009. Oregonians' reasons for requesting physician aid in dying. *Archives of Internal Medicine* 169 (5): 489–492.

Georges, J.-J., B.D. Onwuteaka-Philipsen, M.T. Muller, G. Van Der Wal, A. Van Der Heide, and P.J. Van Der Maas. 2007. Relatives' perspective on the terminally ill patients who died after euthanasia or physician-assisted suicide: A retrospective cross-sectional interview study in the Netherlands. *Death Studies* 31 (1): 1–15.

Heisel, M.J., E. Neufeld, and G.L. Flett. 2016. Reasons for living, meaning in life, and suicide ideation: Investigating the roles of key positive psychological factors in reducing suicide risk in community-residing older adults. *Aging & Mental Health* 20 (2): 195–207.

Kissane, D.W., D.M. Clarke, and A.F. Street. 2001. Demoralization syndrome—A relevant psychiatric diagnosis for palliative care. *Journal of Palliative Care* 17 (1): 12–21.

Kissane, D.W., S. Wein, A. Love, X.Q. Lee, P.L. Kee, and D.M. Clarke. 2004. The demoralization scale: A Report of Its Development and Preliminary Validation. *Journal of Palliative Care* 20(4), 269–276. Retrieved from http://ovidsp.ovid.com/ovidweb.cgi?T=JS&CSC=Y&NEWS=N&PAGE=fulltext&D=med5&AN=15690829.

Kremeike, K., G. Frerich, V. Romotzky, K. Boström, T. Dojan, M. Galushko, H. Pfaff, et al. 2020. The desire to die in palliative care: A sequential mixed methods study to develop a semi-structured clinical approach. *BMC Palliative Care* 19 (1): 1–12.

Krikorian, A., and J.T. Limonero. 2012. An integrated view of suffering in palliative care. *Journal of Palliative Care* 28 (1): 41–49.

LeMay, K., and K.G. Wilson. 2008. Treatment of existential distress in life threatening illness: A review of manualized interventions. *Clinical Psychology Review* 28 (3): 472–493.

Li, M., S. Watt, M. Escaf, M. Gardam, A. Heesters, G. O'Leary, and G. Rodin. 2017. Medical assistance in dying—Implementing a hospital-based program in Canada. *New England Journal of Medicine* 376 (21): 2082–2088.

Libbey, C.L. 2021. *The Relationship Between Fear of Death and Perspectives on Physician-Assisted Suicide Among Healthcare Professionals.* Alliant International University,

Macleod, S. 2011. *The Psychiatry of Palliative Medicine: The Dying Mind*: Radcliffe Publishing.

Maessen, M., J.H. Veldink, L.H. van den Berg, H.J. Schouten, G. van der Wal, and B.D. Onwuteaka-Philipsen. 2010. Requests for euthanasia: Origin of suffering in ALS, heart failure, and cancer patients. *Journal of Neurology* 257 (7): 1192–1198.

Mishara, B.L., and A. Kerkhof. 2018. Canadian and Dutch doctors' roles in assistance in dying. *Canadian Journal of Public Health* 109 (5–6): 726–728. https://doi.org/10.17269/s41997-018-0079-9.

Mount, B.M., P.H. Boston, and S.R. Cohen. 2007. Healing connections: On moving from suffering to a sense of well-being. *Journal of Pain and Symptom Management* 33 (4): 372–388.

Nicolini, M.E., J.R. Peteet, G.K. Donovan, and S.Y. Kim. 2020. Euthanasia and assisted suicide of persons with psychiatric disorders: The challenge of personality disorders. *Psychological Medicine* 50 (4): 575–582.

Noble, S.I., A. Nelson, and I.G. Finlay. 2008. Challenges faced by palliative care physicians when caring for doctors with advanced cancer. *Palliative Medicine* 22 (1): 71–76. https://doi.org/10.1177/0269216307084607.

Noyes, R., Jr., and J. Clancy. 1977. The dying role: Its relevance to improved patient care. *Psychiatry* 40 (1): 41–47.

Nuhn, A., S. Holmes, M. Kelly, A. Just, J. Shaw, and E. Wiebe. 2018. Experiences and perspectives of people who pursued medical assistance in dying: Qualitative study in Vancouver, BC. *Canadian Family Physician* 64 (9): e380–e386.

Ohnsorge, K., C. Rehmann-Sutter, N. Streeck, and H. Gudat. 2019. Wishes to die at the end of life and subjective experience of four different typical dying trajectories. A Qualitative Interview Study. *Plos One* 14 (1): e0210784.

Oldham, R.L., S.K. Dobscha, E.R. Goy, and L. Ganzini. 2011. Attachment styles of Oregonians who request physician-assisted death. *Palliative & Supportive Care* 9 (2): 123–128.

Pearlman, R.A., C. Hsu, H. Starks, A.L. Back, J.R. Gordon, A.J. Bharucha, M.P. Battin, et al. 2005. Motivations for physician-assisted suicide. *Journal of General Internal Medicine* 20 (3): 234–239.

Price, A., W. Lee, L. Goodwin, L. Rayner, R. Humphreys, P. Hansford, M. Hotopf, et al. 2011. Prevalence, course and associations of desire for hastened death in a UK palliative population: A cross-sectional study. *BMJ Supportive & Palliative Care* 1 (2): 140–148. https://doi.org/10.1136/bmjspcare-2011-000011.

Rabkin, J.G., R. Goetz, P. Factor-Litvak, J. Hupf, M. McElhiney, J. Singleton, A.C.S. Group, et al. 2015. Depression and wish to die in a multicenter cohort of ALS patients. *Amyotrophic Lateral Sclerosis and Frontotemporal Degeneration* 16(3–4), 265–273.

Raikin, A. 2022. No other options. *The New Atlantis*, December 16. Retrieved from https://www.thenewatlantis.com/publications/no-other-options.

Rayner, L., J.H. Loge, E. Wasteson, and I.J. Higginson. 2009. The detection of depression in palliative care. *Current Opinion in Supportive and Palliative Care* 3 (1): 55–60.

Rayner, L., A. Price, M. Hotopf, and I.J. Higginson. 2011. The development of evidence-based European guidelines on the management of depression in palliative cancer care. *European Journal of Cancer* 47 (5): 702–712. https://doi.org/10.1016/j.ejca.2010.11.027.

Redelmeier, D.A., K. Ng, D. Thiruchelvam, and E. Shafir. 2021. Association of socioeconomic status with medical assistance in dying: A case–control analysis. *British Medical Journal Open* 11 (5): e043547.

Rietjens, J.A., M. Schermer, A. van der Heide. 2009. Judgement of suffering in the case of a euthanasia request in The Netherlands. *Journal of Medical Ethics, 8*(35), 502–507.

Rodríguez-Prat, A., A. Balaguer, A. Booth, and C. Monforte-Royo. 2017. Understanding patients' experiences of the wish to hasten death: An updated and expanded systematic review and meta-ethnography. *British Medical Journal Open* 7 (9): e016659. https://doi.org/10.1136/bmjopen-2017-016659.

Rosenstein, D.L. 2011. Depression and end-of-life care for patients with cancer. *Dialogues in Clinical Neuroscience* 13(1), 101–108. Retrieved from http://ovidsp.ovid.com/ovidweb.cgi?T=JS&CSC=Y&NEWS=N&PAGE=fulltext&D=med5&AN=21485750.

Sacks, J.L. 2013. Suffering at end of life: A systematic review of the literature. *Journal of Hospice & Palliative Nursing* 15 (5): 286–297.

Saunders, C. 1978. *The Management of Terminal Malignant Disease*. London, UK: Edward Arnold.

Selby, D., and S. Bean. 2019. Oncologists communicating with patients about assisted dying. *Current Opinion in Supportive and Palliative Care* 13 (1): 59–63.

Selby, D., S. Bean, E. Isenberg-Grzeda, B.H.D. Bioethics, and A. Nolen. 2020. Medical assistance in dying (MAiD): a descriptive study from a Canadian tertiary care hospital. *American Journal of Hospice and Palliative Medicine®* 37(1), 58–64.

Simmons, J.-G., G. Reynolds, M. Kekewich, J. Downar, S.R. Isenberg, and D. Kobewka. 2022. Enduring physical or mental suffering of people requesting medical assistance in dying. *Journal of Pain and Symptom Management* 63(2), 244–250, e242.

Smith, K.A., T.A. Harvath, E.R. Goy, and L. Ganzini. 2015. Predictors of pursuit of physician-assisted death. *Journal of Pain and Symptom Management* 49 (3): 555–561.

Stutzki, R., U. Schneider, S. Reiter-Theil, and M. Weber. 2012. Attitudes toward assisted suicide and life-prolonging measures in Swiss ALS patients and their caregivers. *Frontiers in Psychology* 3: 443.

Stutzki, R., M. Weber, S. Reiter-Theil, U. Simmen, G.D. Borasio, and R.J. Jox. 2014. Attitudes towards hastened death in ALS: A prospective study of patients and family caregivers. *Amyotrophic Lateral Sclerosis and Frontotemporal Degeneration* 15 (1–2): 68–76.

Tran, M., K. Honarmand, R. Sibbald, F. Priestap, S. Oczkowski, and I.M. Ball. 2021. Socioeconomic status and medical assistance in dying: A regional descriptive study. *Journal of Palliative Care*, 08258597211053088.

Truchon c. Procureur général du Canada (Procureur général du Canada 2019).

Verhofstadt, M., L. Thienpont, and G.-J.Y. Peters. 2017. When unbearable suffering incites psychiatric patients to request euthanasia: Qualitative study. *The British Journal of Psychiatry* 211 (4): 238–245.

Verschueren, A., G. Kianimehr, C. Belingher, E. Salort-Campana, A. Loundou, A.-M. Grapperon, and S. Attarian. 2019. Wish to die and reasons for living among patients with amyotrophic lateral sclerosis. *Amyotrophic Lateral Sclerosis and Frontotemporal Degeneration* 20 (1–2): 68–73.

Watt, C.L., J. Lapenskie, M. Kabir, G. Lalumiere, M. Dionne, J. Rice, J. Downar, et al. 2022. Accelerated rTMS for existential distress in palliative care: A report of two cases. *Brain Stimulation: Basic, Translational, and Clinical Research in Neuromodulation* 15 (1): 197–200.

Wiebe, E., J. Shaw, S. Green, K. Trouton, and M. Kelly. 2018. Reasons for requesting medical assistance in dying. *Canadian Family Physician* 64 (9): 674–679.

Wilson, K.G., H.M. Chochinov, B.J. de Faye, and W. Breitbart. 2000a. Diagnosis and management of depression in palliative care. In *Handbook of Psychiatry in Palliative Medicine*, 25–49.

Wilson, K.G., J.F. Scott, I.D. Graham, J.F. Kozak, S. Chater, R.A. Viola, D. Curran, et al. 2000b. Attitudes of terminally ill patients toward euthanasia and physician-assisted suicide. *Archives of Internal Medicine* 160 (16): 2454–2460.

Yalom, I.D. 1980. *Existential Psychotherapy*. New York, NY, USA: Basic Books.

John F. Scott, MD, MDiv Associate Professor of Palliative Medicine, Department of Medicine, University of Ottawa, Canada.

Assisted Dr. Balfour Mount in developing the first palliative care unit at McGill University, Montreal. Trained in medicine, theology and clinical epidemiology, Dr Scott pioneered the Division of Palliative Medicine at the University of Ottawa in which he continues to practise.

Mary M. Scott, BA, MSc Masters of Epidemiology from the University of Ottawa, Mary's research experience has focused on Canadian-based end-of-life and palliative care with multiple publications of literature reviews, qualitative studies, and quantitative health services analyses.

Chapter 14
The Assessment and Relief of Suffering in the Shadow of MAID

John F. Scott and Mary M. Scott

Although the world is full of suffering, it is full also of the overcoming of it.
—Helen Keller

Abstract The chapter explores the suffering associated with MAID giving special attention to assessment and the psychological responses elicited in caregivers highlighting the need for all MAID enquiries to activate a period of intense assessment and the provision of detailed treatment alternatives. This chapter calls for a renewed commitment to compassion ('suffering together with') as the communal dynamic to relieve and assuage such suffering. Using the four domains of 'total pain' (Saunders in The management of terminal malignant disease. Edward Arnold, London, UK, 1978), we summarize the evidence for the multiple opportunities available to 'relieve often and comfort always.' Finally, we review evidence on the impact of MAID suggesting that the overall burden of suffering in Canada has increased since legalization.

Keywords (MeSH) Empathy · Needs assessment · Culture · Terminal care

14.1 The Nature of Suffering

The word 'suffering' is complex, evocative, and elusive. While belonging more in the world of philosophy and spirituality than in science or law, suffering remains at the heart of medicine's vocation (Cassell 1998). The Latin root of suffering (sub-below; ferre-to bear) presents the image of a person who is pressed upon by a heavy

J. F. Scott (✉)
University of Ottawa, Ottawa, ON, Canada
e-mail: jscott2@uottawa.ca

M. M. Scott
The Ottawa Hospital Research Institute, Ottawa, ON, Canada
e-mail: mscot070@uottawa.ca

J. Kotalik and D. W. Shannon (eds.), *Medical Assistance in Dying (MAID) in Canada*, The International Library of Bioethics 104,
https://doi.org/10.1007/978-3-031-30002-8_14

burden (Merriam-Webster 2022). Suffering is associated with a threat to integrity or wholeness, usually perceived as external (death, disease, disabling injury, rape, invasion, bereavement, divorce etc.). Research suggests that the essence of suffering is the intrapersonal self-perception rather than the characteristics of the threat itself. Thus, two persons with a terminal disease having equivalent medical and symptom profiles, may have very different experiences of suffering.

The following vignettes illustrate the diversity and complexity of the various manifestations of suffering experienced by sick patients:

Gloria Taylor, first Canadian to receive judicial permission for assisted death (yet died without using assisted death) described her fear of future suffering, "T*here is still a lot of good in my life... I intend to get every bit of happiness I can wring from what is left of my life... but I live in apprehension that my death will be slow, difficult, undignified... wracked with pain...an ugly death. I want the legal right to die peacefully, at the time of my own choosing*" (SCC 2015).

A 62-year-old man with a cancer prognosis of short years explained why he wanted MAID "*No, I'm not suffering, I'm not grieving or depressed and I don't want to talk about that kind of stuff... I have very few symptoms and I trust you guys can control any future pain. But I don't want to wait with this cancer hanging over my head. I want to go out when I am in control. The assessors need me to say that I am suffering, and I will tell them what they need to hear. But, really, we should all have the right to die when we choose*" (personal communication).

A family was asked about the suffering of a woman waiting for MAID "*Is my sister suffering ? Yes, but that started long before her cancer. Throughout her life she struggled with depression, bursts of anger, confused thoughts, isolated, unable to work, several suicide attempts but refused to see a psychiatrist. For years before cancer, she often called me to say she wanted to die. She was a troubled soul*" (personal communication).

In contrast, Dr. Balfour Mount told the story of his 30-year-old patient, whose surgery for testicular cancer was unsuccessful in preventing disease progression. Outgoing, energetic, handsome and a world-class athlete, he had always been a winner but now those close to him were devastated. Just days before he died, despite weakness and cachexia, he did not look back on his experience as one of 'suffering'. "*You know, this last year has been the best year of my life...I had a wonderful life, turned to the outside world. During this last year, I made an inner journey and it was the most amazing trip of my life*" (Mount and Flanders 2003; Gravel 2010).

14.1.1 Provider Response to Suffering: Compassion and the Goals of Medicine

When confronted with the suffering of others, caregivers can experience thoughts and feelings of grief, guilt, helplessness, and disconnection (Back et al. 2015). Whenever health care professionals are called to respond to suffering, evidence indicates high levels of avoidance, denial, intellectualization, emotional disconnection, and intensified attempts to control processes that are not controllable (Chen et al. 2019). These patterns of response are tied to high levels of professional 'burnout' and decreased resilience (Engler-Gross et al. 2020). Signals of this avoidant response include the pathologizing and medicalizing of the normal processes of human life (i.e., birthing, puberty, aging, dying, grieving, suffering) which are conceptualized as external threats, and then treated with technological medical "fixes." These intellectualized responses to suffering have been accompanied by a shift in the understanding of the goals of medicine. In the past, medicine aimed to "cure sometimes, relieve often and comfort always" by accepting limitations imposed by disease but taking pride in its potential to relieve and comfort. Now, medicine has begun to frame suffering in black and white terms embracing the responsibility to eradicate suffering, even at the expense of eradicating the one who suffers. This means providing death as a medicalized treatment. Our reading of the literature suggests this seismic shift does not arise only from a reasoned philosophical analysis but also reflects the unprocessed thoughts and feelings of providers as they struggle with helplessness, guilt, fear, and uncertainty when faced with suffering.

In searching for a more balanced and healthy response to suffering, there has been a renewed interest in the dynamic of compassion in health care. Compassion-building strategies, such as Schwartz Rounds, have demonstrated positive effects on patient care and provider health by encouraging them to process their feelings, reflect on their avoidant behaviours, share the burden of care with others, and set more realistic goals (Flanagan et al. 2020; Sinclair et al. 2016). Compassion is a rich word with a root meaning of 'suffering together with' (Etymology 2022). Compassion assumes a worldview in which suffering and the overcoming of it is shared. A compassionate community is sceptical of radical individualistic self-defining autonomy since it often masks unrecognized avoidance and internalization of past trauma. A compassionate culture seeks ways to relieve and assuage suffering but shares the burden rather than attempting to control or eradicate it.

14.1.2 Assessment of Suffering

Extensive experience exists in developing research tools to categorize and quantify complex subjective parameters such as pain (Bruera et al. 1991; Jones 2013), depression (Bagha et al. 2013; Baile et al. 2011; Gheihman et al. 2016; Scheier et al. 2015) and quality-of-life (Cohen et al. 2017; Scheier et al. 2015). New tools are being developed to measure suffering which may prove important in gathering baseline data on suffering in different populations (Krikorian et al. 2012; Ruijs et al. 2009; Streffer et al. 2009; Mack et al. 2008). These tools also map ebbs and flows and crescendo points, which have been previously described by clinicians and include events such as the relapse of cancer or the early months following a spinal cord injury (Chochinov et al. 1999). It will be important to measure how requests for MAID are correlated with these crescendo points and document the ability of evidence-based short interventions like dignity therapy to prevent, blunt, or shorten these peaks. Even more important may be their use in providing bedside clinicians with a richer set of questions for the difficult conversation toolbox (Eagar et al. 2003). However, most of these tools may prove too onerous for routine bedside assessment of gravely ill patients (Krikorian and Román 2015). The gold standard remains the meticulous, individualized assessment by an experienced multidisciplinary team that includes nurse, social worker, spiritual care practitioner and physician. However, their skills can be honed by systematized checklists, practice guidelines and conversation guides (Howell et al. 2015; Bernacki and Block 2014; Schulz et al. 2007; Lesho 2003; Cassell 1998; Mount et al. 2007; Canadian Association of Psychosocial Oncology 2017).

Deep listening to the story of one who suffers using both verbal and non-verbal cues can focus the professional's ability to communicate empathy, congruence (i.e., 'being real') and 'unconditional positive regard' which is pivotal in establishing a healing connection (Mount and Flanders 2003; Kirschenbaum and Jourdan 2005; Cassell 1999; Rogers 1957). In the clinical encounter between the suffering one ('patient') and caregiver, assessment <u>is</u> therapy.

14.1.3 Assessment of Suffering and Referral for MAID

There is an added urgency when an assessment occurs in the context of a patient request or enquiry about MAID. Analysis of MAID assessor checklists and final consultation notes indicates a dearth of detail and specificity which appears to indicate a lack of thorough, meticulous, and systematic review of patients' sources of suffering. This underlines the need to avoid premature referral to a MAID provider. The two MAID assessments are designed to be external reviews to determine if minimal legal requirements, including capacity and consent, have been met. It assumes that the primary care team has engaged in repeated conversations to both understand the patient distress ('peeling the layers of the onion' of their suffering), and to demonstrate active compassion, including providing strategies to relieve sources of suffering. Too often, this process is truncated by an immediate referral to a MAID provider which may be interpreted by the patient as a recommendation or an indication that other ways to alleviate suffering are not available. Discussing options and providing compassionate care may be particularly important to inform and guide patients and families without prior experience or knowledge of the healthcare system and end-of-life care. Early referral may betray the discomfort many professionals feel about such conversations. Patients requesting MAID who are resistant to discuss feelings or deny specific sources of suffering are particularly challenging for the clinician. There are ways to engage in meaningful discussions, even with resistant patients and family members. For example, patients may 'project' their own feelings unto loved ones and an exploration of interactions between patient and family may open new channels of communication, ultimately assisting the patient to discover new coping mechanisms and relational strengths. Exploring patients' memories of what they are most proud and their hoped-for legacy, may reveal new insights on their suffering and not uncommonly provides the patient with a new perspective on the meaning of their life and death.

14.2 Strategies to Relieve Suffering

> She can take her own life prematurely…or she can suffer until she dies from natural causes. The choice is cruel. (Carter v. Canada—Attorney General 2015)

The Supreme Court of Canada judgement in support of assisted death claims that the suffering person had only two options prior to MAID legislation—to commit suicide when they still were able to do so or to be "condemned to a life of severe and intolerable suffering" until death from natural causes. We propose to demonstrate that the evidence does not support this stark choice between hastened death and intolerable suffering. Research in this field does not substantiate that the suffering of advanced disease is irremediable. Before legislating further expansions of MAID as a method to manage suffering, access must be ensured to interventions and supports known to be effective while continuing to explore innovative and alternate responses to suffering. There is an enormous body of literature on active, positive ways to respond compassionately to relieve suffering (Dunlop 2012; Schulz et al. 2007). Training programs are already available to support providers as they learn to recognize suffering, communicate empathy, encourage patients to share their struggles, and thereby decrease the burden (Schulz et al. 2007; Dunlop 2012; Cassell 1999).

14.2.1 Physical

Paradoxically, we are expanding legalized forms of hastened death for unrelieved suffering at a time when far more options (treatments, medications, technologies, methodologies) to relieve physical symptoms are available than ever before (Bruera et al. 2015). In the last 50 years, palliative care has made huge strides in decreasing physical distress and reshaping the dying process. Yet, despite Canada being one of the global pioneers of palliative care, the Economist Quality of Death report points to gradual deterioration in our global ranking due to problems of access, off-loading of costs to the sick, and low public awareness (Economist 2010). These gaps highlight huge opportunities to decrease end-of-life suffering by recognizing palliative care as a public health priority both locally and globally (Abel and Kellehear 2022).

There are increasingly optimistic possibilities of treating symptoms and enhancing function in non-terminal illness through multidisciplinary pain clinics, rehabilitation and disability care (McMahon et al. 2013; Struhkamp 2005); yet again, the obstacles are not lack of solutions but inadequate resources. Improvements in management of chronic diseases and organ failure have not only extended life but also decreased symptom burden. There is increasing evidence that prognostication tools combined with targeted patient conversations can produce better decisions in the face of serious illness (Bernacki and Block 2014; White et al. 2016). Western medicine rarely places limits on costly and often futile attempts to extend life when death is imminent. Far too many patients receive "meddlesome medicine", forms of overtreatment for grave illness with a short prognosis, sometimes due to inappropriate physician zeal but also driven by radical interpretations of autonomy and fueled by patient misinformation. We ourselves have witnessed, in the last 5 years, multiple patients with serious illness 'saved' from imminent death by aggressive treatment who then turn around and request MAID. To decrease unnecessary suffering, providers must shift attitudes towards natural death by using tools to clarify advance directives and re-shape medical culture and law concerning offering futile treatment.

14.2.2 *Psychological*

Dignity Therapy (Chochinov et al. 2002) and Meaning Therapy (Breitbart et al. 2004; Breitbart and Heller 2003) are two examples of intervention therapies that provide research evidence of positive impacts on mental health, global quality of life and suffering including reversing patients' wish to die. This suggests that a holistic, multi-domain approach to therapy may prove even more fruitful than attempts to target the biochemistry of demoralization and depression through new antidepressants (Rayner et al. 2011), psychedelics (Schimmel et al. 2021), amphetamines (Hardy 2009), ketamine (Goldman et al. 2019) and transcranial magnetic stimulation (Baeken 2018; Watt et al. 2022). The philosophy and techniques of Cognitive Behavioural Therapy (CBT) can support the patient to reframe their thoughts and feelings, shift cognitive distortions, such as fear of inevitable cruel death, and hone coping strategies that refocus on the present (Mannix et al. 2006; Anderson et al. 2008). The Canadian Association of Psychosocial Oncology published a framework for assessing and treating the desire for hastened death as a "manifestation of despair …most often transient, triggered by stressful illness events and responsive to increased psychosocial and supportive care" (Canadian Association of Psychosocial Oncology 2017). Instability with gradual decreases in the desire for hastened death also was reported in a UK palliative care population (Price et al. 2011). Furthermore, there is evidence that serious illness and new disabilities can precipitate a grieving process comparable in shape and intensity to the loss of partner or child (Livneh and Antonak 2005; Barrio et al. 2016; Prigerson and Maciejewski 2008). Many of the psychological dimensions of suffering, including a wish to die, may be integral components of this grief 'work' that is a necessary pathway to healing and re-establishing integrity or coming to a state of acceptance (Prigerson and Maciejewski 2008). A minority of patients will develop symptoms that meet the criteria of clinical depression but many more will have comparable peaks and valleys in their thoughts and feelings as they process their sense of loss. Supportive accompaniment through the crescendos of painful grief has been reported to be followed by a re-establishment of a desire to live until death (Rammant et al. 2021).

14.2.3 Social

Canada has significant gaps in affordable housing options, home care and chronic care facilities. Our sophisticated acute care system saves us from acute complications but then, too often, is unable to provide the ongoing social support that illness or disability demands (Severino 2020). There are media reports of MAID requests arising from gaps in social support or housing (Favaro 2022a, b; Hopper 2022). We must desist from increasingly off-loading community health care costs to vulnerable patients and families. The knowledge that these gaps are contributing to end-of-life suffering opens huge opportunities for governments and professionals to decrease suffering through system re-design and funding. The Compassionate Communities movement (Tompkins 2018) has demonstrated that the entire community can engage in this re-imaging of 'living until we die' including targeted public education initiatives such as the popular book 'With the End in Mind: How to Live and Die Well" (Mannix 2017) and the 'Last Aid Course' which aims to diminish fear and provide practical training to counter the view of dying as cruel and useless (Tompkins 2018; Bollig et al. 2019).

14.2.4 Spiritual and Existential

When faced with a terminal prognosis or any serious illness, many patients report an abrupt shift in their sense of reality and a new interest in exploring a philosophical understanding of human existence (Schulz et al. 2007). From its origins, the hospice literature has emphasized the centrality of 'spiritual pain' and its interaction with physical and other domains of suffering (Saunders 1988). The philosophy and interdisciplinary team techniques of hospice palliative care have been successful in dealing with spiritual pain, which need to be widely taught and incorporated into care for all seriously ill patients (Doka 2011). The work of Chochinov and Brietbart bridge psychological and spiritual by focusing on dignity, meaning, legacy, memories, and connections (Chochinov et al. 2002; Breitbart et al. 2004). Through clinical experience, suffering often elucidates the spiritual, metaphysical and philosophical dimensions of life-purpose, meaning, gratitude, love, fear, guilt, forgiveness, shame, blame, and hope (Breitbart and Heller 2003; Cheavens and Guter 2018). The concept of 'lamentation' can assist in framing and responding to the complex, ambivalent cry for death while accompanying the seriously ill. This deep cry of the human spirit is also heard at a communal level in our struggles with dying and suffering (Scott 1992). Religion and the spiritual wisdom traditions can provide supportive communities, intellectual perspectives to restore meaning and reframe hope for many. Spiritual techniques of meditation, prayer and communal worship can be used to find inner peace or to detach, relinquish, overcome, redeem or transcend suffering (Nelson et al. 2002; Breitbart and Heller 2003).

14.3 The Impact of MAID on Suffering

Media reports and early research point to increased satisfaction in some patients, families, and professional staff because of MAID participation. However, these and other studies also demonstrate that a proportion of family members and caregivers have increased psychological and moral distress because of secondary participation in the MAID experience (Joolaee et al. 2022; Thangarasa et al. 2022; Pesut et al. 2020; Ganzini et al. 2009; Wagner et al. 2012; Dees et al. 2013). MAID is now a dominant focus of media stories about death, at times providing an idealized portrait of the last hours of life for a MAID patient (Grant 2018; Brassolotto et al. 2022). While current research evidence is scant, there are indicators provided by caregivers that the last 2–4 weeks of life for a patient who chooses MAID may be more fraught with distress than initial media reports would indicate, due to increased anxiety and turmoil precipitated by the assessment process itself (Herx et al. 2020; Leiva et al. 2018; Devos 2021; Koch 2022). These patients, as opposed to those who do not opt for MAID, may face opposition or ambivalence from family, friends, and religious and cultural communities. Their fear of losing mental capacity has frequently led to inadequate care, most specifically the inadequate control of pain and other symptoms, because medications are being held back out of fear of somnolence or delirium (Gallagher 2020; Thangarasa et al. 2022). While some patients who prize autonomy may be relieved, others can demonstrate ongoing ambivalence, angst, and anxiety. This can be a time of sudden telescoped grief, but one further complicated by attempts to hide or deny grief by both patient and family in case it demonstrates ambivalence or lack of support.

More importantly, there is a lack of adequate research evidence on the impact of MAID on the majority (over 95%) of patients who die without MAID or continue to live with non-terminal disease or disability without MAID. Anecdotal reports from providers and an analysis of media accounts suggest that since MAID legalization there is increasing fear of the natural process of dying experienced by patients, families and the general public (Brassolotto et al. 2022). Patients may fret over a potential choice that may have no bearing on their circumstances, over increased feelings of guilt about family burden if they choose natural death, and may feel further pressured by providers who offer MAID without the patient broaching the topic (Chochinov 2022). Professionals report increasing frequency of family distress and anger when loss of capacity rules out the MAID option. Clinicians report this distress is based on the family's media-derived belief that natural death will always be painful and that waiting for death is cruel and useless (Herx et al. 2020). These case reports and clinical experience lead us and some other bedside clinicians to conclude that the overall level of suffering and distress in terminal illness has increased since the introduction of MAID. The use of palliative sedation has rapidly increased since MAID legalization in some Canadian centers (Nolen et al. 2022; College De Medecins du Quebec 2022, 2020) following a similar pattern in Netherlands where 18.3% of all deaths in 2015 were with palliative sedation (Rietjens et al. 2019). While designed as a symptom control measure, 'palliative sedation' may be utilized by some physicians as a non-regulated method of hastening death (Scott 2015; Koksvik et al. 2020). There are also reports suggesting attitudes and practice are changing in ICU in that patients withdrawn from life support may be given additional active interventions to hasten their death (Andersen et al. 2022, 2021; Nolen et al. 2022). Thus, the institution of MAID may have a domino effect on the way physicians behave in other end-of-life scenarios.

Even more challenging to measure or document are the shifts in cultural values and norms caused by or associated with MAID that some predict will greatly increase the overall burden of suffering (Snead 2020). These include:

- Increasing support for the concept of rational suicide, the impending expansion of MAID to psychiatric diagnoses and dementia, its impact on suicide prevention, and decreased societal empathy for chronic mental illness and cognitive impairment (Albert Jones 2018).
- Decreasing respect, equality, and funding for disabled, sick and elderly persons (Wallace 2017).
- Increasing fear of death and the belief that the dying process is valueless and undignified, thus undermining the last 40 years of palliative care advocacy.
- Increasing shift in both Law and Medicine towards radical individual autonomy and away from a community-focused 'common good' approach, thereby placing vulnerable, marginalized populations at higher risk of harm since they have far fewer alternative options (Sprung et al. 2018).
- Negative impacts on patients who do not wish for a hastened death including a distrust and fear of palliative care and decreased incentives for governments to improve palliative care resources (Mathews et al. 2021; Herx et al. 2020; Li et al. 2017).

14.4 Conclusion

While respectfully accepting the legal right of the patient to request MAID, health care providers should not hesitate to propose and recommend interventions to relieve distress. Compassion does not mean neutrality. In our two chapters we have reviewed the existing evidence indicating that patients' wish to die may ebb and flow throughout the illness trajectory and may respond to various therapies. This knowledge should lead professionals to actively propose alternatives to MAID. This should not be interpreted as blocking access but exploring the possibility of a solution to minimize a patient's suffering without ending their life, always remaining cognizant that MAID is a final and gravely irreversible action. While honouring the autonomy of the patient, the provider's role is to propose options for relief, support, encouragement and 'suffering together with'. Our stance should be one of active waiting; giving time and permission for the patient's lament to be heard and for the grief work to be accomplished. There is also important 'work' that the professional must accomplish during this period of active waiting, by resisting the temptation to 'fix' the problem in order to decrease their own emotional angst and their sense of powerlessness or vulnerability that can arise when the caregiver must face the uncontrollable.

In Canada, medical intervention has expanded to include MAID but the full impact of this monumental shift in the medical and social mores of our culture is only now being appreciated. Somerville, the Australian-Canadian secular ethicist, calls us to frame MAID less as a triumph of human rights and more as a crisis of the human 'spirit' (Somerville 2014). She describes how 'death-talk' and the push to legalize assisted death may arise from a secular world that is searching for shared meaning. From this perspective, MAID is a signal of societal suffering and a communal lament. We have presented the literature on the end-of-life period to highlight the tumultuous, dynamic, emotional experience of patients while also cataloguing many strategies available to relieve and assuage their distress. Comparable to birthing and aging, the suffering associated with illness cannot be entirely avoided. How we assess and respond to suffering is integral to how our society defines meaning and value. Accompanying a fellow human in their sickness and death is both a privilege and a burden that holds meaning in our social and philosophical understanding of human existence. We must listen deeply to the individual and collective cry 'help me to die', discern its roots, respond with compassion and innovative action, and continue to engage in political dialogue on the values we wish to uphold and protect.

References

Abel, Julian, and Allan Kellehear. 2022. *Public health palliative care: Reframing death, dying, loss and caregiving*. London, England: SAGE Publications.

Albert Jones, David. 2018. Assisted dying and suicide prevention. *Journal of Disability & Religion* 22 (3): 298–316.

Andersen, Sarah K., James Mellett, Marghalara Rashid, Samuel Stewart, Brendan Leier, Derek R. Townsend, and Daniel Garros. 2022. Has medical assistance in dying changed end-of-life care in the ICU? A qualitative survey of Canadian intensivists. *Journal of Critical Care* 72: 154136.

Andersen, Sarah K., Samuel Stewart, Brendan Leier, Lynn E. Alden, Derek R. Townsend, and Daniel Garros. 2021. Hastening death in Canadian ICUs: End-of-life care in the era of medical assistance in dying. *Critical Care Medicine* 50 (5): 742–749.

Anderson, T., M. Watson, and R. Davidson. 2008. The use of cognitive behavioural therapy techniques for anxiety and depression in hospice patients: A feasibility study. *Palliative Medicine* 22 (7): 814–821. https://doi.org/10.1177/0269216308095157.

Back, Anthony L., Cynda H. Rushton, Alfred W. Kaszniak, and Joan S. Halifax. 2015. "Why are we doing this?": Clinician helplessness in the face of suffering. *Journal of Palliative Medicine* 18 (1): 26–30.

Baeken, Chris. 2018. Accelerated rTMS: a potential treatment to alleviate refractory depression. *Frontiers in Psychology*: 2017.

Bagha, S.M., A. Macedo, L.M. Jacks, C. Lo, C. Zimmermann, G. Rodin, and M. Li. 2013. The utility of the Edmonton Symptom Assessment System in screening for anxiety and depression. *European Journal of Cancer Care* 22 (1): 60–69.

Baile, W.F., J.L. Palmer, E. Bruera, and P.A. Parker. 2011. Assessment of palliative care cancer patients' most important concerns. *Supportive Care in Cancer* 19 (4): 475–481. https://doi.org/10.1007/s00520-010-0839-4. http://ovidsp.ovid.com/ovidweb.cgi?T=JS&CSC=Y&NEWS=N&PAGE=fulltext&D=emed13&AN=361540917.

Barrio, Concepcion, Mercedes Hernandez, and Lizbeth Gaona. 2016. The family caregiving context among adults with disabilities: A review of the research on developmental disabilities, serious mental illness, and traumatic brain injury. *Journal of Family Social Work* 19 (4): 328–347.

Bernacki, Rachelle E., and Susan D. Block. 2014. Communication about serious illness care goals: A review and synthesis of best practices. *JAMA Internal Medicine* 174 (12): 1994–2003.

Bollig, Georg, Frans Brandt, Marius Ciurlionis, and Boris Knopf. 2019. Last aid course. An education for all citizens and an ingredient of compassionate communities. Healthcare.

Brassolotto, Julia, Alessandro Manduca-Barone, and Paige Zurbrigg. 2022. Medical assistance in dying: A review of related Canadian News Media Texts. *Journal of Medical Humanities*: 1–20.

Breitbart, William, Christopher Gibson, Shannon R. Poppito, and Amy Berg. 2004. Psychotherapeutic interventions at the end of life: A focus on meaning and spirituality. *The Canadian Journal of Psychiatry* 49 (6): 366–372.

Breitbart, William, and Karen S. Heller. 2003. Reframing hope: Meaning-centered care for patients near the end of life. *Journal of Palliative Medicine* 6 (6): 979–988.

Bruera, Eduardo, Irene Higginson, Charles F Von Gunten, and Tatsuya Morita. 2015. *Textbook of palliative medicine and supportive care*. CRC Press.

Bruera, Eduardo, Norma Kuehn, Melvin J Miller, Pal Selmser, and K Macmillan. 1991. The Edmonton Symptom Assessment System (ESAS): A simple method for the assessment of palliative care patients. *Journal of Palliative Care.*

Canadian Association of Psychosocial Oncology, CAPO. 2017. *Role of psychosocial oncology in medical assistance in dying (MAID): Position statement*. Toronto, ON: CAPO.

Carter v. Canada (Attorney General). 2015. In *Supreme Court Judgments: Constitutional law*: Supreme Court of Canada.

Cassell, Eric J. 1998. The nature of suffering and the goals of medicine. *Loss, Grief & Care* 8 (1–2): 129–142.

———. 1999. Diagnosing suffering: a perspective. *American College of Physicians*.

Cheavens, Jennifer S., and Madison M Guter. 2018. Hope therapy. In *The Oxford handbook of hope*, 133–142.

Chen, Chuqian, Amy YM. Chow, and Suqin Tang. 2019. Professional caregivers' bereavement after patients' deaths: A scoping review of quantitative studies. *Death Studies* 43 (8): 500–508.

Chochinov, H.M., T. Hack, S. McClement, L. Kristjanson, and M. Harlos. 2002. Dignity in the terminally ill: A developing empirical model. *Social Science & Medicine* 54 (3): 433–43. http://ovidsp.ovid.com/ovidweb.cgi?T=JS&CSC=Y&NEWS=N&PAGE=fulltext&D=med4&AN=11824919.

Chochinov, Harvey Max. 2022. *The platinum rule: A new standard for person-centered care*. Mary Ann Liebert, Inc., publishers 140 Huguenot Street, 3rd Floor New

Chochinov, Harvey Max, Douglas Tataryn, Jennifer J. Clinch, and Deborah Dudgeon. 1999. Will to live in the terminally ill. *The Lancet* 354 (9181): 816–819.

Cohen, S Robin, Richard Sawatzky, Lara B. Russell, Javad Shahidi, Daren K. Heyland, and Anne M. Gadermann. 2017. Measuring the quality of life of people at the end of life: The McGill quality of life questionnaire-revised. *Palliative Medicine* 31 (2): 120–129.

College De Medecins du Quebec. 2020. *De la Commission sur les soins de fin de vie: Rapport Annuel D'Activitiés*. https://www.recherchesoinspalliatifs.ca/wp-content/uploads/2020/10/RAA_2019-2020_Commission-soins-fin-de-vie-2020-10-20.pdf.

———. 2022. *Rapport Annuel 2021/2022 Concernant les soins de fin de vie*. Accessed January 15, 2023. http://www.cmq.org/pdf/outils-fin-de-vie/rapa-csfv-2021-2022-20220425.pdf.

Dees, Marianne K., Myrra J. Vernooij-Dassen, Wim J. Dekkers, Glyn Elwyn, Kris C. Vissers, and Chris van Weel. 2013. Perspectives of decision-making in requests for euthanasia: A qualitative research among patients, relatives and treating physicians in the Netherlands. *Palliative Medicine* 27 (1): 27–37.

Devos, Timothy. 2021. *Euthanasia: Searching for the full story: Experiences and insights of Belgian doctors and nurses*. Springer Nature.

Doka, Kenneth J. 2011. Religion and spirituality: Assessment and intervention. *Journal of Social Work in End-of-Life & Palliative Care* 7 (1): 99–109.

Dunlop, Robert. 2012. *Cancer: palliative care.* Springer Science & Business Media.

Eagar, Kathy, Kate Senior, Dave Fildes, Karen Quinsey, Alan Owen, Heather Yeatman, Rob Gordon, and Natasha Posner. 2003. *The caring communities evaluation tool kit: A compendium of tools to aid in the evaluation of palliative care projects.* Wollongong: Centre for Health Service Development, University of Wollongong.

Economist, The. 2010. Ranking care for the dying: Quality of death. https://www.economist.com/news/2010/07/14/quality-of-death.

Engler-Gross, Adi, Gil Goldzweig, Ilanit Hasson-Ohayon, Rony Laor-Maayany, and Michal Braun. 2020. Grief over patients, compassion fatigue, and the role of social acknowledgment among psycho-oncologists. *Psycho-Oncology* 29 (3): 493–499.

Etymology. compassion (n.). *Online Etymology Dictionary.* 2022. Taken from: https://www.etymonline.com/word/compassion#etymonline_v_17258.

Favaro, A. 2022a. Woman with chemical sensitivities chose medically-assisted death after failed bid to get better housing, April 14. https://www.ctvnews.ca/health/woman-with-chemical-sensitivities-chose-medically-assisted-death-after-failed-bid-to-get-better-housing-1.5860579.

———. 2022b. Woman with disabilities approved for medically assisted death relocated thanks to 'inspiring' support, May 28. https://www.ctvnews.ca/health/woman-with-disabilities-approved-for-medically-assisted-death-relocated-thanks-to-inspiring-support-1.5921893.

Flanagan, E., R. Chadwick, J. Goodrich, C. Ford, R. Wickens. 2020. Reflection for all healthcare staff: A national evaluation of Schwartz rounds. *Journal of Interprofessional Care* 34 (1): 140–142. https://doi.org/10.1080/13561820.2019.1636008.

Gallagher, Romayne. 2020. Lack of palliative care is a failure in too many MAiD requests. *Policy Options.*

Ganzini, Linda, Elizabeth R. Goy, Steven K. Dobscha, and Holly Prigerson. 2009. Mental health outcomes of family members of Oregonians who request physician aid in dying. *Journal of Pain and Symptom Management* 38 (6): 807–815.

Gheihman, G., C. Zimmermann, A. Deckert, P. Fitzgerald, A. Mischitelle, A. Rydall, A. Schimmer, L. Gagliese, C. Lo, and G. Rodin. 2016. Depression and hopelessness in patients with acute leukemia: the psychological impact of an acute and life-threatening disorder. *Psycho-Oncology*: 979–989. https://doi.org/10.1002/pon.3940. https://doi.org/10.1002/(ISSN)1099-1611.

Goldman, Nathaniel, Michael Frankenthaler, and Lidia Klepacz. 2019. The efficacy of ketamine in the palliative care setting: A comprehensive review of the literature. *Journal of Palliative Medicine* 22 (9): 1154–1161.

Grant, Kelly. 2018. Medically assisted death allows couple married almost 73 years to die together. *The Globe and Mail.* Accessed March 11, 2022. https://www.theglobeandmail.com/canada/article-medically-assisted-death-allows-couple-married-almost-73-years-to-die/.

Gravel, P. 2010. L'entrevue - La médecine désarmée devant la mort. *Le Devoir*. https://www.ledevoir.com/societe/sante/282228/l-entrevue-la-medecine-desarmee-devant-la-mort.

Hardy, Susan E. 2009. Methylphenidate for the treatment of depressive symptoms, including fatigue and apathy, in medically ill older adults and terminally ill adults. *The American Journal of Geriatric Pharmacotherapy* 7 (1): 34–59.

Herx, Leonie, Margaret Cottle, and John Scott. 2020. The "Normalization" of Euthanasia in Canada: The cautionary tale continues. *World Medical Journal* 66: 28–34.

Hopper, T. 2022. FIRST READING: Fears that the Canadian health system is far too quick to prescribe death. *National Post*, May 3. Accessed May 31, 2022. https://nationalpost.com/news/canada/first-reading-fears-that-the-canadian-health-system-is-far-too-quick-to-prescribe-death.

Howell, D., H. Keshavarz, M.J. Esplen, T. Hack, M. Hamel, J. Howes, J. Jones, M. Li, D. Manii, and D. McLeod. 2015. Pan-Canadian practice guideline: screening, assessment and management of psychosocial distress, depression and anxiety in adults with cancer. *Canadian Partnership Against Cancer, Canadian Association of Psychosocial Oncology.*

Jones, S. 2013. Pain in cystic fibrosis: Assessment. *Pediatric Pulmonology* 48: 162–164. https://doi.org/10.1002/ppul.22896. http://ovidsp.ovid.com/ovidweb.cgi?T=JS&CSC=Y&NEWS=N&PAGE=fulltext&D=emed15&AN=71321309.

Joolaee, Soodabeh, Anita Ho, Kristie Serota, Matthieu Hubert, and Daniel Z. Buchman. 2022. Medical assistance in dying legislation: Hospice palliative care providers' perspectives. *Nursing Ethics* 29 (1): 231–244.

Kirschenbaum, Howard, and April Jourdan. 2005. The current status of Carl Rogers and the person-centered approach. *Psychotherapy: Theory, Research, Practice, Training* 42 (1): 37.

Koch, Tom. 2022. A sceptics report: Canada's five years experience with medical termination (MAiD). HEC Forum.

Koksvik, Gitte Hanssen, Naomi Richards, Sheri Mila Gerson, Lars Johan Materstvedt, and David Clark. 2020. Medicalisation, suffering and control at the end of life: The interplay of deep continuous palliative sedation and assisted dying. *Health.* 1363459320976746.

Krikorian, A., J.T. Limonero, and J. Mate. 2012. Suffering and distress at the end-of-life. *Psycho-Oncology* 21 (8): 799–808. https://doi.org/10.1002/pon.2087. https://doi.org/10.1002/(ISSN)1099-1611.

Krikorian, A., and J.P. Román. 2015. Current dilemmas in the assessment of suffering in palliative care. *Palliative & Supportive Care* 13 (4): 1093–1101.

Leiva, R., M. Cottle, C. Ferrier, S.R. Harding, T. Lau, T. McQuiston, and J.F. Scott. 2018. Euthanasia in Canada: A cautionary tale. *World Medical Journal* 64 (3): 17–23.

Lesho, Emil P. 2003. When the spirit hurts: An approach to the suffering patient. *Archives of Internal Medicine* 163 (20): 2429–2432.

Li, Madeline, Sarah Watt, Marnie Escaf, Michael Gardam, Ann Heesters, Gerald O'Leary, and Gary Rodin. 2017. Medical assistance in dying—implementing a hospital-based program in Canada. *New England Journal of Medicine* 376 (21): 2082–2088.

Livneh, Hanoch, and Richard F. Antonak. 2005. Psychosocial adaptation to chronic illness and disability: A primer for counselors. *Journal of Counseling & Development* 83 (1): 12–20.

Mack, Jennifer W., Matthew Nilsson, Tracy Balboni, Robert J. Friedlander, Susan D. Block, Elizabeth Trice, and Holly G. Prigerson. 2008. Peace, Equanimity, and Acceptance in the cancer experience (PEACE) validation of a scale to assess acceptance and struggle with terminal illness. *Cancer: Interdisciplinary International Journal of the American Cancer Society* 112 (11): 2509–2517.

Mannix, Kathryn. 2017. *With the end in mind: Death, dying, and wisdom in an age of denial.* William Collins.

Mathews, Jean Jacob, David Hausner, Jonathan Avery, Breffni Hannon, Camilla Zimmermann, and Ahmed Al-Awamer. 2021. Impact of medical assistance in dying on palliative care: A qualitative study. *Palliative Medicine* 35 (2): 447–454.

McMahon, Stephen, Martin Koltzenburg, Irene Tracey, and Dennis C Turk. 2013. *Wall & melzack's textbook of pain e-book*. Elsevier Health Sciences.

Merriam-Webster, Dictionary. 2022. Suffer. Merriam-Webster.com.

Mount, Balfour M., Patricia H. Boston, and S. Robin Cohen. 2007. Healing connections: On moving from suffering to a sense of well-being. *Journal of Pain and Symptom Management* 33 (4): 372–388.

Mount, Balfour M., and E.M. Flanders. 2003. Existential suffering and the determinants of healing. *European Journal of Palliative Care* 10 (2 SUPP): 40–42.

Nelson, Christian J., Barry Rosenfeld, William Breitbart, and Michele Galietta. 2002. Spirituality, religion, and depression in the terminally ill. *Psychosomatics* 43 (3): 213–220.

Nolen, Amy, Rawaa Olwi, and Selby Debbie. 2022. Impact of legalization of medical assistance in dying on the use of palliative sedation in a tertiary care hospital: A retrospective chart review. *American Journal of Hospice and Palliative Medicine®* 39 (4): 442–447.

Pesut, Barbara, Sally Thorne, Catharine J. Schiller, Madeleine Greig, and Josette Roussel. 2020. The rocks and hard places of MAiD: A qualitative study of nursing practice in the context of legislated assisted death. *BMC Nursing* 19 (1): 1–14.

Price, A., W. Lee, L. Goodwin, L. Rayner, R. Humphreys, P. Hansford, N. Sykes, B. Monroe, I. Higginson, and M. Hotopf. 2011. Prevalence, course and associations of desire for hastened death in a UK palliative population: A cross-sectional study. *BMJ Supportive & Palliative Care* 1 (2): 140–148. https://doi.org/10.1136/bmjspcare-2011-000011.

Prigerson, Holly G., and Paul K. Maciejewski. 2008. Grief and acceptance as opposite sides of the same coin: Setting a research agenda to study peaceful acceptance of loss. *The British Journal of Psychiatry* 193 (6): 435–437.

Rammant, Elke, Valérie Fonteyne, Vincent Van Goethem, Sofie Verhaeghe, Anneleen Raes, Mieke Van Hemelrijck, Nihal E Mohamed, Karel Decaestecker, and Ann Van Hecke. 2021. *Supportive roles of the health care team throughout the illness trajectory of bladder cancer patients undergoing radical cystectomy: A qualitative study exploring the patients' perspectives.* Seminars in Oncology Nursing.

Rayner, Lauren, Annabel Price, Alison Evans, Koravangattu Valsraj, Matthew Hotopf, and Irene J. Higginson. 2011. Antidepressants for the treatment of depression in palliative care: Systematic review and meta-analysis. *Palliative Medicine* 25 (1): 36–51.

Reid, S. Standart, and J. Scott. 2006. Effectiveness of brief training in cognitive behaviour therapy techniques for palliative care practitioners.*Palliative Medicinel* 20 (6): 579-84. https://doi.org/10.1177/0269216306071058.

Rietjens, Judith AC., Madelon T. Heijltjes, Johannes JM. van Delden, Bregje D. Onwuteaka-Philipsen, and Agnes van der Heide. 2019. The rising frequency of continuous deep sedation in the Netherlands, a repeated cross-sectional survey in 2005, 2010, and 2015. *Journal of the American Medical Directors Association* 20 (11): 1367–1372.

Rogers, Carl R. 1957. The necessary and sufficient conditions of therapeutic personality change. *Journal of Consulting Psychology* 21 (2): 95.

Ruijs, Cees DM., Bregje D. Onwuteaka-Philipsen, Gerrit van der Wal, and Ad.J.F.M. Kerkhof. 2009. Unbearability of suffering at the end of life: The development of a new measuring device, the SOS-V. *BMC Palliative Care* 8 (1): 1–10.

Saunders, Cicely. 1978. *The management of terminal malignant disease.* London, UK: Edward Arnold.

Saunders, C. 1988. Spiritual pain. *Journal of Palliative Care* 4 (3): 29–32.

Scheier, B., A.T. Pearson, J.M. Wilfong, and L.H. Baker. 2015. Symptom assessment in patients with cancer of unknown primary site. *Journal of Clinical Oncology. Conference* 33 (15 Suppl. 1). http://meeting.ascopubs.org/cgi/content/abstract/33/15_suppl/e20696?sid=88d12a62-008f-422c-ba10-4544d9ae8dad.

Schimmel, Nina, Joost J Breeksema, Sanne Y Smith-Apeldoorn, Jolien Veraart, Wim van den Brink, and Robert A. Schoevers. 2021. Psychedelics for the treatment of depression, anxiety, and existential distress in patients with a terminal illness: A systematic review. *Psychopharmacology*: 1–19.

Schulz, Richard, Randy S. Hebert, Mary Amanda Dew, Stephanie L. Brown, Michael F. Scheier, Scott R. Beach, Sara J. Czaja, Lynn M. Martire, David Coon, and Kenneth M. Langa. 2007. Patient suffering and caregiver compassion: New opportunities for research, practice, and policy. *The Gerontologist* 47 (1): 4–13.

Scott, John F. 1992. Lamentation and euthanasia. *Humane Medicine* 8 (2): 116–121.

———. 2015. The case against clinical guidelines for palliative sedation. In *Sedation at the end-of-life: An interdisciplinary approach*, 143–159. Springer.

Severino, S.M. 2020. Is medically assisted dying a choice if persons with disabilities aren't given the necessary supports to live? *Toronto Star.* Opinion. Accessed May 31, 2022. https://www.thestar.com/opinion/contributors/2020/01/28/is-medically-assisted-dying-a-choice-if-persons-with-disabilities-arent-given-the-necessary-supports-to-live.html?rf.

Sinclair, S., J.M. Norris, S.J. McConnell, H.M. Chochinov, T.F. Hack, N.A. Hagen, S. McClement, S.R. Bouchal. 2016. Compassion: A scoping review of the healthcare literature. *BMC Palliative Care* 15 (1), 1–6. https://doi.org/10.1186/s12904-016-0080-0.

Snead, O. Carter. 2020. *What it means to be human: The case for the body in public bioethics.* Boston, USA: Harvard University Press.

Somerville, Margaret. 2014. *Death talk: The case against euthanasia and physician-assisted suicide.* McGill-Queen's Press-MQUP.

Sprung, Charles L., Margaret A. Somerville, Lukas Radbruch, Nathalie Steiner Collet, Gunnar Duttge, Jefferson P. Piva, Massimo Antonelli, Daniel P. Sulmasy, Willem Lemmens, and E. Wesley Ely. 2018. Physician-assisted suicide and euthanasia: Emerging issues from a global perspective. *Journal of Palliative Care* 33 (4): 197–203.

Streffer, Marie-Luise, Stefan Büchi, Hanspeter Mörgeli, Ursula Galli, and Dominik Ettlin. 2009. PRISM (pictorial representation of illness and self measure): A novel visual instrument to assess pain and suffering in orofacial pain patients. *Journal of Orofacial Pain* 23 (2).

Struhkamp, Rita. 2005. Wordless pain: Dealing with suffering in physical rehabilitation. *Cultural Studies* 19 (6): 701–718.

Thangarasa, Tharshika, Sarah Hales, Eryn Tong, Ekaterina An, Debbie Selby, Elie Isenberg-Grzeda, Madeline Li, Gary Rodin, Sally Bean, and Jennifer AH. Bell. 2022. A race to the end: Family caregivers' experience of Medical Assistance in Dying (MAiD)—A qualitative study. *Journal of General Internal Medicine* 37 (4): 809–815.

Tompkins, Bonnie. 2018. Compassionate communities in Canada: It is everyone's responsibility. *Annals of Palliative Medicine* 7 (Suppl 2): S118-s129.

Wagner, Birgit, Valerie Keller, Christine Knaevelsrud, and Andreas Maercker. 2012. Social acknowledgement as a predictor of post-traumatic stress and complicated grief after witnessing assisted suicide. *International Journal of Social Psychiatry* 58 (4): 381–385.

Wallace, C. 2017. Euthanasia a choice for people with disability? It's a threat to our lives. *The Guardian.* https://www.theguardian.com/commentisfree/2017/sep/27/euthanasia-a-choice-for-people-with-disability-its-a-threat-to-our-lives.

Watt, Christine L., Julie Lapenskie, Monisha Kabir, Genevieve Lalumiere, Michel Dionne, Jill Rice, Chelsea Noël, Jonathan Downar, and James Downar. 2022. Accelerated rTMS for existential distress in palliative care: A report of two cases. *Brain Stimulation: Basic, Translational, and Clinical Research in Neuromodulation* 15 (1): 197–200.

White, Nicola, Fiona Reid, Adam Harris, Priscilla Harries, and Patrick Stone. 2016. A systematic review of predictions of survival in palliative care: How accurate are clinicians and who are the experts? *PLOS ONE* 11 (8), e0161407. https://doi.org/10.1371/journal.pone.0161407.

John F. Scott MD, MDiv. Associate Professor of Palliative Medicine, Department of Medicine, University of Ottawa, Canada.

Assisted Dr Balfour Mount in developing the first palliative care unit at McGill University, Montreal. Trained in medicine, theology and clinical epidemiology, Dr Scott pioneered the Division of Palliative Medicine at the University of Ottawa in which he continues to practise.

Mary M. Scott, BA, MSc Masters of Epidemiology from the University of Ottawa, Mary's research experience has focused on Canadian-based end-of-life and palliative care with multiple publications of literature reviews, qualitative studies, and quantitative health services analyses.

Part IV
Mental Disorders and MAID

Chapter 15
Medical Assistance in Dying (MAID) and Suicide: A Community Perspective

Mara Grunau, Robert Olson, and Crystal Walker

Abstract The introduction of Medical Assistance in Dying (MAID) legislation in 2016 has affected the suicide prevention community in significant ways. People considering suicide and people considering MAID when their death is reasonably foreseeable and they are in irremediable physical circumstances may both express a wish to die. These two groups can be considered distinct entities, however. Suicide is a possible yet preventable outcome of mental anguish, burdensomeness, and psychological pain in the context of a painful life. Recovery from suicidal ideation is possible—it is not irremediable. If Bill C-7 (2021), an amendment to MAID that would expand eligibility to those suffering mental illness, comes into effect in March of 2024, the impacts will be severe, as distinctions between suicide and MAID will further erode. Many vulnerable individuals may decide on this MAID option without fully exploring recovery options. Those of us in suicide prevention believe a shift in values must occur to mitigate the impacts of this legislation. This means, ultimately, a belief that mental health care is health care and those suffering mental illness must have open and easy access to care.

Keywords Suicide prevention · Community mental health · Mental health recovery · Suicidal crisis · Priority populations

M. Grunau (✉) · R. Olson · C. Walker
Centre for Suicide Prevention, Calgary, AB, Canada
e-mail: mara@suicideinfo.ca

R. Olson
e-mail: robert@suicideinfo.ca

C. Walker
e-mail: crystal@suicideinfo.ca

J. Kotalik and D. W. Shannon (eds.), *Medical Assistance in Dying (MAID) in Canada*, The International Library of Bioethics 104,
https://doi.org/10.1007/978-3-031-30002-8_15

15.1 Introduction

The landmark passage of Bill C-14, An Act to amend the Criminal Code (medical assistance in dying) in 2016 launched Canada into a club of few members: five countries have assisted dying laws, and two others have states within them which allow it (Australia and the United States). Only two other countries have laws pending (Roehr 2021). Though the intent of Bill C-14 was to balance Canadians' personal autonomy and freedom of choice with the protection of vulnerable people (Bill C-14 2016), the amendments, introduced in 2021 (Bill C-7 2021), tip the scales.

The particular amendment in question, for those of us who work in suicide prevention, is the accessibility to MAID for people whose primary diagnosis is psychiatric.

This change, expected March 2024, will have far-reaching effects. No longer will the concepts of MAID and suicide be distinct. This chapter, from the perspective of a community organization, briefly explains suicide, the suicide landscape in Canada, the historical distinction between suicide and MAID, and how we may mitigate the effects of C-14's amendments in the future.

15.2 Suicide in Canada

Suicide is a mental health and public health tragedy in Canada. It's an issue that needs to be mitigated at both the individual and population levels. People in suicidal crisis are in deep psychological pain—they don't want to die—they want the pain of living to end. As their pain intensifies, their vision constricts until they cannot see any way out of their pain save death. In this state, people are rarely able to think critically or clearly. Their feelings eclipse rational thinking. While suicide can be impulsive in the moment, it rarely comes out of nowhere—suicide intensity (or, acquired capability) builds over time. Likewise, recovery from suicidal crisis also takes time, however, recovery is possible. Few people who experience suicidal crises at some point in their life experience it for the long term. While it is overwhelming and punishing, many people recover and many more learn to manage their crises (Player et al. 2015). The danger lies with the people who may not opt to give recovery a chance. Others may not even understand that recovery is possible if the option for MAID is available, especially in the absence of access to high-quality, individualized mental health care, including adequate psychosocial support.

Every year, about 4000 people die by suicide in Canada. In 2020, the number was 3839, equating to a rate of 10.1 per 100,000 (Statistics Canada 2020). This rate has remained relatively consistent, hovering between 10 and 12 per 100,000 for the last twenty years.

Suicide itself is a result of a complex interplay of many, often interconnected, factors. These factors can range from social, psychological, and cultural, to psychiatric. Suicide's complexity makes its prevention challenging.

It also makes prediction of who will die by suicide imprecise. Many people have thoughts of suicide—researchers estimate the number at 1 in 20. Far fewer make suicidal plans and fewer still attempt. Very few people die by suicide compared to the number of people experiencing suicidal thoughts. A Centers for Disease Control and Prevention report in the United States showed that 10.6 million (4%) of the adult population had suicidal thoughts in the past year. Further, 3.1 million (1.3% of the adult population) had made a suicide plan in the previous year and 1.4 million (0.6% of the adult population) had made a suicide attempt in the past year. Of these, there were 45 861 suicidal deaths in 2019 which would translate to a very small percentage of a population of 330 million (Ivey-Stephenson et al. 2022). There are many theories as to why someone may choose to die by suicide, but none accurately predict who will die by suicide. The suggested risk factors themselves vary so widely from person-to-person that it is impossible to formulate a clear pathway from suicidal ideation to suicide that will apply to everyone.

In addition to suicide risk at the level of the individual, many sub-populations are more susceptible to suicide than others (sometimes called priority populations). These include middle-aged men, older adults, members of the 2SLGBTQ+ community, and Indigenous populations, among others. Each group has unique contributing circumstances which elevate their risk.

Because of the complexity and individuality of suicidality, there are relatively limited universal best practices in its prevention. The World Health Organization groups best practices into 4 categories: open and easy access to health care, responsible media reporting, means restriction or reduction, and education, including research and surveillance (World Health Organization 2014).

Suicide needs to be addressed both by government policy and strategy in a top-down manner in tandem with community organizations incorporating grassroots efforts in a bottom-up trajectory (World Health Organization 2018). No one approach or pathway can possibly tackle its complexity. Suicide prevention efforts must be multifaceted and multi-pronged.

15.2.1 *Suicide and Assisted Dying: Distinct Entities*

People considering suicide and people considering MAID when death is imminent (or when MAID eligibility is restricted to the end of life) may both express a wish to die. These two groups can be considered distinct entities. Suicide is a possible yet preventable outcome of mental anguish, burdensomeness, and psychological pain in the context of a painful life. It requires acquired capability. It is often violent and is usually carried out alone. People considering suicide tend to not enter the medical system and are often misunderstood by it. We work to prevent suicide because people considering suicide do not actually want to die—they want the pain of living to end... suicide is preventable. Recovery from suicidal ideation is possible—it is not irremediable.

Assisted dying, as defined by Bill C-14 in 2016, allows a degree of control over an inevitable, reasonably foreseeable death for people in irremediable physical circumstances. The applicant can have a mental illness but only as a secondary diagnosis, that is, mental illness cannot be the reason for the MAID application. Assisted dying can be carried out alone, however, most people who seek it choose to have it administered in the medical system, as a non-violent medical intervention. It involves decision-making by medical professionals and the option to involve loved ones.

15.2.2 Bill C-7 and Community Suicide Prevention Response

Arguably, the concepts of suicide and MAID remained distinct under C-14. However, the introduction of *Bill C-7: An Act to amend the Criminal Code* (medical assistance in dying) in 2021 (Bill C-7 2021) has blurred this distinction. The federal government was directed by the Superior Court of Quebec to expand the use of MAID. This amendment to the original MAID legislation removed the need for an applicant to establish death as "reasonably foreseeable," and broadened the irremediable circumstances to include mental, not only physical, illness.

Canadian Mental Health Association (CMHA)—National's response was swift. They issued a statement in February 2021 expressing their disappointment with the government's broadened view. They highlighted the failure of our current medical system to meet the mental health needs of Canadians and indicated this decision will only encourage people receiving inadequate care to seek this deadly alternative. Further, they pointed out that the irreversibility and incurability of mental illness is extremely difficult to determine and that those suffering mental illness or suicidality often recover over time (CMHA 2021).

Similarly, the Canadian Association for Suicide Prevention issued a statement that, "encouraged legislators to ensure that MAID does not overlap with what is traditionally understood as suicide and continues to protect those suffering from mental illness from a medically assisted premature death that could be avoided by adequate treatment and care." They acknowledged the two-year exclusion limit with hopes the expert panel list of recommendations (the reason for the hiatus) would contain "proper protective measures" when released in spring 2022. As with CMHA, they believe the protection of those most vulnerable to be of paramount importance (Canadian Association for Suicide Prevention 2021).

Central arguments for and against the inclusion of mental illness for MAID eligibility are mirrored in the central tenets of the original MAID legislation (Bill C-14) itself. The preamble stated that the aim was to find "the most appropriate balance between autonomy of persons who seek medical assistance in dying, on one hand, and the interests of vulnerable persons in need of protection and those of society, on the other" (Bill C-14 2016). We strongly believe the amendment will remove many of the protections for those vulnerable persons by allowing them greater freedoms to access MAID in the absence of improved access to life affirming care and support. As a result, we have grave concerns that people with remediable illnesses and/or suffering will receive MAID inappropriately, thus exacerbating the problem of suicide in Canada, the exact opposite of the outcome we work tirelessly to mitigate.

Brian Mishara, suicidologist and psychologist at Université du Québec à Montréal articulates one "ethical" position, stressing the concern for vulnerable people. He asserts that personal freedoms and autonomy should not be absolute for MAID applicants, citing helmet laws and seat belt laws as everyday examples of how society enacts legislation to protect us from ourselves. He contends "death is no substitute for treatment" as the individual right to decide to die may override an exhaustion of treatment options (Mishara and Weisstaub 2018). Unlike in jurisdictions like the Netherlands where all avenues for treatment must be explored, in Canada this requirement will not be required. An applicant can decide to proceed with MAID with or without a thorough investigation of treatment options. The path that Canada is following is especially concerning as we would not support even the more stringent Netherlands model. This is because a comprehensive exploration of treatment is only one important way to protect people with remediable suffering from dying prematurely by MAID. Treatment exploration, at a minimum, would also require that treatment options are available. Psychosocial supports must be in place to address the societal drivers of suicidality including poverty, accessible housing, meaningful social connections, and marginalization.

15.3 Final Report of the Expert Panel on MAID and Mental Illness

The much anticipated and controversial expert panel's final report on MAID and mental illness was released on May 13, 2022, with two of the original 12 members resigning prior to its release.

The report has 19 recommendations, including an overarching recommendation which informs the subsequent 18: Development of MAID Standards by federal, provincial, and territorial governments with physician and nursing regulatory bodies regarding questions of incurability, irreversibility, capacity, suicidality and structural vulnerabilities. The remaining recommendations cover areas including understanding enduring and intolerable suffering (4); situations of involuntariness (9); training of assessors and providers in specialized topics (15). The implementation of these recommendations would offer health care providers background and context for implementation policies and processes (Health Canada 2022).

A key message from these suggestions appears to be for providers to establish eligibility for MAID on a case-by-case basis but, overall, these recommendations strive to encourage health care providers to offer as comprehensively safe a space as possible. These are non-binding guidelines. Individual health care organizations will be tasked with developing and implementing policies and procedures.

We are disappointed with the report and believe that its recommendations, if implemented, will contribute to the problem of suicide in Canada. We had hoped the report would have recommended halting further expansion of the legislation pending rigorous scientific study of the risks of MAID expansion, specifically those related to suicide, and the degree to which any proposed safeguards might mitigate them. We had also hoped the report would call for research to help determine the reliability of mental health practitioners' assessments of irremediability of suffering from mental illness. Instead, the report assumes outcomes of such research that are not yet grounded in rigorous evidence. Further, it takes for granted that issues arising from MAID applications and assessments can be sorted out by practitioners and/or with consensus recommendations. Finally, its tone implies that the inclusion of mental illness as an eligible criterion is a given. In our view, this is a dangerous mistake.

15.3.1 Actions Going Forward: What Needs to Be Done?

If the last piece of Bill C-7 (2021) comes into effect in 2024, how will we mitigate its effects? How can we best support people with debilitating mental illness and suicidality? How do we know if someone with mental illness who has chosen MAID would not recover at some point in the future? The answer to the last question is certain—we do not have scientific evidence regarding which people may or may not recover. We cannot allow MAID to become an acceptable option for emotional suffering.

Just as those who oppose the use of MAID for physical illnesses tout the need for more robust and accessible palliative care to help ease physical suffering, we need expanded mental health care in Canada to ease people's mental suffering. As stated above, a key best practice in suicide prevention is open and easy access to mental health care. The right care in the right place at the right time needs to be available for people across Canada. This care needs to come from an integration of the medical system and the community. It needs to be person-centered, so that it builds agency, skills, and recovery. It's simplistic to think that this can be merely accomplished with resources. Though dedicated investment is important, what is truly needed is a shift in values. Mental health care is health care.

As Brian Mishara reminds us, death is no substitute for treatment. MAID should not be a substitute for reducing the stigma around suicidality, promoting recovery, and making treatment options familiar and accessible.

People in mental anguish need compassionate care and a societal milieu that addresses their basic human needs. This is how we prevent suicide, and it involves us all.

Acknowledgements We wish to thank Dr. Mark Sinyor, professor of Psychiatry at the University of Toronto and psychiatrist at Sunnybrook Health Sciences Centre, and Dr. Michael Trew, psychiatrist and Senior Medical Director with Addiction and Mental Health for Alberta Health Services for their expertise in the preparation of this chapter.

References

Bill C-14. 2016. An act to amend the Criminal Code and to make related amendments to other acts (medical assistance in dying). 2016. Assented to June 17, 42nd Parliament, 1st session. Retrieved from the Justice Laws website: https://laws-lois.justice.gc.ca/PDF/2016_3.pdf.

Bill C-7. 2021. An act to amend the Criminal Code (medical assistance in dying). Assented to March 17, 43rd Parliament, 2nd session.

Canadian Association for Suicide Prevention. 2021. Statement on recent MAID developments. https://suicideprevention.ca/media/statement-on-recent-MAID-developments/.

Canadian Mental Health Association. 2021. *Statement on Medical Assistance in Dying (MAID)*. https://cmha.ca/brochure/statement-on-medical-assistance-in-dying-MAID/.

Health Canada. 2022. Final report of the expert panel on MAID and mental illness. https://www.canada.ca/en/health-canada/corporate/about-health-canada/public-engagement/external-advisory-bodies/expert-panel-MAID-mental-illness/final-report-expert-panel-MAID-mental-illness.html#glo.

Ivey-Stephenson, A., A. Crosby, J. Hoenig, S. Gyawali, E. Park-Lee, and S. Hedden. 2022. Suicidal thoughts and behaviors among adults aged 18 years-United States, 2015–2019. *Morbidity and Mortality Weekly Report (MMWR)*. https://www.cdc.gov/mmwr/volumes/71/ss/ss7101a1.htm.

Mishara, B., and D. Weisstaub. 2018. Is suicide prevention an absolute? *Crisis*, 39, 313–317.

Player, M., Judy Proudfoot, Andrea Fogarty, Erin Whittle, Michael Spurrier, Fiona Shand, Helen Christensen, et al. 2015. What interrupts suicide attempts in men: A qualitative study. *Plos One*. Retrieved from http://journals.plos.org/plosone/article?id=https://doi.org/10.1371/journal.pone.0128180.

Roehr, B. 2021. Assisted dying around the world. *BMJ*. https://doi.org/10.1136/bmj.n2200.

Statistics Canada. 2020. Table 13-10-0801-01 Leading causes of death, total population (age standardization using 2011 population). https://www150.statcan.gc.ca/t1/tbl1/en/tv.action?pid=1310080001.

World Health Organization. 2014. Preventing suicide. A global imperative. https://www.who.int/publications/i/item/9789241564779.

World Health Organization. 2018. Preventing suicide: A community engagement toolkit. https://apps.who.int/iris/bitstream/handle/10665/272860/9789241513791-eng.pdf?sequence=1&isAllowed=y.

Mara Grunau, B.Ed, MPA is the Executive Director of the Centre for Suicide Prevention (CSP), a branch of the Canadian Mental Health Association in Alberta. Established in 1981, CSP equips Canadians to support people considering suicide.

Under her leadership, CSP has been an active partner in the suicide prevention landscape nationally, provincially, and locally. She is an original, standing member of the National Collaborative for Suicide Prevention where she has also served in a leadership capacity. Mara is and has also been an advisor to many suicide prevention projects and committees.

Mara holds a Bachelor of Education and a Master of Public Administration.

Robert Olson, MLIS As Librarian for CSP, Robert Olson leads the knowledge translation efforts produced by CSP and helps manage the largest English-language library collection on suicide information in the world. From the collection, Olson and team disseminate this information through secondary research publications, including resource toolkits, opinion-editorial articles, special features, fact sheets, infographics, and webinars.

Olson holds an MLIS (Master of Library and Information Science) and has been with CSP since June 2011.

Crystal Walker, BA is Communications Lead at CSP. She holds a Bachelor of Communication Studies and a diploma in Journalism Arts. Crystal has been working in communications and marketing with CSP for over 10 years.

Chapter 16
Unbalanced: Mental Illness, MAID and Medico-legal Principles

David W. Shannon, Alexandra Giancarlo, and Elaine Toombs

Abstract On March 17, 2021 the Parliament of Canada passed Bill C-7, an amendment to the *Criminal Code of Canada,* expanding access to medical assistance in dying (MAID). The law formerly did not allow for anyone to receive MAID where death was not naturally foreseeable. This limitation requiring naturally foreseeable death was removed, now permitting individuals experiencing prolonged suffering solely from mental illness to access MAID. The final version of Bill C-7 allows MAID for mental illness after a "sunset clause" of two years from March 17, 2021. With the upcoming adoption of this change, there will be significant implications for what is generally termed mental health or consent and capacity law, as well as policy in the treatment and behavioural containment of individuals experiencing psychiatric suffering, and/or suicidal ideation. In this chapter the current state of applicable mental health law in the Province of Ontario will be reviewed, covering pressing concerns about the extension of MAID to those solely experiencing mental illness. The perspectives of individuals groups impacted by euthanasia provided to individuals experiencing mental illness in other jurisdictions will be outlined. From this, consistent patterns and or probable outcomes of such a change in the law will be discussed.

Keywords Mental illness · Suicidal ideation · Psychological suffering · Capacity · Involuntary committal

---

D. W. Shannon (✉)
Barrister-and-Solicitor, Shannon Law, Thunder Bay, Toronto, ON, Canada
e-mail: shannonlaw@tbaytel.net

A. Giancarlo
Faculty of Kinesiology, University of Calgary, Calgary, AB, Canada

E. Toombs
Clinical Psychologist, Dilico Anishnawbe Family Care, Toronto, ON, Canada
e-mail: etoombs@lakeheadu.ca

J. Kotalik and D. W. Shannon (eds.), *Medical Assistance in Dying (MAID) in Canada*, The International Library of Bioethics 104,
https://doi.org/10.1007/978-3-031-30002-8_16

16.1 Preface

A patient approaches their family physician, defeated and hopeless. Fifteen years after experiencing a traumatic event, they are still attempting to reduce persistent symptoms of depression, numbness, overwhelming feeling of anger, and distressing flashbacks of the event, including upsetting nightmares that make it difficult to fall asleep each night. They have come to their doctor for an end to their suffering and want to know if they qualify for MAID. It is over a two year wait to see yet another specialized trauma counsellor, and the patient says they are done. No more treatment attempts, I'm ready for this to finally be over. Should they receive MAID?

16.2 Introduction

Extending MAID to those whose underlying medical condition mental illness was mandated to occur in Canada in March 2023. In February 2023 this was delayed until March 17, 2024 (Anderssen 2023). Currently we do not know if the individual described above would qualify for MAID. At the time of writing, there has been no decisions made on who will be eligible for this program nor what safeguards will be in place to ensure patient safety. These expanded eligibility criteria could make the Canadian MAID regime the most permissive in the world (Kirkey 2022).

Since the Supreme Court of Canada's decision in *Carter v. Canada* it has been legal for a qualified physician to provide euthanasia to individuals under strict circumstances. Within the decision, however, the Supreme Court clearly stated that it did not apply to individuals experiencing psychological suffering or mature minors, and applied only to the facts of the case before them (*Carter v Canada* 2015, para. 111). As a response to this decision, the Government of Canada amended the *Criminal Code* in what is known as *Bill C-14* (c.3 s. 241 ((2016)). A key provision of this amendment was that natural death must be foreseeable.

The foreseeability criteria (*Criminal Code* s. 241.2 (2)) was later challenged and deemed to be unconstitutional by the Quebec Superior Court in the decision of Truchon c. Procureur general du Canada 2019 QCCS 3794 (hereinafter referred to as *Truchon)* as an infringement of section 7 of the *Canadian Charter of Rights and Freedoms* as well as section 15, and it could not be justified under section 1, which allows for breaches to be allowed in a "free and democratic society" (*Carter v Canada* 2015; Truchon v Canada 2019). Canada chose not to appeal, and instead further amended the *Criminal Code* with *Bill C-7*, which received Royal Assent March 17, 2021 (Bill C-7 c. 2 (2021)).

Bill C-7 not only removed the natural foreseeability of death clause, but also made a commitment to study expanding the eligibility criteria for MAID to people whose sole underlying condition is mental illness. exclusive of physical suffering. As the preamble states,

> Whereas further consultation and deliberation are required to determine whether it is appropriate and, if so, how to provide medical assistance in dying to persons whose **sole underlying medical condition is a mental illness** in light of the inherent risks and complexity of the provision of medical assistance in dying in those circumstances;
>
> And whereas the law provides that a committee of Parliament will begin a review of the legislative provisions relating to medical assistance in dying and the state of palliative care in Canada in June 2020, which review may include issues of advance requests and requests where mental illness is the sole underlying medical condition

It is expected that this change will become law in March 2023 following "expert review, which will be tasked with considering recommending protocols, guidance and safeguards for MAID for persons suffering from mental illness" (Government of Canada 2021). It does not include neurocognitive and neurodevelopmental disorders, or other conditions that may affect cognitive abilities (Government of Canada 2021).

This chapter will discuss the approach that Ontario law takes when individuals suffering with mental illness are at risk of harming themselves, and the implications for this body of law that extending euthanasia to individuals with mental illness as a criterion will have on its foundational principles. The views of proponents wishing to extend MAID to those experiencing mental illness and the much more complex arguments against extending euthanasia this group will be summarized with specific reference to the Canadian and international experience. We will ask: Is it appropriate to provide MAID to persons whose sole underlying medical condition is a mental illness? If so, what are the inherent risks and complexity, and what would be the impact to the law in Ontario should it be allowed? Ultimately, we will argue that mental illness as a sole criterion is unprincipled, blurs the lines of suicide prevention possibly replacing these efforts with suicide facilitation and ignores mounting legal, social and clinical considerations that must gain paramountcy.

16.3 Existing Criteria for MAID

The criteria for accessing MAID In Canada is comprehensively discussed in other chapters, and therefore only a brief summary will be provided. Also, the central argument of this chapter will surround those with mental illness for whom death is not naturally foreseeable pursuant to Bill C-7. For detailed information regarding the eligibility criteria of MAID, refer to Chaps. 2, 5 and 7 of this book.

16.4 Current Law and Legal Principles Re Psychological Suffering and Self Harm in Ontario: A Reversal of Jurisprudence

A cornerstone of the rationale for extending MAID to mental illness as a sole criterion utilized by proponents is that it respects the dignity of person and individual autonomy. As will be discussed, the body of law built over several decades in Canada runs counter to that principle, should there be a threat to individual or public safety posed by an individual with a mental disorder. The expansion of MAID to persons with mental illness for whom death is not reasonably foreseeable turns mental health law on its head, in that, it ignores the nature of fluctuations in individual capacity to consent to treatment; puts eligibility assessments at risk of manipulation; denies clinical applications that often succeed at reversing mental illness to the point of bringing the 'irremediable' criteria into question; reverses clinical approaches to the concept of wellness; does not consider the limitations of the narrow legal test for capacity to consent to treatment in its current form; runs contrary to traditional Indigenous practices; and denies the unresolved controversy that continues to revolve around this question at both a domestic and international level. Reconciling these flaws and contradictory principals into coherent legislation that balances competing interests within the rubric of law and mental health will create a fractured law that ignores both legal and medical best interests.

The *Mental Health Act* (hereinafter referred to as MHA) and the *Health Care Consent Act* (hereinafter referred to as HCCA) in the Province of Ontario provide good illustrations of how the current law runs counter to extensions of MAID eligibility to those with mental illness, and points to the question, why rush to opening these doors of change to the Criminal Code when doing so in haste may be met with folly? Within the context of self-harm and/or a propensity for self-harm the body of consent and capacity law would appear to run in stark contra-distinction to the federal MAID legislation even with its bundle of so-called safeguards.

Both the MHA and HCCA speak to the criteria and obligations of physicians who assess a person with mental illness for the propensity to self-harm and/or commit suicide, and how capacity to consent to treatment is determined. Each piece of legislation appears to attempt to control self-harming behaviours through an intersecting series of detention and medication, and would appear to suggest that MAID in the context of mental illness runs counter to a physician's duty of care. The legislative focus is one of health and recovery, and it does not contemplate euthanasia as an option (HCCA 1996, s. 1). The current law will be surveyed to demonstrate the radical departure from current legal principles the law would need to take should MAID be extended to psychological suffering.

According to the *MHA* if there is a nexus between an underlying mental disorder and a risk of harm to self or imminent risk of bodily harm, the person will be detained involuntarily in the hospital. If the physician assessing an individual with mental disorder is of the opinion that the conditions of "serious bodily harm to the patient" (MHA S. 20 (a)(i)) or "serious physical impairment of the patient" (MHA S. 20 (a)(i)) then the attending physician shall admit the person as an involuntary patient. Therefore, the legal and ethical dilemmas posed by a physician ignoring the direction of the MHA and recommending what would surely be the ultimate form of bodily harm to self would need to be reconciled with any future MAID law should it be allowed.

The courts have held that the legal test for involuntary committal was one of moderate severity because it affected the liberty interest of the appellant (*Lucas v*

Bardham 1987). In T.W., (2004) although the eligibility criteria for detention in a psychiatric facility has this elevated test, the Consent and Capacity Board interpreted the definition of serious physical impairment broadly by indicating at para. 23 that it includes situations where the patient due to mental disorder failed to: attend adequately to the activities of daily living; seeking medical attention or treatment for physical or mental disorders; comply with drug and treatment regimens outside of hospital, which adversely affected his or her health to a serious degree; or lived in housing or out of doors without adequate support, clothing, sustenance, or assistance. The courts and Consent and Capacity Board have indicated they should be alert to situations where the patient exposed himself or herself to danger due to lack of concentration, infirmity or otherwise in such instances as walking carelessly in traffic, or refusal to use walkers or canes despite their need; pay attention to situations where the patient exposed himself or herself to danger, injury or harm at the hands of others; protect from manipulation by con artists and the like who would leave the patient destitute or reduced resources or otherwise without the necessaries of life. The Consent and Capacity Board also instructed that those with mental disorders must be prevented being manipulated by others into unwanted sexual activities. Attempted suicide and suicidal ideation, as may be expected, is a significant factor in demonstrating a risk of likely or imminent harm to the person. See also: *Azhar v Anderson* (1985); *Blakely v Kingston Psychiatric Hospital* (1995); *Diilanceai v Bell* (1988); *Geiger v Dua* (1994); *Levinskas v Hutson* (1989); *Lucas v Bardham* (1987); *Maracle v. Potopsingh* (1987); *Subotic v Roopchand* (1989); *The Matter of M.M.* (1991); and *The Matter of N.N.* (1996).

16.5 Established Family Intervention: A Justice of the Peace's Order for Psychiatric Examination, and Police Apprehension

Although the courts have indicated that an individual with capacity to consent to treatment may seek MAID without any right of intervention by family or loved ones, this does not square with consent and capacity legislation where a person, due to mental disorder, or a person be at risk of self-harm and or imminent risk of mental or physical deterioration may seek the intervention of the court. Without the inclusion of this model for in MAID procedures where there is a substantial risk of error and safeguards may be severely diluted.

The intervention of family members, loved ones, the police and the courts within the context of mental illness has traditionally been seen as a valid form of evidence showing when or if a person is at risk of self-harm resulting in intervention from the courts and medical practitioners. This may inform appropriate legal standards should MAID be extended to mental illness as a sole criterion for MAID.

A justice of the peace has the authority under s. 16 of the MHA (1990) to order a psychiatric assessment should information upon oath demonstrate to the justice that they have reasonable cause to believe that the person is apparently suffering from a mental disorder of a nature or quality that will likely result in serious bodily harm to the person or another person, or serious physical impairment. The order is sufficient authority, for seven days including the date it was made, for any police officer to whom the order is addressed to take the person who is the subject of the order forthwith to an appropriate place to be detained for examination by a physician (MHA 1990, s.16(3)). This examination will usually occur in the hospital emergency department. The purpose of the examination will be to permit the physician to decide whether to sign a Form 1 authorizing the detention and assessment of the person in hospital for up to 72 hours.

Best-practices have explored the balancing of autonomy and the paternalism of coercion where individual safety is at risk within the mental health context and found that both competing legal interests can find an uneasy truce. They have yet to explore if and how those involuntarily admitted to in-patient facilities can request or consent to MAID, or if specific mandates will omit this population from accessing MAID programs. Further, as the presence of significant suicidality and harm to self can warrant hospitalization for those living with mental illness, the reasoning behind such ideation must be explored by those who are experts within these populations, in the context of assessing MAID and alternative reasons for both living and dying.

16.6 The Legal Definition of Capacity

For an individual to be eligible for MAID they must have capacity to consent to its administration (Criminal Code 1985, s.241.2(1)(b)). The legal test and the limitation of its application within the context of extending MAID to persons with mental illness will be outlined in this section. Emphasis will be placed on the questions of: whether uncoerced and freely informed consent is achieved, whether the criteria of irreversibility can be met, and can it be achieved in a manner that comports with long held legal principles as it applies to individuals with mental illness, albeit capable to consent to treatment who seek to harm themselves viz. MAID? There is a presumption of capacity to consent to treatment that must be applied on an issue specific and individual basis. The test is set out in s.4(1) of the HCCA, which reads in part,

> 4.(1)—A person is capable with respect to a treatment… if the person is able to understand the information that is relevant to making a decision about the treatment, … and able to appreciate the reasonably foreseeable consequences of a decision or lack of decision.
>
> (2) Presumption of capacity—A person is presumed capable with respect to treatment, admission to a care facility and personal assistance services, unless……

Also, pursuant to s.15(1) and (2) of the *HCCA*, capacity is a fluid concept. It can vary over time, and it can also vary in relation to the type of treatment.

The leading case in relation to the law on consent to treatment in Ontario is Starson v Swayze (2003). In it, the majority noted that the HCCA presumes a person capable to decide to accept or reject medical treatment and patients with mental disorders were presumptively entitled to make their own treatment decisions. This presumption could only be displaced by evidence that a person is not able to understand the information that is relevant to making a treatment decision and second, a person must be able to appreciate the reasonably foreseeable consequences of a decision or lack of one as it applies to the proposed form of treatment. Furthermore, it was important that autonomy and self-determination be given priority when assessing individuals in this group.

The court noted that a person need not agree with the diagnosis of an attending physician in order to be able to apply the relevant information to their own circumstances. Also, even if they do not apply the information provided, they must have the ability to appreciate the consequences of a decision (*Starson* 2003, para. 80). Also, "if the patient's condition results in him being unable to recognize that he is affected by its manifestations, he will be unable to apply the relevant information to his circumstances, and unable to appreciate the consequences of his decision" (*Starson* 2003, para. 79). This decision was viewed from the vantage point of an individual seeking autonomy in treatment decisions related to a psychiatric disorder as opposed to an individual with mental illness seeking to terminate their life through MAID. The distinction is discrete, yet it bears consideration by a court at a later date. Capacity to consent to MAID is discussed also in a separate chapter of this book.

16.7 The Limitations of the Legal Definition of Capacity and Irreversible Decline

Given that pathological fixation on death and suicide is central to particular psychiatric disorders, particularly depressive disorders which affect approximately 5.4% of Canadians (Statistics Canada 2013), accurate assessment of capacity, authenticity, autonomy, suicide rationale, and more broadly, how an individual deems their own "life worth living", is necessary (Platt 2021). Accepting an illness as incurable does not necessarily mean that an individual's quality of life is diminished (Wilson 2019), however a core principle of MAID is that one has the capacity to decide one's life trajectory following determination that one's mental illness is "grievous and irremediable".

Capacity assessment for MAID remains a challenging issue, as there are no standardized practices for physicians and nurse practitioners seeking to determine patient capacity to consent to MAID. Other chapters have explored concerns that capacity assessments for MAID remain a challenging issue with a lack of standardized practices. Although Justice Major in the Starson v Swayze (2003, para. 116) case, when quoting the Weisstub report (1990), noted that one should not automatically conflate the presence of mental illness with a presumption of incapacity to consent to treatment, it is unknown how individuals with mental illness will be assessed for capacity during the 90-day wait period between the initial assessment and consent for MAID on the day the procedure occurs. For example, how will they be assessed for capacity at the first assessment, how will they remain capable during the 90-day wait period when they may withdraw consent, and how are they assessed on the final day when MAID is performed? Additionally, the question arises, how does the presence of psychiatric symptoms that interfere with moment to moment capacity, including mania, hallucinations, or delusions, impair one's eligibility for MAID, or how frequently one must be assessed during the mandatory waiting period?

Clinical and legal interpretations of the "irreversible decline" criteria for MAID also warrant careful consideration, particularly in the context of previous treatment attempts and availability of resources in remote and rural care contexts. The definition of irreversible decline, particularly in the context of declining mental illness, cannot be affiliated with simply lack of access to treatment, even over a significant period of decline. Scholars have challenged, in fact, if the concept of incurability even applies in the present context of psychiatric care as psychiatric assessments are known to be unreliable in terms of predicting course and outcomes on an individual patient basis (Dom et al. 2020). Furthermore, a "cure" in cases of psychological suffering is frequently related to factors other than the disorder per se, such as an improvement in financial or employment status. Patients may unexpectedly improve for as-yet-unknown reasons or because of unpredictable factors such as the patient-therapist bond. A third factor to consider in discussing incurability centres on distinguishing between the *availability* of new treatments and the *willingness* of the client to undergo these treatments. For example, in the Belgian case, data shows that most psychiatric patients refuse additional treatments offered, such as electro-convulsive therapy for depression (Dom et al. 2020, 667).

Perpetuating factors of mental illness often relate to access to adequate health care and ability to effectively engage in treatment, as treatment options can be limited across socio-economic disparities and other social determinants of health (Downie and Schuklenk 2021). It is perhaps an inconvenient truth that requests for psychiatric euthanasia have corresponded with the rolling back of social safety nets across many Western countries that have led to increased socio-economic inequalities. The first mention of persons who had mental disorders being euthanized in Belgium is in its 2004–2005 annual report about euthanasia; there were six cases. By its most recent reports, the authors documented 57 and 49 cases in 2018 and 2019, respectively (cited in Haekens 2021, 39). Haekens (2021) argues that psychiatric patients in Belgium receive inadequate care due to overall underfunding (leading to long waiting lists) and social stigma. Psychological suffering, and, therefore, requests for euthanasia based on psychological conditions, cannot be extricated from social determinants of health.

16.8 The Domestic and International Context

According to surveys among clinicians in the Netherlands and Canada, support is higher for assisted dying in physical and terminal illness—and significantly so—than for patients with psychiatric conditions (Nicolini et al. 2020), despite a high degree of comorbidity among physical and mental health conditions (Isenberg-Grzeda et al. 2021). Evidence from countries in which such procedures have been legal for years to decades can be instructive for the Canadian context. It is worth noting at the outset that, perhaps contrary to international media narratives, in countries where euthanasia is legal for psychiatric conditions it remains controversial (Lemmens 2021; Nicolini et al. 2020). Scholars from a range of disciplines including but not limited to psychiatry, psychology, nursing, disability studies, public health, and ethics have elucidated a range of arguments "for" and "against", as well as examined societal consequences and policy implications. Nicolini et al. (2020) argued that "parity" forms the core of the present international debate about psychiatric euthanasia. The crux of the parity argument is that mental and physical suffering should not be differentiated. How might the parity argument come into play in terms of physician-administered death for patients with psychiatric conditions? Consider one of the main arguments against euthanasia for those with mental illness: prognostic uncertainty. Under the principle of parity, the argument for denying MAID is no more valid for someone living with severe depression—in which prognostic unpredictability is high—than it is for someone with cancer. According to the parity argument, broadening euthanasia to those with psychiatric conditions underlying medical condition being a mental illness is "a matter of principle". Hence, it takes an extremely narrow standard where the possibility of false positives should be tolerated to reduce overall suffering (Nicolini et al. 2020, 12).

A second key point of debate hinges on autonomy, one of the guiding principles of life in Western society and one of the guiding principles of modern patient-focused care. In the context of medical ethics, patient autonomy is paramount and it "seeks to ensure the right of a patient to make their own decisions while providing realistic information without undue influence" (Khawaja and Khawaja 2021, 3). In the context of medically assisted dying, however, scholars have raised a range of counterpoints that cloud this seemingly straightforward doctrine. For example, valuing personal autonomy as an unquestioned good above all other considerations precludes contemplating the impact of these practices on broader society, such as the withering of a social contract in which supposedly all people have value, and the ramifications for healthcare professionals on a personal level (Somerville and Ely 2021). In Belgium, which has arguably the most permissible euthanasia legislation, proponents claim that personal freedom is maximized because it is an unforced, individual decision, ensuring "maximum space for personal choice" (Lemmens 2021). But is it? Legally, euthanasia can be performed for non-terminally ill cases such as those with exclusively psychiatric conditions, yet most physicians acknowledge that existential and social concerns cannot be eliminated. That physicians in Belgium who are considering approving a euthanasia request are told to attempt to consult with the patient's family and friends—but only if given approval to do so by the requesting patient—aptly illustrates this tension (Lemmens 2021). Canada would be well advised to learn from these jurisdictions that have jurisprudence, law, and empirically based research historically more advanced than its own.

16.9 Indigenous Perspectives on Psychiatric Suffering and the Contradiction of Bill C-7

Conceptualization of psychological suffering and irremediable decline specific to psychiatric illness must also consider both cultural and contextual understandings of wellness, quality of life, and individual autonomy. Specifically, within Canada, authentic partnership with Indigenous communities as distinct rights-holders is necessary to evaluate how governmental interference in end-of-life decision making may contravene Indigenous conceptualizations of wellness, relational wellbeing, autonomy, and non-interference. In an open letter to the Canadian Psychiatrist Association in response to its (2021) Position Statement, Indigenous representatives were united in their condemnation of the proposed changes to MAID, as they believe this part of the MAID framework is being imposed upon them and is not congruent with Indigenous worldviews (Chaimowitz et al. 2021; Nicholas et al. 2021). Importantly, the authors of their submission raised serious concerns that increasing access to MAID will further undermine Indigenous communities' wide-scale efforts to combat disproportionate suicide rates (Nicholas et al. 2021). The federal government should include consultation with Indigenous leadership before continued consideration of extending MAID to mental illness as a sole criterion.

16.10 Conclusion

In conclusion, the practice of providing euthanasia to people experiencing mental illness as a sole criterion is unprincipled. Such a practice erodes the foundation upon which existing Canadian case law rests: that a person should be vigorously prevented from self-harm by, in the most extreme of cases, involuntary commitment to a psychiatric facility. Additional concerns are posed in reference to assessing individual capacity to consent to MAID, as capacity and reversibility of mental illness is shown to be a fluctuating target. Many mentally ill persons' desire for MAID also cannot be extricated from socio-economic hardship they experience because of Western countries' rolling back of public spending on health and welfare. Finally, the issue of euthanasia for mental suffering is not at all settled in peer countries, where it continues to be controversial despite its being legal for years. As Canada speeds toward extending MAID to persons suffering due to mental illness that is "irreversible" and intolerable to them, profound consideration of regulatory safeguards and effective oversight will be critical should it ever be allowed.

References

Anderssen, Erin. 2023. Federal Government Introduces Legislation to Delay Changes to the Maid Law Until Next Year. The Globe and Mail. https://www.theglobeandmail.com/canada/article-federal-government-introduces-legislation-to-delay-changes-to-the-maid/. Accessed 1 Feb 2023.

Azhar v. Anderson. 1985. Ont. Dist. Ct.

Bill C-7, An Act to Amend the Criminal Code (Medical Assistance in Dying), SC 2021, c 2. https://canlii.ca/t/551nn

Bill C-14, An Act to Amend the Criminal Code and to Make Related Amendments to Other Acts (Medical Assistance in Dying), S.C. 2016, c. 3 at Section 241, Royal Assent 2016. https://www.parl.ca/DocumentViewer/en/42-1/bill/C-14/royal-assent

Blakely v. Kingston Psychiatric Hospital. 1995. O.J. No. 2847 (Gen. Div.).

Carter v. Canada (Attorney General). 2015. SCC 5 (CanLII), [2015] 1 SCR 331.

Chaimowitz, Gary, Alison Freeland, Grainne E. Neilson, Nickie Mathew, Natasha Snelgrove, and Melanie R. Wong. 2021. Position Statement: Medical Assistance in Dying: An Update. Canadian Psychiatric Association.

Criminal Code, RSC 1985.

Diilanceai v. Bell. 1988. O.J. No. 1079 (Dist. Ct.).

Dom, Geert, Heidi Stoop, An. Haekens, and Sigrid Sterckx. 2020. Euthanasia and Assisted Suicide in the Context of Psychiatric Disorders: Sharing Experiences From the Low Countries. *Psychiatria Polska* 54 (4): 661–672.

Downie, Jocelyn, and Udo Schuklenk. 2021. Social Determinants of Health and Slippery Slopes in Assisted Dying Debates: Lessons from Canada. *Journal of Medical Ethics* 47 (10): 662–669.

Geiger v. Dua. 1994. O.J. No. 3175 (Gen. Div.).

Government of Canada. 2021. Canada's New Medical Assistance in Dying (MAID) Law. https://www.justice.gc.ca/eng/cj-jp/ad-am/bk-di.html.

Haekens, An. 2021. Euthanasia for Unbearable Psychological Suffering. In *Euthanasia: Searching for the Full Story. Experiences and Insights of Belgian Doctors and Nurses*, edited by Timothy Devos, 39–47. Leuven, Belgium: Springer.

Health Care Consent Act, 1996, S.O., c. 2, Sched. A.

Isenberg-Grzeda, Elie, Amy Nolen, Debbie Selby, and Sally Bean. 2021. High Rates of Psychiatric Comorbidity Among Requesters of Medical Assistance in Dying: Results of a Canadian Prevalence Study. *General Hospital Psychiatry* 69: 7–11.

Khawaja, Masud, and Abdullah Khawaja. 2021. The Ethics of Dying: Deciphering Pandemic-Resultant Pressures That Influence Elderly Patients' Medical Assistance in Dying (MAiD) Decisions. *International Journal of Environmental Research and Public Health* 18 (16): 8819.

Kirkey, Sharon. 2022. Canada Will Soon Offer Doctor-Assisted Death to the Mentally Ill. Who Should Be Eligible? National Post. https://nationalpost.com/health/canada-mental-illness-maid-medical-aid-in-dying. Accessed 4 Apr 2022.

Lemmens, Willem. 2021. When Conscience Wavers. Some Reflections on the Normalization of Euthanasia in Belgium. In *Euthanasia: Searching for the Full Story*, 25–37. Cham: Springer.

Levinskas v. Hutson. Apr. 6, 1989. (Ont. Dist. Ct).

Lucas v Bardham. Oct. 15, 1987. (Ont. Dist. Ct). In T.W., [2004] O.C.C.B.D. No. 220.

Maracle v. Potopsingh. 1987. (Ont. Dist. Ct)

Mental Health Act, R.S.O. 1990, c. M.7

Nicholas, Graydon, Mary Jane McCallum, Myeegun Henry, Rod McCormick, Neil Belanger, Conrad Saulis, Rosella Kinoshameg et al. 2021. *Open Letter to Canadian Psychiatric Association.*

Nicolini, Marie E., Scott Y. H. Kim, Madison E. Churchill, and Chris Gastmans. 2020. Should Euthanasia and Assisted Suicide for Psychiatric Disorders Be Permitted? A Systematic Review of Reasons. *Psychological Medicine* 50(8): 1241–1256.

Platt, Elyse. 2021. Depression and the Desire to Die–Implications for Medical Assistance in Dying.

Somerville, Margaret, and E. Wesley Ely. 2021. Foreword 2. In *Euthanasia: Searching for the Full Story. Experiences and Insights of Belgian Doctors and Nurses*, edited by Timothy Devos, ix–xv. Leuven, Belgium: Springer.

Starson v Swayze. 2003. *SCC 32, 1 S.C.R 722.*

Statistics Canada. 2013. Canadian Community Health Survey: Mental Health, 2012. *The Daily*. https://www150.statcan.gc.ca/n1/en/daily-quotidien/130918/dq130918a-eng.pdf?st=pvrvPMP2

Subotic v. Roopchand. 1989. *(Ont. Dist. Ct) No. 701/89.*

The Matter of M.M. Feb. 22, 1991. (C.C.B.).

The Matter of N.N. Oct. 6, 1996. (C.C.B.), WT-96/525.

Truchon v. Canada (AG). 2019. *QCCS 3792.*

Weisstub, David N. 1990. *Enquiry on Mental Competency: Final Report*. Toronto: Queen's Printer for Ontario.

Wilson, Emily A. 2019. Medical Assistance in Dying and Mental Illness: When the Illness Hindering Your Autonomy is the Illness You Wish to Be Relieved from. *McGill J.L. & Health* 13: 299.

David W. Shannon LLM, Barrister-and-Solicitor , David is a member of the Law Society Ontario practicing health law in Thunder Bay, Ontario. He received a Masters of Law degree at the London School of Economics, and has continued in his law practice and non-government organization leadership since then. His Dave Shannon Cross-Canada Tour in 1997 gained national media attention as he became the first quadriplegic to trek across Canada in a motorized wheelchair, and in 2009 he set another first by going to the North Pole celebrating the message that seemingly insurmountable barriers can be overcome.

Shannon has won numerous prestigious awards and honours, such as the Order of Ontario and the Order of Canada. His academic and professional body of work has been expansive within the realm of human rights and health law. David has also been a sessional lecturer at Lakehead University and Negahnewin College at both the undergraduate and graduate level as well keynote speaker at numerous conferences across North America.

Alexandra Giancarlo, PhD , is a multi-disciplinary settler scholar. She recently joined the University of Calgary as Assistant Professor in the Faculty of Kinesiology. Previously, she was a Postdoctoral Associate in Indigenous Studies-Sociology at Western University. Passionate about human rights, anti-racism, and cross-cultural learning, her research and advocacy have taken her from post-Hurricane Katrina rebuilding to consulting on ground-breaking Canadian accessibility legislation and to working with black Creole cowboys in rural Louisiana. She has researched legal aspects of Canadian disability policy since 2015, including for the Canadian Access and Inclusion Project (CAIP), a national coalition advising the government on the Accessible Canada Act (2019).

Elaine Toombs, PhD is a Banting postdoctoral fellow at Lakehead University and Thunderbird Partnership Foundation. Her research primarily focuses on First Nations mental health using community-based methods, including how eHealth interventions can be used with Indigenous people seeking treatment for substance use. As a co-investigator of a Canadian Institutes of Health Research project grant, her research examines how adverse childhood experiences may affect psychological and physical health outcomes within a First Nations people in residential treatment. In addition to her academic work, Dr. Toombs is a registered clinical psychologist, and provides assessment, intervention, and consultation services for First Nations children, adolescents, and adults at Dilico Anishinabek Family Care.

Chapter 17
MAID for Persons with Mental Illness as a Sole Eligibility Criterion

Sephora Tang, K. Sonu Gaind, and Timothy Lau

Abstract The practice of Euthanasia and Assisted Suicide (EAS), referred to in Canada as "Medical Assistance in Dying" (MAID), evolved within a span of six years from being initially restricted to individuals dying from a terminal illness, to those who are not dying but suffering intolerably from a disability, and was further projected to include eligibility for non-dying individuals suffering solely from a mental disorder by 2023. The Canadian Federal Government indicated in December of 2022 the intention of delaying expansion of MAID to include mental disorders. The delay was extended to March 17, 2024. The extension of the practice of MAID for Mental Disorders as a Sole Underlying Medical Condition (MD-SUMD) has generated extensive debate in professional bodies advocating for and against the inclusion of mental disorders in the provision of MAID. This chapter provides an overview of the chronology of events in Canada pertaining to the evolution of the public discourse on MAID for MD-SUMD, as well as some of the most important arguments put forth in the study of this issue. The chapter concludes with some recommendations on how this issue could be approached in Canada moving forward.

Keywords Mental disorder · Bill C-14 · Bill C-7 · Irremediability · Standards of care · Suicide · Psychosocial determinants of health · Capacity

17.1 Introduction

This chapter describes how Canada addressed the difficult question of expanding euthanasia and assisted suicide (EAS), referred to as "Medical Assistance in Dying" (MAID) in Canada, for patients suffering solely from a mental disorder. After establishing a chronology of events pertaining to the evolution of the public discourse on

---

S. Tang (✉) · T. Lau
University of Ottawa, Ottawa, ON, Canada
e-mail: sephora.md@gmail.com

K. S. Gaind
University of Toronto and Humber River Hospital, Toronto, ON, Canada

J. Kotalik and D. W. Shannon (eds.), *Medical Assistance in Dying (MAID) in Canada*, The International Library of Bioethics 104,
https://doi.org/10.1007/978-3-031-30002-8_17

MAID for Mental Disorders as the Sole Underlying Medical Condition (MD-SUMC) in Canada, the ensuing chapter takes a closer look at some of the most important arguments put forth by advisory bodies, experts, and discussion papers, and ends with some suggestions on how this issue could be approached in Canada moving forward.

17.2 A Chronology of the Canadian Discussion on MAID for Mental Disorders as the Sole Underlying Medical Condition (MD-SUMC)

EAS in Canada was decriminalized in 2016 through the passage of Bill C-14 for adults meeting eligibility criteria for MAID. This change in law was in response to a 2015 Supreme Court of Canada decision (Carter v Canada 2015) that found the blanket prohibition on assisted suicide contravened the Charter Right to Life of the plaintiffs. In the process of crafting the new law, one of the eligibility criteria, initially upheld as necessary by the Federal Government, was that natural death must be "reasonably foreseeable". This criterion was felt to strike a balance between allowing assisted death to those whose natural course of life was limited in the foreseeable future, and acting as a safeguard for those who might still have years to live. The "reasonably foreseeable death" criterion had the practical effect of excluding most people suffering solely from a mental illness from becoming eligible for MAID, since mental illnesses themselves rarely, if ever, lead to naturally foreseeable death. The eligibility criterion requiring that natural death be "reasonably foreseeable" helped to differentiate between those who would choose MAID to avoid a "painful death" from those wishing to avoid, through suicide, a "painful life" (Sinyor and Schaffer 2020).

After the 2015 Carter vs Canada Supreme Court decision, and prior to the establishment of a legal framework within which to carry out MAID with Bill C-14 in 2016, there was a narrow window of time within which individual applications for MAID could be brought to the judicial courts for independent review for approval. During this period, an application for MAID for sole mental illness was heard by the Court of Appeal of the province of Alberta (Canada vs E.F. 2016). The appellant was a 58-year-old woman with conversion disorder. According to court filings, the psychiatrist in the case opined on the irremediability of the patient's condition, and on the patient's capacity, based solely on chart review without ever seeing or speaking with the patient. The court approved the application, and the person died by MAID. This death raised concerns about the risks of providing MAID without evidence-based standards, and of individual assessors applying unscientific personal judgments in the absence of evidence or standards.

With the advent of Bill C-14, the Canadian government committed to arrange an independent review of several issues including requests for MAID where mental illness is the sole underlying medical condition. The Canadian Council of Academies

(CCA) was tasked with this responsibility, and released their report which summarized the findings of their review in 2018. After nearly a year and a half of expert review of international evidence, rather than consensus, this report captured the divergence in views on this issue (Canadian Council of Academies 2018).

In 2019, the Quebec Superior Court struck out the "reasonably foreseeable natural death" (RFND) eligibility criterion for MAID in Bill C -14 as being unconstitutional (Truchon v Procureur général du Canada 2019). Instead of appealing the decision made by one lower court judge to a higher- level court, the Canadian government embarked upon the process of amending the legislation in Bill C-14 by tabling Bill C-7 in February 2020. The government did not alter the terminology of the procedure for those whose deaths were not reasonably foreseeable and maintained the "Medical Assistance in Dying" phraseology.

In the first iteration of Bill C-7 in February 2020, wording of Bill C-7 excluded mental illness from the definition of "an illness, disease or disability" for the purposes of MAID. According to Justice Minister David Lametti, the issue of expanding MAID to mental illness as a sole eligibility criterion was "too complex to address in the context of a court-imposed time limit" and he suggested that it should instead be addressed during the statutory parliamentary review of the MAID law that was to begin in June 2020 (Government of Canada 2020a, 2020b).

Despite these early precautionary words, some Canadian senators felt strongly that MAID should be provided for sole mental illness. Following Senate consultations in late 2020 and early 2021, the Federal Government reversed its initial position and Parliament voted after minimal debate, and largely along party lines, to provide MAID for mental illness after a "sunset clause" of two years. This sunset clause was framed as allowing development of procedural instructions on how MAID for MD-SUMC would be provided. This, despite multiple expert testimonies at Senate and Parliamentary hearings that advised against expanding MAID for these conditions especially in light of the removal of the "reasonably foreseeable natural death" criterion, and which warned of the concerning lack of evidence guiding such expansion.

A discussion paper titled "*MAID Legislation at a Crossroads: Persons with Mental Disorders as Their Sole Underlying Medical Condition*" was released in January 2020 by the Halifax Group (2020), a select subset of members of the CCA Expert Panel on MAID for MD-SUMC, supporting and providing recommendations on how to implement MAID for MD-SUMC. The Expert Advisory Group (EAG), coordinated by Dr. K. Sonu Gaind, and comprised of experts on Canadian and international MAID policy and practices, and experts on issues of mental health and illness, subsequently released an evidence-based critique in February 2020 of the Halifax Group paper (EAG 2020). Both the Halifax Group and the EAG were multi-disciplinary and included participants of the original CCA taskforce, drawing from the expertise of those in psychiatry, medicine, and law. In addition, the EAG included the contributions of individuals who had lived experience with mental disorders. There was significant divergence in views and conclusions between these two reports. This lack of consensus became a recurring theme in subsequent public discourse as legislators and mental health experts grappled with the implications of expanding MAID to include mental illness as a sole eligibility criterion.

The Canadian Psychiatric Association (CPA), which had in 2016 cautioned against extending MAID to individuals with MD-SUMC due to the lack of evidence in accurately predicting irremediability in mental disorders (CPA 2016), issued a controversial position statement on MAID in February 2020, updated in 2021. While stating that it does not take a position on the legality or morality of MAID, the CPA's position statement nonetheless recommended that patients with psychiatric illness not be "discriminated against" and be given the same options for MAID as all other patients meeting eligibility requirements (CPA 2021). The CPA position statement made no reference to suicide prevention strategies or need for caution in addressing the difficulties in determining eligibility criteria for individuals in an expanded regime, and failed to address the issue of whether mental illnesses could be predicted to be irremediable.

The public, policy makers and media's perception was that the CPA supported expansion of MAID for MD-SUMC (Bryden 2021). The CPA position statement was criticized by members (Gaind 2020), including two past CPA presidents who issued an open letter outlining the CPA's lack of member engagement and lack of evidence-based guidance in forming the statement (McGregor and Gaind 2020). The Canadian Association for Suicide Prevention (CASP) (Krausert 2021), and Canadian Indigenous leaders (2021) likewise expressed alarm that the CPA was taking a position that supported MAID for mental illness while claiming neutrality. Indigenous leaders in Canada expressed further shock when the CPA's November 2021 updated position statement failed again to address the phenomenon of suicide contagion, the importance of suicide prevention, and the risks of suicide to marginalized populations; they expressed concern that the CPA's position statement would adversely affect Indigenous populations in Canada by providing recommendations that perpetuated "policies of privilege". (Ibid).

In November 2020, l'Association des Médecins Psychiatres du Québec (AMPQ) (2020) released their discussion paper which supported expansion of MAID for MD-SUMC and which attempted to establish a framework for a permissive regime for MD-SUMC. The AMPQ noted that a consultation amongst Quebec psychiatrists indicated that 54% of participants considered that MAID should be allowed for patients capable of consent who were suffering from mental illness, under certain conditions. In contrast, a survey of the largest group of psychiatrists in Canada, conducted on its members by the Ontario Medical Association (OMA) Section on Psychiatry in 2021, and which provided context for evolving policy and the "sunset clause", showed that 63% of participants opposed (vs 23% that supported) the position that "*patients with a psychiatric illness should have available the same options regarding MAID as available to all patients, without issues of irremediability in mental illness having been examined*" . (OMA Section on Psychiatry 2021).

In the meantime, the province of Quebec struck a commission to consult on the issue of MAID expansion. The commission released its report in December 2021 and recommended *against* providing MAID for MD-SUMC, and called on the relevant article 26 in Quebec's end of life legislation to be modified accordingly (Commission

Spéciale 2021). The commission cited in the rationale for their conclusions, the lack of consensus amongst experts in addressing fundamental questions relating to specific complexities found amongst mental disorders.

In May 2022, Canada's Ministers of Health and Justice tabled in Parliament the final report of the Expert Panel on MAID and Mental Illness (Gupta et al. 2022). The Expert Panel, appointed by Parliament, was tasked with limiting the scope of its review to "protocols, guidance, and safeguards" to apply to requests for MAID by persons who have a mental illness. Although the Expert Panel reported that consensus on the recommendations was unanimous, two panel members of the original twelve-member panel resigned before the release of the final report, with one contending that the recommendations were inadequate (Kirby 2022), and the other stating that she could not legitimize the process nor condone the conclusions of the Expert Panel which she felt had not meaningfully delivered on its responsibilities (Cohen 2022).

Consensus on MAID for MD-SUMC will not easily be achieved in Canadian Society, if at all. Under such circumstances, and recognizing the diversity of views on a matter affecting the life and death of its citizens, Canada would be wise to use utmost caution in determining its approach to this issue, and to engage in careful study before proceeding further. The Canadian Government indicated in December 2022 their intention to delay the March 2023 start date for including MAID for mental illness (Major 2022), subsequently extending the start date to March 17, 2024.

17.3 The Question of "Irremediability" in Mental Disorders

The difficulties in reliably prognosticating the course of illness and irremediability for mental disorders in any given individual is a key consideration when deliberating on expanding MAID to MD-SUMC, especially since legislative requirements for MAID eligibility require that the illness for which an individual is requesting MAID must be deemed "irremediable".

As with any chronic condition, certain mental illnesses may be deemed incurable, however an incurable illness does not necessarily mean that it is "irremediable", the term that Canadian law requires for eligibility. Some experts have written that there is a crucial difference between an incurable and an irremediable condition; lack of recovery from a mental disorder is not evidence that a condition is "unbearable". Some authors posit that whether a mental illness is truly "irremediable", rather than being very challenging to remediate, is a question that can be tested scientifically, but which remains unanswered (Sinyor and Schaffer 2020). Thus, data needed to inform questions pertaining to the nature of irremediability in mental disorders is lacking in informing public policy.

The 2018 CCA report highlighted the fact that it can be difficult for clinicians to ascertain when they have achieved a reasonable degree of certainty that a mental disorder diagnosis will persist over time, in comparison to the physical conditions that, prior to expansion, underlaid most MAID requests in Canada. The CCA report explained: "*If a person has lived with a mental disorder for a long time, diagnostic change may be less likely. However, diagnostic stability does not guarantee accurate prognostication (i.e., the prediction of the course of the illness over time.)*" (Canadian Council of Academies 2018).

While the Halifax Group (2020) asserted that "*in specific cases, the eligibility criteria of enduring, intolerable and irremediable suffering, and advanced and irreversible decline in capability could be met*", the EAG report on MAID concluded that "*determinations of irremediability and irreversible decline cannot be made for mental illnesses at this time*". (EAG 2020).

Similarly, the Centre for Addiction and Mental Health, an internationally recognized authority, testified before the Canadian Senate during the Legislative Debate on Bill C-7 that "*there are currently no established criteria that define if and when a mental illness should be considered irremediable*". They testified that currently, there is not enough evidence in the field to predict the trajectory of any one person's mental illness and to ascertain irremediability in mental illness. The "irremediable criteria" would therefore be "*open to interpretation*" by the individual MAID assessor, and that any determination of irremediability for mental illness would be "*inherently subjective and therefore arbitrary*". They cautioned that this could put people with mental illness at risk of accessing MAID when they do not, in reality, meet the eligibility criteria (Senate 2020).

The CCA captured the crux of the issue by stating that the issue was "*not whether there are people who have mental disorders that are irremediable but rather whether clinicians can confidently determine whether a particular case is irremediable*" . (Canadian Council of Academies 2018).

The uncertainty in accurately predicting irremediability in mental disorders was the key factor that dissuaded the Quebec Commission from recommending MAID for MD-SUMC in 2021. Given the obvious irremediable nature of MAID, doubts about the incurability of mental disorders led to exercising utmost caution in concluding that it was not appropriate to extend MAID to people whose only medical problem is a mental disorder. While they found that such expansion was favoured by certain groups including the *Collège des Médecins* and the *Association des Médecins Psychiatres du Québec*, they noted that this position was far from a consensus within the medical profession, and lacked evidence (Commission Spéciale 2021).

In the absence of scientific evidence, advocates of MAID for MD-SUMC have justified their position by invoking a values-based argument. While acknowledging that "*diagnosis generally does not guarantee a prognosis, and deciding with certainty on the incurability of a disorder is not only very difficult, but often impossible*", the AMPQ (2020) report nonetheless asserts that "*eligibility criteria must be interpreted in light of the specificity of the mental disorder*" and that "*these criteria are not based only on facts, but also on a person's values*". The AMPQ report opines that it is an "*ethical question*" for assessors to determine the certainty of eligibility for someone requesting MAID for mental illness, while at the same time acknowledging that "*the underlying pathophysiological processes involved in most mental disorders are unknown*" and that "*a person who has recourse to MAID could have regained the desire to live at some point in the future*". (Ibid).

The Final Report of the Expert Panel on MAID and Mental Illness acknowledged that due to the unknown underlying pathology for the majority of mental disorders, the terms "incurable" and "irreversible" used in Canadian MAID legislation are difficult to apply to mental disorders. The Final Report also acknowledged the difficulty, if not impossibility, for clinicians to make accurate predictions about an individual's future in regards to the evolution of that individual's mental disorder. This was a failure of the Expert Panel's mandate to provide definitive protocols for providing death for mental disorders as a sole underlying medical condition. The Final Report stated that evaluating a mental disorder's treatment responsiveness on the basis of past treatments can be done, but that it was an "ethical" choice. It also did not provide objective criteria with which to make these determinations. The Final Report indicated that it is not possible to provide "fixed rules" for how many and what kinds of treatment/intervention attempts are required, nor the time period required to make the determinations of incurability and irreversibility. The Final Report recommended that the determinations of irremediability and incurability be assessed on a "case-by-case" basis, with the requester and assessor coming to a "shared understanding" that the person has a serious and incurable illness, disease or disability. In cases where there is too much uncertainty, the person may not be found eligible for MAID. (Gupta et al. 2022). However, there is nothing to stop patients from finding other assessors who will accede to their request.

It is critically important that public health policy that risks providing death to individuals who would have otherwise recovered should be based on credible evidence and standards, not on arbitrary individual value judgements of assessors (Gaind and Tang 2021). Without objective evidence that determinations of irremediability for mental disorders can be accurately predicted, the determinations would be largely subjective decisions reflecting the underlying philosophy and values of individual assessors.

17.4 Treatment Refusal and Substandard Care

The clause in Canadian legislation that permits individuals to decline treatments they find unacceptable and yet remain eligible for MAID, is particular to Canadian law. As the AMPQ (2020) report states, in Canada "... *a capable person may refuse a treatment considered by the physician to offer a reasonable prospect of relief from his or her suffering*". If standard treatment options do not need to be tried first, this raises challenges as to how one is to determine the irremediability of an illness. Other jurisdictions worldwide require a lack of "reasonable alternatives" prior to MAID being offered. (EAG 2020). In this regard, the clause makes Canada one of the most permissive regimes in the world at the time of this writing.

Prior to the inclusion of mental illness in Bill C-7, Lemmens et al. (2021) had written of the problematic displacement of medical standards of care in pursuit of

MAID under the current legislative framework: "... *since physicians cannot offer treatments that run counter to the standard of care, then surely they cannot, pursuant to the standard of care, offer and provide to the patient the ending of their life—an ultimate and irreversible harm—if they are convinced that other medical procedures or support measures will provide relief*". Lemmens et al. stressed the importance of upholding standards of care as they apply to the mental health context if mental illness is accepted as a basis for MAID requests: "*Imagine what it will mean for the standard of care in mental-health care practice and suicide prevention if a severely depressed person, who is assessed as capable of decision-making, refuses all available treatment and support measures and insists on obtaining MAID instead*". (Ibid).

The Expert Panel on MAID for Mental Illness indicated in their 2022 final report that the "*incurability of a mental disorder cannot be established in the absence of multiple attempts at interventions with therapeutic aims*". (Gupta et al. 2022). The Expert Panel further suggested that a "*capable refusal of treatments with a favourable benefit/burden balance will not lead to automatic access to MAID as in such cases, assessors may not be able to form an opinion about the incurability of the condition or the irreversibility of decline*". (Ibid). At the same time, the Expert Panel indicated that "*incurability and irreversibility do not require that a person has attempted up to that point in time, every potential option for intervention irrespective of the potential harms, nor that a person must attempt interventions that exist but that are inaccessible. The requester and assessors must balance potential benefits and burdens going forward of any remaining interventions*". (Ibid).

Unique to Canada, Canadian legislation lacks the stipulation that euthanasia and assisted suicide be options of last resort after all other treatments deemed appropriate by the patient and their doctor have failed. Without such safeguards, a hopeless suicidal patient could refuse standard treatment options for mental illness (or have lacked access to treatment) and request physician-assisted suicide and be found eligible under the law by a willing practitioner attesting to the eligibility criteria based on their "ethical" belief system. The rate limiting factor to receiving MAID would thus be the accessibility of practitioners willing to endorse such requests. The lack of legislative and evidence-based expectations upholding medical standards in cases of MAID for mental illness is concerning.

17.5 Suicide and MAID for Mental Disorders

In regards to risk of suicide, the preamble to the original Bill C-14 acknowledged that "*vulnerable persons must be protected from being induced, in a moment of weakness, to end their lives*" and that "*suicide is a significant public health issue that can have lasting and harmful effects on individuals, families and communities*". (Bill C-14 2016). In a country where 1 in 5 Canadians suffer from a mental illness in any given year (Smetanin et al. 2011), suicide prevention initiatives received increased attention after the 2012 Mental Health Strategy for Canada highlighted the burden of suicide on individuals, families, and communities, and the need to increase mental health services and suicide prevention awareness (Mental Health Commission of Canada 2012; Government of Canada, Federal Framework for Suicide Prevention 2020b). Suicide is the second leading cause of death among Canadians aged 15–34 (Navaneelan 2017), and suicide among Indigenous populations is disproportionately high with a suicide rate of 78.1/100,000 in 2019, seven times the national average (Cross-Canada Comparison Statistics 2021). The Quebec commission highlighted concerns on the impact that broadening access to MAID for all people with mental disorders would have on the area of suicide prevention. The commission also highlighted the impact of social norms, cautioning that rendering suicide more acceptable, or even "trivialized", could send the message that death is a legitimate or even appropriate option for mental disorders, undermining "decades" of efforts to prevent suicide. (Commission Spéciale 2021).

In the wake of decriminalizing EAS in Canada, Quebec's College of Physicians found it necessary to issue a bulletin in 2016 after emergency physicians reportedly failed to offer life-saving treatments to patients presenting to the hospital following a suicide attempt. The change in Canadian law led to confusion amongst doctors about their moral duty to provide life-saving treatment in emergency settings, even in the case of a toxic overdose from a suicide attempt. The College reminded doctors that it would be negligent to not save life under those circumstances. The bulletin also acknowledged the role of psychiatric illness in suicide, and how difficult it can be at times to assess for capacity to consent to treatment for cases of psychiatric illness (Collège des médecins du Québec 2016).

The change in Canadian law to expand eligibility for EAS to those whose death is not reasonably foreseeable sends implicit messaging to society about the acceptability of suicide as a means to end life suffering. Some individuals have expressed their opinion that permitting individuals with mental illness to request MAID does not impede suicide prevention efforts in circumstances where suicidal desires are present in the context of acute and resolvable stressors (Downie et al. 2021). This opinion fails to consider patients with mental illnesses who suffer from chronic and fluctuating intensities of suicidal ideations for whom suicide prevention methods, namely, restricting access to lethal means, can prolong life and allow them recovery to build a life worth living.

Patients who suffer from personality disorders, such as borderline personality disorder, have DSM-5 diagnostic criteria that include recurrent suicidal behaviours. Formerly thought to be a chronic condition, more recent evidence demonstrates that patients who suffer from borderline personality disorder can get better over time with appropriate therapy (Mehlum et al. 2020). Many suicides are not impulsive, and those who have chronic suicidal but ambivalent thoughts and who may be meticulously planning their suicides, are currently offered protection by a healthcare system designed to assist in saving life. With the advent of MAID in Canada, and the recent removal of the reasonably foreseeable death criterion, the State has removed its ability to safeguard individuals from the enticements of suicide and their own ambivalent suicidal urges.

Under Canadian law, it will be difficult to provide expected standards of medical care when individuals are allowed to refuse standard treatments with the desire to pursue instead an assisted death over therapy options that can sometimes take years, and which may not be readily accessible. Granting an assisted death over therapy options, because a social system has been devised to permit it, is unjust to patients for whom the medical profession has pledged to "first do no harm".

The recognition that media reports on suicide lead to increased suicide rates, and conversely, that media reports on individuals overcoming suicidal crises can lead to lower suicide rates, led to the development of media guidelines on suicide reporting (Sinyor et al. 2018). Amongst the identified harmful elements to avoid suicide contagion in suicide reporting, the guidelines issued by the Canadian Psychiatric Association in 2017 included: avoiding prominent coverage, excessive detail including details or photos of the method, glorifying or glamourizing either the person or the act of suicide in a way that might lead others to identify with them, portraying suicide as achieving results and solving problems, and describing suicidal behaviour as quick, easy, painless, or relieving suffering. Canadian media coverage of MAID deaths since its legalization have espoused the opposite, with the effect of normalizing, and even glamourizing the intervention (Favaro 2022a; Touzin 2019). Media coverage of this kind can be perceived as encouraging individuals to consider an assisted death in lieu of a self-administered suicide. The notion of MAID as an accepted solution to unbearable suffering negatively impacts the efforts of psychiatrists to engage their patients in the often-lengthier process of finding hope and meaning in recovery of chronic illness.

Appearing before the Canadian Senate, Ontario psychiatrist Dr. John Maher testified that the Canadian Law on MAID has already profoundly undermined clinical relationships (Senate 2021). He stated that his patients who are getting better, but can't yet recognize it, ask for MAID and question why they should try to recover when they will soon be able to "choose death" once MAID becomes available to them (Karel 2021). The implicit messaging of legalized assisted-suicide introduces into the clinical relationship the possibility that one's physician can become an instrument to procure, and support, one's death. When death becomes a treatment rather than a consequence of an illness, the role of medicine and of a psychiatrist to prevent suicide, is subverted. For patients who struggle daily with suicidal thoughts and urges, this shift in the clinical relationship can potentially undermine the establishment of recovery-oriented therapeutic goals.

While restricting access to lethal means is a recognized suicide prevention strategy (World Health Organization 2010), there are concerns that providing the means in assisting suicide undermines suicide prevention efforts. The EAG (2020) reported evidence that characteristics of people seeking physician-assisted death for MD-SUMC demonstrate overlap with those contemplating suicide. The AMPQ (2020) discussion paper referenced Dutch guidelines stating that "*the apparent distinction between chronic suicidality and a request for assisted dying can disappear in those cases where a person is able to make a reasoned decision about their condition. However, they call for special attention in situations where a person makes what seems to be a request for euthanasia in the context of a self-destructive, borderline (in the technical sense) relational dynamic*". Despite these precautions, the study by Kim et al. (2016) revealed that 52% or 34/66 cases of the known cases of physician-assisted suicide for MD-SUMC in the Netherlands occurred in patients who had prominent personality disorders which can often be associated with self-destructive tendencies (Mehlum et al. 2020).

The Halifax group (2020) recommended against the inclusion of a "non-ambivalence" safeguard in favour of a "well-considered" clause, yet suggested that "well-considered" should "*not require an assessment of the quality of the decision the person is making…but rather an assessment of the decision-making process to ensure that it is well thought out and not impulsive*". The EAG (2020) argued for the inclusion of a "non-ambivalence" safeguard since ambivalence is recognized as a key feature of suicidal individuals for whom suicide prevention initiatives successfully intervene to keep alive to provide the chance to get better, and evidence shows not all ambivalent suicides are "impulsive". A non-ambivalence criteria had also previously been put forth by the Vulnerable Persons Standard which sought to provide evidence-based safeguards intended to protect Canadian lives with the first iteration of the MAID law (Vulnerable Persons Standard 2017).

From a public policy perspective, expansion of MAID eligibility to include individuals whose deaths are not reasonably foreseeable rested upon the view that MAID did not undermine suicide prevention or cause suicide contagion. In the Quebec Superior Court Case for *Truchon v Procureur général du Canada* (2019), Justice Beaudoin highlighted the "*lack of evidence that MAID undermines suicide prevention or that it has increased suicide rates*" where permitted outside the end-of-life context. Subsequent hearings by the Canadian Senate during the Study of Bill C-7 also highlighted the reported "lack of evidence" for a suicide contagion effect (Senate 2020), and was cited by the Expert Panel report on MAID and Mental Illness (Gupta et al. 2022). However, new evidence has emerged that calls into question the safety of permitting MAID outside the end-of-life context as it pertains to suicide rates. A 2022 study by Oxford Scholar David Jones demonstrated an *increase* in suicide in jurisdictions where EAS have been introduced. Jones (2022) also found that in jurisdictions permitting EAS, there was no reduction in unregulated suicide, relative to the most similar neighbouring jurisdiction that did not permit EAS, and in some cases, there was a relative and/or absolute *increase* in non-assisted suicide. Any country that is truly committed to preventing suicides would seriously consider the implications of this new evidence before introducing permissive EAS laws.

While efforts have been made to strengthen suicide prevention methods in Canada, Canada's expanded MAID law, media reporting, and suggestions by some discussion groups to include less rigorous criteria for eligibility, undermine principles of suicide prevention strategies by publicizing easier access to lethal means of ending one's life with medical assistance. While there may be a belief that certain suicides should not be prevented, and that individuals who have made a "rational" choice for suicide should be permitted to do so with assistance, the difficulty lies in coming to a consensus on which individuals should be assisted to live, and those who should be assisted in ending their life. It is unlikely that a consensus will be reached, and the fate of individual lives under a newly expanded MAID law will likely be determined by individual judgements of assessors making decisions based on individual value judgements. As such, it is imperative that suicide prevention interventions be offered first to *all* individuals presenting with a termination of life request (Mishara and Weisstub 2018).

17.6 Psychosocial Determinants of Health and Limitations of Current Mental Health Resources

Most experts acknowledge psychosocial factors can influence suicidal ideations. The EAG (2020) noted that there is evidence to suggest that the characteristics of individuals seeking euthanasia for mental health reasons differ from those seeking euthanasia for medical reasons, and that it is often the most disenfranchised and marginalized of society that seek physician-assisted suicide for solely mental health reasons. The AMPQ (2020) paper wrote that "*a regime that would allow MAID for persons with MD-SUMC should be accompanied by a commitment from society as a whole to improve access to quality care*". The AMPQ paper acknowledged the problem of scarcity of resources for mental health in Canada, a concern shared by the Quebec Commission in its deliberations: "*The availability of therapeutic interventions can vary considerably depending on where a person lives, the local service organization, and one's individual financial means. Unfortunately, access to mental health services is often suboptimal in several regions in Québec, from first-line care to specialized programs*". (AMPQ 2020). The Halifax Group and some practitioners felt that the scarcity of resources should not be an impediment to the expansion of MAID for MD-SUMC, citing that it has been possible to improve the quality and access to palliative care while implementing MAID when the End-of-Life Care Act was passed (Commission Spéciale 2021). However, as per the 2021 Third Annual Report on MAID in Canada, of 1675 patients who received MAID in 2021 in the absence of palliative care, 201 of those *did not* have *access* to palliative care, and of 4,278 individuals who received MAID in 2021 who required disability support services, 179 did *not* receive needed support services (Health Canada 2022).

In regards to access to psychiatric care, a 2021 study by the Fraser Institute revealed that wait times for psychiatric treatment in Canada ranged from 118 days in Quebec to 477.4 days in New Brunswick, with wait times for urgent care ranging from 2 to 4.5 weeks (Barua and Moir 2021). These wait times do not represent the time required for psychiatric treatments to become effective, which can range from several weeks to years in the case of psychotherapies and recovery-oriented models of care. When juxtaposed against the current legislative requirements for a 90-day assessment period for MAID for those whose deaths are not reasonably foreseeable, and Canada's unique lack of requiring that treatment be tried, the problem of scarce resources becomes evident in the context of current legislation. Psychiatrists surveyed in Quebec felt that a more appropriate wait time to determine eligibility for MAID for MD-SUMC ranged from at least 5 to 10 years (AMPQ 2020). However, explicit eligibility timelines mean very little unless timely mental health and support services can be equitably provided to all Canadians.

17.7 Capacity to Consent

Impaired insight is a common feature of mental illness. Disagreements on patients' capacity to consent to care are common amongst clinicians, and review boards often disagree with physician findings because the matter is often not clear. Patients tend to be found incapable when they decline treatments that are offered; or when they are likely to experience harm that they do not fully appreciate without treatment, that is reversible with treatment. Although the principles of capacity assessments extend beyond MAID, it is critical in the capacity assessment for MAID because the proposed "treatment" in this instance becomes the "cure" that ends life, rather than saves life.

In regards to capacity to consent to MAID for MD-SUMC, the AMPQ (2020) acknowledges the difficulties in assessing decision-making capacity in individuals with MD-SUMC and thus recommends that the assessment take into account several domains in the person's life beyond cognitive criteria such as "*emotional reactions, interpersonal dynamics, and values resulting from the disorder which negatively affect the patient's ability to weigh options and make judgements*". The AMPQ felt a MAID request may not be granted if assessors differ in opinion on capacity assessment and where insurmountable disagreement arise. (Ibid*)*. Similarly, the Halifax group felt that "*where the challenges with respect to capacity are too great, practitioners will not be able to reasonably form the relevant opinion and therefore individuals will not be eligible for MAID* ". (Halifax Group 2020). The Federal Expert Panel echoed these views in their final report: "*It is also possible... assessors will not be able to form an opinion about eligibility for a given case because there is too much uncertainty or complexity .... In such cases, the practitioner cannot find the person eligible for MAID*". (Gupta et al. 2022). However, one assessor's opinion of capacity is independent of another's, and nothing prevents a patient from seeking another provider who will find them capable.

International experience has shown that MAID is granted despite differing opinions in assessors. The CCA report noted that in the Netherlands, there has been an increase in psychiatric EAS cases performed by the formerly named "End-of-Life Clinic" (Levenseindekliniek), now renamed the "Expertisecentrum Euthanasie". According to the CCA report, the Levenseindekliniek organization did not treat patients for illnesses, but evaluated and provided EAS to those whose doctors declined to perform the procedure. One-third of all applicants to the Levenseindekliniek are psychiatric patients. In 2016, 75% of Dutch psychiatric EAS cases were performed by the Levenseindekliniek. In comparison, the Clinic performs approximately 5–8% of the total number of EAS cases in the country. (CCA 2018).

Kim et al. (2016) noted that in psychiatric cases of euthanasia in the Netherlands, physicians consulted extensively, but 11% of cases had no independent psychiatric input, and 24% of cases involved disagreement among consultants. What becomes then of cases that involve disagreements between consultants? In Canada, under the current law, it would be sufficient to find one other physician (who need not have a relationship or in-depth knowledge of the patient) to agree with the finding of capacity to meet eligibility requirements. In other words, even if 98/100 physicians deem a patient to be incapable, it is sufficient to find the 2/100 physicians who agree that the patient has capacity in order to meet the legal requirements. Thus, it is questionable whether the expectation of requiring two consenting opinions supporting MAID will be a meaningful so-called "safeguard" in Canada, especially if individuals are allowed to seek third, fourth, fifth or more opinions to support their request for death when they are not dying.

17.8 Suggested Recommendations

While continuing to prohibit MAID for MD-SUMC would be the surest way to ensure that everyone experiencing mental illness is afforded equal protection under the law, Canadians will continue to grapple with the complexity that MAID for MD-SUMC presents since the passage of Bill C-7 in 2021. Before proceeding further, the Canadian government should delay the start of the implementation of a pre-determined expanded MAID regime that includes MD-SUMC as a sole eligibility criterion in order to adequately study and consider the effects of expansion. The following recommendations are suggested interventions that should be considered during this period of study.

(1). Review Process: The government must complete a 5-year review of the current processes in place to address concerns of abuse and inappropriate application of the law before further expansion. There should also be a review every 5 years thereafter to re-evaluate the regime. Any concerns raised about the potential of abuse in the system should lead to a moratorium and review of the law to prevent further misuse/abuse/unintended consequences, and to ensure appropriate oversight. In addition to reviewing after-the-fact reports of MAID cases, audits should be undertaken to review cases selected for MAID *before* MAID is administered, to assess whether eligibility criteria are being met.

(2). Clearer legislative eligibility criteria: In accordance with the State's responsibility to ensure that the lives of vulnerable and marginalized people are protected, the eligibility criteria for a life-ending process needs to be clarified with stricter legislative criteria that can ensure consistent inter-rater reliability in the identification of eligible patients. The criteria outlined in the current law is purely subjective and reliant on individual choice, without regard to whether medical standards or evidence have been applied prior to granting eligibility for an assisted death. At the very least, the provision of assisted death for MD-SUMC should be deemed an option of last resort. Like in other permissive jurisdictions such as the Netherlands where psychiatric euthanasia is provided, the clinician should have an obligation to confirm that all reasonable treatment options have been exhausted.

(3). Suicide Prevention: Suicide prevention interventions should be provided first in response to all requests for assistance in life termination.

(4). Informed Consent: For true informed consent, patients should not only be *informed* of treatments, but there should be a requirement that standard treatment and supports other than death be *offered* prior to a decision being made on eligibility.

(5). Ensure accessibility of interventions: Treatments options need to be *accessible*. This includes ensuring that psychosocial factors influencing the individual's request for MAID, including food scarcity, appropriate housing, access to medication and psychotherapeutic treatments, social and personal supports, and assisted living options, have been adequately addressed. Assisted death should not be offered as a response to a lack of assistance in living. It is deeply concerning that individuals have already chosen and received death by MAID due to lack of adequate supports in living (CBC News 2019; Favaro 2022b; Fidelman 2016; Noel 2019) and that others have been offered MAID in the face of inadequate living supports even before the law was expanded in 2021 (CTVNews.ca 2018; Gupta 2019). The expansion of the MAID law to include mental disorders in non-dying individuals will significantly increase the numbers of people who could be found eligible. Under the current context, the State has a duty to ensure that equitable access to treatment for mental illnesses is as readily available as services being organized to end lives.

(6). Patient-initiated discussions: In order to preserve the trust that is found implicitly in the therapeutic relationship between a psychiatrist and their patient, psychiatrists in particular should be guarded from having to recommend or approve assisted-dying to their patients unbidden, in order to preserve the therapeutic milieu that should first be oriented towards recovery. To avoid the perception of coercion by an overt recommendation or suggestion for MAID, MAID discussions should only be initiated by the patient.

(7). Scope of psychiatric expertise and role: Psychiatrists with their clinical expertise should be relied upon, when asked, to provide an opinion on the patient's diagnosis, and recommendations for further potential therapies or treatments, as well as an opinion on the patient's prognosis. Clinicians should refer to psychiatric expertise when evaluating eligibility for MAID for MD-SUMC.

(8). Reduce risk of "doctor-shopping": To minimize the risk of "doctor-shopping" to find a clinician ideologically inclined to approve a MAID request, in the event of a disagreement between clinician opinions, the case should be reviewed by a panel of experts whose decision should be final.

(9). Oversight and data collection: Rigorous oversight must be maintained in collecting data on all cases where MAID has been administered for MD-SUMC in order to be able to provide data for future research on understanding the reasons for MAID requests; the social context of the patient who requested MAID; treatments tried, proposed and/or ultimately declined; and details in the evaluation process for those who were approved or declined an assisted death. Such data collection will be critical to refine guidelines provided to clinicians in assessing patients with MD-SUMC, detecting abuse of the law and providing accountability in the process.

(10). Public messaging and media regulations for suicide prevention: From a public health and safety standpoint, the view of suicide as a tragedy to be prevented, not a solution to be facilitated, should be reinforced in public messaging to counter the implicit messaging in MAID law that suicide is an appropriate response to life suffering. Media coverage of MAID stories, and especially news coverage of MAID for mental illness, should follow the same media reporting guidelines as established for suicide.

17.9 Conclusion

The expansion of MAID to include MD-SUMC in Canada is challenged with uncertainty, ethical dilemmas, and the impossibility of ascertaining whether a person requesting MAID truly meets the legislative requirements of irremediability. In the context of a Canadian healthcare system that is still unable to provide timely access to psychiatric care, the risk that individuals could seek and be granted MAID as a response to poor availability of psychiatric care is a serious factor to consider. The danger that an expanded MAID law for mental illness poses for individuals who normally turn to the healthcare system for help and assistance during their most vulnerable period is highlighted by the fact that the medical system is being structured to endorse and facilitate the suicides of non-dying individuals deemed eligible for life-termination. The implications of these changes on suicide prevention initiatives and public messaging on suicide has important and far-reaching consequences on society's view of suicide, particularly in light of the high suicide rates of Canadian Indigenous and other marginalized populations.

Canada's legislative expansion of MAID to include MD-SUMC has raised more questions than solutions, fractured the medical profession, and led to divergent conclusions amongst legislators. How Canada proceeds to manage these questions to implement an expanded MAID regime remains to be seen.

References

AMPQ (Association des Médecins Psychiatres du Québec) 2020. Discussion Paper: Access to Medical Assistance in Dying for People with Mental Disorders. https://ampq.org/wp-content/uploads/2020/12/mpqdocreflexionammenfinal.pdf

Barua, Bacchus, and Mackenzie Moir. 2021. Waiting Your Turn: Wait Times for Health Care in Canada, 2021 Report. Fraser Institute. https://www.fraserinstitute.org/sites/default/files/waiting-your-turn-2021.pdf

Bill C-14. 2016. An Act to Amend the Criminal Code and to Make Related Amendments to Other Acts (Medical Assistance in Dying). First Session, Forty-second Parliament.

Bryden, Joan. 2021. Senators Amend MAID Bill to put 18-Month Time Limit on Mental Illness Exclusion. CTV News. https://www.ctvnews.ca/politics/senators-amend-maid-bill-to-put-18-month-time-limit-on-mental-illness-exclusion-1.5302151. Accessed 31 Jan 2022.

Canada (Attorney General) vs E.F. 2016. Alberta Court of Appeal.

Canadian Council of Academies. 2018. The State of Knowledge on Medical Assistance in Dying Where a Mental Disorder is the Sole Underlying Medical Condition. The Expert Panel Working Group on MAID Where a Mental Disorder is the Sole Underlying Medical Condition, Ottawa, ON.

Canadian Indigneous Leaders. 2021. Indigenous Leaders' Open Letter to CPA. https://static1.squarespace.com/static/5ec97e1c4215f5026a674116/t/619f0a18065082544058cc6e/1637812760998/Indigenous+leaders+open+letter+-+FIN+-+CPA+and+MAID.pdf

Carter v Canada (Attorney General). 2015. Supreme Court of Canada.

CBC News. 2019. B.C. Man with ALS Chooses Medically Assisted Death After Years of Struggling to Fund 24-Hour Care. https://www.cbc.ca/news/canada/british-columbia/als-bc-man-medically-assisted-death-1.5244731. Accessed 9 Feb 2022.

Cohen, Ellen. 2022. Why I Resigned From the Federal Expert Panel on Medical Assistance in Dying. The Globe and Mail. https://www.theglobeandmail.com/opinion/article-expert-panel-maid-mental-illness/. Accessed 17 Dec 2022.

CPA. 2016. Canadian Psychiatric Association Interim Response to Report of the Special Joint Committee on PAD. https://www.cpa-apc.org/wp-content/uploads/2016-04-11-CPA-Letter-Minister-Wilson-Raybould-PAD.pdf

CPA. 2021. Canadian Psychiatric Association: Position Statement: Medical Assistance in Dying.

Collège des médecins du Québec. 2016. Le Geste Suicidaire: Expression Systématique d'un Refus de Traitement? Avis du groupe de travail en éthique clinique. http://www.cmq.org/publications-pdf/p-7-2016-02-01-fr-geste-suicidaire-expression-systematique-refus-traitement.pdf?t=1644117853550

Commission Spéciale sur l'Évolution de la Loi Concernant les Soins de Fin de Vie. 2021. Rapport de la Commission Spéciale sur l'Évolution de la Loi Concernant les Soins de Fin de Vie. Québec: Bibliothèque et Archives nationales du Québec.

Cross-Canada Comparison Statistics. https://www.suicideinfo.ca/local_resource/cross-canada-comparison-statistics/. Accessed 15 Nov 2021.

CTVNews.ca. 2018. Chronically Ill Man Releases Audio of Hospital Staff Offering Assisted Death. CTV News. https://www.ctvnews.ca/health/chronically-ill-man-releases-audio-of-hospital-staff-offering-assisted-death-1.4038841. Accessed 9 Feb 2022.

Downie, Jocelyn, David Wright, and Mona Gupta. 2021. What's the Relationship Between Suicide and MAID? Policy Options Politiques.

EAG. 2020. Canada at a Crossroads: Recommendations on Medical Assistance in Dying and Persons with Mental Disorder-An Evidence-based Critique of the Halifax Group IRPP Report, Toronto, Ontario. https://doi.org/10.13140/RG.2.2.36236.87687 (or) https://doi.org/10.13140/RG.2.2.36236.87687/1 (or) https://www.eagMAID.org/report. Accessed 30 May 2022.

Favaro, Avis. 2022a. A Rare Look at Canada's Growing Demand for Medical Assistance in Dying. CTV News. https://www.ctvnews.ca/w5/a-rare-look-at-canada-s-growing-demand-for-medical-assistance-in-dying-1.5854612. Accessed 18 Apr 2022a.

Favaro, Avis. 2022b. Woman With Chemical Sensitivities Chose Medically-assisted Death After Failed Bid to Get Better Housing. CTV News. https://www.ctvnews.ca/health/woman-with-chemical-sensitivities-chose-medically-assisted-death-after-failed-bid-to-get-better-housing-1.5860579?cache=lxaherxk%3FclipId%3D89619. Accessed 18 Apr 2022b.

Fidelman, Charlie. 2016. Life in Long-term Hospital "Unbearable": Montreal Man with ALS. Montreal Gazette. https://montrealgazette.com/news/local-news/life-in-long-term-hospital-unbearable-montreal-man-with-als. Accessed 9 Feb 2022.

Gaind, Karandeep Sonu. 2020. What Does Irremediability in Mental Illness Mean? The Canadian Journal of Psychiatry 65(9): 604–606. https://doi.org/10.1177/0706743720928656. https://pubmed.ncbi.nlm.nih.gov/32441132/

Gaind, K. Sonu, and Sephora Tang. 2021. MAID for Mental Illness Opens Dangerous Doors. The Hamilton Spectator. https://www.thespec.com/opinion/contributors/2021/02/16/maid-for-mental-illness-opens-dangerous-doors.html. Accessed 26 Mar 2022.

Government of Canada. 2020a. Bill C-7: An Act to Amend the Criminal Code (Medical Assistance in Dying). Publication No. 43–1-.C7-E. Library of Parliament Legislative Summary of Bill C-7. https://www.justice.gc.ca/eng/trans/bm-mb/other-autre/c7/summary-resultats.html#s2-2

Government of Canada, The Federal Framework for Suicide Prevention. 2020b. Working Together to Prevent Suicide in Canada. Progress Report. https://www.canada.ca/content/dam/hc-sc/documents/services/publications/healthy-living/2020b-progress-report-federal-framework-suicide-prevention/2020b-progress-report-federal-framework-suicide-prevention-eng.pdf

Gupta, Rahul. 2019. Confined to a Toronto Nursing Home Bed 24 Hours a Day, Tommy Sec Wants to Die. Toronto Star. https://www.thestar.com/news/gta/2019/11/18/confined-to-a-toronto-nursing-home-bed-24-hours-a-day-tommy-sec-wants-to-die.html. Accessed 9 Feb 2022.

Gupta, Mona, Rose M. Carter et al. 2022. Final Report of the Expert Panel on MAiD and Mental Illness. Ottawa, Ontario: Health Canada.

Halifax Group. 2020. MAID Legislation at a Crossroads: Persons with Mental Disorders as Their Sole Underlying Medical Condition. IRPP Report. Montreal: Institute for Research on Public Policy.

Health Canada. 2022. Third Annual Report on Medical Assistance in Dying in Canada. Ottawa, Ontario: Government of Canada. https://www.canada.ca/content/dam/hc-sc/documents/services/medical-assistance-dying/annual-report-2021/annual-report-2021.pdf

Jones, David. 2022. Euthanasia, Assisted Suicide, and Suicide Rates in Europe. Journal of Ethics in Mental Health 11.

Karel, Richard. 2021. Updated Physician-Aid-in-Dying Law Sparks Controversy in Canada. Psychiatric News. https://doi.org/10.1176/appi.pn.2021.6.31. Accessed 7 Feb 2022.

Kim, Scott Y.H., Raymond G. De Vries, and John R. Peteet. 2016. Euthanasia and Assisted Suicide of Patients With Psychiatric Disorders in the Netherlands 2011 to 2014. *JAMA Psychiatry* 73(4): 362–368. https://doi.org/10.1001/jamapsychiatry.2015.2887.

Kirby, Jeff. 2022. MAID Expert Panel Recommendations Are Inadequate, Contends Panel Member Who Resigned. Hill Times. Accessed June 16, 2022. https://www.hilltimes.com/2022/06/16/MAID-expert-panel-recommendations-are-inadequate-contends-panel-member-who-resigned/367356

Krausert, Sean. 2021. Letter to CPA Re: CPA's Discussion Paper on Medical Assistance in Dying. https://static1.squarespace.com/static/5ec97e1c4215f5026a674116/t/619bc64ac482815740760931/1637598795070/Letter+to+CPA+re+Discussion+Paper+on+MAID+%282021+11+20%29.pdf. Accessed 16 Jun 2022.

Lemmens, Trudo, Mary Shariff, and Leonie Herx. 2021. How Bill C-7 Will Sacrifice the Medical Profession's Standard of Care. Policy Options Politiques. https://policyoptions.irpp.org/magazines/february-2021/how-bill-c7-will-sacrifice-the-medical-professions-standard-of-care/

Major, Darren. 2022. Government Seeking Delay to Maid Expansion That Would Cover Mental Illness. CBC News. https://www.cbc.ca/news/politics/maid-deadline-extension-mental-illness-1.6687429. Accessed 17 Dec 2022.

McGregor, Fiona, and K. Sonu Gaind. 2020. Open Letter from CPA Past-Presidents Outlining Concerns About CPA Processes and 2020 Position Statement. https://www.webcampaign.org/cpa-positions. Accessed 28 Dec 2022.

Mehlum, L., C. Schmahl, A. Berens, S. Doering, J. Hutsebaut, A. Kaera, U. Kramer, P.A. Moran, B. Renneberg, J.S. Ribaudi, S. Simonsen, M. Swales, S. Taubner, and E. di Giacomo. 2020. Euthanasia and Assisted Suicide in Patients with Personality Disorders: A Review of Current Practice and Challenges. *Borderline Personality Disorder and Emotion Dysregulation* 7: 15. https://doi.org/10.1186/s40479-020-00131-9. https://pubmed.ncbi.nlm.nih.gov/32742662

Mental Health Commission of Canada. 2012. Changing Directions, Changing Lives: The Mental Health Strategy for Canada. Calgary, Alberta.

Mishara, B.L., and D.N. Weisstub. 2018. Is Suicide Prevention an Absolute? Considerations When Medical Assistance in Dying Is an Option [Editorial]. *Crisis: the Journal of Crisis Intervention and Suicide Prevention* 39 (5): 313–317. https://doi.org/10.1027/0227-5910/a000568.

Navaneelan, Tanya. 2017. Suicide Rates: An Overview. Statistics Canada. https://www150.statcan.gc.ca/n1/pub/82-624-x/2012001/article/11696-eng.htm#cvsd. Accessed 16 Jun 2017.

Noel, Melanie. 2019. Dénoncer avant de mourir. La Tribune Numérique. https://www.latribune.ca/2019/11/30/denoncer-avant-de-mourir-video-6b5b5b9901c42d9660ec5ba19a0eda78?nor=true. Accessed 9 Feb 2022.

OMA Section on Psychiatry. 2021. Medical Assistance in Dying Survey of OMA Section on Psychiatry Members. Ontario: OMA.

Senate. 2020. Evidence to Examine the Subject Matter of Bill C-7, an Act to Amend the Criminal Code (MAID). Ottawa, Ontario: The Standing Senate Committee on Legal and Constitutional Affairs.

Senate. 2021. Bill C-7, An Act to Amend the Criminal Code (MAID) 43rd Parliament, 2nd Session. https://sencanada.ca/en/Content/Sen/Committee/432/LCJC/55130-e

Sinyor, Mark, Ayal Schaffer, Marnin J. Heisel, André Picard, Gavin Adamson, Christian P. Cheung, Laurence Y. Katz, Rakesh Jetly, and Jitender Sareen. 2018. Media Guidelines for Reporting on Suicide: 2017 Update of the Canadian Psychiatric Association Policy Paper. *The Canadian Journal of Psychiatry* 63(3): 182–196. https://doi.org/10.1177/0706743717753147.

Sinyor, Mark, and Ayal Schaffer. 2020. The Lack of Adequate Scientific Evidence Regarding Physician-assisted Death for People with Psychiatric Disorders is a Danger to Patients. Canadian Journal of Psychiatry 65(9): 607–609. https://doi.org/10.1177/0706743720928658. https://pubmed.ncbi.nlm.nih.gov/32452224/

Smetanin, Paul, David Stiff, Carla Briante, Carol E. Adair, Sheeba Ahmad, and Minhal Khan. 2011. The Life and Economic Impact of Major Mental Illnesses in Canada: 2011 to 2041. RiskAnalytica, on behalf of the Mental Health Commission of Canada.

Touzin, Caroline. 2019. Aide médicale à mourir: Je pars sereine. La Presse. https://www.lapresse.ca/actualites/sante/201902/13/01-5214532-aide-medicale-a-mourir-je-pars-sereine.php. Accessed 18 Apr 2022.

Truchon v Procureur general du Canada. 2019. QCCS 3792.

Vulnerable Persons Standard. 2017. vps-npv.ca. https://static1.squarespace.com/static/56bb84cb01dbae77f988b71a/t/59ce974012abd96a0e1facfd/1506711361151/VPS+Standard+%26+Compliance+Table+-+Sept+2017.pdf. Accessed 9 Feb 2022.

World Health Organization. 2010. *Towards Evidence-based Suicide Prevention Programmes*. Manila, Philippines: WHO, Regional Office for the Western Pacific.

Chapter 18
MAID for Mental Illness: What Exactly is Being Consented To?

John Maher

Abstract The Canadian MAID framework for mental illness will permit suicide facilitation under the guise of medical practice. MAID for mental illness is being operationalized by the government within a medical framework where it does not belong ethically or clinically. Persons seeking MAID for mental illness will be choosing suicide. The critical role for psychiatrists involved in the care of such patients is determining whether they are mentally capable of refusing continuing offers of psychiatric treatment. Suicide prevention is a social imperative that is undermined by the legalization of suicide assistance. MAID assessors and providers will be providing suicide facilitation while not appreciating the complexities and effectiveness of tertiary psychiatric care, and while placing the lives of Canadians at risk.

Keywords Medical assistance in dying · Euthanasia · Physician assisted suicide · Capacity · Consent

18.1 Introduction

Background: The author of this chapter was asked to discuss and provide advice on obtaining consent for MAID from applicants with mental disorders. Carefully examining this issue he concluded that there is no scientifically grounded and clinically sound method for MAID assessors and providers to obtain consent for MAID. Furthermore, persons interested in MAID need to receive a consultation and follow-up from a psychiatrist. This opinion is based on the author's accumulated clinical, research and educational experience from caring for people with mental health problems for 28 years. The author is a psychiatrist who also obtained a BA in philosophy and MA in medical ethics. For almost 20 years he has been a consultant psychiatrist

---

J. Maher (✉)
Barrie & South Georgian Bay ACT Teams, Canadian Mental Health Association, Barrie, ON, Canada
e-mail: jmaher@cmhastarttalking.ca

J. Kotalik and D. W. Shannon (eds.), *Medical Assistance in Dying (MAID) in Canada*, The International Library of Bioethics 104,
https://doi.org/10.1007/978-3-031-30002-8_18

for Assertive Community Treatment Teams, and in that capacity he takes care of individuals with the most serious mental illnesses including those who are involved with the criminal justice system.

Overview: This chapter presents some relevant facts about suicide and shows how MAID is based largely on a false narrative, using distorted language. The main part of the chapter will explain the conviction that MAID is not a medical act and that there is no conceptually coherent process of informed consent that includes a choice of death. MAID is a choice for suicide. The key clinical question is whether the person interested in MAID is mentally capable of choosing to refuse psychiatric treatment and thereafter opt for facilitated suicide (which has been falsely labelled as a medical act in Canada). Only an appropriately experienced psychiatrist can make such a capacity determination. There is in fact no valid medical process that constitutes a capacity assessment for MAID in and of itself.

Suicide Facts: As the Canadian Supreme Court has highlighted, we citizens are already free to independently plan and complete our own suicides (Supreme Court of Canada 2015). It is an existential decision, not a medical decision. Doctors causing certain death is not, and never has been considered ethical medicine.

All suicide is tragic. Of all the people who attempt suicide, 23% will try again and 7% will complete suicide (Harvard School of Public Health 2022). 75% of those who do complete suicide have thoughtfully planned their deaths. Of particular relevance to the question of MAID for mental illness, 90% of the 4000 people who kill themselves in Canada every year have diagnosed mental illness (Mental Health Commission of Canada 2022; Statistics Canada 2017). Helping people suffering with mental illness to complete suicide is not the goal of psychiatric medicine. Suicide prevention and recovery are the only ethically defensible goals of good mental health care. Suicide prevention, as a social imperative, is in fact being undermined by our government offering death to the most vulnerable and undertreated among us.

18.2 False Narratives

The debate in Canada about MAID for mental illness rests on false premises. As Foucault would say, the false production of "knowledge" has distorted the information flow to the point that the public conversation has become ideological and not reasoned.

"Medical Assistance in Dying" as a euphemism for both "Physician Assisted Suicide" and "Euthanasia" has served its purpose in misdirecting people away from careful ethical analysis and towards public media driven acceptance of a facile and simplistic argument about the primacy of autonomy over the preservation of life.

The World Medical Association (WMA) has soundly condemned physician assisted suicide and euthanasia as immoral acts by doctors (WMA 2022). The Canadian Mental Health Association (CMHA 2021) and the Centre for Addiction and Mental Health (CAMH 2017) have stated that MAID for mental illness is not a clinically acceptable practice.

The unjustifiably delimited parameters of the debate have been shaped by MAID activists and constrained by the legal language (e.g., "irremediability") of the court decisions (Supreme Court of Canada 2015; Truchon 2019) and the ensuing legislation (Bill C-14 2016; Bill C-7 2021).

In Canadian law, MAID is defined as an exemption to the criminal code to allow health care practitioners to commit homicide or assist in the suicide of a patient (Bill C-14 2016). MAID was adopted by the federal government and forced onto our medical framework despite it being incongruent with the operations of medicine and its profound commitment to the preservation of life. MAID is not medicine.

Medicine is based on best evidence gathered from research coupled with clinical expertise. This provides the basis of patient care, medical decision making and informed consent. Scientific medicine is based on average responses, risks, and harms, and requires that patients only be offered treatments that have gone through a rigorous process of study (Association of Faculties of Medicine of Canada 2021).

The principle of autonomy in medicine has been understood as the patient's ability to engage in shared decision making of the well understood treatments offered by the physician, not the making of unilateral demands that must be acquiesced to (Sandman and Munthe 2009). Asking a doctor to kill you has never been an appropriate expression of autonomous decision making within the ethical framework of a clinical relationship.

18.3 MAID is Not a Medical Act

In Canada, the Quebec and federal governments unilaterally and legislatively declared MAID a medical act for reasons of expediency, legitimization, and because they wanted gatekeepers for MAID.

Let's distinguish between two parts in the MAID process as it has been imposed on the medical profession in Canada. Part One is the assessment. Part Two is the act of introducing the lethal poison into the person's body.

Let's start with Part Two. How do people actually die? In the United States and Switzerland people are only allowed to swallow lethal pills as the means of death (BMA 2021). A protocol requiring an IV is not necessary for causing death. Swallowing pills in itself is not a medical act. I can take pills to die on my own time, in the place of my own choosing, accompanied by whomever I wish. That process need not require that medical personnel be involved at all.

But of course, a prescription for the pills had to be provided under the respective countries' legislative frameworks (BMA 2021). Some will say that prescribing the pill is a necessarily medical act. And yet, once someone has had all the boxes checked off on the Bill C-7 checklist for MAID in Canada (they say they have suffered enough, there is progressive decline, and the suffering is irremediable) there is no medical skill required in prescribing the right dose of poison, monitoring for side-effects, or providing follow-up care for the condition being treated (the usual components of medical care). A government office could courier you the pill and make you sign for the package. Or, under a different operational framework, it could just be shipped directly from the manufacturer if you show your MAID permit. The point is that *deciding who can get the pill* is the issue at hand and subsequently framing it as a "prescription" is like calling rat poison a medical prescription for rats.

In Canada, 99% of people get MAID through an intravenous (IV) injection of poison, that is, voluntary euthanasia (Government of Canada 2020). It would seem obvious that inserting an IV is a medical act and therefore MAID is a medical act. Yet, if as a lay person I take a 4-hour phlebotomy course on how to find a vein and inject someone with drugs, does my subsequently doing so make it a medical act? The key point is that it doesn't need to be administered by health care providers, or via an IV route of administration, because the exact same endpoint can be achieved with simple swallowing. If we remove doctors from Part Two of the process a person can still die on their own terms. But Part One is needed to get to Part Two. A legal hurdle, the MAID assessment required to get the poison, has been deemed a medical act by legislative fiat. Is it actually that?

Why are doctors and nurse practitioners assessing and providing MAID in Canada? There are many reasons: (1) It artificially transforms the suicide support into the semblance of a medical procedure; (2) It makes a person feel cared for by a doctor; (3) It validates the person's choice to die through an external, supposedly objective and expert opinion that nothing more could have been done to relieve their suffering; (4) It provides a person with a kind of moral exoneration against a social or personal values backdrop that might otherwise make them feel ashamed for giving up or choosing suicide; (5) It comes with the guarantee a person won't "botch" the suicide and end up worse off than before, or having to face loved ones or forced hospitalization after a "failed" attempt; (6) It reflects the distinct psychology, moral values, and libertarian ideology of certain doctors who are willing to do what the majority of their professional colleagues will not.

With respect to this last point about doctors we wish to ask, "Whose needs are being met?" This seems to be an important part of the analysis of the dynamics and psychological complexity of some of the end-of-life dramas that are being played out and glorified. This is not an expression of doubt about altruistic motivations of the players, but the interpersonal context must be evaluated, especially when one considers the role of physicians in multiple ethical perversions in history under the guise of altruism (e.g., eugenics, research without consent, etc.). And when doctors who are already personally doing hundreds of MAID procedures publicly report in many settings (social media, parliamentary committees) how wonderful and moving it is *for them*, it drives the questions of motivation, objectivity, and bias (Green 2022; Holland 2021; Wiebe 2022). When a small minority is willing to act in ways counter to the long established ethical and medical practice norms of the absolute majority there should be very careful consideration of the reasons why and whether they are ethically defensible.

It has been argued that what makes Part Two a medical act is the emotional accompaniment and support that only a doctor can give. In response we will say that, given the patent absurdity of any such claim that physicians must of necessity be involved for a suicide to be subjectively satisfactory, it is not worth arguing the point further. The protocol could equally be called, "Mom's Assistance in Dying", "Partner's Assistance in Dying", or "Best Friend's Assistance in Dying". It might in fact be a more intimate and less intrusive experience without the doctor's presence.

If a person will not go ahead and swallow a pill without a doctor's presence it raises the very disturbing question of whether the act is fully autonomous. Or whether a doctor has recognized their role in pushing a person towards the suicide decision or their part in sustaining the conviction to go through with it. Is a doctor needed like a final shot of whisky for courage? Is it the direct and continuing proof that a decision to die is the right one and a person need feel no shame at ending it all?

What the claim about the need for a doctor's involvement does raise is the profoundly serious ethical question about doctors being responsible for suicide inducement. The very act of being a MAID assessor and being present to help with suicides is a public declaration of support for suicide and, whether acknowledged or not, directly undermines social and medical community efforts at suicide prevention. Doctors cannot both support suicide and suicide prevention simultaneously in any logically consistent way. Hence the repeated assertions by MAID activists that MAID is not suicide.

The Canadian Association for Suicide Prevention (CASP) has declared that all MAID in the context of someone who is not dying is suicide (Krausert 2021). All suicide is the result of people believing nothing can ever get better and that suffering cannot be reduced. Suicide is clinically defined as taking steps to arrange your own death. At its conceptual root, the MAID doctor has become the sanitized gun in a cleaner process with a false veneer of social sanction. MAID for mental illness is ethically indistinguishable from unassisted suicide and the recent Gupta report (Gupta et al. 2022) has acknowledged this, while simultaneously, and astonishingly, saying it doesn't matter. That is tantamount to saying suicide prevention doesn't matter. Which is in fact the ideological position of extreme libertarianism.

The rejoinder from MAID providers is that MAID for mental illness will only be done with people who have "irremediable" mental illness, as per court and legislative (Bill C-7 2021) directives. However, both sides in this debate are saying there is a lack of science to support predictions of irremediability (AMPQ 2020; EAG 2020; Gaind 2020; Gupta et al. 2022). The MAID advocates are comfortable with best guesses based on *past* treatment failures, but medicine/science uses evidence/data to predict prognosis or *future* outcomes.

Support for unscientific assessments that are just value-based guesswork (rooted in the doctor's personal ethics) is indefensible when the patient is actually seeking your professional opinion on whether they can get better. It is analogous to saying to a patient, "I think you may have terminal cancer", without being able to stage or identify the cancer type, or say whether it is even a treatable cancer, and then offering MAID under those conditions of uncertainty. And making the offer while the patient is led to believe you have certain knowledge, namely that they can never get better. This is clinical deceit and does not meet even a basic standard of medical care.

Much has been written about the question of irremediability of mental illness. No doubt, some people may have irremediable mental illness. The empirical reality, despite indefensible claims by some outlier psychiatrists, is that there is no clinical or scientific way to definitively identify which particular patients those are (EAG 2020; Gaind 2020). And unlike physical medicine, there are endless treatment options for mental illness that will never be fully exhausted. Some people get better after one year and some people get better after 10 years. The question is whether the person will persist in trying different treatments. The experience of subspecialist psychiatric teams is that suffering can always be ameliorated but Canadians wait years for these underfunded services. We will offer death while only 1 in 3 Canadians get the mental health care they need in either a timely way, or at all (Mental Health Commission of Canada 2022).

18.4 Since MAID is not a Medical Act It is not Possible to Assess Capacity for MAID Decisions

There is a very substantial medical ethics literature on informed consent and issues related to capacity assessments in medicine. Questions of interrater reliability, physician skill and experience and bias, patient comprehension and emotions, coercion, and the adequacy of legal or clinical frameworks have been explored in great detail.

Everyone agrees that most people living with mental illness are perfectly capable of making their own treatment decisions. Everyone also agrees that there are some people who by virtue of illness (e.g., psychosis, depression induced suicidal thinking) can't make their own reasoned treatment decisions and a process of substitute or supported decision making is an ethical imperative. Where matters are complicated is with people who fall in the middle; people whose decisional capacity is uncertain and who therefore may not be able to provide fully informed consent.

It is common for patients to appear as though they understand and appreciate treatments being offered while concealing psychotic processes or disease induced despair. Someone may happily take their antipsychotic medication and seem fully capable while concealing their belief that if they don't take the medication that demons will kill their family. Patients with anorexia are classic examples of patients who meet normative clinical tests for capacity but who literally starve themselves to death. Hopelessness leads to assumptions of certainty ("I will never get better") without factual validation.

Different psychiatrists have different skill sets and levels of experience. They also have biases like everyone else. Unfortunately, "there tends to be low interrater reliability among clinician judgments of decisional capacity" of psychiatric patients generally (Palmer 2007). This should engender grave concern because a substantial error rate will allow incapable patients to choose a pathway to assisted suicide. There is, not surprisingly, high interrater reliability with patients on inpatient units where symptomatology is more severe and clearly expressed (Cairns et al. 2018; Okai et al. 2007). In any setting, interrater reliability can be improved (up to 0.8) with the use of some of the many available structured assessment tools (Palmer 2007). However, these are not commonly used in the quotidian work of psychiatry.

Interrater reliability therefore can and does vary substantially. Of note is the fact that in addition to worrisome variable interrater reliability among psychiatrists, one can reasonably expect dramatic variance among the 85,000 non-psychiatrist doctors and nurse practitioners eligible to provide MAID who both lack psychiatric expertise and are not trained to use psychiatric capacity assessment tools. Regardless, what the capacity assessor variance also means is that incapable patients in the gray zone will, with minimal effort or persistence, likely be able to find a different psychiatrist or MAID assessor who declares them capable.

Charland et al. (2016) and multiple commentators have written thoughtfully about the issues specifically related to capacity determinations with MAID for mental illness. Framing a request for MAID as a matter of medical consent and capacity once again serves to falsely characterize it as a medical act. But if it is not a medical act, as we contend, then what is actually being sought?

What is being sought by the person is simply assistance with suicide. What is being sought is expedient help from a member of the very large group of people in Canada (over 90,000 doctors and NPs) who have been given an extraordinary legal exemption under the Criminal Code homicide law. If anyone else helps with the suicide it is punishable by up to 14 years in prison.

It is really just one human being asking another willing human being with a licence to end lives to help them. That licence is restricted to the medical profession because we expect and trust them to exercise good judgement. Is that trust well founded? When a prolific Canadian MAID provider recently testified before a Canadian parliamentary committee (Wiebe 2022) that she would provide MAID *while a patient is on a waitlist* for potentially completely effective psychiatric care it begs disturbing questions about the lack of oversight in Canada. Not all doctors will respect the lax law and abuses leading to premature and unnecessary deaths are inevitable under the current legislative framework.

Informed consent requires that a patient be offered substantial and material information about a proposed procedure and the risks of accepting or refusing treatment. The greater the risks the higher the bar for ensuring full comprehension and supporting the person through the decision-making process. In all of medicine, one of the greatest risks that might attend any proposed treatment is death. A person may be told, for example, that there is a 30% chance of death from cardiac surgery. A person starting an antipsychotic might be told there is a 1 in one hundred million chance of death. With MAID, there is no patient evaluation of the risk of death. Or of any other possible harms, side-effects, or sequelae. With death as the intended outcome there can be no discussion of risks or consequences. This has never occurred in the history of medicine. It is not the same as stopping or withdrawing ineffective treatment. It is not the same as a competent patient refusing even easily administered life sustaining treatment (e.g., Jehovah's Witness and blood transfusion). It is choosing death. And frankly, and paradoxically, death is one of the easiest things to understand.

There is no traditional or conceptually coherent process of informed consent happening with a choice for death. In the case of MAID for mental illness, informed consent would pertain to *a choice to refuse any further psychiatric treatment options.* And so, the capacity assessment that is taking place is related exclusively to the psychiatrist's effort to ensure that the patient understands and appreciates the consequences of accepting or refusing additional psychiatric treatments. Insofar as potentially effective treatment is refused because of the seduction of immediate relief of suffering through MAID the psychiatrist must make some judgement about whether the present decision to refuse treatment is based on full understanding and appreciation. *It is not an assessment about capacity for choosing MAID.* It is the specialist psychiatrist assessment of whether someone is suicidal, whether decision making is impaired by brain disease, and whether the person can provide fully informed consent with respect to proffered treatment options. Of particular note, these evaluations are best accomplished within the privileged position of an ongoing therapeutic relationship. A person's values and authentic desires are often only revealed over time and are often impossible to determine with a onetime assessment by an "independent" psychiatrist.

Insofar as biopsychosocial treatment options in psychiatry are limitless, the key and only question is whether the person is capable of choosing psychiatric treatment refusal. And given that this is specialized psychiatric treatment being offered, the assessment must be made exclusively by the psychiatrists offering it. To become a psychiatrist requires the same residency training period (5 years) as a surgeon. We frown on people operating on themselves. Remarkably, MAID advocates are claiming that assessments by family doctors, nurse practitioners, or any other non-psychiatrists are adequate. This claim devalues the training and skillset required for psychiatric care, and if a patient is seeking death, then by definition the person does not have stable or well treated psychiatric illness and needs specialist evaluation. In this regard, even with the false claim that persons with irremediable illness are identifiable, the government has failed catastrophically with its lack of explicit requirements that a person with serious and persistent mental illness have received at least standard psychiatric treatment trials, and that psychiatric specialist and sub-specialist services must actually be available in a timely way. That is to say, other doctors offering death before psychiatrists can offer treatment is a perversion of psychiatric care. It strikes us as another example of systemic stigmatization. It undermines hope and is an ablest message to people with mental illness that their lives are not worth living.

18.5 Conclusion

The entire Canadian MAID framework facilitates suicide under the guise of medical practice. It is an insidious and disturbing invention of a government ignoring its duty to preserve life as was affirmed unequivocally in the Canadian Supreme Court "Carter" decision (2015). And as the Parliamentary Budget Officer (2021) has declared, it will save a lot of money.

Many other libertarian idealogues besides activist doctors would be happy to step forward and assist people with suicides. In fact, some groups have been trying to do this for years but had no legal exemption for doing so.

The government wants gatekeeping because society will reject full laissez faire support for suicide facilitation. Thus, the government has falsely ascribed the moral high ground and probity to over 90,000 Canadians without individually assessing their biases, judgement, ability to choose wisely, or their commitment to suicide prevention, not to mention their actual understanding of the complexity of tertiary psychiatric care. They are granted the astonishing liberty to approve and assist with suicide planning and completion. Our Canadian judges, who are not even allowed to render capital punishment judgements, have their careers and lives extensively vetted and only become judges following a remarkably detailed screening process. And yet, in Canada, if my application to medical or nursing school has been accepted I can become a health professional who ends lives through MAID. It should go without saying that good marks do not necessarily equate with wisdom or good judgement or moral propriety.

Anyone objecting to MAID for mental illness is labelled by MAID expansionists as an extremist or as callously indifferent to the suffering of others despite the fact we have dedicated our lives to relieving that very suffering. And their refrain has been echoed by politicians. The public debate has become ideological and divorced from serious ethical analysis. The clear majority of Canadian psychiatrists object to MAID. Their expertise has been devalued and their understanding of the profound impact of the introduction of MAID on the clinical encounter is being ignored. Some patients are now refusing effective psychiatric treatments to pursue MAID. Some patients now want easy death over the protracted work of recovery. The rending of the clinical encounter, and the tragic reality that many will die who would have gotten better, is a moral travesty.

References

AMPQ (Association Des Médecins Psychiatres Du Québec) 2020. Access to Medical Assistance in Dying For People with Mental Disorders: Discussion Paper. https://ampq.org/wp-content/uploads/2020/12/mpqdocreflexionammenfinal.pdf

Association of Faculties of Medicine of Canada. 2021. Chapter 5: Assessing Evidence and Information. https://phprimer.afmc.ca/en/part-ii/chapter-5/. Accessed 21 Jun 2022.

Bill C-7. 2021. An Act to Amend the Criminal Code (Medical Assistance in Dying). Parliament of Canada, Second Session, Forty-third Parliament. https://parl.ca/DocumentViewer/en/43-2/bill/C-7/royal-assent. Accessed 21 Jun 2022.

Bill C-14. 2016. An Act to Amend the Criminal Code and to Make Related Amendments to Other Acts (Medical Assistance in Dying). Parliament of Canada, First Session, Forty-second Parliament. https://www.parl.ca/DocumentViewer/en/42-1/bill/C-14/royal-assent. Accessed 21 Jun 2022.

BMA. 2021. Physician-assisted Dying Legislation Around the World. https://www.bma.org.uk/media/4402/bma-where-is-pad-permitted-internationally-aug-2021.pdf. Accessed 21 Jun 2022.

Cairns, R., C. Maddock, A. Buchanan, A.S. David, P. Hayward, G. Richardson, G. Szmukler, and M. Hotopf. 2018. Reliability of Mental Capacity Assessments in Psychiatric In-patients. *The British Journal of Psychiatry* 187(4): 372–378. https://doi.org/10.1192/bjp.187.4.372

CAMH. 2017. Policy Advice on Medical Assistance in Dying and Mental Illness. https://www.camh.ca/-/media/files/pdfs---public-policy-submissions/camh-position-on-mi-MAID-oct2017-pdf.pdf. Accessed 21 Jun 2022.

Charland, Louis C, Trudo Lemmens, and Kyoko Wada. 2016. Decision-making Capacity to Consent to Medical Assitance in Dying for Persons with Mental Disorders. *Journal of Ethics in Mental Health* Open Volume: 1–14. https://papers.ssrn.com/sol3/papers.cfm?abstract_id=2784291. Accessed 21 Jun 2022.

CMHA. 2021. Statement on Medical Assistance in Dying (MAID). https://cmha.ca/brochure/statement-on-medical-assistance-in-dying-MAID/. Accessed 21 Jun 2022.

EAG. 2020. Canada at a Crossroads: Recommendations on Medical Assistance in Dying and Persons with Mental Disorder-An Evidence-based Critique of the Halifax Group IRPP Report, Ontario, Toronto.

Gaind, K.S. 2020. What Does "Irremediability" in Mental Illness Mean? *The Canadian Journal of Psychiatry* 65 (9): 604–606. https://doi.org/10.1177/0706743720928656.

Government of Canada. 2020. Second Annual Health Canada Report on MAID in Canada 2020, Section 3.1. https://www.canada.ca/en/health-canada/services/medical-assistance-dying/annual-report-2020.html#3_1. Accessed 21 Jun 2022.

Green, Stefanie. 2022. This is Assisted Dying. https://www.stefaniegreen.com/author/. Accessed June 21, 2022.

Gupta, Mona, Rose M. Carter, et al. 2022. Final Report of the Expert Panel on MAiD and Mental Illness. Ottawa, Ontario: Health Canada.

Harvard School of Public Health. 2022. Mean's Matter: Attempter's Longterm Survival. https://www.hsph.harvard.edu/means-matter/means-matter/survival/. Accessed 21 Jun 2022.

Holland, Timothy. 2021. Senate of Canada. https://sencanada.ca/en/Content/Sen/Committee/432/LCJC/11EV-55129-E. Accessed 21 Jun 2022.

Krausert, Sean. 2021. Letter to CPA Re: CPA's Discussion Paper on Medical Assistance in Dying. https://static1.squarespace.com/static/5ec97e1c4215f5026a674116/t/619bc64ac482815740760931/1637598795070/Letter+to+CPA+re+Discussion+Paper+on+MAID+%282021+11+20%29.pdf. Accessed June 21, 2022.

Mental Health Commission of Canada. 2022. Resources. https://mentalhealthcommission.ca/resources/. Accessed 21 Jun 2022.

Okai, D., G. Owen, H. McGuire, S. Singh, R. Churchill, and M. Hotopf. 2007. Mental Capacity in Psychiatric Patients: Systematic Review. *The British Journal of Psychiatry* 191 (4): 291–297. https://doi.org/10.1192/bjp.bp.106.035162.

Palmer, Barton W. 2007. Assessment of Decisional Capacity. *Psychiatric Times* 24(13): 38.

Parliamentary Budget Officer. 2021. Cost Estimates for Bill C-7 MAID. https://www.pbo-dpb.gc.ca/web/default/files/Documents/Reports/RP-2021-025-M/RP-2021-025-M_en.pdf. Accessed 21 Jun 2022.

Sandman, L., and C. Munthe. 2009. Shared Decision-making and patient autonomy. *Theoretical Medicine and Bioethics* 30(4): 289–310. https://doi.org/10.1007/s11017-009-9114-4.

Statistics Canada. 2017. Health at a Glance, Suicide Rates: An Overview. https://www150.statcan.gc.ca/n1/pub/82-624-x/2012001/article/11696-eng.htm. Accessed 21 Jun 2021.

Supreme Court of Canada. 2015. Carter v. Canada (Attorney General). https://scc-csc.lexum.com/scc-csc/scc-csc/en/item/14637/index.do

Truchon v Procureur General Du Canada. 2019. The Superior Court of Quebec. (CanLII). https://canlii.ca/t/j2bzl. Accessed 21 Jun 2022.

Wiebe, Ellen. 2022. AMAD Committee. https://parl.ca/DocumentViewer/en/44-1/AMAD/meeting-9/evidence. Accessed 21 Jun 2022.

World Medical Association. 2022. WMA Declaration on Euthanasia and Physician-assisted Suicide. https://www.wma.net/policies-post/declaration-on-euthanasia-and-physician-assisted-suicide/. Accessed 21 Jun 2021.

John Maher, MA, MD, FRCPC Barrie & South Georgian Bay ACT Teams, Ontario, Canada.

Dr. Maher did his medical training at McMaster University, psychiatry residency at the University of Ottawa, and is a Fellow of the Royal College of Physicians and Surgeons of Canada. He has a BA in philosophy and MA in medical ethics.

He has been an Assertive Community Treatment (ACT) Team consulting psychiatrist for 20 years. He previously did adult and emergency psychiatry, was director of a personality disorders program, and provided consultations to transitional youth services, a mental health support court, and a college mental health service.

He was founder and Executive Director of the Trillium Childhood Cancer Support Centre, founding board member of the Canadian Candlelighters Childhood Cancer Foundation, Assistant Director of the National Cancer Control Task Force, Executive Director of Cancer Canada, and founder and Executive Director of Families and Children Experiencing AIDS (FACE AIDS).

He is currently Editor-in-Chief of the Journal of Ethics in Mental Health and President of the Ontario Association for ACT & FACT (OAAF).

Part V
Disability Perspectives, Human Rights and MAID

Chapter 19
MAID to Die by Medical and Systemic Ableism

Heidi Janz

Abstract This chapter explores the concept of ableism (that is, prejudice and discrimination based on disability), and the impact that ableism has had, and continues to have, on the implementation and expansion of Canada's MAID regime. More specifically, it considers the ways in which systemic ableism limits social and economic opportunities for people with disabilities, sometimes to the degree that they are compelled to seek an assisted death because they cannot access sufficient assistance to live. The chapter concludes with reflections on what the expansion of MAID in the context of the COVID-19 pandemic reveals about the depth to which systemic ableism is ingrained in Canadian society, and the existential threat that this poses for people with disabilities in Canada.

Keywords Ableism · Disability discrimination · Social determinants of health · Medical assistance in dying · Medical ableism

19.1 Introduction

On March 17, 2021, Bill C-7 became law in Canada, expanding eligibility for Medical Assistance in Dying to people with disabilities who are not at end-of-life. While proponents of the so-called "right to die" heralded this development as an advancement of the autonomy of Canadians, disability-rights advocates mourned the passing of this bill into law. These disability-rights advocates recognized that, contrary to the claims of its champions, Bill C-7 would not advance the human rights of people with disabilities; rather, it would enshrine into law the quintessential ableist stereotype that it's better to be dead than disabled. In this chapter, I will explore the concept of ableism (that is, prejudice and discrimination based on disability), and the impact that ableism has had, and continues to have, on the implementation and expansion of Canada's MAID program. More specifically, I will consider the ways in which

H. Janz (✉)
Associate Adjunct Professor, University of Alberta, Edmonton, AB, Canada
e-mail: hjanz@ualberta.ca

J. Kotalik and D. W. Shannon (eds.), *Medical Assistance in Dying (MAID) in Canada*, The International Library of Bioethics 104,
https://doi.org/10.1007/978-3-031-30002-8_19

systemic ableism limits social and economic opportunities for people with disabilities, often to the degree that they are compelled to seek an assisted death because they cannot access sufficient assistance to live. I will conclude this chapter by reflecting on what the expansion of MAID in the context of the COVID-19 pandemic reveals about the depth to which systemic ableism is ingrained in Canadian society, and the existential threat that this poses for people with disabilities in Canada.

19.2 What is Ableism?

Ableism can be defined as prejudice and discrimination against people with disabilities based on the belief that typical abilities are superior. At its heart, ableism is rooted in the assumption that disabled people require 'fixing' and defines people by their disability. Like racism and sexism, ableism classifies entire groups of people as 'less than', and perpetuates harmful stereotypes, misconceptions, and generalizations about people with disabilities (Eisenmenger 2019). Clear, yet unacknowledged, evidence of ableism can be seen in everything from the way in which buildings are built to exclude people who do not walk or see (Livingston 2000), to the way in which students with disabilities are taught that it's better to look, move, and behave as much like their nondisabled peers as possible (Hehir 2002), to the way in which terms associated with disability are used as insults in everyday language (Aaron 2015). Unlike racism or sexism, however, ableism remains, in the words of Canadian disability scholar, Gregor Wolbring, "one of the most societally entrenched and accepted isms" (Wolbring 2008). Ableism becomes systemic as patterns of discriminatory and exclusionary practices against people with disabilities are entrenched in sectors which are fundamental to societal membership, sectors such as education, employment, and housing. Limited access and opportunities in one of these key sectors, such as education, often automatically leads to further restrictions in employment opportunities, which, in turn, limits housing options (Emerson et al. 2011; Frier et al. 2018; Haigh et al. 2019).

19.3 The Unique Role of Medical Ableism in Perpetuating Systemic Ableism

One key sector in which ableism is most deeply rooted is healthcare. Indeed, disability scholar Fiona Kumari Campbell draws a direct link between ableism and a biomedical paradigm, in that she defines ableism as "a network of beliefs, processes and practices that produces a particular kind of self and body (the corporeal standard) that is projected as the perfect, species-typical and therefore essential and fully human. Disability then, is cast as a diminished state of being human" (Campbell 2001). Campbell delineates an inherent link between this deficit-based construction of disability and a "biomedicalist stance", which, since the Age of Reason, has "played a critical intervening role in the lives of people with disability and people with anomalous bodies or mentalities. Medicine has operated as the primary paradigm not only for the treatment of disabled bodies but has also shaped the way decision makers, legislators, families and society in general *think* about and *sense* disability" (Campbell 2009).

It is, therefore, quite troubling that, while Medicine has traditionally been viewed as a benevolent discipline in which every human life is valued equally, without any form of prejudice or discrimination, the reality is that, as individuals, physicians are not immune to the influence of dominant societal understandings of, and attitudes towards, individuals and groups of people deemed to be Others (Moskowitz et al. 2012).

Despite the major role that medical professionals play in the lives of people with disabilities, research indicates that people with disabilities frequently encounter discrimination in healthcare settings, even in ordinary, non-pandemic times (McMillan 2016). Studies have consistently shown that, as a group, health care professionals tend to substantially underestimate the quality of life of people with disabilities (Kothari and Kirschner 2006; Landry 2022; Macauley 2019; Reynolds 2017). Most recently, a survey of more than 700 U.S. physicians, across several specialties, found that just over 82 percent of physicians reported that people with disabilities have a significantly worse quality of life than nondisabled people (Iezzoni et al. 2021).

The novel coronavirus, which, for the better part of two years, put "normal" life across the planet on pause, has revealed in the starkest of terms the real and present danger that medical ableism poses for people with disabilities. Just as COVID-19 is not confined by national borders, the prevalence of medical ableism that the pandemic has revealed likewise transcends geographical boundaries. As Panocchia et al. observe, "Although different countries have proposed different guidelines to manage this emergency, these proposed criteria do not sufficiently consider people living with disabilities. People living with disabilities are therefore at a higher risk of exclusion from medical treatments as physicians tend to assume they have poor quality of life" (Panocchia et al. 2021). In the same vein, Ignagni et al. identify an ableist, utilitarian mindset, which they call "eugenic logic", underpinning most worldwide healthcare responses to COVID-19, such as Critical Care Triage Protocols which list pre-existing disability as an exclusion criterion for receiving Critical Care in the event that the system is overwhelmed, and Critical Care resources need to be rationed (Ignagni et al. 2020).

As healthcare professionals, MAID assessors are not immune to the prevalent influence of medical ableism rooted in utilitarian, eugenic logic (Gill 2000). This becomes particularly significant in light of the fact that studies consistently demonstrate that healthcare professionals who provide care to disabled individuals are often more pessimistic and upset about their charges' prognoses than the affected individuals themselves (Bodenhamer et al. 1983; Cushman and Dijkers 1990; Ernst 1987). Some may argue that this is essentially a moot point because people with disabilities meet with MAID assessors only after they have already applied for MAID. However, such an argument fails to take into consideration the fact that, among other things, MAID assessors are tasked with determining "the greatest source of suffering" that is prompting a person's request for MAID (Health Canada 2021). MAID assessors afflicted with undiagnosed ableism are likely to automatically view the existence of disability, in and of itself, as the greatest source of suffering. They thus risk conflating suffering stemming from remediable social conditions, such as poverty and the lack of affordable, accessible housing and community-based, user-directed personal support services, with the "irremediable" presence of disability. In a recent article published in *The New Atlatntis*, investigative journalist, Alexander Raikin, reports that, "Based on vague criteria and with precious little oversight, the MAID system has given enormous latitude to euthanasia providers to make the judgment calls about who should be helped to live and who should be helped to die. The result is a system that is highly efficient at finding reasons patients should qualify, not reasons they shouldn't" (Raikin 2022).

19.4 Systemic Ableism and Restricted Access to the Social Determinants of Health for People with Disabilities

While medical ableism routinely restricts disabled people's access to medical treatments and resources, systemic ableism routinely restricts the access that people with disabilities have to Social Determinants of Health (SDH). SDH are defined as the non-medical factors that influence health outcomes. They are the conditions in which people are born, grow, work, live, and age, and the wider set of forces and systems shaping the conditions of daily life. These forces and systems include economic policies and systems, development agendas, social norms, social policies and political systems (World Health Organization n.d.). SDH for people with disabilities are generally understood as including **income and social protection, affordable accessible housing along with community-based, user-directed disability support services, and social inclusion and non-discrimination.** What follows is a brief overview of the level of access that Canadians with disabilities generally have to these Social Determinants of Health, and a consideration of the ways in which systemic ableism creates barriers to accessing these Social Determinants of Health for people with disabilities. Embedded in this overview will be a discussion of a few recent cases in which people with disabilities were compelled to seek MAID because systemic ableism caused them to be denied access to the Social Determinants of Health.

19.4.1 Income and Social Protection

Statistics Canada reports that, in 2017, persons with disabilities faced lower employment rates, particularly for Canadians with very severe disabilities, and those with lower levels of education (Wall 2017). About 59% of working-age adults with disabilities were employed compared with around 80% of those without disabilities. However, as severity of disability increased, the percentage of those employed fell from 76% among those with mild disabilities to 31% among those with very severe disabilities. In addition, among working age adults, 28% of those with more severe disabilities were living below Canada's official poverty line, compared with 14% of those with milder disabilities and 10% of those without disabilities.

Manitoba is the top-performing province on income of people with disabilities—the average income of people with disabilities is 76.5 per cent of the income of people without disabilities. In Ontario and Nova Scotia, the income of people with disabilities is 74.8 and 74.4 per cent of the income of people without disabilities, respectively. P.E.I. (73.3 per cent) does better than the national average (72.7 per cent), while Newfoundland and Labrador matches the national average. There is a slight difference between the performance of New Brunswick (71.9 per cent) and Saskatchewan (71.7 per cent) on this indicator. The income of people with disabilities in B.C. and Quebec is 70.7 and 70.5 per cent that of people without disabilities, respectively. Alberta is the worst-performing province—the income for people with disabilities is 66.1 per cent that of people without disabilities (Conference Board of Canada 2017).

19.4.2 Affordable and Accessible Housing Along with Community-based, User-run Disability Support Services

The availability of affordable and accessible housing along with community-based, user-run disability support services is arguably the most critical Social Determinant of Health in terms of its impact on decisions by people with disabilities to end their life via MAID. Adults with disabilities younger than 65 years are the fastest growing population within public long-term care (LTC) (Canadian Institute for Health Information 2000). LTC homes in Canada, commonly referred to as nursing homes, are intended for individuals who do not require hospital care but need additional care supports above and beyond what is available through home care programs or assisted-living housing (Barber et al. 2020; Goffin 2017; Haiven 2019; Kaldy 2012; Kerr 2017). About 10% of people living in LTC in Canada are under age 65 (Hay and Chaudhury 2015). Across Canada, thousands of people well below old age are housed in institutions like nursing homes. In Ontario, for example, more than 4,700 long-term care residents, or 12 percent of the total LTC population, were under the age of 65 in April 2021 (Gillmore 2021). This, despite the fact that a growing body of research indicates that people with all types of disabilities, even those considered "severely disabled", benefit by living in typical housing in the community, as opposed to living in an institution or an institution- like congregate living facility (Kanter 2014).

A Hobson's choice is defined as a situation in which it seems that you can choose between different things or actions, but there is really only one thing that you can take or do (Hobson's Choice, n.d.). Since the 2021 expansion of eligibility for MAID to include people with illnesses and disabilities who are not at end-of-life, increasing numbers of ill and disabled people are turning to MAID as a Hobson's choice—that is, death by MAID is their only remaining alternative to abject poverty and/or incarceration in a long term care facility.

Among these are:

Chris Gladders, age 35, who died by MAID in January, 2021. He was battling Fabry's disease, a genetic condition which affects the body's ability to break down a specific fatty acid and causes a number of side-effects. He had two daughters: Hailee, 13, and Savannah, 5. He lived in a long term care facility in Niagra, Ontario. His brother reported that at the time of his assisted death, "The bedding hadn't been changed for weeks. There was feces on the bed. There was urine on the bed. There was urine and feces on the floor, the room was absolutely disgusting". And that the day before his death "He pulled the call bell beside his bed. I was on the phone with him for 40 min and nobody answered that bell. That was his last night" (Hristova 2021).

Equally disturbing is the case of "**Sophia**".

"**Sophia**" was the preferred pseudonym of a 51-year-old Ontario woman who had Multiple Chemical Sensitivities, a chronic condition. She underwent a medically-assisted death after her desperate search for affordable housing free of cigarette smoke and chemical cleaners failed. "The government sees me as expendable trash, a complainer, useless and a pain in the ass", 'Sophia' said in a video filmed on Feb. 14, 2022, eight days before her death. She died after a frantic effort by friends, supporters and even her doctors to get her safe and affordable housing in Toronto. She also left behind letters showing a desperate two-year search for help, in which she begs local, provincial and federal officials for assistance in finding a home away from the smoke and chemicals wafting through her apartment. Four Toronto doctors were aware of Sophia's case and they also wrote to federal housing and disability government officials on her behalf. In that letter the doctors confirmed that her symptoms improved in cleaner air environments and asked for help to find or build a chemical-free residence. "We physicians find it UNCONSCIONABLE that no other solution is proposed to this situation other than medical assistance in dying", they wrote (Favaro 2022).

And finally, there is the equally troubling and tragic story of **Sathya Dhara Kovak**.

Sathya Dhara Kovak was a Winnipeg woman who died by MAID on October 3, 2022. Kovac lived with amyotrophic lateral sclerosis (ALS). Her condition was worsening, but she felt she had more life to live—just not enough home care support to do so. "Ultimately it was not a genetic disease that took me out, it was a system", Kovac wrote in an obituary to loved ones. Her obituary continues, "There is desperate need for change. That is the sickness that causes so much suffering. Vulnerable people need help to survive. I could have had more time if I had more help" (KOVAC SATHYA–Obituaries–Winnipeg Free Press Passages n.d.).

There, *is*, indeed a desperate need for change. In the words of Canadian disability scholar, Jerome Bickenbach, when an individual chooses death as the only viable way to escape an intolerable situation partly brought on by the social environment, it seems "perverse and unfair to say that this is an expression of self-determination or autonomy" (Bickenbach 2012).

19.4.3 Social Inclusion and Non-discrimination

The term *social inclusion* is used by policy makers and professionals as a synonym for social cohesion, social integration, and social participation, or, as the opposite of social exclusion (Rimmerman 2013). However, in her qualitative meta-analysis of social inclusion of people with disabilities, Sarah A. Hall found the following six common elements of social inclusion of people with disabilities: (1) being recognized and accepted as an individual apart from their disability; (2) having personal relationships with family, friends, and acquaintances; (3) being involved in recreation, leisure, and other social activities; (4) having appropriate living accommodations; (5) having employment, and (6) having appropriate formal and informal supports (Hall 2009). Csiernik et al. observe that, "In Canada, the lack of national housing, income support, and mental health policies has further added to the basic problems related to the issue of mental health and social inclusion" (Csiernik et al. 2021). This statement is not just applicable to people with mental health issues, rather, it is an accurate articulation of the barriers to social inclusion facing all people with disabilities in Canada due to systemic ableism. Thus, as we have seen throughout this chapter, entrenched systemic ableism and concurrent lack of national housing, income support and community-based disability support programs at the federal and provincial level remain formidable barriers to the social inclusion of people with disabilities in Canada. Consequently, disabled people are left very much on the margins of Canadian society.

19.5 Conclusion: What the COVID-19 Pandemic and the Expansion of Eligibility for Medical Assistance in Dying Reveal about Systemic Ableism in Canada

The Parliamentary and Senate Committee studies of Bill C-7, the proposed expansion of eligibility for MAID, took place in the fall and winter of 2020–2021, while the entire country was under pandemic lockdown. The breathtaking irony of this situation is that, at the very time when government and public health officials were urging Canadians to take precautions and even to make sacrifices for the sake of protecting the lives of vulnerable people, some of these same government officials were working to push through a bill which would make many of these same vulnerable people eligible for a state-sanctioned death. The kind of logical and ethical contortions that this required had to be of Olympic calibre, matched only by the logical and ethical contortions that it took to quote from the UN Convention on the Rights of Persons with Disabilities in the Preamble to Bill C-7, when three UN Human Rights experts declared Canada's MAiD regime to be in violation of that very Convention. Thus, it is clear that systemic ableism in Canada is, not just alive, but thriving, with the passing of Bill C-7, and current efforts to further expand eligibility for MAID to include people with a sole diagnosis of mental illness and so-called "mature minors", including kids with disabilities. Sadly, this will inevitably result in a growing number of Canadians with disabilities losing their lives to ableism disguised as autonomy.

References

Aaron, Jessi Elana. 2015. Forever Crooked: How Everyday Language Reflects Negative Attitudes About the Physically Disabled. The Conversation. http://theconversation.com/forever-crooked-how-everyday-language-reflects-negative-attitudes-about-the-physically-disabled-38881

Barber, Brittany V., Lori E. Weeks, Natasha A. Spassiani, and Brad A. Meisner. 2020. Experiences of Health and Aging for Younger Adults in Long-term Care: A Social-ecological Multi-method Approach. *Disability & Society* 36(3): 1–20.

Bickenbach, Jerome. 2012. Ethics, Disability and the International Classification of Functioning, Disability and Health. *American Journal of Physical Medicine & Rehabilitation* 91 (13): S163–S167.

Bodenhamer, Eugenia, Jeanne Achterberg-Lawlis, George Kevorkian, Anne Belanus, and Jeanne Cofer. 1983. Staff and Patient Perceptions of the Psychosocial Concerns of Spinal Cord Injured Persons. *American Journal of Physical Medicine* 62(4): 182–193.

Campbell, Fiona Kumari. 2001. Inciting Legal Fictions: Disability's Date with Ontology and the Ableist Body of the Law. *Griffith Law Review* 10(1): 42–62.

Campbell, Fiona Kumari. 2009. Medical Education and Disability Studies. *Journal of Medical Humanities* 30(4): 221–235. https://doi.org/10.1007/s10912-009-9088-2.

Canadian Institute for Health Information. 2000. When a Nursing Home is Home: How Do Canadian Nursing Homes Measure up on Quality? Ottawa, ON, Canada: Canadian Institute for Health Information.

Conference Board of Canada. 2017. Income of People with Disabilities–Society Provincial Rankings–How Canada Performs. https://www.conferenceboard.ca/hcp/provincial/society/disability-income.aspx

Csiernik, Rick, Cheryl Forchuk, Katherine Turner, Libbey Joplin, Ruth Schofield, and Carolyn Gorlick. 2021. Creating Social Exclusion: The Disconnection Between Housing, Income Support, and Mental Health Policies. In *Poverty, Mental Health, and Social Inclusion*, 32.

Cushman, Laura A., and Marcel P. Dijkers. 1990. Depressed Mood in Spinal Cord Injured Patients: Staff Perceptions and Patient Realities. *Archives of Physical Medicine and Rehabilitation* 71: 191–196.

Eisenmenger, Ashley. 2019. Ableism 101–What It Is, What It Looks Like, and How to Become a Better ally. Access Living. https://www.accessliving.org/newsroom/blog/ableism-101/

Emerson, Eric, Ros Madden, Hilary Graham, Gwynnyth Llewellyn, Chris Hatton, and Janet Robertson. 2011. The Health of Disabled People and the Social Determinants of Health. *Public Health* 125(3): 145–147.

Ernst, Frederick A. 1987. Contrasting Perceptions of Distress by Research Personnel and Their Spinal Cord Injured Subjects. *American Journal of Physical Medicine and Rehabilitation* 66 (1): 12–15.

Favaro, Avis. 2022. Woman with Chemical Sensitivities Chose Medically-assisted Death after Failed Bid to Get Better Housing. CTV News. https://www.ctvnews.ca/health/woman-with-chemical-sensitivities-chose-medically-assisted-death-after-failed-bid-to-get-better-housing-1.5860579

Frier, Amanda, Fiona Barnett, Sue Devine, and Ruth Barker. 2018. Understanding Disability and the 'Social Determinants of Health': How Does Disability Affect Peoples' Social Determinants of Health? *Disability and Rehabilitation* 40 (5): 538–547. https://doi.org/10.1080/09638288.2016.1258090.

Gill, Carol J. 2000. Health Professionals, Disability, and Assisted Suicide: An Examination of Relevant Empirical Evidence and Reply to Batavia. *Psychology, Public Policy, and Law* 6 (2): 526–545.

Gillmore, Meagan. 2021. Some Young Adults with Disabilities Are Stuck in Long-Term care. They Say That's discrimination. Without Proper Support to Live at Home, They Have No Choice but to Stay in "Medical Prison Cells". Broadview. https://broadview.org/young-people-with-disabilities-long-term-care/

Goffin, Peter. 2017. Thousands of Under-65 Adults with Physical Disabilities are Being Forced into Ontario Nursing Homes: Ministry Data. The Toronto Star. https://www.thestar.com/news/gta/2017/07/09/thousands-of-under-65-adults-with-physical-disabilities-are-being-forced-into-ontario-nursing-homes-ministry-data.html

Haigh, Fiona, Lynn Kemp, Patricia Bazeley, and Neil Haigh. 2019. Developing a Critical Realist Informed Framework to Explain How the Human Rights and Social Determinants of Health Relationship Works. *BMC Public Health* 19 (1571): 1–12.

Haiven, Judy. 2019. OPINION: Nursing Homes No Place for Severely Disabled. Saltwire: Halifax, Powdered by The Chronicle Herald. https://www.saltwire.com/halifax/opinion/opinion-nursing-homes-no-place-for-severely-disabled-287769/

Hall, Sarah A. 2009. The Social Inclusion of People with Disabilities: A Qualitative Meta-Analysis. *Journal of Ethnographic & Qualitative Research* 3(3): 162–173.

Hay, Kara, and Habib Chaudhury. 2015. Exploring the Quality of Life of Younger Residents Living in Long-term Care Facilities. *Journal of Applied Gerontology* 34 (6): 675–690.

Health Canada. 2021. Medical Assistance in Dying: Implementing the Framework. Frequently asked questions. https://www.canada.ca/en/health-canada/services/medical-assistance-dying/guidance-reporting-summary/implementing-framework.html

Hehir, Thomas. 2002. Eliminating Ableism in Education. *Harvard Educational Review* 72(1): 1–33. https://doi.org/10.17763/haer.72.1.03866528702g2105.

Hobson's Choice. n.d. In *Cambridge Advanced Learner's Dictionary & Thesaurus*.

Hristova, Bobby. 2021. Niagara MPP Calls for Province to Take Over 'Disgusting' Greycliff Manor after 35-Year-Old Dies. CBC News. https://www.cbc.ca/news/canada/hamilton/greycliff-manor-covid19-1.5891730

Iezzoni, Lisa I., Sowmya R. Rao, Julie Ressalam, Dragana Bolcic-Jankovic, Nicole D. Agaronnik, Karen Donelan, Tara Lagu, and Eric G. Campbell. 2021. Physicians' Perceptions of People with Disability and Their Health Care. *Health Affairs* 40(2): 297–306. https://doi.org/10.1377/hlthaff.2020.01452.

Ignagni, Esther, Eliza Chandler, and Loree Erickson. 2020. Crips and COVID in Canada. iHuman. https://www.sheffield.ac.uk/ihuman/disability-and-covid-19-global-impacts/crips-and-covid-canada

Kaldy, Joanne. 2012. Riding the Wave of Younger Residents in Long-term Care. *The Consultant Pharmacist®* 27(8): 538–543.

Kanter, Arlene S. 2014. The Development of Disability Rights Under International Law: From Charity to Human Rights. New York: Routledge.

Kerr, Sheri. 2017. Report and Recommendations: A Place to Call Home: A Continuing Care Symposium for Younger Adults with Disabilities Held May 5, 2016—Edmonton, Alberta. https://mssociety.ca/library/document/gzovOjV6GeJHNI8pScqm40MiQuZLaW7x/original.pdf

Kothari, Sunil, and Kristi L. Kirschner. 2006. Abandoning the Golden Rule: The Problem with "Putting Ourselves in the Patient's Place". *Topics in Stroke Rehabilitation* 13 (4): 68–73. https://doi.org/10.1310/tsr1304-68.

Kovac Sathya Dhara–Obituaries–Winnipeg Free Press Passages. n.d. https://passages.winnipegfreepress.com/passage-details/id-311052/KOVAC_SATHYA?fbclid=IwAR3g9BBlzc5CqpU8AF93Zu346irJ-PiNBbOqVg0Ut44yV8tvN8D1woRi25A. Accessed 12 Nov 2022.

Landry, Lydia Nunez. 2022. Chronic Illness, Well-Being, and Social Values. In *The Disability Bioethics Reader*, 156–69. Routledge.

Livingston, Kathy. 2000. When Architecture Disables: Teaching Undergraduates to Perceive Ableism in the Built Environment. *Teaching Sociology* 28 (3): 182–191. https://doi.org/10.2307/1318988.

Macauley, Robert. 2019. To Trach or Not to Trach, That Is the Question. *Paediatric Respiratory Reviews* 29: 9–13. https://doi.org/10.1016/j.prrv.2018.05.004.

McMillan, C., et al. 2016. Physician Perspectives on Care of Individuals with Severe Mobility Impairments in Primary Care in Southwestern Ontario, Canada. *Health and Social Care in the Community* 24 (4): 463–472.

Moskowitz, Gordon B., Jeff Stone, and Amanda Childs. 2012. Implicit Stereotyping and Medical Decisions: Unconscious Stereotype Activation in Practitioners' Thoughts About African Americans. *American Journal of Public Health* 102 (5): 996–1001. https://doi.org/10.2105/AJPH.2011.300591.

Panocchia, Nicola, Viola D'ambrosio, Serafino Corti, Eluisa Lo Presti, Marco Bertelli, Maria Luisa Scattoni, and Filippo Ghelma. 2021. COVID-19 Pandemic, the Scarcity of Medical Resources, Community-centred Medicine and Discrimination against Persons with Disabilities. *Journal of Medical Ethics* 47(6): 362–366.

Raikin, Alexander. 2022. No Other Options. The New Atlantis. https://www.thenewatlantis.com/publications/no-other-options

Reynolds, Joel M. 2017. 'I'd Rather Be Dead than Disabled'—The Ableist Conflation and the Meanings of Disability. *Review of Communication* 17(3): 149–163. https://doi.org/10.1080/15358593.2017.1331255.

Rimmerman, Arie. 2013. Social Inclusion of People with Disabilities: National and International Perspectives. Human Rights Review 16(4): 397–399.

Wall, K. 2017. Low Income Among Persons with a Disability in Canada. Insights on Canadian Society. Statistics Canada. https://www150.statcan.gc.ca/n1/en/pub/75-006-x/2017001/article/54854-eng.pdf

Wolbring, Gregor. 2008. The Politics of Ableism. *Development* 51 (2): 252–258.

World Health Organization. n.d. Social Determinants of Health. https://www.who.int/westernpacific/health-topics/social-determinants-of-health. Accessed on 27 Dec 2021.

Heidi Janz, Ph.D. (She/Her/Hers) is a Core Faculty Member and Assistant Adjunct Professor with the John Dossetor Health Ethics Centre at the University of Alberta. Her areas of specialization include Disability Ethics, Critical Disability Studies, and Research-Based Drama. She is also an active disability-rights advocate at the national level. In her "other life", she is a writer/playwright and filmmaker. Her creative work focuses on making the experiences of people with disabilities accessible to audiences made up of both people with disabilities and people who are temporarily able-bodied. Heidi Janz also has cerebral palsy.

Chapter 20
Assisted Life Before Assisted Death: Disability Discomfort Regarding MAID

Tim Stainton

Abstract This chapter considers the vulnerability of disabled persons within Canada's Medical Assistance in Dying regime. It briefly outlines the history of disabled persons opposition and the failure of both government and the media to respond to these concerns. It then examines some of the contextual and historical issues which underlay disabled persons concerns. The main part of the chapter considers three key factors which create this vulnerability and may lead to disabled persons seeking MAID due to preventable socio-economic factors: the threat of or current institutionalization; a lack of access to sufficient and acceptable disability supports; and, poverty. Each of these aspects of vulnerability are illustrated by a number of case examples. Finally, the chapter discusses the absence of any significant safeguards to protect disabled persons and the failure of current monitoring and reporting regimes to identify cases of MAID motivated by the factors noted above.

Keywords Assisted suicide · Disability · Institutionalization · Disability supports · Poverty · Vulnerability

20.1 Introduction

On March 17th 2021 Bill C7, *An Act to amend the Criminal Code (medical assistance in dying)* received royal assent after the Government invoked closure to cut off debate despite the fact that a Senate amendment would significantly expand MAID by allowing mental illness as the sole underlying condition,[1] a significant expansion

[1] The Senate amendment would require the Government to allow access to MAID where mental illness was the sole underlying condition within 18 months, the amendment was revised to allow a two year window for introduction.

T. Stainton (✉)
University of British Columbia, Vancouver, BC, Canada
e-mail: Timothy.Stainton@ubc.ca

J. Kotalik and D. W. Shannon (eds.), *Medical Assistance in Dying (MAID) in Canada*, The International Library of Bioethics 104,
https://doi.org/10.1007/978-3-031-30002-8_20

to the original intent of C7 (*Government Bill (House of Commons) C-7 (43–2)–Royal Assent–an Act to Amend the Criminal Code (Medical Assistance in Dying)–Parliament of Canada* 2021). Even though the Government's own five year review required by the original Act had not taken place, the Government claimed widespread consultation had taken place and that this change reflected the majority opinion of Canadians. The Bill was near universally opposed by Canadian disability rights groups, disabled persons organizations and their allies including BIPOC, Indigenous and disabled women's organizations. If that were not enough to give pause, it was also condemned by the UN Special Rapporteurs on the Rights of Disabled Persons, the UN Independent Expert on the Enjoyment of all Human Rights by Older Persons and, the UN Special Rapporteur on Extreme Poverty and Human Rights (British Columbia Aboriginal Network on Disability Society et al. 2021). The Liberal Government and their allies however chose to ignore these concerns and hurried the Bill through. To add insult to injury, they even reference Canada's commitment to the UN Convention on the Rights of Person's with Disabilities and its own commitment to disability inclusion in the preamble of the Bill. It is hard to imagine a similar dismissal of an equity seeking group's concerns if we were discussing any other equity group were they to voice unified concern over a bill and were backed up by UN human rights experts. And yet the Liberal Government somehow believed, and correctly it would seem, that the concerns of the disability community would not resonate beyond that community and it was safe to ignore them without damage to their political fortunes.

Sadly this mirrors the general response to disabled persons concerns over MAID since the early days of the push for assisted suicide and its expansion in Canada. This is likely in part due simply to a lack of awareness of the concerns of disabled persons given the paucity of coverage in the Canadian media. It also reflects a deeply rooted, and often unconscious, ableism which permeates Canadian and other societies. This chapter will consider one specific set of concerns the disability community has regarding assisted suicide, broadly, the idea that as Canada rushes to an ever more liberal regime of assisted death, it has failed to provide disabled persons with the necessities for an assisted life (Stainton 2019; Tuffrey-Wijne et al. 2018).

Before we begin it needs to be acknowledged that disability covers a vast range of conditions and presentations from sensory impairments, physical and mental disabilities and ranging from lifelong, traumatic, degenerative or episodic impairments. To some degree the implications of MAID will vary across this range but for our purposes we will use the approach adopted by the UN Convention on the Rights Persons with Disabilities which emphasizes the barriers disabled persons face in achieving full and equal participation in society[2] rather than specific impairments.[3] Further, as individuals, disabled Canadian's like the rest of the population, hold a range of personal

[2] *Persons with disabilities include those who have long-term physical, mental, intellectual or sensory impairments which in interaction with various barriers may hinder their full and effective participation in society on an equal basis with others* (United Nations Convention on the Rights of Persons with Disabilities, Article 1).

[3] The issue of mental illness and assisted suicide raises a number of unique issues and deserves a fuller treatment then can be provided here though much of what is contained in this paper applies to mental illness.

views regarding assisted suicide. Some will oppose it on conscience or religious grounds while other will support it. Our concern here is not with MAID per se, but rather how MAID as currently constituted increases the vulnerability of disabled Canadian's and how it will lead to unwanted and unnecessary shortening of lives due to poverty, ableism and, a lack of appropriate and sufficient disability supports.

One final note concerns the rather unique position of disability in the Canadian assisted suicide context. Virtually unique amongst jurisdiction that allow some form of assisted suicide, Canada's law explicitly includes disability as one of the conditions that makes a person eligible for MAID. The Act states:

(2) *A person has a grievous and irremediable medical condition only if they meet all of the following criteria:*
(a) *they have a serious and incurable illness, disease or* ***disability***;
(b) *they are in an advanced state of irreversible decline in capability*;
(c) *that illness, disease or* ***disability*** *or that state of decline causes them enduring physical or psychological suffering that is intolerable to them and that cannot be relieved under conditions that they consider acceptable(emphasis added) (Government Bill (House of Commons) C-14 (42–1)–Royal Assent–an Act to Amend the Criminal Code and to Make Related Amendments to Other Acts (Medical Assistance in Dying)–Parliament of Canada*, 2015).

This then makes disabled Canadian's the only Canadian's who can access MAID without actually being ill. While prior to the amendments under C7 the 'reasonably foreseeable death' criteria provided a certain level of protection, the elimination of this requirement essentially means the mere fact of having a disability and that you feel you are suffering is sufficient to qualify for MAID. What much of the focus of this chapter will be is the nature and cause of the suffering which when combined with the presence of a disability is sufficient grounds to access MAID regardless of whether the suffering is an inevitable result of the disability rather than being a result of a lack of support, poverty, or ableism.

20.2 Some Context

Before we consider the specific areas of concern a little context is useful to fully grasp the disability community's fears and the precarity of life with a disability in Canada. Disabled people in Canada are some of the most marginalized citizens in the country. Income levels of those with more severe disabilities are less than half that of the population without disabilities with 28.6 percent of working age adults with disabilities living in poverty compared to 10% of the general population (Morris et al. 2018). Canadians with a disability, both women and men, were almost twice as likely to be victims of a violent crime as Canadians who do not have a disability. Women with disabilities and persons with mental health disabilities were particularly at risk of violence (Cotter 2018). Add to this poor access to appropriate housing, high rates of institutionalization and the poor access to disability supports, the general picture of being disabled in Canada is not one conducive to living a full and meaningful life.

The lack of appropriate supports also creates an increased reliance on family to provide supports. In the second annual report on medical assistance in dying in Canada 35.9% of people seeking MAID reported perceived burden to family, friends or caregivers as a source of suffering leading to their decision to access MAID (Health Canada 2021). Feelings of burdensomeness observed in persons with disabilities has been associated with suicidal ideation or attempts (Khazem et al. 2015; Russell et al. 2009). This is to a large degree a function of the presence or absence, and the adequacy of, disability supports.

In their submission to the UN Human Right Committee, the major Canadian disability rights organizations note: *Under Article 6, this Committee (UNHRC) has recognized the rights of persons with disabilities to "the effective enjoyment of the right to life on an equal basis with others". States are required to provide "measures of protection" to guarantee this right, including "the provision of reasonable accommodation when necessary to ensure the right to life, such as ensuring access of persons with disabilities to essential facilities and services..." This Committee has interpreted Article 6 of the ICCPR with reference to Article 10 of the CRPD, which provides that "States Parties reaffirm that every human being has the inherent right to life and shall take all necessary measures to ensure its effective enjoyment by persons with disabilities on an equal basis with others". It is clear that the right to life includes obligations on states to ensure access to the social conditions necessary for meaningful enjoyment of the right* (British Columbia Aboriginal Network on Disability Society et al. (2021). Canada is far from meeting its obligation under the Convention of the Rights of Persons with Disabilities.

20.3 History

To fully understand the disability concerns over assisted suicide some historical context is useful. The Eugenics movement of the late 19th and early to mid-twentieth centuries was to a large degree focused on the control, and in many cases, elimination of those consider 'unfit'. People affected by disabling conditions, socially, physically, and cognitively, became a primary target of eugenics inspired policy throughout Europe and North America (Hawkins 1997). The eugenics movement reached its nadir with the Nazi *Aktion T4* program, a precursor and testing ground for the wider holocaust to come. *Aktion T4* led to the state sanctioned physician led murder of some 300,000 people with disabilities, many on the grounds of ending suffering (Hohendorf 2016).

While tempting to dismiss this as simply part of the tyranny of National Socialism, it is important to note this was simply the extreme, and to some degree logical, outcome of eugenic beliefs widely held in the west including prominent figures such as Tommy Douglas the 'father' of medicare and, Winston Churchill (Marshall and Robertson 2006). Nor can we claim that similar beliefs do not still exists. In 2016

26-year-old Satoshi Uematsu murdered 19 disabled people in the care home where he worked in Japan. In explaining his motives he notes 'I envision a world where a person with multiple disabilities can be euthanized' (Willitts 2016). In a letter he attempted to deliver to the Speaker of the lower house he called for 'a revolution', demanding that all disabled people be put to death through 'a world that allows for mercy killings' (Associated Press 2016).

20.4 'New Eugenics'

While this extreme example may be easy to dismiss as a tragic act by a disturbed individual there are other examples of what has been termed the 'new' or the 'quiet' eugenics (Reinders et al. 2019). Current trends in pre-natal testing indicate a strong negative view towards having a child with a disability (Skotko 2009). The potential to 'eliminate Down's Syndrome' through pre-natal testing (PNT) and termination is now being discussed widely as a very positive development both with regards to the elimination itself and the potential cost savings which might be realized (Quinones and Lajka 2017).

In 2002, the Groningen Protocol (GP) for neonatal euthanasia was developed in the Netherlands with the intent to regulate the practice of actively ending the life of newborns. Significant numbers of these cases involve neonates with non-life threatening, medically treatable conditions and disabilities (Verhagen and Sauer 2005). In their review of all 22 cases reported to the district attorneys' offices in the Netherlands from 1998–2005 Verhagen and Sauer found that all involved spina bifida. They report that the considerations used to decide on euthanizing included: expectation of extremely poor quality of life (suffering) in terms of functional disability, pain, discomfort, poor prognosis, and hopelessness; predicted lack of self-sufficiency; expected hospital dependency; and, long life expectancy. In all cases these were largely third party, subjective determinations of perceived future quality of life. It is not an unreasonable proposition that similar consideration would influence the practice of MAID.

20.5 Ethical Rationale

The idea that disability inevitably leads to a life of suffering and poor quality is a key belief that underpins the view that providing access to MAID for disabled persons is both a compassionate and in many cases, the morally correct thing to do. This is most evident in, but not exclusive to, certain strains of utilitarian ethics. Singer's views about disabled persons moral status and the ethics of euthanasia are perhaps best known (Singer 1975, 1993).[4] He is far from alone however. John Harris writes with regards to prenatal testing and elimination of disabled fetuses that 'where we know that a particular individual will be born 'deformed' or 'disfigured' ...the powerful motive that we have to avoid bringing gratuitous suffering into the world will surely show us that to do so would be wrong'. He goes on to state that in the case of severe disability 'we should give them a humane death by legalizing euthanasia in such cases' (Harris 1998).

As Tuffrey-Wijne et al. note: 'the fact that the disability itself, rather than an acquired medical condition, can be accepted as a cause of suffering that justifies euthanasia is deeply worrying' (Tuffrey-Wijne et al. 2019). Where this becomes most concerning is when it is operationalized through approaches like Quality Adjusted Life Years (QALYs) to determine what, if any, interventions offer the best cost–benefit outcome. Leading bioethicists have endorsed the view that utilitarianism requires discrimination against the disabled in the allocation of health care resources based on the maximization of quality adjusted life years (McKie et al. 1998; Stein 2002). As Hilliard states, 'Consistent with the utilitarian ethic, state sanctioned killing of those deemed to have "lost their dignity" is hailed as a "good"' (Hilliard 2011). Barrie notes, 'problems (with QALY)' relate closely to the debate over euthanasia and assisted suicide because negative QALY scores can be taken to mean that patients would be 'better off dead' (Barrie 2015).

20.6 Assisted Life or Assisted Death?

While there are range in inter-related concerns the disability community has around assisted suicide, this chapter will focus on what is perhaps the key area of concern, that is, how disabled persons may consider assisted suicide not because they want to die, but rather they are tired of fighting for the means to live a life of meaning and quality. We will consider three specific issues: institutionalization; a lack of community supports; and, poverty.

20.6.1 Institutionalization

There have been a number of cases that seem to support the concern that people with disabilities are opting for MAID rather than live in an institution which they consider dehumanizing and incompatible with a life of meaning and quality. Archie Rolland died by assisted suicide in 2016. A press report at the time noted: 'It's not the illness that's killing him', Rolland said in a series of emails with the Montreal Gazette. He's tired of fighting for compassionate care' and that life in long term care

[4] Singer has consistently expressed his views on the morality of killing disabled persons and the relative moral status of some disabled persons and certain animals.

was 'unbearable' (Fidelman 2016). M. Truchon, plaintiff in the Quebec case which led to Bill C7, noted it is the nature of the care he is being offered which is, at least in part, behind his suffering: 'At a news conference… Mr. Truchon had an assistant read a statement explaining that he couldn't face the prospect of life confined to an institution' (Tu Thanh Ha 2017).

The testimony of Québécois Jonathan Marchand to the Senate Standing Committee on Legal and Constitutional Affairs powerfully sums up the situation many disabled persons find themselves confronted with:

> *I was prepared to do anything to get out of this medical hell, but just like Jean Truchon, I was denied the home care support that I needed... I was told that it was a political issue as living in the community with the necessary support is not a right in Canada. After two and a half years in the hospital, I ended up in a long-term care facility.... I gave up and sank into depression. I was ashamed to live in this ghetto. Without humanity and freedom, life no longer has any meaning. I regretted having refused euthanasia. I simply wanted to live with my partner, work and have a normal social life. I wanted to die. ... I discovered that about 70% of people with severe disabilities live in institutions in Quebec. The others cling to living at home, but often find themselves also isolated. Many have committed suicide or have accepted euthanasia to avoid suffering my fate. My disability is not the cause of my suffering, but rather the lack of adequate support, accessibility, and the discrimination I endure every day…Why is it so hard to be seen and heard when we want to live?* (Marchand 2021).

According to the 2016 census data, over 65,000 people between the ages of 18 and 64 live in health care facilities in Canada with the vast majority of that number in nursing homes and seniors facilities, mostly people with disabilities (Government of Canada 2017). The failings of our long term care system is well documented, but even in the best of circumstances these are not suitable homes for working age people with disabilities. There is no reason inherent to their disability that anyone cannot be supported in a suitably adapted home. The reason working age people are institutionalized is based not on best practice or the needs or desires of the individual, but on outdated policy and financial constraints. There are few greater incentives for people to access MAID than the prospect of a life in an institution, an outcome which is wholly socially determined and wholly preventable.

20.6.2 Lack of Appropriate or Acceptable Disability Supports

Related to institutionalization is the lack of acceptable disability supports available in the community. Access to community supports, even if sufficient supports are theoretically available, is usually met with significant delays (Direct Funding Ontario n.d.). With regards to the relatively straight forward basic equipment, over 1.5 million persons with disabilities in Canada aged 15 years and over had an unmet need for an aid or device. Of these, 1 million indicated that cost was the reason for their unmet need (Morris et al. 2018).

The disturbing case that seems to indicate a direct example of how MAID was presented as an alternative to acceptable disability supports is that of Roger Foley, a

42 year old man with a degenerative condition. After a lengthy hospital stay, staff are sent, unrequested by Foley, to see if he has an interest in pursuing MAID. Not once, but twice. Foley secretly taped the conversations, which he subsequently released publicly (Chronically Ill Man Releases Audio of Hospital Staff Offering Assisted Death 2018). During one conversation he is informed that if he refuses the home care option available to him he will be charged 'north of $1500 a day' if he remains in hospital. Foley indicates that the current option for home care is unacceptable to him as the government-selected home care provider had previously left him in ill health with injuries and food poisoning. When asked if he thinks of harming himself Foley replies he is "always thinking I want to end my life" but quickly clarifies this by saying if he had self-directed support, a service available but not offered to Foley, he would be fine.

While the exact nature and motivations for the hospital's actions are still unclear, what is clear is that Foley was under pressure to leave the hospital without the type of support he desired or, to consider MAID. The tapes also raise the question of whether MAID was seen as a way to manage the cost the hospital was incurring due to Foley's refusal to leave.

41 year old Sean Tagert, a man with ALS, died by MAID in August of 2019. He was quite explicit that his reason for choosing assisted death was his inability to secure sufficient home care funding in order live a life he considered worth living. A mere 4 h more of home care then the health authority was willing to offer was the difference between life or death for Mr. Tagert (*B.C. Man with ALS Chooses Medically Assisted Death after Years of Struggling to Fund 24-h Care* 2019). All he wanted was to remain in his home which had the necessary adaptations and to be able to spend time with his young son in their home. Instead he was offered the unacceptable option of ending his days in a long term care home, effectively cut off from any normal relationship his son.

'Sophia' died by MAID in February of 2022. She was diagnosed with Multiple Chemical Sensitivities which meant she required accommodation that protected her from airborne chemicals from smoke, perfume etc. Living on disability benefits she was unable to afford a single family home and was unable to secure housing that would keep her safe and healthy. Ultimately this led to her decision to access assisted suicide as she could no longer bear the isolation and constant threats to her health. (Blog, n.d.).[5]

Both of these latter two cases speak to the completely preventable deaths caused by a structural failure of systems which are unable, or more accurately unwilling, to provide the means for an assisted life, but are quite willing to provide an assisted death in its stead. Clearly in Sophia's case poverty was also an underlying cause of her suffering which led to her decision to access MAID.

[5] Sophia is a pseudonym used to protect her identity. She left a statement on her situation and reasons for accessing MAID before she passed with the **GRIM:** The **GR**ievous & **I**rre**M**ediable Project which seeks to document disability related MAID cases. See also: https://aseq-ehaq.ca/en/news/blog/.

20.6.3 Poverty

As noted above, working age adults with severe disabilities are far more likely to live in poverty than those without disabilities. The rate of low income for working age adults with more severe disabilities was double that of working age adults with milder disabilities and nearly triple that for those with no disability (Morris et al. 2018). This not surprising since no province or territory provides a disability benefit which would provide an income above the poverty line with rates as low as $850 a month in some jurisdictions (Canadian Institute for Inclusion and Citizenship 2021). When one considers the extra cost that many disabled people face these rates become even more shocking and highlight the daily struggle many face to simply survive. Poverty and other socio-economic deprivations experienced by disabled people are strongly correlated with suicide and suicidal ideation (Burrows et al. 2010; Curtis et al. 2013). A recent study on suicide and disability notes that while rates are higher than the general population this can in part be explained by social adversity, including food insecurity and low sense of community belonging (McConnell et al. 2016).

One recent case brings this concern into tragic perspective. 'Madeline'[6] has been clear that it is poverty and a lack of financial support that has led her to apply for MAID. While struggling to survive through a GoFundMe campaign, she is aware this is only a temporary stopgap in what seems an inevitable journey towards MAID.

> *I am dying, and my death is completely preventable.*
>
> *My deterioration has two causes. The first is the poverty: the lack of care and financial support for someone like me on disability assistance. The second is the abdication of care, as I battle a 40-year war with a post-viral syndrome, the treatments for which my provincial government refuses to pay...*
>
> *I've started the medical assistance in dying (MAID) process... not because I want to die, but because I'll have to choose between a quick death and a slow, torturous death in a care home—separated from my community, my friends, and my volunteering (which is my purpose).*
>
> *I've chosen the quick death, if you can call it a choice* (The Disabled Are Choosing to Die Because They Can't Afford to Live, n.d.)

'Denise', in a case remarkably similar to that of 'Sophia', the 31 year old women came close to also accessing MAID due to her inability to secure appropriate housing and other supports. She was more fortunate than 'Sophia' in that a GoFundMe campaign that has now has amassed over $65,000 allowed her to find alternative, though temporary, housing which alleviated her suffering. In an interview she notes: "These are strangers saying they do not want this to happen". "I am no longer focusing on just survival", "...Mentally, I am more clear to put things in place to put a more liveable life" (Woman with Disabilities Approved for Medically Assisted Death Relocated Thanks to "Inspiring" Support 2022). Clearly her suffering which motivated her decision to consider MAID was in large part a result of her inability to

[6] Madeline is a pseudonym she uses protect her identity.

find appropriate support and was remediable. Unfortunately this took a GoFundMe campaign in the absence of government support-Canada it would seem is content to provide an assisted death but not the means for an assisted life.

'Kat', a women in her 30s with significant health challenges was approved for MAID. She notes: "I thought, 'Goodness, I feel like I'm falling through the cracks so if I'm not able to access health care am I then able to access death care?' And that's what led me to look into MAID and I applied last year" (*"Easier to Let Go" without Support: B.C. Woman Approved for Medically Assisted Death Speaks Out* 2022). She goes on to state: "From a disability and financial perspective as well, I can't afford the resources that would help improve my quality of life. Because of being locked in financially as well and geographically, it is far easier to let go than keep fighting" (*"Easier to Let Go" without Support: B.C. Woman Approved for Medically Assisted Death Speaks Out* 2022). Fraser Health inadvertently acknowledges the role poverty plays in this decision noting in her documentation: "patient has an extensive medical chart" and that "there were no other treatment recommendations or interventions that were suitable to the patient's needs or to her financial constraints" (*"Easier to Let Go" without Support: B.C. Woman Approved for Medically Assisted Death Speaks Out* 2022). This is a stark admission of the role poverty may play in MAID decision making and the complete absence of any mechanism to address socio-economic factors which may prompt MAID decisions in our current MAID regime, even when known and acknowledged.

It has been argued that socio-economic factors do not play a significant role in decisions to access MAID (Downar et al. 2020; Redelmeier et al. 2021). The two studies cited both looked at Ontario cohorts before the reasonable foreseeability clause was removed from the legislation. As such all of those in the cohort would have been at the end of life and would not have included cases such as 'Sophia' or 'Madeline'. Additionally, both studies indicate even in this cohort a significant minority do access MAID. It is also worth noting that even a small minority accessing MAID due to preventable factors such as poverty should be cause for significant concern. In short, this data does little to allay the disability related concerns noted above.

20.7 Conclusion

While there are many concerns regarding disability and assisted suicide, this chapter has focused on those which stem from the structural barriers faced by disabled persons in Canada which may lead them to seek MAID rather than continue to suffer from preventable hardships. In a context of widespread ableism within society generally, and more acutely within healthcare settings, it is difficult to see how the current MAID regime can safeguard disabled persons effectively from an unwanted and untimely death (Drainoni et al. 2006; Gill 2000, 2010; Iezzoni et al. 2021). Indeed with the prevailing assumptions around inevitable suffering within many ethical streams and the negative attitudes in health care many may be subtly, or overtly, counselled to seek an assisted death rather than struggle to obtain an assisted life[7] (Chronically Ill Man Releases Audio of Hospital Staff Offering Assisted Death 2018).

With the removal of the reasonable foreseeability criteria, there are few, if any safeguards against the type of concerns discussed above (Stainton 2019). From the early days of the MAID debate, disability groups and others advocated for a 'vulnerability assessment' to be included in the process (Canadian Association for Community Living 2016; CCD Submission to Special Joint Committee on Physician Assisted Dying | Council of Canadians with Disabilities, n.d.; Read the Standard, n.d.). This would encompass a full psycho-social assessment by a qualified professional such as a social worker. This would at least help to uncover cases where remediation of suffering through non-medical interventions is possible as in many of the cases discussed here. Current safeguards for a patient whose natural death is not reasonably foreseeable do require that the patient be informed of other means available to them, including counselling, mental health supports, disability supports, community services and palliative care; offered consultation with relevant professionals. There are of course serious limitations to how effective this is as a safeguard. Firstly the assessors may not be particularly familiar with the alternatives available to the person as social support issues are generally not areas of core medical expertise. A more fundamental issues is that even if appropriate supports are identified, access to them remains difficult, if not impossible, in many case. Many of the cases noted above are quite clear about what supports would alleviate their suffering but in none of these cases has this resulted in them being provided beyond reliance on GoFundMe campaigns. Disabled people are well aware of what they need-the issue is that these are not readily available, unlike MAID. As such, even a comprehensive vulnerability assessment remains a hollow safeguard.

Monitoring also remains problematic as little in the current regime would help to identify the number of persons seeking MAID due to suffering caused not by their underlying condition, but due to the social and economic fallout related to that condition. While the current monitoring system does identify persons who use disability supports this is largely irrelevant in the current context for a number of reasons. First it is such an imprecise term covering everything for aides and devices–walkers, hearing aids, cleaning assistance to full 24/7 attendant care supports. More importantly, it fails to answer the critical question of whether the person had sufficient disability supports to allow them to live a life of meaning and quality. In many ways this data on disability supports will feed a false narrative that given they had disability supports the absence of such was not a factor in their decision to access MAID. In order to get the kind of information that would help us to understand the extent to which socio-economic factors are a significant element in disabled person's decision making with regards to MAID a far more detailed and nuanced reporting would be required.

Narrative data which would help us to better understand the above issue is completely absent from the current system of reporting and monitoring. The Dutch

[7] The Foley case is perhaps the best known example.

RTE system requires a certain number of case reports, detailed narratives of the individual's situation and the process of assessment and approval. These are publicly available which allows researchers and others to do detailed analyses. Tuffrey-Wijne et al. (2018; 2019) and colleagues were able to search these documents in order to identify persons with intellectual disabilities (ID) and/or Autism Spectrum Disorder who had been euthanized, data which is not collected in the general demographics. Without this resource there would be no way of knowing if people with ID/ASD had been euthanized nor allow an assessment of how issues such as of consent were dealt with (an issue the authors found significant problems with) or the broader context of the requests. This demonstrates why such data is crucial to getting a full and complete picture of MAID and to fully analyze the extent disabled and vulnerable people are accessing MAID and the underlying reasons.

With limited ability to even know the extent to which disabled people are accessing MAID due to socially determined suffering, it is easy for this issue to simply continue under the radar. What is more disturbing is, as we saw with Bill C7, even when people are made aware of the concerns of the disability community it prompts little response or attention. Until Canada lives up to its commitments under the CRPD and ensure the means for an assisted life in Canada, MAID will continue to present a grave danger to disabled people.

References

B.C. Man with ALS Chooses Medically Assisted Death After Years of Struggling to Fund 24-Hour Care. 2019. CBC. https://www.cbc.ca/news/canada/british-columbia/als-bc-man-medically-assisted-death-1.5244731

Barrie, S. 2015. QALYs, Euthanasia and the Puzzle of Death. *Journal of Medical Ethics* 41 (8): 635–638. https://doi.org/10.1136/medethics-2014-102060.

Blog. n.d. ASEQ-EHAQ. https://aseq-ehaq.ca/en/news/blog/. Retrieved 14 Jun 2022.

British Columbia Aboriginal Network on Disability Society (BCANDS), Council of Canadians with Disabilities, DisAbled Women's Network Canada, Inclusion Canada, and Vulnerable Persons Standard to the Human Rights Committee. 2021. On the List of Issues Prior to Reporting for Canada to be adopted during the 132nd Session of the Human Rights Committee (28 June to 23 July 2021). https://tbinternet.ohchr.org/_layouts/15/treatybodyexternal/Download.aspx?symbolno=INT%2FCCPR%2FICS%2FCAN%2F44924&Lang=en

Burrows, S., N. Auger, M. Roy, and C. Alix. 2010. Socio-economic Inequalities in Suicide Attempts and Suicide Mortality in Québec, Canada, 1990–2005. *Public Health* 124 (2): 78–85.

Canadian Association for Community Living. 2016. Assessing Vulnerability in a System of Physician-assisted Death in Canada.

Canadian Institute for Inclusion and Citizenship. 2021. Supporting Siblings: A New Path to Financial Security.

CCD Submission to Special Joint Committee on Physician Assisted Dying | Council of Canadians with Disabilities. n.d. www.ccdonline.ca. http://www.ccdonline.ca/en/humanrights/endoflife/SJCPAD-28jan2016. Retrieved 15 Jun 2022.

Chronically Ill Man Releases Audio of Hospital Staff Offering Assisted Death. CTV National News. 2018. The recordings and commentary can be found at: https://www.ctvnews.ca/health/chronically-ill-man-releases-audio-of-hospital-staff-offering-assisted-death-1.4038841

Cotter, A. 2018. Violent Victimization of Women With Disabilities. 2014. statcan.gc.ca; Government of Canada, Statistics Canada. https://www150.statcan.gc.ca/n1/pub/85-002-x/2018001/article/54910-eng.htm

Curtis, B., C. Curtis, and R.W. Fleet. 2013. Socio-economic Factors and Suicide: The Importance of Inequality. *New Zealand Sociology* 28 (2): 77–92.

de Bruin, Gerald Tabitha, and Gerald Robertson. 2006. Eugenics in Canada. In The Canadian Encyclopedia. https://www.thecanadianencyclopedia.ca/en/article/eugenics

Direct Funding Ontario. n.d. Waiting List: How the Waiting List Works. https://www.dfontario.ca/application/waiting-list.html. Accessed 1 Apr 1 2022.

Downar, J., R.A. Fowler, R. Halko, L.D. Huyer, A.D. Hill, and J.L. Gibson. 2020. Early Experience with Medical Assistance in Dying in Ontario, Canada: A Cohort Study. *Canadian Medical Association Journal* 192 (8): E173–E181. https://doi.org/10.1503/cmaj.200016.

Drainoni, M.-L., E. Lee-Hood, C. Tobias, S.S. Bachman, J. Andrew, and L. Maisels. 2006. Cross-disability Experiences of Barriers to Health-care Access: Consumer Perspectives. *Journal of Disability Policy Studies* 17 (2): 101–115. https://doi.org/10.1177/10442073060170020101.

"Easier to Let Go" Without Support: B.C. Woman Approved for Medically Assisted Death Speaks Out. 2022. British Columbia. https://bc.ctvnews.ca/easier-to-let-go-without-support-b-c-woman-approved-for-medically-assisted-death-speaks-out-1.5937496

Fidelman, C. 2016. Life in Long-term Hospital "Unbearable": Montreal Man with ALS. Montreal Gazette. https://montrealgazette.com/news/local-news/life-in-long-term-hospital-unbearable-montreal-man-with-als

Gill, C.J. 2000. Health Professionals, Disability, and Assisted Suicide: An Examination of Relevant Empirical Evidence and Reply to Batavia. *Psychology, Public Policy, and Law* 6 (2): 526–545. https://doi.org/10.1037/1076-8971.6.2.526.

Gill, C.J. 2010. No, We Don't Think Our Doctors Are Out to Get Us: Responding to the Straw Man Distortions of Disability Rights Arguments Against Assisted Suicide. *Disability and Health Journal* 3 (1): 31–38. https://doi.org/10.1016/j.dhjo.2009.10.003.

Government Bill (House of Commons) C-7 (43–2)–Royal Assent–An Act to Amend the Criminal Code (Medical Assistance in Dying)–Parliament of Canada. 2021. parl.ca. https://parl.ca/DocumentViewer/en/43-2/bill/C-7/royal-assent

Government Bill (House of Commons) C-14 (42–1)–Royal Assent–An Act to Amend the Criminal Code and to Make Related Amendments to Other Acts (Medical Assistance in Dying)–Parliament of Canada. 2015. parl.ca. https://www.parl.ca/DocumentViewer/en/42-1/bill/C-14/royal-assent

Government of Canada, S. C. 2017. 2016 Census of Canada: Data tables–Type of Collective Dwelling (16), Age (20) and Sex (3) for the Population in Collective Dwellings of Canada, Provinces and Territories, 2016 Census–100% Data. www12.statcan.gc.ca. https://www12.statcan.gc.ca/census-recensement/2016/dp-pd/dt-td/Rp-eng.cfm?TABID=2&Lang=E&APATH=3&DETAIL=0&DIM=0&FL=A&FREE=0&GC=0&GID=1234492&GK=0&GRP=1&PID=109537&PRID=10&PTYPE=109445&S=0&SHOWALL=0&SUB=0&Temporal=2016&THEME=116&VID=0&VNAMEE=&VNAMEF=&D1=0&D2=0&D3=0&D4=0&D5=0&D6=0

Harris, J. 1998. *Clones, Genes, and Immortality: Ethics and the Genetic Revolution*. Oxford: Oxford University Press.

Hawkins, M. 1997. *Social Darwinism in European and American Thought 1860–1945: Nature as Model and Nature as Threat*. London: Cambridge University Press.

Health Canada. 2021. Second Annual Report on Medical Assistance in Dying in Canada. 2020. Ottawa. https://www.canada.ca/en/health-canada/services/medical-assistance-dying/annual-report-2020.html

Hilliard, M.T. 2011. Utilitarianism Impacting Care of Those with Disabilities and Those at Life's End. *The Linacre Quarterly* 78 (1): 59–71. https://doi.org/10.1179/002436311803888474.

Hohendorf, G. 2016. "Death as a Release From Suffering"—The History and Ethics of Assisted Dying in Germany Since the End of the 19th Century. *Neurology, Psychiatry and Brain Research* 22 (2): 56–62. https://doi.org/10.1016/j.npbr.2016.01.003.

Khazem, L.R., D.R. Jahn, K.C. Cukrowicz, and M.D. Anestis. 2015. Physical Disability and the Interpersonal Theory of Suicide. *Death Studies* 39 (10): 641–646.

Iezzoni, L.I., S.R. Rao, J. Ressalam, D. Bolcic-Jankovic, N.D. Agaronnik, K. Donelan, T. Lagu, and E.G. Campbell. 2021. Physicians' Perceptions of People With Disability and Their Health Care. *Health Affairs* 40 (2): 297–306. https://doi.org/10.1377/hlthaff.2020.01452.

Marchand, J. 2021. Testimony of Jonathan Marchand, Senate, Standing Committee on Legal and Constitutional Affairs, Evidence, 43–2. 2021. https://sencanada.ca/en/Content/Sen/Committee/432/LCJC/10ev-55128-e

McConnell, D., L. Hahn, A. Savage, C. Dubé, and E. Park. 2016. Suicidal Ideation Among Adults with Disability in Western Canada: A Brief Report. *Community Mental Health Journal* 52 (5): 519–526. https://doi.org/10.1007/s10597-015-9911-3.

McKie, J., J. Richardson, Peter D Singer, and Helga Kuhse. 1998. The Allocation of Health Care Resources: An Ethical Evaluation of the 'QALY' Approach. Aldershot, England; Brookfield, USA: Ashgate Publishing Limited.

Morris, S., Gail Fawcett, Laurent Brisebois, and Jeffrey Hughes. 2018. A Demographic, Employment and Income Profile of Canadians with Disabilities Aged 15 Years and Over, 2017. statcan.gc.ca; Government of Canada, Statistics Canada. https://www150.statcan.gc.ca/n1/pub/89-654-x/89-654-x2018002-eng.htm

Quinones, Julian, and Arijeta Lajka. 2017. "What Kind of Society Do You Want to Live In?": Inside the Country Where Down Syndrome Is Disappearing. CBS News. https://www.cbsnews.com/news/down-syndrome-iceland/

Read the Standard. n.d. Vulnerable Persons Standard. http://www.vps-npv.ca/readthestandard. Retrieved 15 Jun 2022.

Redelmeier, Donald A., Kelvin Ng, Deva Thiruchelvam, and Eldar Shafir. 2021. Association of Socioeconomic Status with Medical Assistance in Dying: A Case–control Analysis. BMJ Open 11: e043547. https://doi.org/10.1136/bmjopen-2020-043547.

Reinders, J., T. Stainton, and T.R. Parmenter. 2019. The Quiet Progress of the New Eugenics. Ending the Lives of Persons With Intellectual and Developmental Disabilities for Reasons of Presumed Poor Quality of Life. *Journal of Policy and Practice in Intellectual Disabilities* 16 (2): 99–112. https://doi.org/10.1111/jppi.12298.

Russell, D., R.J. Turner, and T.E. Joiner. 2009. Physical Disability and Suicidal Ideation: A Community-Based Study of Risk/Protective Factors for Suicidal Thoughts. *Suicide & Life-Threatening Behavior* 39 (4): 440–451.

Singer, P. 1975. *Animal Liberation*. New York: Harper Collins.

Singer, P. 1993. *Practical Ethics*, 2nd ed. Cambridge: Cambridge University Press.

Skotko, B.G. 2009. With New Prenatal Testing, Will Babies With Down Syndrome Slowly Disappear? *Archives of Disease in Childhood* 94 (11): 823–826. https://doi.org/10.1136/adc.2009.166017.

Stainton, T. 2019. Disability, Vulnerability and Assisted Death: Commentary on Tuffrey-Wijne, Curfs, Finlay and Hollins. *BMC Medical Ethics* 20: 89. https://doi.org/10.1186/s12910-019-0426-2.

Stein, M.S. 2002. Utilitarianism and the Disabled: Distribution of Resources. *Bioethics* 16 (1): 1–19. https://doi.org/10.1111/1467-8519.00264.

The Associated Press. 2016. Police Search Home of Suspect in Japan Stabbing Spree. CBC News. https://www.cbc.ca/news/world/japan-stabbing-search-1.3696644

The Disabled Are Choosing to Die Because They Can't Afford to Live. n.d. www.ipolitics.ca. https://www.ipolitics.ca/news/the-disabled-are-choosing-to-die-because-they-cant-afford-to-live. Retrieved 14 Jun 2022.

Tu Thanh Ha. 2017. Montrealers File Civil Suit over Assisted-dying Laws. The Globe and Mail. https://www.theglobeandmail.com/news/national/montrealers-file-legal-action-contesting-restrictions-on-assisted-dying/article35308414/

Tuffrey-Wijne, I., Leopold Curfs, Ilora Finlay, Sheila Hollins. 2018. Euthanasia and Assisted Suicide for People with an Intellectual Disability and/or Autism Spectrum Disorder: An Examination of Nine Relevant Euthanasia Cases in the Netherlands (2012–2016). BMC Medical Ethics 19: 17. https://doi.org/10.1186/s12910-018-0257-6.

Tuffrey-Wijne, I., L. Curfs, I. Finlay, and S. Hollins. 2019. "Because of His Intellectual Disability, He Couldn't Cope". Is Euthanasia the Answer? *Journal of Policy and Practice in Intellectual Disabilities* 16 (2): 113–116. https://doi.org/10.1111/jppi.12307.

Verhagen, E., and P.J.J. Sauer. 2005. The Groningen Protocol—Euthanasia in Severely Ill Newborns. *New England Journal of Medicine* 352 (10): 959–962. https://doi.org/10.1056/nejmp058026.

Willitts, P. 2016. Disability Hate Leads to Mass Murder in Japan. Global Comment. https://globalcomment.com/disability-hate-leads-to-mass-murder-in-japan/

Woman with Disabilities Approved for Medically Assisted Death Relocated Thanks to 'Inspiring' Support. 2022. CTV News. https://www.ctvnews.ca/health/woman-with-disabilities-approved-for-medically-assisted-death-relocated-thanks-to-inspiring-support-1.5921893

Tim Stainton, BSW, MSW, PhD Tim Stainton holds both a BSW and MSW as well as a PhD in Political Theory and Social Policy from the London School of Economics. He is Co-Director of the Canadian Institute for Inclusion and Citizenship and a Professor in the School of Social Work at the University of British Columbia. He has been active in the disability rights movement for over 40 years. He was a member of the Council of Canadian Academies expert review panel on MAID and has written and lectured widely on issues related to assisted suicide.

Chapter 21
Implementing Medical Assistance in Dying in Canada: Implications for People with Intellectual and Developmental Disabilities

William F. Sullivan and John Heng

Abstract The Canadian law on Medical Assistance in Dying (MAID) is an attempt to balance various constitutional rights of individuals with society's interest in protecting vulnerable groups from harm. It aims for this protection primarily by enacting procedural safeguards. Little is known regarding how these safeguards will be implemented when the Canadian law is extended to include eligibility for MAID for those with medical conditions (including disabilities) in which death is not reasonably foreseeable. In this chapter, we explore the implications of this legislative extension on people with intellectual and developmental disabilities (IDD) in two ways: first by reviewing case studies from the Netherlands involving euthanasia of people with IDD, and second by examining tools for healthcare practitioners that have been developed to implement the Canadian law on MAID. We conclude, for various reasons, that procedural safeguards alone are ineffective in protecting people with IDD from the harm of an unintended or premature death.

Keywords Intellectual and developmental disabilities · Autism · Medical Assistance in Dying · Procedural safeguards · Implementation · Vulnerability · Complexity · Health inequities

21.1 Introduction

Since Canada's law regulating 'medical assistance in dying' (MAID) was enacted in June 2016, the number of people within the Canadian health system who die in this way has steadily increased each year (Health Canada 2021). The criteria of eligibility for MAID have also broadened incrementally. As a result of such changes, the

W. F. Sullivan (✉)
University of Toronto, Toronto, ON, Canada
e-mail: bill.sullivan@utoronto.ca

J. Heng
The King's University College of Western University, London, ON, Canada
e-mail: john.heng@uwo.ca

J. Kotalik and D. W. Shannon (eds.), *Medical Assistance in Dying (MAID) in Canada*, The International Library of Bioethics 104,
https://doi.org/10.1007/978-3-031-30002-8_21

Canadian healthcare system, whose goals are to promote and restore health, prevent illness, and palliate pain and distress, is increasingly overlapping with Canada's death system. The latter has expanded now to include persons being eligible to request that their lives be terminated even when they do not have a terminal illness or 'reasonably foreseeable natural death'. The Canadian law on MAID also introduces the notion that a *disability* can be considered for requests for MAID from a person with such a disability if it causes this person intolerable suffering that cannot be relieved by any means acceptable to the person (House of Commons Canada 2020). This potentially includes lifelong disabilities such as intellectual and developmental disabilities (IDD). Moving forward, what will be the implications for people IDD of such ambiguities when MAID is implemented in healthcare systems? The aim of this chapter is to address this question.

In this chapter, we will first summarize what is known regarding health inequities experienced by people with IDD. Health 'inequities' or 'disparities' stem from social, economic, environmental, and organizational factors. These inequities result in inadequate or inappropriate health care and related supports for certain members in society and fewer opportunities that such people have for achieving optimal health (Weinstein 2017). Health inequities, therefore, can result in poorer health and increased experiences of trauma for people with IDD compared with other people in the community (Cook and Hole 2021; Lunsky et al. 2013; Sullivan et al. 2018). Such facts are part of the context of vulnerability that must be considered concerning why people with IDD might be at greater risk of requesting MAID.

We will next review the studies that have been published in peer-reviewed medical journals regarding people with IDD or those with similar impairments of cognitive and/or social functioning in the Netherlands who have received 'euthanasia' or 'assisted suicide' (what 'MAID' is called there). There are no comparable studies on how many Canadians with IDD have requested and received MAID and under what circumstances. In the Netherlands, as in Canada, physicians are given the role of assessing eligibility for and carrying out euthanasia or assisted suicide. The Dutch studies raise poignant clinical and ethical questions regarding implementing euthanasia when healthcare practitioners who are unfamiliar with people with IDD and their complex needs play such a key role.

We will then examine a typical published tool to assist implementing the Canadian law on MAID in clinical settings—the *Medical Assistance in Dying Resource* (Centre for Effective Practice 2021). Based on our analysis, we conclude that, although there are some variations in how euthanasia is regulated in the Netherlands and Canada, similar clinical and ethical issues that have been raised by the Dutch studies on euthanasia involving people with IDD or similar impairments of cognitive and social functioning can also arise in Canada.

21.2 People with IDD in the Canadian Healthcare System

Like everyone else in society, people with IDD have a range of capabilities and limitations. Their disability and vulnerability in the healthcare system arise in the context of complex interacting individual, environmental, and societal factors.

IDD are lifelong challenges to cognitive, adaptive, and social skills that result from differences in neuropsychological development that become manifest before adulthood (Schalock et al. 2021). These challenges vary in type and severity among people with IDD. The cause of IDD is unknown in roughly half of those who have been diagnosed with IDD. But people with IDD are not always diagnosed as such. They can often go unrecognized in the healthcare and other systems. This can happen especially when challenges to cognitive, adaptive, and social functions in a person are mild or the person has concurrent mental health conditions. Mental illness is more common among people with IDD than those without IDD (Einfeld et al. 2011).

Known causes of IDD can be genetic (e.g., Down syndrome or Fragile-X Spectrum Disorder), environmental (e.g., Fetal Alcohol Spectrum Disorder), prenatal and perinatal (e.g., some instances of Cerebral Palsy) or could be linked to multiple causes (e.g., Autism Spectrum Disorder or ASD). Although some persons with an ASD do not meet cognitive and language impairment criteria for IDD, they can have significantly impaired adaptive and social functioning. People with ASD share many similar challenges relating to their health and health care as people with cognitive and language impairments.

Despite their multiple challenges, people with IDD can cope well and be resilient, be supported to lead lives that are purposeful, and develop social relationships that are mutually enriching and fulfilling (Brown 2020). However, the challenges that people with IDD experience are often associated with, and compounded by, negative social, economic, organizational, and ecological factors. These include stigma, isolation, abuse, poverty, poor nutrition, inadequate social supports, inappropriate housing, lack of opportunities for education, work, and unsafe environments.

Within healthcare systems, such members of society often encounter negative bias and discrimination due to their disability. These were especially manifested during the COVID-19 pandemic (Sullivan et al. 2022). People with IDD experience multiple and compounding health inequities (Krahn et al. 2006). Healthcare practitioners in Canada are generally unfamiliar with people with IDD, who comprise about 1% of the population, and can fail to recognize and address the complex health and health-related needs of such patients (Sullivan et al. 2018). Some of these health needs can be different from those of their other patients (Grier et al. 2018; Tuffrey-Wijne et al. 2014). Adjustments to practice that are necessary to address these health and health-related needs are frequently not offered (Hollins and Tuffrey-Wijne 2013). As a result, many people with IDD have poor health and, as a group, have high rates of premature deaths (Hosking et al. 2016; Lunsky et al. 2022). People with IDD also experience high rates of trauma and are at higher risk for suicidal behaviours than those in the general population (Giannini et al. 2010).

At the end of life, people with IDD are often not offered palliative care and other supports. When adapted to the needs of people with IDD, however, such care and supports can be beneficial for them (Adam et al. 2020). Systemic barriers to palliative care for people with IDD have been documented and should be overcome (McNamara et al. 2020).

The health inequities experienced by people with IDD are part of the context of vulnerability that must be considered concerning why such a group might be at risk of inappropriately receiving MAID. This risk increases when there are no adequate alternatives to addressing health and health-related needs of people with IDD, and to increasing coping and resilience of such people in many communities. It is a grave injustice to people with IDD to recognize their equal 'right to liberty' to other Canadians to request MAID without first addressing their fundamental right to health, life, and development, and to equitable health care and related supports.

21.3 Studies Concerning Euthanasia or Assisted Suicide in the Netherlands Involving People with IDD or Similar Impairments of Cognitive or Social Functioning

There are no Canadian studies on how many people with IDD have requested and received MAID since the Canadian law on MAID came into effect. In the Netherlands, which has a similar legal framework for euthanasia and assisted suicide to Canada's in many respects, detailed case summaries describing these practices are made available to the public online (https://english.euthanasiecommissie.nl/judgments). This has generated research regarding people with IDD or similar impairments of cognitive or social functioning that can be examined for relevance to the Canadian context.

The *Termination of Life on Request and Assisted Suicide (Review Procedures) Act*, which legally recognizes and regulates these practices in the Netherlands, has been in effect since 2002. This Dutch law stipulates that all physicians who carry out requests by patients for euthanasia must ensure that 'due care' has been taken. Regional Euthanasia Review Committees (*Regionale Toetsingscommissie Euthanasie* or RTE) must be notified of each occurrence of euthanasia or assisted suicide. Studies show that not all such occurrences are reported, although the majority are (Onwuteaka-Philipsen et al. 2012; Rurup et al. 2008). The RTE reviews retrospectively whether due care has been taken according to the following six criteria (Regional Euthanasia Review Committees 2018):

That the reporting physician judged that:

- The patient's request had been voluntary and well-considered;
- The patient's suffering had been unbearable, with no prospect of improving;
- The patient's situation and prognosis had been disclosed to the patient;
- The patient and the physician together had reached the conclusion that there was no reasonable alternative in the patient's situation.

Also, the reporting physician must verify that:

- One other independent physician had been consulted (although the advice of this physician may be disregarded);
- Medication had been administered with due care and attention in terminating the patient's life or assisting in the patient's suicide.

As Miller and Kim (2017) noted, the Dutch legal framework for medical assistance in dying "places tremendous trust in its physicians" (p. 11). In their study of 32 of 33 cases between 2012–2016 in which RTEs concluded that due care criteria were not met, most infractions (69%) concerned the last two criteria (i.e., whether certain procedures were followed) rather than the physician's assessment of the patient's eligibility for medical assistance in dying. None of the findings of infraction resulted in serious legal penalties for the physicians involved.

Kim and colleagues (2016) analyzed 66 reported cases of euthanasia or assisted suicide involving people with psychiatric disorders between 2012–2016. In these cases, many physicians reported that they found assessing decision-making capacity and 'unbearable suffering' in patients with such disorders complex and challenging. Multiple assessments by different healthcare providers were often needed. Disagreements among these providers sometimes arose. Psychiatrists were not always sought or available for input into such assessments. Many of these instances of euthanasia or assisted suicide ended up being carried out by a physician associated with organizations advocating for these practices.

Recent studies of euthanasia and assisted suicide in the Netherlands involving people with dementia (Mangino et al. 2020) and people with personality disorders (Nicolini et al. 2020) reached similar conclusions regarding the considerable variability of assessments among physicians of decision-making capacity or 'unbearable suffering' in relation to the same patient. Mangino and colleagues found that, in practice, a lower threshold of cognitive function was often accepted by the Dutch RTEs for determining the capacity of people with dementia to make a decision regarding euthanasia than the 'well-considered' standard for people without dementia. That is, a physician's assessment that a patient with dementia was capable of making a decision regarding euthanasia was judged to meet standards of due care if it had been based on such indicators as the patient's persisting in making a request for euthanasia, and the physician's interpreting the patient's body language or wish for euthanasia as expressed in the patient's advance directive (regardless of the patient's current circumstances). Mangino et al. (2020) also found, however, that many such assessments of decision-making capacity, which require being familiar with a patient over time, were being made by physicians who were new to the patient.

Tuffrey-Wijne et al. (2018) examined case summaries published by Dutch RTEs for reported deaths by euthanasia between 2012–2016 of people identified as having or likely to have IDD. These same researchers subsequently published a study of two cases involving the euthanasia of people with IDD between 2017–2018 (Tuffrey-Wijne et al. 2019). These studies raise important and concerning questions regarding assessing 'unbearable suffering' that has 'no prospect of improving' in people with lifelong disabilities with impaired functioning in communication and socialization, such as those with IDD. None of the cases examined described how the physicians involved made such assessments, which can be complex and require adjustments in communication and engaging people close to the patient who understand and know the patient well. Many of the people with IDD who received euthanasia appear to have been assessed alone. Physicians also did not appear to probe or question whether a patient with IDD's persisting in requesting euthanasia or refusing therapies and other interventions as alternatives to euthanasia really was an expression of the patient's will, as they might have in the case of any other patient making such requests. In many of the examined cases, the suffering described by the assessing physician was *explicitly attributed to the patient's IDD*, such as difficulty accepting or coping with changes to life circumstances, dependency on caregivers, and inability to have meaningful social relationships. In all the examined cases, the reporting physician had concluded that the patient's mental suffering or challenges in coping or adapting 'had no cure' or was 'palliative' (despite the patient being otherwise healthy in some cases). These findings point to the problem of physicians assessing 'unbearable suffering' and that there is 'no prospect of improving' in people with IDD when they are unfamiliar with people with IDD and/or have unrecognized biases that result in negative attitudes towards people with cognitive impairments (Chicoine et al. 2022).

It is important also to highlight that people with IDD, especially those with mild cognitive impairments, can often not be identified as such or are misdiagnosed and receive inappropriate treatments. Schmahl and colleagues (2021) discussed a case of a 68-year-old woman with unrecognized IDD in the Netherlands who persistently requested euthanasia because of unbearable suffering due to 'extreme tiredness' and 'racing thoughts' that she was experiencing over many years. She had been diagnosed more than 20 years ago with a bipolar disorder and was prescribed psychotropic medications. Schmahl and colleagues were part of an outpatient geriatric psychiatry team that was consulted regarding the patient's eligibility for euthanasia. On psychiatric examination, this team found no signs in this patient of mood disturbances or psychotic features that would indicate the presence of a bipolar disorder in the woman. They discontinued her psychotropic medications and improved management of her diabetes and chronic obstructive pulmonary disease (COPD). The woman had no adverse physical or psychological side effects from discontinuing her medications, and her physical functioning improved significantly. Cognitive tests led to the woman being diagnosed with IDD and either concurrent ASD or attention deficit hyperactivity disorder (ADHD). The woman subsequently participated in a psycho-social intervention that resulted, after a year, in her developing an improved outlook on life. She no longer expressed a wish for death. Schmahl and colleagues concluded that the woman's IDD and concurrent ASD or ADHD did not, by themselves, entail that her suffering could not be ameliorated. Moreover, much of this suffering appeared to have been caused by the negative iatrogenic effects of her psychotropic medications that had been inappropriately prescribed because she was misdiagnosed with bipolar disorder.

In their case review, Tuffrey-Wijne and colleagues (2019) also discussed the ethical issue, raised in the previous section, that unequal opportunities for health care and social supports of people with IDD can play a role in influencing assessments by physicians that the suffering of people with IDD has no prospect of improving. For instance, these researchers discussed the case of a Dutch man in his 50s with IDD and ASD who received euthanasia when his struggles with mental health intensified after the death of one of his parents and led to emerging suicidal tendencies. The man lacked appropriate housing and bereavement supports adapted to his needs. The reporting physician had attributed the 'hopeless' situation of this patient to the *patient's* apparent inability to cope with various adversities because of IDD, and not to *unavailable supports* for people with IDD relative to others in the community. It is a social injustice that lives of people with IDD are ended because a community does not offer adequate and appropriate systems of supports that are adapted to the needs of people with IDD.

Objections might be raised that the Dutch studies reviewed above are descriptive case studies and are limited in the extent to which they can be generalized to show trends. The strength of case studies, however, is that they reveal the uncertainties and complexities of real-life clinical situations. They also show examples of a range of relevant factors and considerations (Morgan 2019). They can reveal helpful contextual information. Case studies can also provide evidence for further exploration regarding groups in healthcare and related systems that have distinctive needs or groups that are under-represented in mainstream research, such as people with IDD. When combined with robust and transparent analytical methods like the directed content analysis method used in many of the above studies, such case studies can generate reliable knowledge (Assarroudi et al. 2018).

21.4 People with IDD and Implementing MAID in Canada

It could also be objected that conclusions from the studies above cannot be applied straightforwardly to the Canadian context. There are, however, many similarities between the Dutch and Canadian legal frameworks on medical assistance in dying, and hence, many similar clinical and ethical issues are very likely to arise in implementing these laws in both countries. One important similarity is that, in both legal regimes, healthcare practitioners (in the Netherlands, physicians, but in Canada, also nurse practitioners) have a central role in assessing eligibility for medical assistance in dying. This entails that euthanasia and assisted suicide have become regarded as medical interventions that will continue to be normalized in these healthcare systems over time. The central role of healthcare practitioners also entails that, in practice, they are the ones who must deal with the complexities involved in implementing MAID. Healthcare practitioners are entrusted with considerable discretion in assessing the eligibility of persons for medical assistance in dying and carrying it out, and as in the Netherlands, there is little scope within the legal regime for MAID in Canada for the judgments and conduct of practitioners to be challenged. Rather, this lack of oversight is likely to be a greater ethical problem in Canada.

In the Netherlands, there is at least a system of oversight in the role of RTEs in reviewing the decisions and conduct of physicians involved in euthanasia of patients.

There is detailed and accessible reporting of these findings, on the level of individual cases, to the public. As noted above, however, the Dutch review process is retrospective—after euthanasia or assisted suicide has been carried out. Findings that due care criteria had not been met by physicians in this context are generally on the grounds of failing to follow certain procedures rather than of poor judgment in assessing eligibility. Infractions rarely result in negative professional or legal consequences for physicians. In Canada, the legal framework for MAID gives physicians and nurse practitioners even greater leeway. Practitioners are protected from legal prosecution if they carry out MAID 'with reasonable knowledge, care and skill' according to the rules laid out by the law on MAID. But there is no Canadian system for reviewing that this has been the case. The federal government's reports regarding MAID, from data collected from practitioners, are quantitative and aggregate. There is no descriptive case-level information, as in the Dutch reports.

One tool that is consulted by Canadian healthcare practitioners is the *Medical Assistance in Dying Resource* (Centre for Effective Practice 2021)—hereafter referred to as the *Resource*. The *Resource* (one of several such tools adopting the same approach) maps decision-making for healthcare practitioners regarding MAID in the format of an 'algorithm' that lays out various decision points and alternative action trajectories or pathways. The emphasis of the *Resource* is on translating Canada's law on MAID into easy-to-follow steps for physicians and nurse practitioners. Its goal is to clarify legally required procedures and not to offer guidance on the clinical and ethical *grounds* of decisions regarding a given request for MAID. More importantly, for the purpose of this chapter, there is no scope in this generic tool to alert physician and nurse assessors, at various decision points, to considerations that might pertain to certain patient groups, such as people with IDD, whose circumstances and needs are complex and might differ from those of other patient groups. There are also no directions to screen for risk factors that might make such persons vulnerable in their decision making regarding MAID.

Consider, therefore, how using the generic, algorithmic approach of the *Resource* and other such tools might have clinically and ethically problematic implications for a person with IDD requesting MAID. When any patient requests MAID, whether or not that patient's death is reasonably foreseeable, the *Resource* directs healthcare practitioners to 'discuss' and 'explore' the motivation and the suffering behind the request for MAID and to 'inform' the patient of all available treatment and care options including MAID. Discussing and exploring the motivation and the suffering behind a person's request for MAID is important because, as Branigan (2015)'s review reveals, there can be multiple reasons behind a patient's communicating a desire for hastened death, the vast majority of which can be addressed by means acceptable to the patient.

The Dutch studies discussed above illustrate that there are risks that opportunities to address suffering and improve the lives of people with IDD are missed when physicians are unfamiliar with IDD or with the person with IDD. It is concerning that, in many of the Dutch cases, the nature and intractability of a person with IDD's suffering are assumed to be fixed and attributed to the person's disability. It is also troubling that, in many cases, the physician discussing and exploring the suffering of a person with IDD who requested euthanasia had no prior relationship with the person with IDD, and that no close family member or other caregiver or friend of this person was contacted or asked for input. Under Canadian law, such risks are magnified for two reasons. First, in the Netherlands, the patient requesting euthanasia or assisted suicide and the physician must *jointly* conclude that the patient's suffering is unlikely to improve. In Canada, patients requesting MAID are *the sole judge* that their suffering is intolerable to them and that this suffering cannot be relieved under conditions that the patient considers acceptable. Secondly, in Canada, if neither of the two healthcare practitioners who must concur on a patient's eligibility for MAID have expertise in the medical condition causing the patient's suffering, they are required by law to consult a clinician with such expertise. But, as we have seen in the Dutch studies, it is conceivable that healthcare practitioners will not recognize the presence of IDD in certain patients or they might not appreciate the need to seek such consultations if they hold negative or unfounded views regarding people with IDD. Moreover, unlike in the Netherlands where certain primary care physicians and nurses receive specialist training and have practices focused on the health care of people with IDD, there is no such area of expertise in Canada.

There are no guidelines in the Canadian law on MAID for assessing a patient's capacity for making a decision regarding MAID. The *Resource* interprets the law as follows: 'The requirements to undergo MAID *are the same as* [emphasis added] the requirements for any healthcare treatment. The use of existing procedures for capacity assessments are encouraged'. Little is known, however, regarding how people with IDD perceive and understand death or a procedure such as MAID which results in death. On this question, we cannot straightforwardly and uncritically extrapolate from research on promoting greater participation of people with IDD in decision making. Those studies concern options for enhancing *living*. The increasing entanglement and ambiguity in the Canadian healthcare system between treatments that promote and restore health, prevent illness, and palliate pain and distress, on the one hand, and MAID, on the other, can be confusing. There is a need for further investigations to determine how this entanglement and ambiguity affects people with IDD.

The typical clinical and legal approaches to assessing a patient's capacity to make specific healthcare decisions in Canada requires demonstrating that the patient is capable *independently* of understanding and appreciating information communicated by the healthcare practitioner. All of us benefit when healthcare practitioners communicate effectively. Additionally, we might depend to varying degrees on advice and support from close and trusted people in our lives (e.g., family members, other caregivers, and friends) to help us reach significant healthcare decisions. It should not be assumed, however, that most healthcare practitioners know what the communication needs of people with IDD are or how to adjust their ways of communicating with people with IDD as needed. Having an ongoing professional relationship with the person with IDD and being familiar with this person's communication capabilities and needs are crucial. The input and assistance that caregivers and others close to the person with IDD can give are also important to enhance communication, as well as to elicit and clarify the person with IDD's goals and to decide on healthcare options that best align with this patient's goals (Bekkema et al. 2014; Sullivan et al. 2019).

For healthcare practitioners, trying to interpret a person with IDD's request for MAID without being familiar with that person or having input from those who are familiar with the person is challenging. Misinterpreting what a person with IDD truly wants can also put the person with IDD at risk of unintended harm. It is significant that the *Resource* highlights the lack of a formal process under the Canadian law on MAID to inform close family members or other caregivers that their loved one has requested MAID. Rather, this is left to the discretion of the healthcare practitioner discussing and assessing eligibility for MAID to bring up with the person with IDD. However, it should be standard practice for healthcare practitioners, when a request for MAID has been made by a person with IDD, to ask whether that person would like a close family member, other caregiver, friend or regular family physician to be informed and involved in discussions.

Healthcare practitioners should also be aware of and screen for risks that compromise the voluntariness of decisions when a person with IDD requests MAID. These could be different or more prevalent among people with IDD than people without IDD. For instance, it is important to screen for factors such as suggestibility, learned compliance to authorities in the person's life (e.g., healthcare providers or caregivers), low self-esteem, a sense of being overly burdensome to caregivers, impulsive behaviours, compulsive fixations, heightened withdrawal or flight responses to trauma, and limited ability to envision alternatives (Heng and Sullivan 2018). These vulnerabilities often can be addressed through various therapies that can diminish suicidal ideation or self-harming behaviours and inducements to pursue MAID.

There can also be systemic factors that reduce the voluntariness of a person with IDD's decision to pursue MAID. The *Resource* directs the patient's healthcare practitioner to 'discuss available treatments and care including MAID'. But in Canada, as in many places around the world, there are significant gaps in services and appropriate supports for people with IDD. Hence, basic counselling adapted to the needs of the person with IDD who is requesting MAID might be 'unavailable'. This would render the 'choice' of MAID non-voluntary.

21.5 Conclusion

The Canadian law on MAID is an attempt to balance various constitutional rights of individuals with society's interest in protecting vulnerable groups from harm. Whether the law has achieved this judicious balance is a topic that goes beyond the scope of this chapter. On examining actual practices in implementing MAID, however, we conclude that the vulnerability of people with IDD cannot simply be protected by regulatory safeguards without addressing deeper systemic issues in Canadian health care that relate to the care of people with IDD.

In our view, despite the promise that procedural safeguards at the level of national legislation can robustly and effectively protect vulnerable groups, such as people with IDD or those with similar cognitive and/or social impairments, these regulatory safeguards often fail in practice. This will persist as long as (1) healthcare practitioners who assess and carry out MAID are not aware of and attentive to the complexities involved in assessing and addressing decision-making vulnerabilities and suffering in people with IDD, and (2) the current healthcare system is not integrated with other systems to meet the health and related needs of people with IDD adequately and appropriately. The suffering of people with IDD is most often due to their experiences of being devalued and lacking meaningful participation in society, and poor health and trauma caused by inadequate or inappropriate health care and related supports. It is also due to unequal access to such services as comprehensive palliative and end-of-life care. We question whether implementing MAID in such a context does offer people with IDD real choices regarding alternatives to MAID.

The considerable and central role that physicians and nurse practitioners have, under Canadian law, to assess patients' eligibility for MAID and to carry out MAID introduces ambiguity to their primary commitment to promote the health and well-being of people with IDD. The vulnerability experienced by people with IDD, as a result, affects not only exceptional cases but all people with IDD. Such a law does not promote their equal standing in Canadian society but instead violates their basic rights, under international law, to life, health, good care, and development (United Nations 2010, art. 10, 25).

We conclude that people with IDD will be at even greater risk of vulnerability to premature death through MAID in a healthcare system in which MAID is increasingly accepted as a *medical* intervention. This vulnerability is compounded in a society, and a medical culture, that too readily regards people with disabilities, such as those living with IDD, as having potentially 'grievous and irremediable' *medical* conditions.

References

Adam, E., K.E. Sleeman, S. Brearley, K. Hunt, and I. Tuffrey-Wijne. 2020. The Palliative Care Needs of Adults With Intellectual Disabilities and Their Access to Palliative Care Services: A Systematic Review. *Palliative Medicine.* 34 (8): 1006–1018. https://doi.org/10.1177/0269216320932774.

Assarroudi, A., F. Heshmati Nabavi, M.R. Armat, A. Ebadi, and M. Vaismoradi, M. 2018. Directed Qualitative Content Analysis: The Description and Elaboration of Its Underpinning Methods and Data Analysis Process. *Journal of Research in Nursing (JRN)* 23(1): 42–55. https://doi.org/10.1177/1744987117741667

Bekkema, N., A.J.E. de Veer, C.M.P.M. Hertogh, and A.L. Francke. 2014. Respecting Autonomy in the End-of-life Care of People With Intellectual Disabilities: A Qualitative Multiple-case Study. *Journal of Intellectual Disability Research (JIDR)* 58 (4): 368–380. https://doi.org/10.1111/jir.12023.

Branigan, M. 2015. Desire for Hastened Death: Exploring the Emotions and the Ethics. *Current Opinions in Supportive and Palliative Care* 9 (1): 64–71. https://doi.org/10.1097/SPC.0000000000000109.

Brown, I. 2020. Life Planning For People with Neurodevelopmental and Intellectual Disability: Effective Support, Quality of Life, and Community Engagement. Preface. *Annali Dell'istituto Superiore Di Sanita* 56 (2): 168–170. https://doi.org/10.4415/ANN_20_02_06.

Centre for Effective Practice. 2021. Medical Assistance in Dying Resource. Toronto: Centre for Effective Practice. http://cep.health/MAID/

Chicoine, C., E.E. Hickey, K.L. Kirschner, and B.A. Chicoine. 2022. Ableism at the Bedside: People with Intellectual Disabilities and COVID-19. *Journal of the American Board of Family Medicine (JABFM)* 35 (2): 390–393. https://doi.org/10.3122/jabfm.2022.02.210371.

Cook, S., and Rachelle Hole. 2021. Trauma, Intellectual and/or Developmental Disability, and Multiple, Complex Needs: A Scoping Review of the Literature. Research in Developmental Disabilities 115: 103939. https://doi.org/10.1016/j.ridd.2021.103939.

Einfeld, S.L., L.A. Ellis, and E. Emerson. 2011. Comorbidity of Intellectual Disability and Mental Disorder in Children and Adolescents: A Systematic Review. *Journal of Intellectual and Developmental Disability* 36 (2): 137–143. https://doi.org/10.1080/13668250.2011.572548.

Giannini, M.J., B. Bergmark, S. Kreshover, E. Elias, C. Plummer, and E. O'Keefe. 2010. Understanding Suicide and Disability Through Three Major Disabling Conditions: Intellectual Disability, Spinal Cord Injury, and Multiple Sclerosis. *Disability and Health Journal* 3 (2): 74–78. https://doi.org/10.1016/j.dhjo.2009.09.001.

Greenspan, S., G. Loughlin, and R.S. Black. 2001. Credulity and Gullibility in People with Developmental Disorders: A Framework for Future Research. *International Review of Research in Mental Retardation.* 24: 101–135. https://doi.org/10.1016/S0074-7750(01)80007-0.

Grier, E., D. Abells, I. Casson, M. Gemmill, J. Ladouceur, A. Lepp, U. Niel, S. Sacks, and K. Sue. 2018. Managing Complexity in Care of Patients with Intellectual and Developmental Disabilities: Natural Fit for the Family Physician as an Expert Generalist. Canadian Family Physician. 64(Supplement 2): S15–S22. https://www.cfp.ca/content/64/Suppl_2/S15.long

Health Canada. 2021. Second Annual Report on Medical Assistance in Dying in Canada 2020.Ottawa: Government of Canada. https://www.canada.ca/en/health-canada/services/medical-assistance-dying/annual-report-2020.html

Heng, J., and W.F. Sullivan. 2017. Ethics of Decision Making and Consent in People with Intellectual and Developmental Disabilities. In A Comprehensive Guide To Intellectual And Developmental Disabilities, 2nd ed, ed. Wehmeyer, M.L., I. Brown, M. Percy, W.L.A. Fung, and K.A. Shogren, 655–664. Baltimore, MD: Paul H. Brookes.

Ho, A., J.S. Norman, S. Joolaee, K. Serota, L. Twells, and L. William. 2021. How Does Medical Assistance in Dying Affect End-of-life Care Planning Discussions? Experiences of Canadian Multidisciplinary Palliative Care Providers. *Palliative Care and Social Practice* 15: e26323524211045996. https://doi.org/10.1177/26323524211045996.

Hollins, S. and I. Tuffrey-Wijne. 2013. Meeting the Needs of Patients with Learning Disabilities. *BMJ*, no. 346: f3421. https://doi.org/10.1136/bmj.f3421.

Hosking, F.J., I.M. Carey, S.M. Shah, T. Harris, S. DeWilde, C. Beighton, and D.G. Cook. 2016. Mortality among Adults with Intellectual Disability in England: Comparisons with the General Population. *American Journal of Public Health* 106 (8): 1483–1490. https://doi.org/10.2105/AJPH.2016.303240.

House of Commons Canada. 2020. Bill C-7 [now, Legislation on MAID]. Ottawa: Government of Canada. https://www.parl.ca/Content/Bills/431/Government/C-7/C-7_1/C-7_1.PDF

Kim, S.Y.H., R.G. De Vries, and J.R. Peteet. 2016. Euthanasia and Assisted Suicide of Patients With Psychiatric Disorders in the Netherlands 2011 To 2014. *JAMA Psychiatry* 73 (4): 362–368. https://doi.org/10.1001/jamapsychiatry.2015.2887.

Krahn, G.L., L. Hammond, and A. Turner. 2006. A Cascade of Disparities: Health and Health Care Access for People With Intellectual Disabilities. *Mental Retardation and Developmental Disabilities Research Reviews* 12 (1): 70–82. https://doi.org/10.1002/mrdd.20098.

Lunsky, Y., J.E. Klein-Geltink, E.A. Yates, eds. 2013. Atlas on the Primary Care of Adults with Developmental Disabilities in Ontario. Toronto, ON: Institute for Clinical Evaluative Sciences and Centre for Addiction and Mental Health. https://www.camh.ca/-/media/files/pdfs---reports-and-books---research/hcardd_atlas---rev_web_2014-pdf.pdf

Lunsky, Y., M.-C. Lai, R. Balogh, H. Chung, A. Durbin, P. Jachyra, A. Tint, J. Weiss, and E. Lin. 2022. Premature Mortality in a Population-based Cohort of Autistic Adults in Canada. Autism Research 15(8): 1550–1559. https://doi.org/10.1002/aur.2741.

Mangino, D.R., M.E. Nicolini, R.G. De Vries, and S.Y.H. Kim. 2020. Euthanasia and Assisted Suicide of Persons with Dementia in the Netherlands. *The American Journal of Geriatric Psychiatry* 28 (4): 466–477. https://doi.org/10.1016/j.jagp.2019.08.015.

McNamara, B., A. Same, and L. Rosenwax. 2020. Creating Person-Centred Support for People with Intellectual Disabilities at the End of Life: An Australian Qualitative Study of Unmet Needs and Strategies. *Journal of Intellectual Disabilities.* 24 (4): 543–558. https://doi.org/10.1177/1744629518823887.

Miller, D.G., and S.Y.H. Kim. 2017. Euthanasia and Physician-assisted Suicide Not Meeting Due Care Criteria in the Netherlands: A Qualitative Review of Review Committee Judgements. *British Medical Journal Open* 7 (10): e017628. https://doi.org/10.1136/bmjopen-2017-017628.

Morgan, M.S. 2019. Exemplification and the Use-values of Cases and Case Studies. *Studies in the History and Philosophy of Science.* 78: 5–13. https://doi.org/10.1016/j.shpsa.2018.12.008.

Nicolini, M.E., J.R. Peteet, G.K. Donovan, and S.Y.H. Kim. 2020. Euthanasia and Assisted Suicide of Persons with Psychiatric Disorders: The Challenge of Personality Disorders. *Psychological Medicine* 50 (4): 575–582. https://doi.org/10.1017/S0033291719000333.

Onwuteaka-Philipsen, B.D., A. Brinkman-Stoppelenburg, C. Penning, et al. 2012. Trends in End-of-life Practices Before and After the Enactment of the Euthanasia Law in the Netherlands From 1990 to 2010: A Repeated Cross-sectional Survey. *Lancet* 380 (9845): 908–915.

Regional Euthanasia Review Committees [The Netherlands]. 2018. Euthanasia code. https://www.euthanasiecommissie.nl/

Rurup, M.L., H.M. Buiting, H.R.W. Pasman, P.J. van der Maas, Agnes van der Heide, and B.D. Onwuteaka-Philipsen. 2008. The Reporting Rate of Euthanasia and Physician-assisted Suicide: A Study of the Trends. *Medical Care* 46(12): 1198–1202. https://doi.org/10.1097/MLR.0b013e31817d69e8

Schalock, R.L., R. Luckasson, and M.J. Tassé. 2021. *Intellectual Disability: Definition, Classification, and Systems of Supports*, 12th ed. Washington, DC: American Association on Intellectual and Developmental Disabilities.

Schmahl, O., R.O. Voshaar, A. van de Poel-Mustafayeva, and R. Marijnissen. 2021. Request for Euthanasia by a Psychiatric Patient With Undetected Intellectual Disability. *British Medical Journal Case Reports* 14 (8): e239862. https://doi.org/10.1136/bcr-2020-239862.

Sullivan, W.F., and J. Heng. 2018. Supporting Adults with Intellectual and Developmental Disabilities to Participate In Health Care Decision Making. *Canadian Family Physician* 64 (Supplement 2): S32–S36.

Sullivan, W.F., H. Diepstra, J. Heng, S. Ally, E. Bradley, I. Casson, B. Hennen, M. Kelly, M. Korossy, K. McNeil, D. Abells, K. Amaria, K. Boyd, M. Gemmill, E. Grier, N. Kennie-Kaulbach, M. Ketchell, J. Ladouceur, A. Lepp, Y. Lunsky, S. McMillan, U. Niel, S. Sacks, S. Shea, K. Stringer, K. Sue, S. Witherbee. 2018. Primary Care of Adults with Intellectual and Developmental Disabilities: 2018 Canadian Consensus Guidelines. Canadian Family Physician 64(4):254–279. PMID: 29650602; PMCID: PMC5897068.

Sullivan, W.F., J. Heng, K. McNeil, M. Bach, M. Henze, A. Perry, and J. Vogt. 2019. Promoting Health Care Decision-making Capabilities of Adults With Intellectual and Developmental Disabilities. *Canadian Family Physician* 65 (Suppl): S27–S29.

Sullivan, W.F., P. Björne, J. Heng, and R. Northway. 2022. Ethics Framework and Recommendations to Support Capabilities of People With Intellectual and Developmental Disabilities (IDD) During Pandemics. *Journal of Policy and Practice in Intellectual Disabilities* 19 (1): 116–124. https://doi.org/10.1111/jppi.12413.

Tuffrey-Wijne, I., L. Goulding, N. Giatras, E. Abraham, S. Gillard, S. White, C. Edwards, and S. Hollins. 2014. The Barriers to and Enablers of Providing Reasonably Adjusted Health Services to People with Intellectual Disabilities in Acute Hospitals: Evidence from a Mixed-methods Study. BMJ Open 4(4): e004606. https://doi.org/10.1136/bmjopen-2013-004606.

Tuffrey-Wijne, I., L. Curfs, I. Finlay, and S. Hollins. 2018. Euthanasia and Assisted Suicide for People With an Intellectual Disability and/or Autism Spectrum Disorder: An Examination of

Nine Relevant Euthanasia Cases in the Netherlands (2012–2016). *BMC Medical Ethics*. 19 (1): 17. https://doi.org/10.1186/s12910-018-0257-6.
Tuffrey-Wijne, I., L. Curfs, I. Finlay, and S. Hollins. 2019. "Because of His Intellectual Disability, He Couldn't Cope". Is Euthanasia the Answer? Journal of Policy and Practice in Intellectual Disabilities 16(2): 113–116.
United Nations. 2010. Convention on the Rights of Persons with Disabilities. Geneva: United Nations. https://www.un.org/development/desa/disabilities/convention-on-the-rights-of-persons-with-disabilities/convention-on-the-rights-of-persons-with-disabilities-2.html
Weinstein, J.N., et al. 2017. Communities in Action: Pathways to Health Equity. https://www.ncbi.nlm.nih.gov/books/NBK425845/

William F. Sullivan, MD, CFPC (F), PhD William F. Sullivan is a physician with a Ph.D. in Philosophy (Bioethics). He was a member of the St. Michael's Hospital Academic Family Health Team, Unity Health, Toronto, and is an Adjunct Professor of Family and Community Medicine at the University of Toronto. His clinical and academic focus is on integrating ethics and primary care of people who are vulnerable in healthcare systems, especially adults with intellectual and developmental disabilities (IDD). Dr. Sullivan's collaborative research develops guidelines on complex ethical and clinical questions, based not only on integrating expertise from different relevant fields, but also engaging with researchers with distinct skills that are needed for comprehensive investigations. He has been appointed to the Joseph P. Kennedy Chair at the Kennedy Institute of Ethics, Georgetown University, Washington, D.C., commencing in 2023.

John Heng John Heng is a tenured professor at King's University College of Western University, London, Ontario, Canada. He teaches in the Departments of Philosophy, Thanatology, and Disability Studies, and serves as Chair of the Department of Philosophy.

Chapter 22
The Psychosocial Aspects of Adapting to Traumatic Non-life-threatening Disability

Irmo Marini and Laura Villarreal

Abstract Historical attitudes and treatment of persons with disabilities has not been kind. This population has endured institutionalization, torture, sterilization, and forcible euthanasia over the centuries. Presently, persons with disabilities are thought to be helpless, incapable, pitied, and invisible in society, treated as second-class citizens. Statistically, persons with disabilities are highly likely to be unemployed and on government assistance, less likely to be socially engaged in the community, and still deal with physical barriers despite legislation encouraging equal access. Despite second-class citizenry, the majority of people with disabilities generally report having a good quality of life and life satisfaction. Those who sustain a traumatic injury typically grieve the loss of function and lifestyle, but generally transition to a state of acceptance and adaptation within a year. The Criminal Code of Canada was recently amended to permit a family physician the power to grant persons with non-life-threatening disabilities to be euthanized if they so desire without such safeguards as a mandated psychological evaluation. A minority of persons with disabilities living under dismal circumstances in poverty due to lack of government resources may indeed opt for Medical Assistance in Dying (MAID) not due to their disability, but rather lack of governmental or related support to ensure a better quality of life. These issues are explored in this chapter.

Keywords Quality of life · Stages of adjustment · Adaptation to loss · Integrated adaptation model · Social injustice · Lives of despair · Deaths of despair

---

I. Marini (✉) · L. Villarreal
The University of Texas Rio Grande Valley, Edinburg, TX, USA
e-mail: irmo.marini@utrgv.edu

L. Villarreal
e-mail: laura.villarreal02@utrgv.edu

J. Kotalik and D. W. Shannon (eds.), *Medical Assistance in Dying (MAID) in Canada*, The International Library of Bioethics 104,
https://doi.org/10.1007/978-3-031-30002-8_22

22.1 The Psychosocial Aspects of Adapting to Traumatic Non-life-threatening Disability

The decision to end one's life for what is perceived as unbearable and unending suffering is not only a complex and multifaceted decision, but in the end, an irreversible one once the perceived sufferer has been pronounced dead. The right to die has been debated endlessly for decades among physicians, bioethicists, mental health professionals, activists on both sides, and ultimately politicians and the courts. The decision to take one's life approved with medical assistance is opined upon by numerous authorities in this book, within each of their areas of expertise. The focus of this chapter is to explore the psychosocial aspects of adapting to a traumatic but non-life-threatening disability, by examining all facets of why a person with a disability would otherwise want to end their life examining the topic from a holistic perspective. As such, we briefly explore the historical treatment of persons with disabilities and societal attitudes toward disability, how such attitudes have impacted the psychosocial adjustment for this population, adjusting or adapting to disability, the multifaceted reasons as to why a person with a disability would want to end his or her life, how the medical community perceives disability, and loopholes in MAID in its current legislative form.

22.2 Historical Treatment of Disability and Its Impact on Those with Disabilities

The psychological and sociological impact of oppression, social injustice, and dehumanizing of minorities in every country and culture has in many countries been brought to light in recent decades such as the atrocities against Indigenous peoples in Canada and African Americans in the United States. However, the plight of persons with disabilities has received much less attention across the globe, and in many instances, they continue to be perceived and treated as second-class citizens in overt and covert ways (Marini 2018c). Without understanding how persons with disabilities have been treated and perceived by society in the past, we are left with presuming why some with disabilities may want to end their life based on preconceived notions that their life is not worth living. Although for a small minority of those with disabilities this may be true, for the vast majority, empirical research suggests most with disabilities would disagree with this assumption (Baynton 2001; Diener and Lucas, 1999; Marini 2018a; Seligman 2002).

22.2.1 A Brief History of Disability Treatment

Dating back to Plato and Aristotle who sanctioned infanticide for children born with a disability in fourth century B.C., the treatment of those with disabilities has included everything from torture, possessed by demons and exorcised, burned at the stake, incarceration, mocked, and objects of pity (Coco 2010; Marini 2018b; Smart; 2009). Centuries later, the cruelty towards those with disabilities became less common, but they were still often shut-ins, placed in asylums, and otherwise kept out ashamed to be out in public. The psychosocial impact these negative societal attitudes had on persons with disabilities were those of shame, embarrassment, loneliness, social isolation, low self-esteem, and unworthy of love (DiTomasso and Spinner 1997; Gallagher 1995; Li and Moore 1998).

In the latter half of the nineteenth century led by Charles Darwin, Herbert Spencer, and other scholars, the science of eugenics, natural selection, and survival of the fittest became of great interest and debate among scholars, physicians, politicians, and the like. The debate revolved around the irrational fear and presumption that if persons who had some physical and/or mental defect were allowed to procreate with able-bodied partners or among themselves, it would ultimately dilute and weaken the human species to a point of extinction (Carlson 2009; Gallagher 1995; Largent 2002). This eventually led in the early twentieth century to the US beginning a forced sterilization program of those with various disabilities and perceived degenerates. Other US laws forbade persons considered to be developmentally disabled, have epilepsy, or some other mental disability from getting married (Gallagher 1995; Marini 2018b). Some states even passed what came to be known as the "ugly act" the late nineteenth century where persons with visible disabilities were forcibly removed from public site off the streets. In Chicago, some 30,000 individuals with disabilities and the poor were confined to the Cook County poorhouse which was known to have deplorable living conditions (Coco 2010).

The eugenics movement culminated prior to the second world war in Germany, when Hitler ignited similar forced sterilization of German citizens with disabilities, ultimately leading to his secret order to deliver what was termed *final medical assistance* to an estimated 300,000 citizens with disabilities (Bayton 2001; Gallagher 1995) to whom they referred to as *useless eaters.* The program secretly ran between 1939—1941 after outrage from German citizens who realized Hitler and the Nazis were exterminating their own people (Gallagher 1995).

Proponents of Social Darwinism and eugenics had come full circle beginning with sterilization, laws forbidding marriage, and to the eventual extermination of those deemed hereditarily unfit. The central questions arising from this historical perspective is with regards to whether some individuals with disabilities perceive their life as not worth living due to ongoing discrimination, oppression, poverty, and negative attitudes imposed by society (Marini 2018c). At present, attitudes towards persons with disabilities have improved in the past century, but are still largely perceived as helpless, incapable, asexual, pitied and admired, ultimately continuing to be subjected to inequitable treatment in employment, housing, and community participation (Coco 2010; Graf et al. 2009; Olkin 1999; Smart 2009).

22.3 The Psychosocial Adaptation of Persons with Disabilities

There are numerous theories regarding how individuals with traumatic disabilities respond or adapt to their condition (Bishop 2005; Kendall and Buys 1998; Livneh and Antonak 1997; Marini 2018a; Olkin 1999; Vash and Crewe 2004). Livneh (1991) found there were over 40 models of adaptation to disability, most originating from the stage theory of grief after a loss by Elizabeth Kubler Ross (1969), who observed hundreds of terminally ill patients noting that the majority typically transitioned through five stages of adjustment and eventually emotionally and intellectually come to terms with their imminent death over time. The other major school of thought originated from Lewin's (1935) theory of personality development based on the inter-dynamics of our individual personality traits (e.g., intellect, locus of control, humor, coping skills, temperament) interacting with the outside world and how we perceive others view and treat us. Lewin's theory has been adapted to include organic factors such as disability (Dembo et al. 1956; Trieschmann 1988). As such, behavior is a function of our personality traits, organic factors, and environmental influences.

The stage theory of grieving the loss of a loved one is typically described as having five linear stages that most individuals transition through in a linear fashion. Kubler-Ross' (1969) original five stages for loss of a loved one or our own imminent death included the following:

Shock and Denial—confused, avoidant, fear and denying that the loved one is deceased.

Anger—irritation, shame, and anger at the person or situation for having died.

Bargaining—bargaining with God or a higher power for some type of payment to bring the loved one back.

Depression—the realization that the loved one is gone and not coming back.

Acceptance—the individual emotionally and intellectually stops looking back yearning for the loved one and begins to embrace the future of life without their loved one.

More recent studies involving the grieving process, define with the concept of complex grief (Boelen 2016). It was included in the Diagnostic and Statistical Manual-5 (DSM-5) (American Psychiatric Association [APA] 2013) as Persistent Complex Bereavement Disorder. It is described differently than acute grief that is endorsed after one or more years after the loss when the bereaved continues to experience disbelief about the irreversibility of the loss, feels as if the death occurred yesterday, and the bereaved continues to yearn with all the intense feelings, thoughts, and emotions experienced with acute grief. This definition indicates that although a grieving loss is experienced differently and for varying lengths of time by most individuals, the DSM-5 acknowledges approximately one-year as a natural grieving length of time.

Ott et al. (2007) and Stroebe et al. (2007) cite the finding that those who experience ongoing yearning, unacceptance of loved ones lost, and disbelief over six months following the death, begin to engage in maladaptive bereavement behaviors. These may include not taking care of oneself, poor quality of life, higher suicide rates and self-neglect medical impairments. For persons with disabilities, their grief is initially from the loss of physical and/or mental functions, way of life, and overall change of lifestyle status.

The stages of adjustment to a traumatic disability proposed by Livneh (1991) however, noted a change in the five stages of adjustment. He hypothesized that some individuals do not always transition to final acceptance, and can sometimes get stuck, regress, or remain in a state of complicated grief. Livneh's five stages of adaptation included:

Initial Impact—includes two sub-stages, shock, and anxiety. It is observed as including fear, confusion, poor concentration, anxiety, crying, and panic.

Defense Mobilization—this stage also includes two sub-stages, bargaining and denial. Bargaining with God or a higher power for complete recovery from traumatic injuries as well as denial of the severity and permanency of the loss of bodily functions from the trauma.

Initial Realization—this stage has two-substages, mourning/depression and internalized anger. The injured individual begins to realize that their disability is permanent and will either grieve and/or become clinically depressed over this realization. Not all individuals experience a clinical depression, but all will grieve the loss of function. During this stage, individuals will also internalize their anger and blame themselves, ask "why me" and contemplate suicide.

Retaliation—this stage is observed as the patient's anger is turned outward to loved ones, medical staff, and God for not being cured. He or she becomes frustrated, irritable, and angry, and this is taken out on others around him or her.

Reintegration—stages comparable to Kubler Ross's stage of acceptance. It is marked by acknowledging the permanency of the situation and looking forward to the future instead of dwelling on the past. The individual emotionally and intellectually can come to terms with his or her disability self and contemplates his or her future living with a disability.

There is also a need to distinguish between grief and clinical depression. An individual can grieve their loss of function, but it is distinctly different than also being diagnosed with clinical depression (Prigerson and Maciejewski 2006) as these authors found that over half of the reported symptoms were contributed to grief but not depression. Although most individuals will experience some form of grief, they do not all become clinically depressed. Several researchers have found that symptoms of depression (e.g., withdrawal, crying, poor appetite, sleep disturbance, memory problems and loss of interest) will overlap with grief initially. Kendler et al. (2003) found that at one-year post-loss, only 16% showed signs of clinical depression. More recently, Mash et al. (2013) reported that 16% of participants who had lost a loved one between 1 to 2 years, met the criteria for maladaptive grief, and 34% met the criteria for moderate to severe depression. Overall factors in these studies considered time since loss, type of loss, resilience, and coping strategies. Such statistics become important when considering MAID criteria, in that individuals' who are experiencing maladaptive grief or meets criteria for clinical depression, arguably both should receive psychotherapy and psychopharmacology to see if they respond well and can then still decide whether they want to assist in sending their life. This cannot be determined by their family practitioner.

Finally, regarding stage theory, there is general agreement among these researchers empirically and theoretically that suggests most individuals will grieve the loss or the loss of function for a varying length of time depending on the factors discussed (e.g., coping strategies, personality characteristics, importance of functions lost such as vision). This appears to be in the range of 6 to 12 months of displaying the symptoms noted. After this period, grief and/or depression become maladaptive and persistent requiring psychotherapy and/or prescription medication. Some individuals will seek out or be referred for treatment, others will deny or ignore it and potentially become worse. This knowledge becomes important when we discuss the Medical Assistance in Dying (MAID) Canadian Supreme Court decision later.

The second school of thought regarding adjustment or adaptation to a traumatic disability, is propounded by those theorists who postulate that human response to disability is more complex. They emphasize that successful adjustment involves not only one's personality traits, but also the type of disability and functions lost (mobility, sensory, cognitive, or emotional), as well as the interaction that occurs between the individual in his or her environment (community and family support, societal attitudes, cultural influences, physical access barriers, income status, available resources, etc.) (Kendall and Buys 1998; Trieschmann 1988; Vash and Crewe 2004). Adaptation does not simply occur in a vacuum of one's unique traits and coping skills, but also encompasses how he or she perceives support or lack of it from his or her immediate environment. Overall, the greater one perceives he or she is alone in navigating the world with little or no support, the greater the likelihood he or she may not want to continue living.

A key aspect to one's adjustment within this model is how the outside world views and treats persons with disabilities. Although societal attitudes toward disability have markedly improved from centuries ago, many persons with disabilities still experience negative attitudes, discrimination, oppression, and social injustice (DiTomasso and Spinner 1997; Graf et al. 2009; Li and Moore 1998; Pinon et al. 2021).

Li and Moore (1998) surveyed 1,266 persons with disabilities regarding their experiences in their communities, noting that aside from family and friend emotional support, they perceived that negative societal attitude blocked their ability to fully adjust. Similarly, DiTomasso and Spinner (1997); Hopps et al. (2001) found that their participants with disabilities who experienced negative attitudes of others, endorsed greater feelings of loneliness, anxiety, social isolation, and poor social skill development. Graf et al. (2009) found that 63 individuals who wrote 100 words or less about their lived experience with disability shared common themes of daily anger and frustration due to physical access barriers and negative attitudes.

More recently, Kellett et al. (2021) assessed 1,566 individuals with various physical, cognitive and emotional disabilities transitioning from an institution into community living situations and evaluated their quality of life and life satisfaction at six, 12, and 24 month-time intervals. The authors found that these individuals with disabilities reported a higher quality of life and life satisfaction the longer they were out of the institution and integrated into the larger community. Reaffirming earlier studies regarding community attitudes and resources, the authors further found that those participants who reported problems with healthcare coverage, poor transportation options, lower income, and poor assistive technology assistance claimed greater dissatisfaction with their life circumstances. However, when community resources and assistance were available, participants reported a better quality of life and life satisfaction.

In summarizing this section, two key factors should stand out. The first is the fact that immediately after a traumatic and permanent disabling incident, many individuals with disabilities will initially be distressed and grieve their loss of lifestyle and function. Although these chaotic and uncertain cognitions, emotions and behaviors will initially temporarily disrupt their life, most will generally transition through stages of adjustment over varying individualized periods of time that experts describe as several months to a year. After 6 to 12 months of what experts believe to be a normal grieving process, persistent and complex grief then is diagnosed as maladaptive. Second, aside from the coping process an individual is undergoing after a traumatic disability, how he or she experiences and perceives his or her environment in terms of equity in services, treatment, and societal attitudes also plays a significant role in whether, and when an individual adjusts to disability.

22.4 Reasons as to Why Some with a Disability May Want to End Their Life

There are various reasons why some persons with non-life-threatening disabilities may desire to be euthanized, but statistically, the majority do not (Baynton 2001; Diener and Lucas 1999; Marini 2018a; Seligman 2002). Of the minority group who may consider suicide or PAS, we have already briefly discussed the negative societal influences that contribute to a poorer quality of life and life satisfaction. Some of these reasons are facilitated by social inequities and degrading and dehumanizing treatment of others as well as one's living conditions. As Kellett et al. (2021) found, some people with disabilities lack the resources, finances, transportation, adaptive technology, and healthcare, that would permit them to live a better life.

The desire to die varies considerably depending on one's severity of disability and their living circumstances such as living in poverty. This is a topic of complex and multifaceted set of factors that is beyond the scope of this chapter. Nevertheless, several important factors are discussed. The suicide rate among persons with spinal cord injuries (SCI) for example is higher than that of the general population, however not solely due to the disability itself (DeVivo and Stover 1995). In a series of longitudinal prospective studies (Krause and Crewe 1987; Krause et al. 1990, 1997; Krause and Kjorsvig 1992), these researchers explored the differences in earlier versus later mortality rates among persons with SCI using a longitudinal pre-and post-survey methodology. Variables included medical stability, employed versus unemployed, psychosocial adjustment, social engagement activities, and life satisfaction rated by psychologists. Findings indicated that the longer survival group were typically employed, engaged in more social activities, and were emotionally adjusted. Those who died earlier had reported more boredom, loneliness, depression, and lack of transportation.

Similarly, and perhaps living under more dire circumstances, are the life circumstances of many United States (US) veterans. Kamdar et al. (2021) accessed the National Health and Nutrition Examination Survey 2007–2016 database for 2,630 veterans exploring the risk of depression and suicidal ideation resulting from food insecurity in the US. The authors noted the suicide rate for veterans in the US has steadily increased since 2005, and although veterans account for only 8% of the US population, they disproportionately account for 13.5% of those who die from suicide. Although the reasons why veterans consider or complete suicide is multifaceted (mental illness, chronic pain, homelessness, etc.), food insecurity has been noted as a main contributor and correlates highly with homelessness (Hoffberg et al. 2018). Kamdar and associates (2021) found that veterans with less than a high school education were statistically more likely to consider suicide and have depression living with food insecurity.

There are numerous real-life examples of people with disabilities wanting to die because of their restricted living options (Stramondo 2021). Stramondo argues that when a society restricts the rights and privileges of those with disabilities leaving them with few if any choices and an otherwise unacceptable quality of life, some individuals may choose to be euthanized or consider suicide. For example, a young person with high-level tetraplegia requiring a 24/7 live-in attendant in Tennessee with insufficient government assistance or program has no choice but to live in a nursing home despite their age. However, in Colorado where financial resources and those with similar disabilities who can cost share for 24/7 attendant care are able to live independently in the community in an apartment. The positive impact on one's quality of life are immeasurable and previously discussed in the Kellett et al. (2021) study.

Finally, Na et al. (2022) accessed data from the National Epidemiologic Society on Alcohol and Related Conditions Wave III studying populations with substance use disorder (SUD), suicide attempt, and each condition individually compared to those with neither condition. The authors note a dramatic increase in US mortality rates among Caucasians with no religious affiliation without a college degree due to drug overdose, suicide, and alcohol-related liver disease. Case and Deaton (2020) refer to this growing phenomenon as “deaths of despair” from living “lives of despair”. The authors note other factors including the high cost of medical care, childhood trauma and history of psychiatric disorder, and low socioeconomic status as also main contributors to suicide and drug overdose deaths. Na and colleagues (2022) found a significant increase in SUD related deaths in this population, noting dismal living circumstances and no hope for succeeding in life. Tilstra et al. (2021) also found a growing trend in deaths of despair due to the opioid epidemic for those with similar demographic characteristics.

Despite some of the worst living conditions however, Stramondo (2021) emphasizes the concept of ‘adaptive preferences’ regarding quality of life, noting that when individuals have limited options to improve their life circumstances, they often adapt accordingly to that lifestyle over time. The other argument he cites is that able-bodied bioethicists evaluate the lives of those with disabilities more negatively, whereas persons with disabilities self-assess their life circumstances more positively (Singer 1995; Terlazzo 2016; Tsai et al. 2021). Thus, even for many of those with disabilities who are living in dismal situations or surroundings, the majority tend to adapt to that situation if their basic living conditions are met.

22.5 How the Medical Community Perceives Disability

It would be erroneous to assume that all healthcare practitioners are well versed with the lifestyle circumstances and adjustment factors for those with disabilities. Carmona et al. (2010) cited The Office of the Surgeon General (2005) report regarding the need for medical practitioners to acquire additional competencies to be better equipped with the tools to screen, diagnose, and treat the whole person with a disability as a whole. This is not surprising as prior studies have revealed that medical students do not feel sufficiently competent in addressing people with disabilities’ unique needs and circumstances (Iezzoni et al. 2021; Jacobson and Gammon 1997; Long-Bellil et al. 2011). Medical training regarding disability is essential to improving quality health care and providing a comprehensive and holistic approach to caring for people with disabilities.

Related to improving training for practitioners and students around chronic illness and disability, is the recognition of one’s own biases towards this population. One example of the common myths and misconceptions noted by Dunnum (1990), is the finding that many non-disabled persons expressed the view that people with SCI are not able to lead a normal life and would otherwise be better off not living. In the same light, research also indicates that healthcare providers will commonly be more inclined to object to artificial ventilation, give intubation and allow death for high level SCI patients (Gardner et al. 1985; Gerhart et al. 1994).

Similarly, Albrecht and Devlieger (1999) described the disparate viewpoints of physicians and healthcare workers who perceive people with disabilities have an unsatisfying quality of life, whereas over 50% of those with disabilities describe having a good or excellent quality of life (Dunn and Brody 2008; Hart et al. 2003; McCord et al. 2016). These insider experiences reported by those with disabilities are generally starkly different from outsider perceptions by those without disabilities. For example, Gerhart et al. (1994) found that only 18% of emergency room medical personnel would be glad to survive a severe SCI when compared to 92% of individuals living with a severe SCI.

In the Gerhart et al. (1994) study, the authors surveyed 153 emergency room physicians, nurses, residents, and technicians regarding level of care that should be provided to persons with high-level SCI, additional responses were solicited. Among the emergency medical responders, 41% reported they thought resuscitation of such patients was too aggressive, and 28% expressed that SCI patients' quality of life should be considered before attempting resuscitation measures. In addition, 22% reported they would not want resuscitation measures, and 23% of the remaining would only want pain relief, but not lifesaving measures. The majority of emergency responders perceived persons with severe SCI would have a poor quality of life with no explanation why.

Similarly in recent studies, Iezzoni et al. (2021) found that 82.4% of practicing US physicians perceived that people with a severe disability have a worse quality of life. Furthermore, the authors found that only 40% of practitioner participants felt very confident in their ability to provide the same quality care to patients with disabilities as they do for patients without a chronic disability. Relatedly, Asch and Ubel (1997) further argue the incentives some physician's practice when deciding the level of care to provide patients with various conditions. Specifically, the authors discuss the rationing care ordering fewer tests, less expensive tests, less expensive treatment to persons with chronic, but not terminal disabilities. The health insurance industry, government subsidized insurance, and individualized physician philosophy of delivering healthcare, all may play a role in the quality of care or lack of it for people with chronic disabilities (Batavia 1997).

These concerns become only too obvious in massive casualty situations such as the COVID-19 pandemic where emergency room physicians must triage and decide who receives lifesaving measures versus those, they perceive do not. This became apparent in several provinces in Canada in 2020 where they put out a list of patients with certain conditions who should be the last to receive lifesaving medical care. The Ontario Health (2020) ministry published a set of guidelines for ER physicians on the clinical triage protocol for the Covid pandemic. In this document, Appendix A provides a frailty criteria scale on a nine-point scale being 1 = very fit to 9 = terminally ill for physicians to follow. Ranking seventh for example on the scale are patients who are "severely frail–completely dependent for personal care from whatever cause (physical or cognitive). Even so, they seem stable and not a high risk of dying within six months" (p 10). This is level 7/9 of triage lifesaving measures initiated or not if for example a very fit patient requires a ventilator and the choice with someone who is not terminally ill but has a lesser chance of surviving even with a ventilator. Appendix B provides physicians with a verbatim script of responses they can memorize to offer patients and their family members on what type of care the patient will receive. For patient's not considered for lifesaving measures, one scripted response is "I regret to inform you that we are unable to offer you intensive care treatments at this time" (p 12). Despite a seemingly medically objective set of guidelines for physicians to follow during such an event, with the empirical based evidence regarding physician attitudes towards individuals with disabilities, their own subjective opinions may inherently be biased in a negative way.

Aside from a perceived poor quality of life of people with disabilities, physicians can also possess incorrect assumptions about patients' preferences and beliefs, thus hindering the ability to provide quality health care (Iezzoni and O'Day 2006; Iezzoni et al. 2021). Some physicians may also assume that patients with disabilities probably endure a great deal of pain and suffering, justifying a desire as to why some will opt for PAS. Disability advocates however advocate that it is more the loss of dignity, and fear of loss of bodily functions as well as the uncertainty of what life will be like living with a disability that is more the issue, particularly in the early stages of adapting to their trauma (Golden and Zoanni 2010; Leiby 1996).

Concerns arise when this uncertainty is unseen or overlooked by a physician and consequently leads to assisted suicide being considered more seriously as a good faith or reasonable option (Hendin and Foley 2008). However, other research findings studying physician perspectives surrounding assisted suicide for people with disabilities varies. Levy et al. (2013) found that psychiatrists were more conservative in their opinions regarding physician assisted suicide than physicians in other specialties. Golden and Zoanni (2010) cite the paucity of empirical research demonstrating that physicians' opinions are influenced by societal attitudes, however one cannot ignore the similarity in beliefs regarding disability. Nevertheless, physicians much like laypersons are not immune to projecting how horrendous their lives would become if they became paralyzed or lost their sight, that they would also consider suicide.

Radtke (2005), a physician, shares his own experience of acquiring trigeminal neuralgia and multiple sclerosis that caused him severe pain that does not respond to morphine. He discussed his transition adapting to his condition and now encourages medical students and other physicians to improve palliative care and enhance the quality of life of patients with disabilities. He also addressed how physicians should incorporate obtaining knowledge and addressing patient's financial needs as well as their spiritual and emotional state toward a more holistic way to treat patients with chronic disabilities.

22.6 Loopholes in MAID in Its Present Legislative Form

Having explored some of the history of disability treatment over the centuries, current societal attitudes and its ramifications on the psychosocial adjustment or adaptation to disability, as well as healthcare provider attitudes, the MAID Criminal Code (Criminal Code of Canada 1985) amendments have some notable flaws and may itself be inherently bias. First, since the American Psychiatric Association has distinguished a relative time-period of adaptive grief to loss at no more than 12 months, the Criminal

Code should allot for a minimum 12 month waiting period and then 90 days thereafter. This contrasts with the current 90-day wait period for MAID with no further psycho-social intervention.

Second, having learned the inherent bias many nondisabled individuals including physicians have toward the perceived poor quality of life or life satisfaction of persons with disabilities, having a single decision point physician as the sole gatekeeper in granting permission to be euthanized is concerning. It is common for individuals to project how they would feel if they became paralyzed, blind, etc., and the insider versus outsider perspective has typically been vastly divergent as noted in the empirical based literature provided.

Third, without a mandatory psychological evaluation to support someone with a disability is not clinically depressed, and of sound mind, PAS for someone will be premature if counseling and psychopharmacology treatment is not first tried. Fourth, if an individual wants PAS because he or she is under duress because of dismal community resources that could be remedied (i.e., government subsidized low-income housing), approving PAS for someone will again be premature. Finally, the inherent negative bias likely possessed by able-bodied politicians, judges, as well as medical providers cannot be understated in making decisions regarding who should live and who should not. Without the inclusion of individuals with disabilities to have a voice in any advisory or policy-setting forum, society will continue to get it wrong, and persons with disabilities will continue to be the invisible ones without a voice.

References

Albrecht, G.L., and P.J. Devlieger. 1999. The Disability Paradox: High Quality of Life Against All Odds. *Social Science & Medicine* 48 (8): 977–988.

American Psychiatric Association. 2013. Diagnostic and Statistical Manual of Mental Disorders, 5th ed. Washington, DC: Author.

Asch, D.A., and P.A. Ubel. 1997. Rationalizing by Any Other Name. *The New England Journal of Medicine* 336 (23): 1668–1671.

Batavia, A.I. 1997. Disability and Physician-assisted Suicide. *The New England Journal of Medicine* 336 (23): 1671–1673.

Baynton, D.C. 2001. Disability and the Justification of Inequality in American History. In *The New Disability History: American Perspectives*, ed. P.K. Longmore and L. Umansky, 33–57. New York: New York University Press.

Bishop, M. 2005. Quality of Life and Psychosocial Adaptation to Chronic Illness and Acquired Disability: A Conceptual and Theoretical Synthesis. *Journal of Rehabilitation* 71 (2): 5–13.

Boelen, P.A. 2016. Improving the Understanding and Treatment of Complex Grief: An Important Issue for Psychotraumatology. *European Journal of Psychotraumatology* 7: 32609. https://doi.org/10.3402/ejpt.v7.32609.

Carlson, E.A., and P.A. Lombardo. 2009. Review of Three Generations, No Imbeciles: Eugenics, the Supreme Court, and Buck v. Bell. *The Quarterly Review of Biology* 84 (2): 178–180.

Carmona, R.H., M. Giannini, B. Bergmark, and J. Cabe. 2010. The Surgeon General's Call to Action to Improve the Health and Wellness of Persons with Disabilities: Historical Review, Rationale, and Implications 5 Years After Publication. *Disability and Health Journal* 3 (4): 229–232.

Case, A., and A. Deaton. 2020. *Deaths of Despair and the Future of Capitalism*. Princeton, NJ: Princeton University Press.

Coco, A.P. 2010. Diseased, Maimed, Mutilated: Categorizations of Disability and Ugly Law in Late Nineteenth-century Chicago. *Journal of Social History* 44 (1): 23–37.

Dembo, T., G.L. Leviton, and B.A. Wright. 1956. Adjustment to Misfortune: A Problem in Social-psychological Rehabilitation. *Artificial Limbs* 3 (2): 4–62.

DeVivo, M.J., and S.L. Stover. 1995. Long-term Survival and Causes of Death. In *Spinal Cord Injury: Clinical Outcomes From the Model Systems*, ed. S.L. Stover, et al., 289–316. Aspen Publishers.

Diener, E., and S. Lucas. 1999. Personality and Subjective Well-being. In *Well-being: The Foundations of Hedonic Psychology*, ed. E. Diener and N. Schwarz, 213–229. Russell Sage Foundation.

DiTomasso, E., and B. Spinner. 1997. Social and Emotional Loneliness: A Re-examination of Weiss' Typology of Loneliness. *Personality and Individual Differences* 22 (3): 417–427.

Dunnum, L. 1990. Life Satisfaction and Spinal Cord Injury: The Patient Perspective. *The Journal of Neuroscience Nursing* 22 (1): 43–47.

Dunn, D.S., and C. Brody. 2008. Defining the Good Life Following Acquired Physical Disability. *Rehabilitation Psychology* 53 (4): 413–425. https://doi.org/10.1037/a0013749.

Gallagher, H.G. 1995. By Trust Betrayed: Patients, Physicians, and the License to Kill in the Third Reich (rev. ed.). Arlingto, VA: Vandamere Press.

Gardner, B.P., F. Theocleous, J.W. Watt, and K.R. Krishnan. 1985. Ventilation or Dignified Death for Patients with High Tetraplegia. *British Medical Journal* 291 (6509): 1620–1622.

Gerhart, K.A., J. Koziol-McLain, S.R. Lowenstein, and G.G. Whiteneck. 1994. Quality of Life Following Spinal Cord Injury: Knowledge and Attitudes of Emergency Care Providers. *Annals of Emergency Medicine* 23 (4): 807–812.

Golden, M., and T. Zoanni. 2010. Killing Us Softly: The Dangers of Legalizing Assisted Suicide. *Disability and Health Journal* 3 (1): 16–30.

Graf, N.M., I. Marini, and C. Blankenship. 2009. 100 Words About Disability. *Journal of Rehabilitation* 75 (2): 25–34.

Hart, T., J. Whyte, M. Polansky, S. Millis, F.M. Hammond, Sherer, et al. 2003. Concordance of Patient and Family Report of Neurobehavioral Symptoms at 1 Year After Traumatic Brain Injury. *Archives of Physical Medicine and Rehabilitation* 84 (2): 204–213.

Hendin, H., and K. Foley. 2008. Physician-assisted Suicide in Oregon: A Medical Perspective. *Michigan Law Review* 106 (8): 1613–1640.

Hoffberg, A.S., E. Spitzer, J.L. Mackelprang, S.A. Farro, and L.A. Brenner. 2018. Suicidal Self-directed Violence Among Homeless US Veterans: A Systematic Review. *Suicide and Life-threatening Behavior* 48(4): 481–498.

Hopps, S.L., M. Pepin, I. Arseneau, M. Frechette, and G. Begin. 2001. Disability-related Variables Associated with Loneliness Among People with Disabilities. *Journal of Rehabilitation* 67 (3): 42–48.

Iezzoni, L.I., and B. O'Day. 2006. More than Ramps: A Guide to Improving Health Care Quality and Access for People with Disabilities. *Oxford University Press*. https://doi.org/10.1093/acprof:oso/9780195172768.001.0001.

Iezzoni, L.I., S.R. Rao, J. Ressalam, D. Bolcic-Jankovic, N.D. Agaronnik, K. Donelan, T. Lagu, and E.G. Campbell. 2021. Physicians' Perceptions of People with Disability and Their Health Care. *Health Affairs* 40 (2): 297–306.

Jacobson, E.W., and W. Gammon. 1997. Using Standardized-patient Instructors to Teach Students About the Needs of Patients With Disabilities. *Academic Medicine* 72 (5): 442.

Kamdar, N.P., M.L. Horning, J.C. Geraci, A.W. Uzdavines, D.A. Helmer, and N.E. Hundt. 2021. Risk for Depression and Suicidal Ideation Among Food Insecure US Veterans: Data From the National Health and Nutrition Examination Study. *Social Psychiatry and Psychiatric Epidemiology* 56 (12): 2175–2184.

Kellett, K., K. Ligus, and J. Robison. 2021. "So Glad to Be Home": Experiences of Money Follows the Person Participants After Transitioning Out of an Institution. *Journal of Disability Policy Studies* 33 (2): 122–132. https://doi.org/10.1177/10442073211043519.

Kendall, E., and N. Buys. 1998. An Integrated Model of Psychosocial Adjustment Following Acquired Disability. *Journal of Rehabilitation* 64 (3): 16–20.

Kendler, K.S., J.M. Hettema, F. Butera, C.O. Gardner, and C.A. Prescott. 2003. Life Event Dimensions of Loss, Humiliation, Entrapment, and Danger in the Prediction of Onsets of Major Depression and Generalized Anxiety. *Archives of General Psychiatry* 60 (8): 789–796.

Krause, J.S., and N.M. Crewe. 1987. Prediction of Long-term Survival of Persons with Spinal Cord Injury: An 11-year Prospective Study. *Rehabilitation Psychology* 32 (4): 205–213.

Krause, J.S., J.M. Saari, and D. Dykstra. 1990. Quality of Life and Survival After Spinal Cord Injury. *SCI Psychosocial Process* 3: 4–8.

Krause, J.S., and J.M. Kjorsvig. 1992. Mortality after Spinal Cord Injury: A 4-year Prospective Study. *Archives of Physical Medicine and Rehabilitation* 73 (6): 558–563.

Krause, J.S., M. Sternberg, J. Maides, and S. Lottes. 1997. Mortality After Spinal Cord Injury: An 11-year Prospective Study. *Archives of Physical Medicine and Rehabilitation* 78 (8): 815–821.

Kubler-Ross, E. 1969. *On Death and Dying*. Scribner.

Largent, M. 2002. The Greatest Curse of the Race: Eugenic sterilization in Oregon 1909–1983. *Oregon Historical Quarterly* 103 (2): 188–209.

Leiby, R. 1996. Whose Death Is It Anyway? The Kevorkian Debate: It's a Matter of Faith, in the End. Washington Post.

Levy, T.B., S. Azar, R. Huberfeld, A.M. Siegel, and R.D. Strous. 2013. Attitudes Towards Euthanasia and Assisted Suicide: A Comparison Between Psychiatrists and Other Physicians. *Bioethics* 27 (7): 402–408.

Lewin, K. 1935. *A Dynamic Theory of Personality*. McGraw-Hill.

Li, L., and D. Moore. 1998. Acceptance of Disability and Its Correlates. *The Journal of Social Psychology* 138 (1): 13–25.

Livneh, H. 1991. On the Origins of Negative Attitudes Toward People with Disabilities. In *The Psychological and Social Impact of Disability*, ed. R.P. Marinelli, and A.E.D. Orto, 111–138. New York: Springer.

Livneh, H., and R.F. Antonak. 1997. Psychosocial Adaptation to Chronic Illness and Disability. Aspen Publishers.

Long-Belil, L.M., K.L. Robey, C.L. Graham, P.M. Minihan, S.C. Smeltzer, and P. Kahn. 2011. Teaching Medical Students About Disability: The Use of Standardized Patients. *Academic Medicine* 86 (9): 1163–1170.

Marini, I. 2018a. Societal Attitudes and Myths About Disability: Improving the Social Consciousness. In *Psychosocial Aspects of Disability: Insider Perspectives and Strategies for Counselors*, 33–62. New York: Springer.

Marini, I. 2018b. The History of Treatment Toward People with Disabilities. In *Psychosocial Aspects of Disability: Insider Perspectives and Strategies for Counselors*, 2nd ed, ed. I. Marini, N.M. Graf, and M.J. Millington, 3–32. New York: Springer.

Marini, I. 2018c. Social Justice, Oppression, and Disability: Counseling Those Most in Need. In *The Psychological and Social Impact of Illness and Disability*, 7th ed, ed. I. Marini, and M. Stebnicki, 503–519. Springer.

Mash, H.B.H., C.S. Fullerton, and R.J. Ursano. 2013. Complicated Grief and Bereavement in Young Adults Following Close Friend and Sibling Loss. *Depression and Anxiety* 30 (12): 1202–1210.

McCord, C.E., T.R. Elliott, J.W. Berry, A.T. Underhill, P.R. Fine, and M.H.C. Lai. 2016. Trajectories of Happiness 5 Years Following Medical Discharge for Traumatic Disability: Differences Between Insider and Outsider Perspectives. *Journal of Happiness Studies* 17 (2): 553–567.

Na, P.J., E.A. Stefanovics, T.G. Rhee, and R.A. Rosenheck. 2022. "Lives of Despair "At Risk for "Deaths of Despair': Tracking and Under-recognized, Vulnerable Population. *Social Psychiatry and Psychiatric Epidemiology* 57: 1123–1134. https://doi.org/10.1007/s00127-022-02218-w.

Olkin, R. 1999. *What Psychotherapists Should Know About Disability*. New York: Guilford Press.

Ontario Health. 2020. Clinical Triage Protocol for Major Surge in COVID Pandemic. https://med.uottawa.ca/pathology/sites/med.uottawa.ca.pathology/files/clinical_triage_protocol_for_major_surge_in_covid_pandemic_-_march_28_20205.pdf

Ott, C.H., R.J. Lueger, S.T. Kelber, and H.G. Prigerson. 2007. Spousal Bereavement in Older Adults: Common, Resilient, and Chronic Grief with Defining Characteristics. *The Journal of Nervous and Mental Disease* 195 (4): 332–341.

Pinon, R.M., I. Marini, and D.L. Antol. 2021. Every Day Lived Experiences of Persons with Physical Disabilities: Implications for Rehabilitation Counselors. *Directions in Rehabilitation Counseling* 31 (10): 17–30.

Prigerson, H.G., and Paul K. Maciejewski. 2006. A Call for Sound Empirical Testing and Evaluation of Criteria for Complicated Grief Proposed for DSM-V. *Omega: Journal of Death and Dying*, 52(1): 9–19.

Radtke, R. 2005. A Case Against Physician-assisted Suicide. *Journal of Disability Policy Studies* 16 (1): 58–60.

Seligman, M.E.P. 2002. *Authentic Happiness: Using the New Positive Psychology to Realize Your Potential for Lasting Fulfillment*. New York: Free Press.

Singer, P. 1995. *Rethinking Life and Death*. Oxford: Oxford University Press.

Smart, J. 2009. Disability, Society, and the Individual. Austin, TX: Pro-ED Publishing, Inc.

Stramondo, J.A. 2021. Bioethics, Adaptive Preferences, and Judging the Quality of a Life with Disability. *Social Theory and Practice* 47 (1): 199–220.

Stroebe, M., H. Schut, and W. Stroebe. 2007. Health Outcomes of Bereavement. *The Lancet* 370 (9603): 1960–1973.

Terlazzo, R. 2016. Conceptualizing Adaptive Preferences Respectively: An Substantial Account. *Canadian Journal of Philosophy* 45 (2): 179–196.

The Criminal Code of Canada (R.S.C., 1985, c. C-46) at Section 241.

Tilstra, A.M., D.H. Simon, and R.K. Masters. 2021. Trends in "Depths Of Despair" Among Working-aged White and Black Americans, 1990–2017. *American Journal of Epidemiology* 190 (9): 1751–1759. https://doi.org/10.1093/aje/kwab088.

Trieschmann, R.B. 1988. Spinal Cord Injuries: Psychological, Social, and Vocational Rehabilitation, 2nd ed. New York: Demos.

Tsai, J., N. Jones, D. Szymkowiak, and R.A. Rosenheck. 2021. Longitudinal Study of the Housing and Mental Health Outcomes of Tenants Appearing in Eviction Court. *Social Psychiatry and Psychiatric Epidemiology* 56 (9): 1679–1686.

Office of the Surgeon General (US); Office on Disability. 2005. U.S. Department of Health and Human Services. The Surgeon General's Call to Action to Improve the Health and Wellness of Persons with Disabilities. US Department of Health and Human Services, Office of the Surgeon General. PMID: 20669510.

Vash, C.L., and N.M. Crewe. 2004. *Psychology of Disability*, 2nd ed. New York: Springer.

Irmo Marini, PhD, CRC, CLCP is originally from Thunder Bay, Ontario and is currently a professor in the School of Rehabilitation at the University of Texas Rio Grande Valley. Dr. Marini has authored over 100 peer-reviewed journal publications, seven co-authored and co-edited texts, and over 55 book chapters. He has received numerous institutional and national distinguished career awards teaching, research, and service. Dr. Marini is the former Chair of the Commission on Rehabilitation Counselor Certification as well as former president of the American Rehabilitation Counseling Association. Dr. Marini is CEO of Marini & Associates, rehabilitation consultant experts in personal injury litigation.

Laura Villarreal, MS, CRC is from Monterrey, Mexico and grew up in the Rio Grande Valley. She is a PhD student in Rehabilitation Counseling at the University of Texas at Rio Grande Valley. She is a Certified Rehabilitation Counselor and currently works as an ADA Accommodation Specialist for Sedgwick. She is also a research and teaching assistant at the university. She has a special interest in forensic rehabilitation work and provides consulting work as a rehabilitation consulting expert in litigated cases involving personal injury.

Chapter 23
Canada's Medical Assistance in Dying Law and the Rights to Life and Equality at International Law

Kerri Joffe and Roberto Lattanzio

Abstract In international law, the right to life has been described as the most fundamental human right, without which other core human rights cannot be enjoyed. This chapter explores the right to life, together with the right to equality, with a focus on three international instruments; two United Nations treaties, the *International Covenant on Civil and Political Rights* and the *Convention on the Rights of Persons with Disabilities*, and a regional treaty for the Americas, the *American Declaration on the Rights and Duties of Man*. This Chapter analyzes Canada's MAiD legislation from a disability rights perspective, within the context of the right to life and the right to equality as understood in these international instruments. It reviews implications for Canadian government, policy and lawmaking.

Keywords International law · Right to life · Right to equality · Systemic inequality · Safeguards · Discrimination · Disability rights · Domestic law

23.1 Introduction

The protections inherent in the right to life and the right to equality are fundamental within international human rights law. The United Nations Human Rights Committee has characterized the right to life as the supreme right from which no derogation is permitted even in times of public emergency (UN Human Rights Committee 2019b). The right to life and the right to equality are foundational rights that are necessary for the enjoyment and fulfillment of other civil, political, economic, social and cultural rights.

On February 3, 2021, while Canadian lawmakers debated Bill C-7, a bill that proposed to expand eligibility for medical assistance in dying (MAiD) to people

K. Joffe (✉) · R. Lattanzio
ARCH Disability Law Centre, Toronto, Canada
e-mail: kerri.joffe@arch.clcj.ca

R. Lattanzio
e-mail: robert.lattanzio@arch.clcj.ca

J. Kotalik and D. W. Shannon (eds.), *Medical Assistance in Dying (MAID) in Canada*, The International Library of Bioethics 104,
https://doi.org/10.1007/978-3-031-30002-8_23

whose death was not reasonably foreseeable, three United Nations experts expressed serious concerns to the Government of Canada that the Bill violated the rights to life and equality in international law (United Nations 2021). They stated that, "provisions contained in Bill C-7 would appear to violate the right of persons with disabilities to life" (United Nations 2021, 4). Further they expressed, "grave concern that provisions contained in the Bill may be contrary to Canada's international obligations to respect, protect and fulfil the core right of equality and non-discrimination of persons with disabilities" (United Nations 2021, 4). These concerns were echoed and underscored by disability rights advocates from across Canada, who conveyed to lawmakers the grave and deeply discriminatory impacts they feared Bill C-7 would have on persons with disabilities. For example, in a letter to Senators, Aislinn Thomas wrote:

> [p]oliticians seem to choose not to acknowledge that the majority of disabled people in Canada are also poor, racialized, struggling to survive, and not afforded access to the resources needed to live, let alone thrive. ...offering early death for disabled people in this context is not just morally reprehensible ... It is also a loss. ... Disabled lives, like all lives, have inherent value. (Thomas 2021)

Before the House of Commons committee studying Bill C-7, Dr. Heidi Janz testified that, "[t]he removal of 'reasonably foreseeable' natural death as a limiting eligibility criterion for the provision of MAiD will result in people with disabilities seeking MAiD as an ultimate capitulation to a lifetime of ableist oppression" (Canada 2020).

Despite these warnings from the United Nations and disability communities, on March 17, 2021, Parliament passed Bill C-7, bringing into law Canada's current medical assistance in dying regime. The current regime significantly expanded eligibility, beyond what was permitted when MAiD was first legalized throughout Canada in 2016. In 2016, Canada's *Criminal Code* was amended to provide a legal exception to the prohibition on assisted suicide that still applies (Criminal Code 1985). In certain circumstances, the law permits physicians and nurse practitioners to administer a substance that will cause a person's premature death, and pharmacists, family members or others to assist without facing criminal charges, provided that the person meets certain eligibility criteria. When first legalized, it was clear that MAiD was limited to end of life situations: the law authorized MAiD only for persons who were suffering with serious diseases, illnesses or disabilities whose death was reasonably foreseeable (Criminal Code 1985). However, with the passage of Bill C-7, MAiD is now legally permitted for persons suffering with serious diseases, illnesses or disabilities who are in the process of dying (i.e., whose death is reasonably foreseeable) and for persons who are not near the end of their natural life and who meet the remaining eligibility criteria (Criminal Code 1985) (See Appendix 1.1).

Building upon the concerns outlined by the UN, this chapter examines whether and how Canada's existing MAiD regime violates the right to life and equality for persons with disabilities, at international law. In particular, this analysis is grounded and framed through a disability rights and intersectional lens, and draws on international human rights obligations that must inform Canadian public policy and lawmaking.

23.2 The Right to Life in International Law

Canada is a State Party to a number of international human rights treaties under which it has legal obligations to protect the right to life, including the *Universal Declaration of Human Rights* (UDHR), the *International Covenant on Civil and Political Rights* (ICCPR) and the *Convention on the Rights of Persons with Disabilities* (CRPD) (Universal Declaration of Human Rights 1948; International Covenant on Civil and Political Rights 1966; Convention on the Rights of Persons with Disabilities 2006). From a disability rights perspective, the CRPD is arguably the most significant international human rights treaty. The CRPD elaborates human rights and freedoms in the context of disability. It provides for civil, political, economic, social and cultural rights for persons with disabilities and establishes obligations that states must fulfill in order to ensure the full and equal enjoyment of these rights.

Article 3 of the UDHR provides that, "everyone has the right to life, liberty and security of the person." Article 6(1) of the ICCPR states that, "[e]very human being has the inherent right to life. This right shall be protected by law. No one shall be arbitrarily deprived of his life." The UN Human Rights Committee, in its General Comment on Article 6, has explained that the right to life encapsulates the entitlement of individuals to enjoy a life with dignity and to be free from acts or omissions that are intended or may be expected to cause their unnatural or premature death.

Building upon the right to life articulated in the UDHR and the ICCPR, Article 10 of the CRPD provides for the right to life for persons with disabilities:

> States Parties reaffirm that every human being has the inherent right to life and shall take all necessary measures to ensure its effective enjoyment by persons with disabilities on an equal basis with others.

The CRPD highlights that for persons with disabilities, the right to life is fundamentally interrelated with the right to equality: one cannot exist without the other. Indeed, the UN Human Rights Committee has interpreted the right to life contained in the ICCPR with reference to the right to life contained in the CRPD, emphasizing that persons with disabilities must have effective enjoyment of the right to life <u>on an equal basis with others</u> [emphasis added] (UN Human Rights Committee 2019b). To this end, states are required to provide "measures of protection" to guarantee this right, including "the provision of reasonable accommodation when necessary to ensure the right to life, such as ensuring access of persons with disabilities to essential facilities and services" (UN Human Rights Committee 2019b). The right to life includes obligations on states to ensure access to the essentials of life such as food, shelter, and water, as well as measures to combat stigmatization associated with disabilities which can undermine access to health care (UN Human Rights Committee 2019b). It is clear that at international law, the right to life is deeply interconnected with the right to equality, and that it includes obligations on states to ensure access to the social conditions necessary for meaningful and dignified enjoyment of the right.

Free choice is a central pillar of the right to life as it applies to assisted dying. The right to life requires that a person freely choose assisted death, without any coercion or external pressure. To ensure that this condition is met, international law requires that states create and implement safeguards. The UN Human Rights Committee's General Comment on the right to life explains that assisted dying regimes must include "robust legal and institutional safeguards to verify that medical professionals are complying with the free, informed, explicit and unambiguous decision of their patients, with a view to protecting patients from pressure and abuse" (UN Human Rights Committee 2019b). Consistent with the General Comment, the UN Human Rights Committee has recommended that states strengthen their assisted dying safeguards to bring them into compliance with the right to life (UN Human Rights Committee 2009; UN Human Rights Committee 2019a).

In addition to Canada's legal obligations to protect and promote the right to life under the international human rights treaties described above, Canada is also a member of the Organization of American States (OAS) and is thereby bound to respect the rights and duties set out in the *American Declaration on the Rights and Duties of Man* (American Declaration) (American Declaration on the Rights and Duties of Man 1948). The American Declaration is one of the core legal instruments within the inter-American system for the protection of human rights. Article I of the American Declaration provides that, "[e]very human being has the right to life, liberty and the security of his person" (American Declaration on the Rights and Duties of Man 1948). Consistent with the United Nations, the inter-American system has interpreted the right to life as the most fundamental right and as necessary to the enjoyment of other human rights (Inter-American Commission on Human Rights 2003, 2018) (See Appendix 1.2). As a result, states have positive obligations "to guarantee the creation of the conditions that may be necessary in order to prevent violations of such inalienable right" (Inter-American Court of Human Rights 2006).

23.2.1 Canada's MAiD Legislation and the Right to Life

The Government of Canada has consistently taken the position that Canada's MAiD regime appropriately balances the interests of autonomy (people should have the right to choose MAiD, whether or not their death is reasonably foreseeable) with the need to prevent suicide and protect vulnerable persons from ending their lives prematurely. This rationale is a cornerstone of the Government's response to the concerns expressed by the three UN experts (Government of Canada 2021). According to the Government, both vulnerable people whose deaths are reasonably foreseeable and vulnerable people who are not near death are adequately protected via safeguards in the legislation. For people whose deaths are reasonably foreseeable, the legislation includes safeguards such as requiring a written request for MAiD signed by one independent witness; two independent physicians or nurse practitioners must assess the person and confirm that all eligibility requirements are met; the person must be informed that they can withdraw their request; and the person must be given an opportunity to withdraw immediately before MAiD is administered (although this final consent can be waived) (Criminal Code 1985). For people who are not dying, the legislation further requires a mandatory 90-day assessment period, a requirement that a practitioner with expertise in the condition that is causing the suffering be consulted as part of the eligibility assessment, a requirement that the person be informed of the available supports and services, and that reasonable means of alleviating suffering be discussed and considered (Criminal Code 1985).

Are these legislative requirements sufficient to meet the standard of robust legal and institutional safeguards that ensure people have free, unencumbered choices about assisted dying, in accordance with the international right to life? From a disability rights perspective, the only answer can be no. To understand why, it is critical to appreciate the context in which MAiD operates. To be eligible for MAiD, whether at the end of life or not, Canada's legislation requires that a person have a serious and incurable illness, disease or disability. As a result, it is only persons with disabilities (including illnesses and diseases) who meet the remaining eligibility requirements who can request MAiD. The choice to do so must, therefore, be understood within the context in which persons with disabilities live.

In Canada, persons with disabilities experience a myriad of systemic civil, social and economic inequalities. Persons with disabilities are more likely to experience poverty, inadequate housing or homelessness, unmet healthcare needs, and ableist barriers that prevent them from living dignified lives in the community (See Appendix 1.3). Persons with disabilities who are further marginalized by their racialized, Indigenous, gender, sexual orientation or other identity status are often less likely to be able to access appropriate services and supports (See Appendix 1.4). The Supreme Court of Canada has recognized the long history of discrimination and exclusion experienced by persons with disabilities in Canada (*Eldridge v British Columbia (Attorney General)* 1997).

For some people with disabilities, this context profoundly limits the choices available to them. Tragically, the media, family, friends and community-based disability organizations have documented cases of persons with disabilities resorting to MAiD because they cannot get affordable, accessible housing, despair in poverty, or cannot access the services and supports necessary to live in the community (*Several Cases of Concern* 2021). In particular, institutionalization, poor conditions in institutions, and lack of alternatives to institutional life have been reported as reasons for people with disabilities seeking MAiD (See Appendix 1.5). In testimony before Canada's Senate, disability rights activist Jonathan Marchand explained that, "[t]here can be no death with dignity and freedom of choice as long as we are forced to live in institutions, made to feel like burdens, while we face discrimination and systemic violence at all levels" (Canada 2021). For many people with disabilities who lack meaningful alternatives to assisted death, choice is illusory.

In their February 2021 communication to Canada, the three UN experts recognized that many persons with disabilities in Canada are not in a position to exercise free, unconstrained choice when making decisions about MAiD, writing:

> It is not beyond possibility that, if offered an expanded right as per Bill C-7, persons with disabilities may decide to end their lives because of broader social factors, including loneliness, social isolation and lack of access to quality support services. Indeed, persons with disabilities, particularly older persons with disabilities, may be vulnerable to explicit or

implicit pressures arising from their context, including expectations from family members, financial pressures, cultural messages. (United Nations 2021)

Given these systemic inequalities, and the resulting constraints on choice that may induce people with disabilities to resort to MAiD, it is absolutely imperative that there be effective safeguards to ensure that decisions about MAiD are truly free and informed, not pressured or coerced. Indeed, for a number of years the United Nations has called upon Canada to strengthen its safeguards. In her report to the Human Rights Council at its 43rd session, the former UN Special Rapporteur on the Rights of Persons with Disabilities called on Canada to enact adequate safeguards to ensure that persons with disabilities do not resort to MAiD because of the absence of community-based alternatives and palliative care (UN Human Rights Council 2019). A similar recommendation was made to Canada by the CRPD Committee in its 2017 Concluding Observations (UN Committee on the Rights of Persons with Disabilities 2017). The Committee expressed its concern about the adoption of MAiD based on disability and recommended that Canada ensure that people who seek MAiD at the end of life have access to alternative courses of action and the supports necessary to live a full life in the community (UN Committee on the Rights of Persons with Disabilities 2017). To be effective, safeguards must provide real options that actually ameliorate physical and socio-economic suffering, allow for meaningful alternatives to MAiD, and facilitate free, informed choice.

The safeguards built into Canada's existing MAiD regime fall short of the standard set out under the right to life. For persons whose natural death is reasonably foreseeable, the legislation only requires that they be *informed* about interventions and supports that might alleviate their suffering. For persons who are not dying, the legislation further requires that they are offered consultations with relevant professionals who provide services or care that might relieve their suffering and that these are given serious consideration. Despite these requirements, there is no explicit assurance that such interventions are actually available in a manner and form that meet people's needs and offer meaningful options for a dignified life in the community. There is no guaranteed right to access palliative care, in-home supports, alternatives to institutions or other disability services and supports in Canada. The legislation does not address what happens when critical housing, services and supports are not available, have excessively long waitlists or are only offered at levels that do not meet actual needs. While it is certainly important that people requesting MAiD have information about the means available to relieve their suffering, the mere provision of information about available services does not translate into people having real options that actually ameliorate their suffering. Providing information is not sufficient to ensure that people with disabilities who are suffering due to systemic civil, social, and economic inequality can make free and informed decisions about MAiD, unencumbered by external pressures. In this way, Canada's MAiD regime falls short of the necessary institutional and legal safeguards to meet the requirements of the right to life.

23.3 Canada's MAiD Legislation and the Right to Equality in International Law

Like the right to life, the right to equality is a particularly important human right because it serves as the foundation for other rights and freedoms. Article 5 of the CRPD provides that:

> ...all persons are equal before and under the law and are entitled without any discrimination to the equal protection and equal benefit of the law. States Parties shall prohibit all discrimination on the basis of disability and guarantee to persons with disabilities equal and effective legal protection against discrimination on all grounds. (Convention on the Rights of Persons with Disabilities 2006)

In its guidance document on the meaning and interpretation of Article 5, the UN CRPD Committee explained that the right to equality should account for the human rights model of disability—"that disability is a social construct and impairments must not be taken as a legitimate ground for the denial or restriction of human rights" (UN Committee on the Rights of Persons with Disabilities 2018).

Contrary to this UN guidance, Canada's MAiD legislation treats people with disabilities who are suffering as a result of their disability differently than non-disabled people who are suffering. Canadian disability rights advocates have consistently advocated that the 2021 expansion of MAiD legislation to persons with disabilities who are suffering but are not dying violates the right to equality and constitutes discrimination on the basis of disability. It must be remembered that all people experience suffering at some point in life, and for many this suffering is at times sustained and intolerable. Governments often respond with policies and programs aimed at remediating suffering, such as social supports and suicide prevention programs. MAiD is not, for good reason, counted among such social remedies. Nevertheless, under Canada's current law, an exception exists for persons with disabilities who suffer intolerably: they *are* eligible to receive MAiD. Only suffering that is caused by disability can render a person eligible for MAiD. No other group of people can receive MAiD due to suffering as a result of an immutable personal characteristic. In this way, the 2021 amendments to Canada's MAiD legislation target disability-related suffering as an eligibility criterion for MAiD. It is this distinction, based on the protected ground of disability, that discriminates.

Underlying this distinction in the law between those who are suffering as a result of their disability and those who are suffering for other reasons is the discriminatory view that living with a disability may be worse than death. It is not intolerable suffering that makes a person eligible for MAiD, but rather intolerable suffering caused by a disability. It follows that according to the law's logic, there must be something about disability that is bad enough to justify the state's intervention to hasten death. By targeting disability-related suffering for assisted death, MAiD legislation signals in Canadian public policy that the lives of people with disabilities who desire an end to suffering are less valuable or worthy than others. In turn, this devaluing of disabled lives reinforces negative stereotypes, thereby contributing to systemic ableism and negative social perceptions of persons with disabilities in Canada.

These concerns were outlined in the February 2021 communication to Canada by the three UN experts, who stated that Canada's current MAiD law risks, "reinforcing (even unintentionally) ableist and ageist assumptions about the value or quality of life of persons with disabilities and older persons with or without disabilities" (United Nations 2021). Further they stated that, "[f]rom a disability rights perspective, there is a grave concern that, if assisted dying is made available for all persons with a health condition or impairment, regardless of whether they are close to death, a social assumption might follow (or be subtly reinforced) that it is better to be dead than to live with a disability" (United Nations 2021) (See Appendix 1.6). They found that "[b]y expanding access to medical assistance in dying based on disability .... the policy would ... have a potentially discriminatory impact on persons with disabilities and older persons who are not at the end of their life or nearing death from natural causes" (United Nations 2021).

International law recognizes that central to the right to be free from discrimination is the need to combat negative stereotypes which undermine equality. Article 26 of the ICCPR requires states:

> ... to take affirmative action in order to diminish or eliminate conditions which cause or help to perpetuate discrimination prohibited by the Covenant. For example, in a State where the general conditions of a certain part of the population prevent or impair their enjoyment of human rights, the State should take specific action to correct those conditions. (UN Human Rights Committee 1989)

Article 8 of the CRPD requires states to "promote positive perceptions and greater social awareness towards persons with disabilities" and to "combat stereotypes, prejudices and harmful practices" relating to persons with disabilities. The three UN experts warned Canada that:

> ... Article 8 envisages a positive policy eco-system that takes great care to ensure that otherwise sound and well-intentioned legislation does not indirectly pivot on, or subtly reinforce, ableist assumptions in society. It is hard to see how a legislative proposal that extends a right to medically assisted dying to persons with disabilities who are not themselves close to death could send a signal that is compatible with Article 8 (obligations to combat ableism) combined with Article 5 (obligation to secure equal respect for rights) of the CRPD. (United Nations 2021)

Contrary to the right to equality, Canada's MAiD legislation serves to reinforce rather than counteract damaging social forces of systemic ableism and negative social perceptions of persons with disabilities. These societal implications are particularly concerning, in light of the existing ableism and systemic inequality experienced by persons with disabilities in Canada described above. These discriminatory impacts of the law further demonstrate the manner in which the law violates the right to equality.

23.4 The Rights to Life and Equality at International Law: Implications for Canadian Policy and Lawmaking

As a State Party to the international human rights treaties discussed above, Canada has undertaken to uphold and comply with the rights and freedoms set out therein. It is important to note, however, that Canada takes a dualist approach, meaning that generally further steps must be taken to incorporate ratified international treaties into Canadian law before the treaties will have domestic legal effect (De Mestral and Fox-Decent 2008). None of the international human rights treaties discussed above have been wholly incorporated into Canadian domestic law. Even so, Canadian law recognizes that signing or ratifying an international treaty signals governments' intent to comply with the rights and obligations in that treaty (Sullivan 2014). Canadian courts and tribunals may draw upon the values, principles and rights in these treaties to interpret Canadian domestic legislation (*Quebec (Attorney General) v 9147-0732 Québec Inc* 2020; *R v Hape* 2007). Indeed, the Supreme Court of Canada recently reaffirmed that the *Canadian Charter* should generally be presumed to provide protection at least as great as that afforded by similar provisions in international human rights documents which Canada has ratified (*Quebec (Attorney General) v 9147-0732 Québec Inc* 2020). Thus, signing and ratifying international human rights treaties is a significant step, and all levels of government in Canada must take seriously their obligations to comply with and implement the rights set out in these treaties.

Canada's expanded MAiD regime under Bill C-7 was developed, according to the Government, as a response to the Quebec Superior Court's 2019 *Truchon v Canada (AG)* decision (*Truchon v Canada* 2019; Government of Canada n.d.). In that decision the Quebec lower court found that the eligibility requirement for death to be reasonably foreseeable was unconstitutional in that it violated the *Canadian Charter* protection against deprivations of life, liberty and security of the person (section 7) and the right to equality (section 15). Rather than defend its legislation by appealing the decision to the Quebec Court of Appeal, the Government instead introduced Bill C-7, which now allows anyone suffering with a serious disability or illness to resort to MAiD regardless of whether they are at the end of life. However, as a country that has bound itself to uphold and comply with international human rights norms, this was too narrow an approach. Instead, due weight should have been given to the rights set out in the UDHR, ICCPR, CRPD, American Declaration and other relevant international human rights treaties to which Canada is a State Party. When developing public policy, legislating, or amending Canadian laws and regulations, governments in Canada *must* conduct a robust critical analysis to ensure compliance with Canada's international human rights obligations. Anything less risks derogating from fundamental human rights requirements.

Similarly, governments in Canada must heed warnings from UN experts and other international human rights bodies that new legislation or policy threatens to violate Canada's international law obligations and undermine human rights. In the case of Bill C-7, some of these warnings came during the legislative process, while the Bill was being debated in Parliament, when there was still time to amend its discriminatory provisions. Even before the development of Bill C-7, Canada received criticism from the former UN Special Rapporteur and the UN CRPD Committee concerning the lack of adequate safeguards to protect vulnerable people with disabilities from being induced to resort to MAiD. During the course of the development of Canada's MAiD regime, disability rights advocates and disability communities put forward policy alternatives to address the lack of adequate safeguards. For example, many advocated for much more robust tools and comprehensive protocols to assess eligibility for MAiD. This would require, among other things, that health care teams involved with assessing eligibility for and providing MAiD collect additional information regarding a range of factors, including social deprivation factors like intolerable living conditions in prisons, institutions, hospitals or long-term care facilities, that contribute to the person's suffering and motivation to request MAiD (Canadian Association for Community Living 2016). Warnings and critique from international human rights bodies should give rise, at the very least, to a pause in the development of the law, further study, and serious consideration of alternative policy options on the part of Government and lawmakers.

Within international systems, there are various tools to address circumstances when a state's legislation, policy or activities derogate from international human rights treaty obligations. These include processes whereby individuals or groups of individuals whose rights have been violated by a state may, under certain circumstances, bring their concerns to an international body for adjudication. States' compliance with international law is monitored and reviewed using tools such as period UN treaty-body review procedures; systemic investigations into allegations that a state has committed grave human rights violations; country and thematic reports; communications from designated experts and Special Rapporteurs; and other mechanisms. To date, the UN has repeatedly criticized Canada's MAiD legislation, whether through UN experts expressing serious concerns that Canada's MAiD legislation violates the right to life and equality in international law, the former UN Special Rapporteur asking Canada to adopt real alternatives to MAiD, or the UN CRPD Committee calling Canada to task. As long as Canada's MAiD legislation continues to have a discriminatory impact on people with disabilities, we can expect that Canada will continue to be criticized within international human rights systems.

23.5 Conclusion

In this chapter we have argued that Canada's existing MAiD legislation does not comply with its international legal obligations to protect the right to life and to guarantee the right to equality enshrined in the *Universal Declaration of Human Rights,* the *International Covenant on Civil and Political Rights*, the *Convention on the Rights of Persons with Disabilities* and the *American Declaration on the Rights and Duties of Man.* Contrary to the right to life, legislative safeguards have not been effective in protecting persons with disabilities from resorting to MAiD because of suffering related to their systemic civil, social, and economic inequalities. Tragically, the number of documented cases of persons with disabilities resorting to MAiD due to a lack of real alternatives to address their suffering continues to grow. There is a very urgent need to create robust safeguards that actually ameliorate people with disabilities' suffering and provide real, meaningful alternatives to MAiD at the end of life. The introduction of new safeguards cannot, however, remedy the 2021 amendments which allow MAiD for people with disabilities who are suffering but are not at the end of their lives. As we have argued, these amendments violate the right to equality by targeting disability-related suffering for assisted dying, thereby reinforcing negative stereotypes about the value of disabled lives. They are fundamentally discriminatory.

It is imperative that all levels of government in Canada earnestly consider and implement the international human rights obligations that our country has voluntarily assumed. These obligations should not be an after-thought; instead, they must become part of the development of domestic legislation and policy. Now is the time for governments to act to bring Canada's MAiD legislation into compliance with its international human rights obligations.

Acknowledgements We gratefully acknowledge assistance from Keshbir Brar and Jack Fallon, students at Osgoode Hall Law School.

Appendix

1.1	*Criminal Code*, RSC 1985, c C-46, s 241.2(1)–241.2(2). For both categories, the legislation requires that certain threshold requirements must be met—for example, the person must be 18 years old or older, and must have made the request voluntarily. The person must also be provided with information about ways to alleviate suffering, give informed consent, make a written request that is signed, dated, witnessed, and supported by an opinion of a second independent medical practitioner, and receive an opportunity to withdraw their request. Where death is not reasonably foreseeable, the legislation includes additional requirements: consulting with a third independent practitioner if needed; informing the person of available medical and disability services and supports; and offering consultations with service providers; ensuring that available means to relieve suffering are discussed and considered; and providing a waiting period of 90 days. See *Criminal Code* ss 241.2(1)–241.2(9)
1.2	IACHR, Merits Report No 121/18, Case 10.573, *José Isabel Salas Galindo and others* (United States) 5 October 2018 at para 337. IACHR, Merits Report No 97/03, Case 11.193, *Gary T Graham (Shaka Sankofa)* (United States) 29 December 2003 at para 26 ("[T]he right to life is widely recognized as the supreme right of the human being, respect for which the enjoyment of other rights depends"); Inter-American Court of Human rights, *Case of Ximenes-Lopes v Brazil* (4 July 2006), Series C, No 149 at para 124 [*Ximenes-Lopes*]; Inter-American Court of Human Rights, *Case of the Sawhoyamaxa Indigenous Community* (29 March 2006), Series C, No 146 at paras 150–152; Inter-American Court of Human Rights, *Case of the Indigenous Community Yakye Axa* (17 June 2005) Series C, No 125 at paras 161–162

(continued)

(continued)

1.3	Persons with disabilities between the ages of 25–64 years were less likely to be employed (59%) as compared to those without disabilities (80%). Stewart Morris et al., "A demographic, employment and income profile of Canadians with disabilities aged 15 years and over, 2017" (2017), Statistics Canada (online): https://www150.statcan.gc.ca/n1/pub/89-654-x/89-654-x2018002-eng.htm. Similarly, persons with disabilities had a lower median after-tax income as compared to persons without disabilities. These findings are consistent with the 2014 Longitudinal and International Study of Adults, which demonstrated that 23% of adults with a disability in Canada between 25 to 64 were low income, as compared to 9% of adults without a disability: Katherine Wall, "Low income among persons with a disability in Canada" (11 August 2017), online: Statistics Canada https://www150.statcan.gc.ca/n1/pub/75-006-x/2017001/article/54854-eng.htm. Estimates suggest that as many as 45% of the overall homeless population in Canada have physical and mental health disabilities. Homeless Hub, "Poverty" online: <http://homelesshub.ca/about-homelessness/education-training-employment/poverty>. Overall, 17% of people with disabilities in Canada live in households that are in core housing need, with people with disabilities living alone (36%) and in lone parent households (32%) experiencing even higher rates: Canada Mortgage and Housing Corporation, "Population with Disability in Housing Need by Sex and Living Arrangement" 31 March 2018), online: https://www.cmhc-schl.gc.ca/en/professionals/housing-markets-data-and-research/housing-data/data-tables/household-characteristics/population-with-disability-housing-need-by-sex-living-arrangement. In particular, 46% of women who reported being homeless also have a disability: DisAbled Women's Network of Canada, "More Than A Footnote: A Research Report on Women and Girls with Disabilities in Canada" (2019) at 69, online: https://dawncanada.net/media/uploads/news_data/news-279/more_than_a_footnote_research_report.pdf. Persons with disabilities experience long wait times for affordable accessible housing and some people—especially those labelled with intellectual disabilities, mental health disabilities, or cognitive disabilities—are frequently housed in institutions rather than in the community. For example, almost 600 people live in large provincial institutions for people with disabilities in Alberta, Manitoba, Nova Scotia, and Québec: Alzheimer Society of Canada et al., "Meeting Canada's Obligations to Affordable Housing and Supports for People with Disabilities to Live Independently in the Community, Submission to Committee on the Rights of Persons with Disabilities for the 17th Session" (March 2017), online: https://www.ohchr.org/Documents/Issues/Housing/Disabilities/CivilSociety/Canada-ARCHDisabilityLawCenter.pdf. In the Canadian province of Nova Scotia for example, people with disabilities have been on the waitlist for community based residential services for nearly 20 years: *MacLean v Nova Scotia* CA No 486952 (NSCA) (Factum of the Appellants at para 48), online: https://www.disabilityrightscoalitionns.ca/wp-content/uploads/2020/03/DRC-Factum-March-6-CA-486952.pdf. In the most dire situations, people have been held in psychiatric facilities as they wait: *MacLean v Nova Scotia (Attorney General) (No. 2)*, 2019 CanLII 145764 at paras 31–34, online: https://humanrights.novascotia.ca/sites/default/files/editor-uploads/maclean_et_al_decision.pdf. In the province of Ontario, it is common practice for people with intellectual and developmental disabilities to be added to a waitlist for supportive housing as soon as they turn 18, knowing that they will likely be waiting for decades: Jason Viau, "'It scares me:' Adults with disabilities still waiting decades for supportive housing" (CBC News, 12 August 2019), online: https://www.cbc.ca/news/canada/windsor/supportive-housing-delays-continue-1.5241574. Rebecca Casey, "Disability and unmet health care needs in Canada: A longitudinal analysis" (2015), 8 Disability Health Journal 173 at 177–181. Former Special Rapporteur on the Rights of Persons with Disabilities, Catalina Devandas Aguilar, observed that the Canadian health care system incorporates systemic ableism that denies needed supports and undermines quality of life for disabled persons: UNHRC "Report of the Special Rapporteur on the Rights of Persons with Disabilities" (17 December 2019) UN Doc A/HRC/43/41 at paras 26–38. During the COVID-19 pandemic, for example, this ableism has manifested in policies that discriminate against persons with disabilities by deprioritizing persons with disabilities for acute medical care as a triage strategy: ARCH Disability Law Centre, "ARCH Disability Law Centre's Submissions Regarding Ontario's Triage Protocol Draft dated July 7, 2020 Following Meetings with Bioethics Table" (1 September 2021), online: https://archdisabilitylaw.ca/resource/arch-submissions-detailing-its-position-on-the-inclusion-of-the-clinical-frailty-scale-in-ontarios-triage-protocol/

(continued)

(continued)

1.4	DisAbled Women's Network of Canada, "More Than A Footnote: A Research Report on Women and Girls with Disabilities in Canada" (2019) at 69, online: https://dawncanada.net/media/uploads/news_data/news-279/more_than_a_footnote_research_report.pdf at 75–76; DisAbled Women's Network of Canada, "Parliamentary Brief: Health Issues for LGBTQ2 People with Disabilities" (2 May 2019), online: https://www.ourcommons.ca/Content/Committee/421/HESA/Brief/BR10503899/br-external/DisAbledWomensNetworkOfCanada-e.pdf; AB Taylor et al., "Being Safe, Being Me 2019: Results of Canadian Trans and Non-binary Youth Health Survey" (2020), Vancouver, Canada: Stigma and Resilience Among Vulnerable Youth Centre, University of British Columbia, online: https://apsc-saravyc.sites.olt.ubc.ca/files/2020/03/Being-Safe-Being-Me-2019_SARAVYC_ENG-1.pdf. In a recent Canadian study, trans and non-binary respondents reported a variety of disability identities. These disability identities include psychiatric survivor, mad or person with a mental illness (43%), neurodivergent (30%), and disabled or living with a disability (19%). 45% of study respondents reported unmet health needs in the last year and 12% reported that they avoided seeking emergency room care because of their trans or non-binary identity: Trans PULSE Canada Team, "Trans PULSE Canada Report No 1 of 10" (10 March 2020), online: https://transpulsecanada.ca/wp-content/uploads/2020/03/National_Report_2020-03-03_cc-by_FINAL-ua-1.pdf; Testimony of Tyler White, Siksika Health Services & Suzanne L Stewart, Waakebiness-Bryce Institute for Indigenous Health and Associate Professor Dalla Lana School of Public Health, University of Toronto, Senate, Standing Committee on Legal and Constitutional Affairs, *Evidence*, 43–2 (2 February 2021), online: https://sencanada.ca/en/Content/Sen/Committee/432/LCJC/11ev-55129-e
1.5	In a Facebook post written in August 2020, Jennifer Turton-Molgat wrote: "My mom, the most positive, resilient, loving person I know has requested an assisted suicide. She wants to end her life, not because she is in pain or intolerable discomfort, but because she is being held prisoner in her long-term care home and has lost her will to live." Jennifer Turton-Molgat, Facebook Post dated 13 August 2020, online: https://www.facebook.com/jturtonmolgat/posts/10157236222225925; Arleen Reinsborough expressed a wish to apply for MAiD to avoid having to go to a long-term care home. Carmela Fragomeni, "Hamilton senior in unbearable pain wants assisted dying to save her from nursing home" (Toronto Star, 20 January 2019), online: https://www.thestar.com/news/canada/2019/01/20/hamilton-senior-in-unbearable-pain-wants-assisted-dying-to-save-her-from-nursing-home.html?rf
1.6	This same concern was raised by former Special Rapporteur on the Rights of Persons with Disabilities, Catalina Devandas Aguilar: UNHRC, "Visit to Canada: Report of the Special Rapporteur on the Rights of Persons with Disabilities" (19 December 2019) UN Doc A/HRC/43/41/Add.2 UN at paras 68–69 & 99

References

American Declaration on the Rights and Duties of Man 1948 (Resolution XXX, adopted by the Ninth International Conference of American States, Bogotá, Colombia). http://www.oas.org/en/iachr/mandate/Basics/american-declaration-rights-duties-of-man.pdf

Canada, Parliament. 2020. House of Commons. Standing Committee on Justice and Legal Affairs: Evidence of Heidi Janz. 2nd sess., 43rd Parliament, 2020. Committee Report 6. https://www.ourcommons.ca/DocumentViewer/en/43-2/JUST/meeting-6/evidence

Canada, Parliament. 2021. Senate. Standing Committee on Legal and Constitutional Affairs: Evidence Testimony of Jonathan Marchand. 2nd sess., 43rd Parliament, 2021. https://sencanada.ca/en/Content/Sen/Committee/432/LCJC/10ev-55128-e.

Canadian Association for Community Living. 2016. Assessing vulnerability in a system for physician-assisted death in Canada. https://static1.squarespace.com/static/56bb84cb01dbae77f988b71a/t/570811ebe32140e265b04aed/1460146676021/CACL+Vulnerability+Assessment+Apr+8+2016+-+Final.compressed.pdf

Convention on the Rights of Persons with Disabilities 2006, 2515 UNTS 3 (entered into force 3 May 2008, Convention ratified by Canada on 11 March 2010, Optional Protocol ratified by Canada on 3 December 2018). https://www.un.org/development/desa/disabilities/convention-on-the-rights-of-persons-with-disabilities.html.

Criminal Code, Statues of Canada 1985, c C-46. https://laws-lois.justice.gc.ca/eng/acts/c-46/

De Mestral, Armand, and Evan Fox-Decent. 2008. Rethinking the relationship between international and domestic law. *McGill L.J.* 53:573 at 577 & 581.

Eldridge v British Columbia (Attorney General) [1997] 3 SCR 624 at para 56. https://scc-csc.lexum.com/scc-csc/scc-csc/en/item/1552/index.do

Government of Canada. n.d. Legislative Background: Bill C-7: Government of Canada's Legislative Response to the Superior Court of Quebec *Truchon* Decision. https://www.justice.gc.ca/eng/csj-sjc/pl/ad-am/c7/p2.html

Government of Canada. 2021. Response of Canada to the Joint Communication from the Special Rapporteur on the Rights of Persons with Disabilities, the Independent Expert on the Enjoyment of all Human Rights by Older Persons, and the Special Rapporteur on Extreme Poverty and Human Rights. https://spcommreports.ohchr.org/TMResultsBase/DownLoadFile?gId=36253. Accessed 29 March 2022.

Inter-American Court of Human Rights. 2006. *Case of the Sawhoyamaxa Indigenous Community* (29 March 2006), Series C, No 146. https://www.corteidh.or.cr/docs/casos/articulos/seriec_146_ing.pdf. Accessed 29 March 2022.

Inter-American Commission on Human Rights. 2003. Merits Report No 97/03, Case 11.193, *Gary T Graham (Shaka Sankofa)* (United States). http://www.cidh.org/annualrep/2003eng/usa.11193.htm. Accessed 29 March 2022.

Inter-American Commission on Human Rights. 2018. Merits Report No 121/18, Case 10.573, *José Isabel Salas Galindo and others* (United States). https://www.oas.org/en/iachr/media_center/PReleases/2018/258.asp. Accessed 29 March 2022.

International Covenant on Civil and Political Rights 1966, 999 UNTS 171, Can TS 1976 No 47 (entered into force 23 March 1976, Convention ratified by Canada May 1976). https://www.ohchr.org/en/instruments-mechanisms/instruments/international-covenant-civil-and-political-rights

Quebec (Attorney General) v 9147-0732 Québec Inc [2020] SCC 32 at para 30. https://scc-csc.lexum.com/scc-csc/scc-csc/en/item/18529/index.do

R v Hape [2007] SCC 26 at para 53. https://scc-csc.lexum.com/scc-csc/scc-csc/en/item/2364/index.do

Several Cases of Concern Medical Assistance in Dying & the Inducements of Social Deprivation. 2021. https://static1.squarespace.com/static/56bb84cb01dbae77f988b71a/t/615716183c30976483303da6/1633097241515/Updated+Cases+of+Concern+.pdf. Accessed 29 March 2022.

Sullivan, Ruth. 2014. Sullivan on the Construction of Statutes 6th ed. *LexisNexis* 567–568.

Thomas, Aislinn. 2021. Letter to the Senate. https://sencanada.ca/content/sen/committee/432/LCJC/Briefs/Brief_AislinnThomas_e.pdf Accessed 29 March 2022.

Truchon v Canada (AG) [2019] QCCS 3792.

Universal Declaration of Human Rights 1948, G.A. res. 217A (III), U.N. Doc A/810. https://www.un.org/en/about-us/universal-declaration-of-human-rights

United Nations (UN). 2021. Mandates of the Special Rapporteur on the rights of persons with disabilities; the Independent Expert on the enjoyment of human rights by older persons; and the Special Rapporteur on extreme poverty and human rights. UN Doc OL CAN 2/202. https://spcommreports.ohchr.org/TMResultsBase/DownLoadPublicCommunicationFile?gId=26002. Accessed 29 March 2022.

United Nations Committee on the Rights of Persons with Disabilities. 2017. Concluding Observations on the Initial Report of Canada. UN Doc CRPD/C/CAN/CO/1. https://undocs.org/CRPD/C/CAN/CO/1. Accessed 29 March 2022.

United Nations Committee on the Rights of Persons with Disabilities. 2018. General comment No. 6 on equality and non-discrimination CRPD/C/GC/6. https://undocs.org/CRPD/C/GC/6. Accessed 29 March 2022.

United Nations Human Rights Committee. 1989. General Comment No 18 ICCPR General Comment No 18. https://www.refworld.org/docid/453883fa8.html. Accessed 29 March 2022.

United Nations Human Rights Committee. 2009. Concluding Observations of the Human Rights Committee: Netherlands UN Doc CCPR/C/NLD/CO/4. https://documents-dds-ny.un.org/doc/UNDOC/GEN/G09/445/62/PDF/G0944562.pdf?OpenElement. Accessed 29 March 2022.

United Nations Human Rights Committee. 2019a. Concluding observations on the fifth periodic report of the Netherlands" UN Doc CCPR/C/NLD/CO/5. https://documents-dds-ny.un.org/doc/UNDOC/GEN/G19/249/80/PDF/G1924980.pdf?OpenElement. Accessed 29 March 2022.

United Nations Human Rights Committee. 2019b. General Comment No 36, Article 6 (Right to Life) CCPR/C/GC/36. https://documents-dds-ny.un.org/doc/UNDOC/GEN/G19/261/15/PDF/G1926115.pdf?OpenElement. Accessed 29 March 2022.

United Nations Human Rights Council. 2019. Visit to Canada: Report of the Special Rapporteur on the Rights of Persons with Disabilities. UN Doc A/HRC/43/41/Add.2 UN. https://documents-dds-ny.un.org/doc/UNDOC/GEN/G19/348/81/PDF/G1934881.pdf?OpenElement. Accessed 29 March 2022.

Kerri Joffe is a human rights lawyer at ARCH Disability Law Centre. She has been involved in disability rights litigation at various tribunals and courts, including the Supreme Court of Canada. Kerri has presented law reform and policy submissions about disability law issues to legislative committees, governments, administrative bodies and the United Nations Committee on the Rights of Persons with Disabilities. She has authored law reform reports for the Law Commission of Ontario, the Canadian Human Rights Commission and the Government of Canada. Kerri has delivered extensive public legal education to diverse disability communities, and has guest lectured on many disability rights topics. Before joining ARCH, Kerri worked on housing rights, social assistance programs, immigration and refugee issues, human trafficking, and as a law clerk to Superior Court judges. Kerri holds degrees in law and social work from McGill University.

Roberto Lattanzio is the Executive Director of ARCH Disability Law Centre. Robert has been Executive Director since 2015, and first joined ARCH in 2003. He has acted as legal counsel in test case litigation at all levels of court, including the Supreme Court of Canada, and has made law reform submissions to various levels of government, committees, and administrative bodies. Robert has been appointed to numerous advisory committees, and has written and guest-lectured on diverse topics including equality and human rights law, administrative law, education law, capacity and supported decision making, disability rights, legislative reform, and social science evidence. He also has delivered extensive rights education training to communities of persons with disabilities and continuing legal education to the legal profession. Robert received his LL.B and B.C.L. law degrees with distinction from McGill University in 2003. Robert has a long standing interest in disability rights and social justice, with extensive work and volunteer experience within disability communities and other equity seeking communities.

Part VI
Social and Cultural Issues of the MAID Programme

Chapter 24
Examining Indigenous Perspectives on Medical Assistance in Dying (MAID)

Dwight Newman and Gabrielle Robitaille

Abstract This chapter draws attention to Indigenous perspectives on MAID policies, reviewing the public record on Indigenous perspectives introduced into parliamentary discussions and debates. That overview identifies a number of qualitatively significant issues raised. The chapter goes on to show that Health has tended to set aside those issues in its ongoing review of MAID implementation and that real engagement with Indigenous peoples on MAID has not even begun. The chapter suggests that the broader concerns raised within Indigenous perspectives on MAID do not sit easily with narrower evidence-based approaches to health policy but that normative and legal instruments in relation to Indigenous rights do have implications of a new need to engage with these perspectives in ongoing review of MAID. There will need to be further attention to Indigenous perspectives on MAID.

Keywords Medical assistance in dying (MAID) · Indigenous peoples · Indigenous rights · United Nations Declaration on the Rights of Indigenous Peoples (UNDRIP) · Indigenous health

24.1 Introduction

Indigenous communities have distinctive perspectives on medical assistance in dying (MAID) policies, informed both by different worldviews and cultural values and by different lived experiences of health care systems and colonialism generally. To take just one example as an initial illustration of the point, some such different perspectives came to the fore in the original parliamentary debate on Bill C-14 in 2016. This was legislation passed through Parliament to facilitate physician-assisted

D. Newman (✉) · G. Robitaille
University of Saskatchewan, Saskatoon, Canada
e-mail: dwight.newman@usask.ca

G. Robitaille
e-mail: gabrielle.robitaille@usask.ca

J. Kotalik and D. W. Shannon (eds.), *Medical Assistance in Dying (MAID) in Canada*, The International Library of Bioethics 104,
https://doi.org/10.1007/978-3-031-30002-8_24

suicide almost simultaneously with a suicide crisis on First Nations reserves. Robert-Falcon Ouellette, an Indigenous Liberal MP from Winnipeg, would speak against and vote against his own government's legislation based both on his spiritual beliefs and based on concerns about the impact of the legislation in the context of communities facing suicide crises (Barrera 2016).

In the context of Indigenous peoples in another part of the world, with their own distinctive cultures and worldviews, issues related to distinctive Indigenous perspectives on MAID have received some attention in debates of recent years on Voluntary Assisted Dying (VAD) in Australia, even while many of those debates have proceeded with very little perspective by policymakers on reasons for views held by the public generally (Kresin et al. 2021). There had been some early attention to issues of potential impacts of such legislation on Indigenous communities (e.g. Collins and Brennan 1997), and debates of recent years have seen ongoing reference to these issues, including through the intervention of key Indigenous leaders such as Senator Pat Dodson who spoke out in 2019 about his concerns that the adoption of VAD legislation could make Indigenous Australians reluctant to access health care services (Walker 2019).

In this chapter, we overview the public record on Indigenous perspectives on MAID legislation in Canada, focusing particularly on recorded comments within the discussions of the House of Commons and the Senate on the two legislative steps, Bill C-14 in 2016 and Bill C-7 in 2020–21. A review of this record in Part II of this chapter shows qualitatively significant comments on various distinctive perspectives and implications of MAID legislation in the context of Indigenous communities. As we discuss in Part III, our review suggests that these discussions have revealed issues that have been flagged but then largely been shelved, with ongoing review of the legislation being content simply to note lack of data on Indigenous contexts and to make relatively non-committal statements of the need at some point to engage in more consultation with Indigenous communities. In our brief analytical comment in Part IV, we suggest that the effective exclusion of Indigenous perspectives may be partly connected to efforts at evidence-driven policy that does not readily have room for values-based perspectives or Indigenous worldviews—thus raising issues of to what degree current approaches will ever have room for the sort of consultation with Indigenous communities to which they refer—and we raise the prospect of future issues concerning the intersection of MAID legislation with key normative instruments on Indigenous rights, notably the *United Nations Declaration on the Rights of Indigenous Peoples (UNDRIP).*

24.2 Discussions in the House of Commons and the Senate

As both iterations of the MAID legislation moved through the discussions of the House of Commons and the Senate on Bill C-14 in 2016 and Bill C-7 in 2020–21, various issues were raised as to the potential impact upon Indigenous communities and individuals. The discussions highlighted concerns specific to implementing a MAID program within Indigenous communities.

The implications of inadequate health care services as it pertains to MAID's introduction into these communities were discussed, particularly the insufficiency of health care services in remote and Indigenous communities as well as systemic racism within healthcare. Discussions highlighted that there are cultural issues in the implementation of MAID programs in these communities, including vulnerabilities that MAID could exacerbate and complexities facing a culturally safe introduction of MAID programs. Broader issues surrounding MAID's introduction were also raised, such as the widening of the jurisdictional gaps in Indigenous health care and the implications of inadequate consultation in Indigenous communities.

24.2.1 Health Care Inequities

A concern evident throughout discussions for both Bills was the reality of insufficient health services in Indigenous communities, especially remote and Northern communities. A lack of access to various forms of health care is problematic when the ethical delivery of MAID is contingent upon consent. The lead physician for what is now the Indigenous Health Program in Alberta spoke to this point during the Senate's pre-study of Bill C-7, where he stated that "agency requires health systems … to provide the infrastructure for other types of health and wellness" (Canada, Parliament, Standing Senate Committee on Legal and Constitutional Affairs, *Evidence*, 43rd Parl, 2nd sess, 26 November 2020, 5:98, Dr. Alika Lafontaine).

One viewpoint that has been presented in these debates is that Indigenous communities would be vulnerable to forms of coerced consent due to these inadequacies. Parliamentarians have heard a concern that those who would otherwise not seek MAID in Indigenous communities would do so to ease suffering caused by the insufficient provision of palliative care or other health services (Canada, Parliament, *House of Commons Debates*, 43rd Parl, 2nd sess, 2 December 2021, 2235, Mr. Brad Vis). The worry in expanding access to MAID services in the face of the inequities in access across the country arose from the sole emphasis upon individual choice in the legislation when "individuals will have unequal options to choose from" (Canada, Parliament, *Debates of the Senate*, 43rd Parl, 2nd sess, 16 February 2021, 1005, Kim Pate).

The possibility of higher uptake in MAID services in Indigenous communities due to these limited options is complicated by social determinants of health. Alongside the inequity in the services themselves, representatives have raised the concurrent issues of housing, clean water, food, and other support services that lead to worse health outcomes in Indigenous communities (Canada, Parliament, Standing Senate Committee on Legal and Constitutional Affairs, *Evidence*, 43rd Parl, 2nd sess, 3 February 2021, 12:67, Dr. Lisa Richardson). The competing causes of health issues faced in Indigenous communities creates a sensitive landscape for the introduction of a service such as MAID.

24.2.2 Implications of Systemic Racism

The concern about a potential of higher uptake in MAID services in Indigenous communities has also been connected to the systemic racism within the Canadian health care system. Indigenous representatives and senators felt a general unease with the introduction of these services in the face of abuses of medical procedures in Indigenous communities or issues such as forced sterilization (Canada, Parliament, Standing Senate Committee on Legal and Constitutional Affairs, *Evidence*, 43rd Parl, 2nd sess, 3 February 2021, 12:66–12:67, Dr. Lisa Richardson). Many expressed caution in the possibility that such issues would permeate into MAID's implementation considering the delicacies inherent in the practice.

The medical abuses that have taken place in Indigenous communities have also led to a distrust in medical services more generally. A further potential consequence flagged throughout the discussions was that the introduction of MAID would heighten distrust in Indigenous communities and exacerbate barriers in the provision of medical services due to that distrust. Transparency was stated to be essential to prevent the further deterioration of Indigenous peoples' relationship with governmental health care (Canada, Parliament, Standing Senate Committee on Legal and Constitutional Affairs, *Evidence*, 43rd Parl, 2nd sess, 2 February 2021, 11:82, Suzanne L. Stewart).

24.2.3 Tensions with Suicide Prevention

Numerous cultural issues in implementation were raised throughout the legislation's development. The discussion surrounding the health care challenges faced in Indigenous communities was heavily connected to the suicide epidemic faced by many of those communities. Mr. Robert-Falcon Ouellette cited this as a significant reason why he was opposed to the implementation of Bill C-14, drawing on his own and his community's experiences. He was particularly concerned that young people witnessing MAID being utilized would instigate further suicidality in Indigenous communities (Canada, Parliament, *House of Commons Debates*, 42nd Parl, 1st sess, 2 May 2016, 1830–1835). Bill C-7's introduction of mental illness as a sole qualifying condition for MAID led to similar concerns. Parliamentarians have also heard Indigenous representatives express worries that the implementation would be counterproductive to the ongoing suicide prevention efforts in Indigenous communities, and that it would send the message that some "should receive suicide assistance while others suicide prevention" (Canada, Parliament, *House of Commons Debates,* 43rd Parl, 2nd sess, 2 December 2020, 2235–2240, Mr. Brad Vis).

24.2.4 Cultural Conscience Objections

The implementation of MAID programs in Indigenous communities is further complicated by the cultural and spiritual objections to MAID in Indigenous communities. While also being concerned with the suicide epidemic, Mr. Robert-Falcon Ouelette also opposed Bill C-14 due to being spiritually opposed to the practice (Canada, Parliament, *House of Commons Debates*, 42nd Parl, 1st sess, 2 May 2016, 1830–1835). An open letter signed by 15 representatives of different Indigenous Nations written by Tyler White expressed that Bill C-7 "goes against many of [their] cultural values, belief systems, and sacred teachings" (Canada, Parliament, *House of Commons Debates,* 43rd Parl, 2nd sess, 2 December 2020, 2235–2240, Mr. Brad Vis). A Denesuline Elder appearing before the Senate committee in 2021 reflected sentiments of the importance of Denesuline spiritual law, stating that the legislation had not been designed in consideration of these belief systems (Canada, Parliament, Standing Senate Committee on Legal and Constitutional Affairs, *Evidence*, 43rd Parl, 2nd sess, 2 February 2021, 11:78–11:81, François Paulette).

Alongside the broad issue of MAID being antithetical to many Indigenous groups' spiritual beliefs, there have also been associated issues raised concerning conscientious objections to participating in a MAID program. The representative of the Canadian Indigenous Nurses Association requested that the legislation include explicit protections for Indigenous health care workers to refrain from providing MAID services for cultural or spiritual reasons. She stated that the context of many Indigenous communities, especially remote or Northern communities, is that the few nurse practitioners authorized to participate in MAID are also members of that community. There is thus the potential that these Indigenous nurse practitioners face either punishment from their workplace from refraining from the procedure or punishment from their communities for "taking a life" (Canada, Parliament, Standing Senate Committee on Legal and Constitutional Affairs, *Evidence*, 43rd Parl, 2nd sess, 26 November 2020, 5:96–5:97, Marilee Nowgesic). Though the issue of inadequate protection within the legislation for moral objections was raised by numerous communities, the cultural context of Indigenous communities contains additional nuances.

24.2.5 Cultural Competency

Due to many of the other factors concerning the implementation of MAID in the Indigenous context, both representatives and Senators have referenced the importance of cultural competency in implementing MAID. In remote and Northern communities, those likely to be eligible or considered for MAID are often unilingual Elders, and there are thus nuanced issues in communicating significant concepts such as consent or the concept of MAID more generally into languages such as Inuktitut (Canada, Parliament, *Debates of the Senate*, 43rd Parl, 2nd sess, 16 February 2021,

1014, Dennis Glen Patterson). The language barrier is further complicated by the understandings of death and life within Indigenous communities, leading to difficulties in differentiating MAID from suicide and adequately communicating that distinction.

Language and cultural barriers led to pushes for the legislation itself to have specific provisions of implementing these services in different cultural contexts. Implementing "culturally safe health care" within the MAID legislation and the distribution of health care was highlighted as a significant consideration for the legislation when there is currently a lack of "cultural competence" within medicine in Indigenous communities (Canada, Parliament, Standing Senate Committee on Legal and Constitutional Affairs, *Evidence*, 43rd Parl, 2nd sess, 26 November 2020, 5:25, Senator McCallum). It has been emphasized that the introduction of MAID must come within a culturally safe environment, which should include concurrent ceremonies and other considerations (Canada, Parliament, Standing Senate Committee on Legal and Constitutional Affairs, *Evidence*, 43rd Parl, 2nd sess, 26 November 2020, 5:95, Dr. Cornelia Wieman).

24.2.6 Jurisdictional Issues

The implementation of MAID within Indigenous communities also highlighted an ongoing issue of jurisdiction in the provision of health care for Indigenous individuals and communities. Though the provinces generally have jurisdiction over the administration of health care, this is complicated by the federal government's jurisdiction over matters relating to Indigenous peoples. There was concern that the ongoing issue of jurisdiction between the provinces and the federal government would be worsened by the MAID legislation.

Senators raised the issue that in the midst of the amendments posed by Bill C-7, there continued to be jurisdictional disputes over which government "should fund the necessary health services for Indigenous persons with disabilities" (Canada, Parliament, Standing Senate Committee on Legal and Constitutional Affairs, *Evidence*, 43rd Parl, 2nd sess, 1 February 2021, 10:75, Yvonne Boyer). In discussion for both Bills, it was suggested that these issues should be resolved prior to the implementation of MAID to avoid exacerbating the vulnerabilities and issues in health care administration in Indigenous communities. This is particularly due to the delicacies and nuance needed in administering the MAID program, and that the continued dispute would worsen the health care inequities that pose challenges to MAID implementation.

24.2.7 Inadequate Consultation

The most common grievance throughout the legislation's development was the lack of adequate consultation with Indigenous peoples, especially considering their rights to consultation. At the initial stages of Bill C-7 in January 2020, Parliament undertook community consultations to gauge opinions on the prospective amendments to the MAID legislation. There was only one roundtable meeting that was specifically oriented to Indigenous representatives, which included 10 participants and did not include any Métis or Inuit representatives. Though there was a generalized online consultation beyond these roundtable meetings, many Indigenous communities remained not engaged. Many remote or Northern communities were stated to not "even know about the law" and its developments (Canada, Parliament, *Debates of the Senate*, 43rd Parl, 2nd sess, 16 December 2020, 760–762, Mary Jane McCallum).

Considering the aforementioned issues and nuances that MAID poses to Indigenous communities and individuals, developing the legislation without adequate consultation to those communities was stated to be an oversight throughout discussions on both Bills. The potential legal implications are also significant in terms of Canada's obligations to Indigenous peoples. Senator Dennis Glen Patterson emphasized that the lack of consultation in Nunavut in particular violates the terms of their modern treaty that requires Inuit participation in the development of social and cultural policies (Canada, Parliament, *Debates of the Senate*, 43rd Parl, 2nd sess, 11 February 2021, 945–946). Other representatives noted that a lack of consultation on a major medical development that could affect "traditional life practices" could also be a violation of their domestic and international rights as recognized in commitments such as the *United Nations Declaration on the Rights of Indigenous Peoples* (Canada, Parliament, Standing Senate Committee on Legal and Constitutional Affairs, *Evidence*, 43rd Parl, 2nd sess, 2 February 2021, 11:115–116, Dr. Janet Smylie). The impacts of the legislation upon Indigenous communities have thus been suggested to pose broader legal questions.

24.2.8 Ongoing Submissions at the Special Joint Committee

The work of the Special Joint Committee on Medical Assistance in Dying, which brings together Members of Parliament and Senators, continued past specific decisions on legislative amendments. Some Indigenous witnesses appearing before this Committee have made reference to lack of past engagement with Indigenous peoples. For example, appearing on November 4, 2022, Neil Belanger, the Chief Executive Officer of Indigenous Disability Canada, made reference to Health Canada's own reports as showing there had not been engagement with Indigenous peoples or perspectives, and he asked "why Canada, this committee or anyone would presume to have the authority to make recommendations or implement actions for the expansion of state-assisted death and suicide for indigenous children, indigenous persons with disabilities and indigenous persons with mental illness as a sole condition, without first engaging the very people whom these proposed changes will target" (Canada, Parliament, Special Joint Committee on Medical Assistance in Dying, *Evidence*, 44th Parl., 1st sess., 4 November 2022, Neil Belanger).

24.3 Treatment of Indigenous Perspectives on MAID in Ongoing Review Contexts

The various rounds of discussion and debate on MAID legislation have thus seen various distinctive perspectives of and impacts on Indigenous communities raised. However, there has been very little detailed study of MAID and Indigenous communities in Canada, something both referenced in and reflected by initial governmental reviews of MAID.

Health Canada's *First Annual Report on Medical Assistance in Dying in Canada, 2019* did not even mention Indigenous communities or individuals (Health Canada 2020), and the *Second Annual Report on Medical Assistance in Dying in Canada, 2021* mentioned Indigenous communities only in a brief reference in the conclusion calling for the collection of data (Health Canada 2021: 17). The *Third Annual Report on Medical Assistance in Dying in Canada, 2021* did commit to the collection of Indigenous-specific data commencing in 2023 and reporting on that commencing in 2024 (Health Canada 2022b: 14). That last report also saw a direct recognition of failings on Indigenous engagement and need to do better: "Health Canada recognizes the need to enhance engagement with Indigenous peoples on MAID and will undertake engagement in partnership with and respecting the priorities of Indigenous peoples to better understand distinctions-based views and perspectives about MAID, including on access to culturally safe MAID services, and potential avenues for conducting needed research" (Health Canada 2022b: 15).

How Health Canada sees engagement with Indigenous peoples on MAID, though, contains some contradictory statements. In the portion of the *Third Annual Report* reviewing the history of MAID legislation, there was reference to "broad consultations with […] Indigenous peoples" in advance to the 2021 amendments (Health Canada 2022b: 11), something of which there is no evidence, other than one footnote indicating that there had been "one roundtable focused on Indigenous perspectives to inform the legislative amendment process" (Health Canada 2022b: 10). As noted earlier in our chapter, that particular roundtable involved extremely limited participation from only some Indigenous perspectives, so that it could serve as evidence of "broad consultations" is questionable and raises genuine questions about the accuracy of these elements of the *Third Annual Report*. The May 2022 *Final Report of the Expert Panel on MAiD and Mental Illness*, reflecting a more independent review related to the same processes on MAID legislation, stated that "engagement with Indigenous peoples in Canada concerning MAiD has yet to occur" (Health Canada 2022a: 35), a statement consistent with the Minister of Health's statement in

October 2022 referring to Health Canada having reached out to national Indigenous organizations to begin discussing potential engagement (Minister of Health 2022: 5–6).

These ongoing review contexts have not seen any deep engagement with Indigenous issues in relation to MAID. If anything, the Health Canada *Annual Reports* have tended to initially ignore and then to skirt over the broader set of issues raised by Indigenous communities in relation to MAID, finally simply calling for data collection. Here, Health Canada portrays Bill C-7 as having provided an innovative framework: "To address concerns regarding the potential for some vulnerable groups to be over-represented in MAID cases, or possibly denied access, Bill C-7 requires that data be collected with respect to race, Indigenous identity and disability. Expanding the MAID monitoring regime in this way will help to determine the presence of any inequalities or disadvantages in Canada's MAID regime based on these or other characteristics" (Health Canada 2021: 17). The statement ultimately reflects quite technocratic engagement with data about overrepresentation and/or denial of access, suggesting a focus on quantitative data about use of MAID rather than any sort of broader engagement with the more qualitatively complex array of issues raised in parliamentary discussions and debates leading up to the legislation.

The May 2022 *Expert Panel* report in the context of extension of MAID to contexts of mental illness does have a slightly longer discussion in relation to Indigenous issues. It references Indigenous peoples' unique perspectives on death, past and ongoing harms of colonialism that may pressure some Indigenous individuals into choosing MAID, non-homogeneity of viewpoints that also include some Indigenous individuals choosing MAID for similar reasons as non-Indigenous individuals, and broader issues of equity and self-determination (Health Canada 2022a, b: 35). This *Expert Panel* thus includes amongst its recommendations one recommendation specific to Indigenous contexts: "Consultation between health regulatory bodies in each province and territory with First Nations, Métis, and Inuit peoples must aim to create practice standards with respect to MAiD MD-SUMC, and MAiD more generally, that incorporate Indigenous perspectives and are relevant to their communities" (Health Canada 2022a, b: 73).

The May 2022 *Expert Panel* reflects the use of experts external to Health Canada and thus injects some new perspective. In the absence of it, one would surely see Health Canada's engagement with Indigenous issues in the annual reviews as quite anemic. If the *Expert Panel* recommendations foster actual engagement by health regulatory bodies with Indigenous peoples, there may be a new path achieved. However, non-committal references to consultation are nonetheless limited in effect. The recommendation itself does not have legal force, has not been adopted into legislative frameworks, and could yet not be realized. The recognition in 2022 of an essentially complete lack of engagement thus far does not bode well, even while there must be hopes and efforts for new paths together.

24.4 Analysis and Conclusions

Considering these materials, we would raise two points, one concerning the reasons why things have developed as they have and the other concerning normative frameworks that may affect matters going forward. First, in a context where MAID legislation has now been assumed and evidence-based approaches to specific questions of implementation are the order of the day, the broader array of complex questions raised as Indigenous perspectives on MAID do not fit easily into ongoing discussions. There are fundamental questions raised about a system when the system itself is presumed. There are basic questions raised about MAID when MAID is now assumed. There are questions raised about who is to be in charge, which those presently in charge are relatively disinclined to pursue. Even full-fledged engagement of regulatory authorities with Indigenous peoples raises uncomfortable questions about how to accommodate and/or incorporate philosophically different values-based views on MAID and Indigenous worldviews reflecting different philosophical values and different perspectives on Western biomedicine. In some senses, Indigenous perspectives that have been part of MAID discussions and debates to some degree have been amongst those most challenging for those leading the implementation of MAID legislation, making it easier to shelve these issues that do not fit easily into technocratic or evidence-based approaches.

Second, however, those approaches are unlikely to remain acceptable or even permissible. The 2007 *UNDRIP*, has now been incorporated into Canadian federal law in certain ways by the 2021 *United Nations Declaration on the Rights of Indigenous Peoples Act*, S.C. 2021, c. 14 (*UNDRIPA*). The *UNDRIP* is a key normative framework for rights both of Indigenous peoples (as collective entities) and Indigenous people (Indigenous individuals), with a complex interplay of individual and collective rights present in the instrument (Newman 2021). The health rights within the *UNDRIP* still need more attention than they have received, but they include both rights of Indigenous peoples in relation to self-determination in a health context, including to maintenance of distinctive health traditions, and rights of Indigenous individuals, within this context, to the highest possible attainable standard of health. The *UNDRIPA* commits Canada to pursuing, while consulting and cooperating with Indigenous peoples, "all necessary measures" to achieve the consistency of all federal legislation with the *UNDRIP*. That statutory commitment has implications for all legislation, including for the MAID legislation, which may thus legally need to be subject to further review in relation to the *UNDRIPA* and the *UNDRIP.* In October 2022, the Minister of Health acknowledged that the *UNDRIPA* might need to inform future engagement and consultation processes, though without any acknowledgment of the specific requirements within the *UNDRIPA* to achieve consistency of Canadian legislation with the *UNDRIP* (Minister of Health 2022: 6). In addition to other factors tending toward greater Indigenous engagement over time, Canada's commitment to this international normative framework can imply legal issues over time that push for an engagement with the complex Indigenous perspectives on MAID that have been present in discussions and debate but that have not received sufficient attention or sufficient engagement. The need for fuller engagement with Indigenous peoples and Indigenous worldviews may raise broader issues about MAID and the ways in which it is implemented and reviewed, with Indigenous issues having relevance in various ways to everyone.

References

Barrera, Jorge. 2016. Indigenous liberal MP voting against government's assisted dying bill. *APTN News*, 20 April 2016. https://www.aptnnews.ca/national-news/indigenous-liberal-mp-ouellette-voting-against-governments-assisted-dying-bill/

Canada, Parliament, *Debates of the Senate*, 43rd Parl, 2nd sess, 2020–21.

Canada, Parliament, *House of Commons Debates*, 42nd Parl, 1st sess, 2016.

Canada, Parliament, *House of Commons Debates*, 43rd Parl, 2nd sess, 2020–21.

Canada, Parliament, Standing Senate Committee on Legal and Constitutional Affairs, *Evidence*, 43rd Parl, 2nd sess, 2020–21.

Collins, John J., and Frank T. Brennan. 1997. Euthanasia and the potential adverse impacts for northern territory Aborigines. *The Lancet* 349 (9069): 1907–1908.

Health Canada. 2020. *First annual report on medical assistance in dying in Canada, 2019* (June 2020).

Health Canada. 2021. *Second annual report on medical assistance in dying in Canada, 2020* (June 2021).

Health Canada. 2022a. *Final report of the expert panel on MAiD and mental illness* (May 2022).

Health Canada. 2022b. *Third annual report on medical assistance in dying in Canada, 2021* (July 2022).

Kresin, Tracee, et al. 2021. Attitudes and arguments in the voluntary assisted dying debate in Australia: What are they and how have they evolved over time? *International Journal of Environmental Research and Public Health* 18 (23): 12327.

Minister of Health. 2022. Letter of Minister of Health to Joint Chairs of the Special Joint Committee on Medical Assistance in Dying (20 October 2022).

Newman, Dwight. 2021. Peoples and persons in the UNDRIP. In *Struggles for recognition: Cultural pluralism and rights of minorities*, ed. Oscar Pérez de la Fuente et al., 239–247. Madrid: Dykinson.

Walker, Jamie. 2019. Pat Dodson's warning on legalised Euthanasia. *The Weekend Australian*, 5 October 2019.

Dwight Newman, KC is Professor of Law and Canada Research Chair in Indigenous Rights in Constitutional and International Law at the University of Saskatchewan. He completed his doctoral studies at Oxford University in Legal Philosophy and holds seven degrees in all, in economics, finance, theological studies/history of Christianity, law, and legal philosophy. He has published over a hundred articles or book chapters and fifteen books on various topics in constitutional law, international law, legal theory, and Indigenous rights law. He has been a recent visitor at Cambridge, Oxford, Princeton, Université de Montréal, and University of Western Australia. He is a member of the bars of Ontario and Saskatchewan and carries on a selective practice in constitutional law.

Gabrielle Robitaille is a J.D. Candidate and Research Assistant at the University of Saskatchewan College of Law. She is originally from Caronport, Saskatchewan. Prior to her legal studies, she studied interdisciplinary Humanities with a Philosophy concentration at Briercrest College. She has previously published in *Current Developments in Arctic Law*.

Chapter 25
Patient-Physician Relationship and MAID: Trust and Autonomy

Louisa Pedri

Abstract Canada's Medical Assistance in Dying (MAID) Law and practice will likely have a negative impact on the therapeutic value of the patient-physician relationship (PPR). It is affecting the keystone of the relationship, trust. As the number of MAID cases continue to grow and MAID is becoming "normalized," those in the population who would never wish to receive MAID are losing trust in the profession and its members because it is perceived that the profession has abandoned some of its core values. Respect for a patient's autonomy, which has been supporting the patient-physician relationship for at least 50 years is being overpowered by a new and radical autonomy, the role of which is to uphold an individual's legal right (under certain conditions) to access MAID, a medical intervention that the profession is responsible for carrying out. This new element of duty of practice is not in accord with the physicians' code of ethics and with most patients' expectations. If the PPR is to maintain its relational strength and optimal therapeutic value, the newly imposed physician's role will have to be seriously examined and addressed by the professional body.

Keywords Patient-physician relationship · Patient-caregiver relationship · Trust · Autonomy · Radical autonomy · Medical assistance in dying · Assisted suicide · Voluntary euthanasia · Standard of care · Ethics · Ethical

25.1 Introduction

Canada legalized medically assisted suicide (AS) and voluntary euthanasia (VE) under the umbrella term, Medical Assistance in Dying (MAID) in 2016 (Government of Canada 2022). In 2021, over 10,000 people received MAID and the total number of deaths attributed to MAID stand over 30,000 (Health Canada 2022).

L. Pedri (✉)
Centre for Health Care Ethics, Lakehead University, 955 Oliver Road, Thunder Bay, Ontario P7B 5E1, Canada
e-mail: lpedri@lakeheadu.ca

J. Kotalik and D. W. Shannon (eds.), *Medical Assistance in Dying (MAID) in Canada*, The International Library of Bioethics 104,
https://doi.org/10.1007/978-3-031-30002-8_25

In this paper, the author takes the position that the patient-physician relationship (PPR), an ethical relationship of dynamic exchange of energies between physicians and patients, likely will be affected as members of the medical profession in Canada continue to be involved in causing death to persons who qualify for MAID. This phenomenon will be explored from several perspectives. It will look at how the PPR's core values of autonomy and trust were expressed pre-MAID times, and how, within the landscape of MAID, they have been modified by a slight semantic shift. Finally, the author will report on some early MAID experiences of Canadian clinicians, which poignantly supports the assertion that the PPR has been negatively affected. This impact may necessitate a paradigm shift within the profession's ethos to maintain the PPR's value within the healing arts. While the term PPR will be used throughout, it is recognized that much of the discussion will be also relevant to the patient-nurse practitioner.

25.2 PPR Trust—An Art in the Making

Scholarship is not lacking in what constitutes this unique relationship between a physician and a patient. It has been evolving since Hippocrates (c. 460 BCE–c. 375 BCE) first defined the role of the physician toward the patient. The striving for the ideal and successful PPR has necessitated a concerted effort by philosophers, bioethicists and the medical profession to periodically make proposals informed by emerging norms and principles, the dynamics of culture, societal expectations, advances in scientific knowledge, availability of technology, human rights issues, and most importantly, the application of ethical scrutiny (Pellegrini 2017).

One singular fact that ought to motivate the profession to continually strive for its perfection is the realization that in and of itself, the PPR can significantly augment the prescribed medical care and advance the health of the patient (ABIM Foundation 2021; Olaisen et al. 2020; Steward 1995).

Various PPR models have emerged, evolved, were combined, recombined, re-labelled and some just phased out (Marcum 2008). The best-known framework was formulated by Emanuel and Emanuel (1992), known as the Four Models of the Physician–Patient Relationship.

On one end of the spectrum, these authors describe the historic, paternalistic model where the patient accepts the physician's professional authority exercised with the care of a good father, desiring the best for his children, who follows his advice and directions. On the other end is the informative model where physicians provide their patients with all the available facts about the disease and treatment, and it is the responsibility of the patient to select the intervention which the physician will execute. In between, are the interpretative and deliberative models, each taking into account, in varying degrees of weight, the physician and patient goals, the physician's obligations, the patient's values and the degree of autonomy that both parties require. Presently, the most promoted model appears to be the shared decision-making model where physicians bring into the discussion their professional knowledge, and patients aware of their personal values and beliefs make the decision together (Stigglebout 2012).

Two other models need mentioning. The first is the Instrumental model where the physician aims for some goal independent of the patient's, such as the good of society or advancement of scientific knowledge and the patient's values are irrelevant. This model is an aberration unless an informed patient, who is not under duress agrees to such a relationship (Emanuel and Emanuel 1992).

Another, more current model, deemed more egalitarian and encompassing the ethical principle of taking equal interests into equal consideration is known as the Window-Mirror model. Its designers state that it sharpens the focus of both patients and physicians and it illuminates the unmet interests of both parties in the context of a continuing relationship. In contrast to other classic PPR models, the Window-Mirror model gives an equal role to the interdependent and equal moral interests of the patient and physician (Buetow and Elwyn 2018).

These various models of PPR are mentioned to make the point that each of them incorporates autonomy and trust as a necessary condition for the PPR to exist and to be effective.

25.3 Trust—An Ongoing Concern

Among the important attributes of the PPR are those that also partake of the accepted codes of professional ethics, such as veracity, deep listening, confidentiality, care, compassion, open communication, justice, respect for autonomy, adherence to beneficence and non-maleficence. Combined and practiced, these attributes engender mutual trust between the patient and physician. Trust can present as an elusive, ephemeral ideal, yet it is a powerful reality that builds, strengthens, and ultimately binds relationships; an essential component for a therapeutic PPR. One knows what to expect (Ipsos 2021; Pearson et al. 2000; Pellegrini 2017).

25.3.1 MAID Offspring: Two Classes of Physicians, Two Classes of Patients

To provide access to MAID, which entails ending the life of a patient (who requests it and meets the legal requirements), the courts and government conscripted the medical profession to carry out this intervention, while at the same time allowing for individual conscientious objectors. In 2021 less than 2% of all Canadian physicians provided MAID to one or more persons (Health Canada 2022). We can estimate that an additional 2–3% of Canadian physicians were involved in providing assessment of a person's eligibility for MAID. This means that about 5% of Canadian physicians are likely MAID-willing while 90–95% are MAID-averse. Of course, these classes are not fixed, physicians may change their position toward MAID based on a particular patient they care for, or on the basis of accumulated experience. Unfortunately, given the high level of privacy and confidentiality surrounding the MAID program, neither colleagues, nor patients have any effective way of identifying physicians who are MAID-willing; some patients will know only when their interview for their MAID intervention is taking place.

If a patient needs to enter a new PPR because of admission to the intensive care unit (ICU) or because of surgery, or simply to obtain a replacement for their retired family physician, it is unlikely that they will know the new physician's orientation towards MAID. This lack of knowledge can greatly impact patients, who do not aspire to undergo MAID. Let us consider this scenario representing the two classes of patients.

John and Mary are admitted to the ICU with complicated health issues that would qualify both for MAID. Both are well informed about this intervention. John is an advocate of MAID and knew that when the time came, he would apply for it. Mary, on the other hand, has always desired to receive whatever treatments were available to help her live her life to a natural end, as would be made possible in the hospice or palliative care unit.

In their immediate encounter with the attending physician, John and Mary's perceptions, expectations, concerns and feelings of trust will be substantially different from one another. In her sickly and vulnerable state, Mary feels powerless, anxious and fearful in this encounter because she does not know if this new physician is MAID-averse-willing-or-enthusiast. Everything she hears during the consultation is filtered through her doubt and suspicion that this stranger may nudge her towards MAID (as has been reported by various news sources). Her lack of trust is barring the formation of a healthy PPR with this new doctor.

John has no concerns about the MAID status of the physician. After their consultation, he makes his preferences known and he finds everyone to be obliging and helpful. A MAID team will visit him and provide him with the information he needs to start the process. He will complete the application, then be visited by a MAID assessor, who will ask him questions. If he qualifies, a time will be set up for the MAID provider to visit and end his life.

In this scenario, John and Mary represent two categories of patients that are products of the MAID law; those who wish to avail themselves of MAID, and those who do not want to avail themselves of MAID. The John types seem to be able to maintain their trust, while the Mary types seem to have it compromised. Why is this happening and why does it matter?

25.3.2 The Dissolution of Trust

The legalization of MAID did not translate into its universal acceptance in theory, practice, ideology or morality; just the opposite, it appears to have created division and stands to threaten the PPR. The insertion of MAID into medical care has caused two, distinct, fluid groups of physicians and patients, those who embrace MAID and are confident it will be carried out, and those who do not embrace MAID and feel demoralized or threatened by it. The latter group will experience a lack of trust, which will impede the formation of a therapeutic PPR. Since trust is an indispensable component of the PPR, it is important to unpack its operational mode within each group.

It appears that the word, 'trust', although it maintains lexical integrity, possesses two distinct foci. For John the focus (or signifier) for trust is MAID, whereas, for Mary, it is the PPR. In John's case, trust is a conviction that one's decision to obtain either AS or VE, will be realized. Trust is a felt response, a confidence, a surety that John will obtain what he is entitled by law to receive.

In Mary's case, trust must evolve from an encounter with the physician, within which certain ethical principles (as defined by the professional's code practice) are activated as communication takes place. For her, trust is not a given or an expectation, rather it develops as physician and patient exchange mutually necessary information with honesty and respect, and that in turn gives rise to a therapeutic relationship.

Because it is a doctor who is performing MAID, one may be inclined to think that John's trust emerges from the PPR, but this cannot be, because for John, the physician represents a trusted agent, whose role is to oblige his legal right to an assisted death. The intent of John's rapport to his physician is the fulfillment of a duty that originates in the law, and does not represent the initiation or the maintenance of a therapeutic relationship, or PPR.

In her first encounter with the physician, Mary's intent is to establish a relationship of trust that will assure her that the physician has her best medical interests in mind. But she finds herself unable to initiate a workable PPR because most of the ingredients are missing - her perception of the institution's lack of transparency, not being forthright with essential information, and uncertainty in intentions. Mary is bereft. She feels trust's absence and is overwhelmed with fear and anxiety.

The issue of trust as experienced by Mary is grounded in the historic, recognized value of PPR, and Mary's inability to even begin to shape a PPR because of fears, angst and diffidence is a serious assault on this therapeutic modality. The problem exemplified by Mary, and her representative group must be quickly addressed with the intention to avoid and/or repair the damage that is threatening to erode the PPR. To summarize, one can look at the meaning and use of the word, 'trust' to inform us as to why it does matter that two classes of patients and physicians are forming, and why this is a threat to the revered PPR. Even though one can agree that the lexical or literal meaning of 'trust' is maintained in the above examples, one can see that semantic ambiguity arises because the word has been used in two very different contexts. Within the context of MAID, as a legal right, 'trust' appears to maintain its operational validity, whereas, within the context of the relational encounter of the PPR, 'trust' cannot maintain its core value and significance, thereby rendering the PPR moribund. This requires further analysis, but if contextual use points to the problem, then for the sake of the integrity of the PPR, the problem ought to be addressed and resolved by the same agency that caused the problem. The action of reparation matters because without trust, there is no PPR, and without PPR, we will have destroyed an effectively operating institution that is +2500 years old.

25.4 Respect for Autonomy—Shifting Ground

Legalization of medically assisted death has been promoted by some as a triumph for patient autonomy. Yet, this new autonomy, as will be shown, lies outside the boundaries of the PPR. In order to provide evidence for this assertion we will compare the conventional autonomy of Pre-MAID time and Post-MAID Autonomy.

25.4.1 Pre-MAID Autonomy

The principle of respect for autonomy as concerns the human body was already promoted by John Stuart Mill in 1859, affirmed in common law by Judge Cardozo in 2014 (quoted by Jonsen et al. 1982), and introduced into medical care after the middle of the last century by the first generation of bioethicists (Beauchamp and Childress 2021) in order to empower patients to make choices and take actions based on their beliefs and values (Coggon and Miola 2011). The main instrument for autonomy became informed consent, but this autonomy also played a major role in a PPR model based on shared decision-making, namely, the deliberative model. The physician is expected to recommend a medical intervention that appears best for the patient, or a number of interventions for the patient to consider, and subsequently to autonomously refuse or accept. In offering their treatment, physicians are guided by a "standard of care" established by the profession and maintained on the basis of best scientific evidence (Hebert 1996).

Conventionally understood, autonomy, as a bioethics principle within the PPR is directly connected to the ongoing relationship between patients and physicians. For the physician to respect the autonomy of their patient means to view the patient as a moral subject with agency and not a thing to be acted upon or manipulated. To respect a patient's autonomy means that the physician is committed to listen, to respect, to take into consideration whatever the patient is saying, as well as to provide necessary medical knowledge and expertise for the duration of their relationship. In other words, autonomy acts as a signpost to remind the physician that the individual in their presence has intrinsic value.

25.4.2 Post-MAID Autonomy

The MAID Law, which has been founded in the Canadian Charter of Rights and Freedoms, introduced, what some may call, a 'radical autonomy' (See McGuinty

Chap. 1), which empowers qualifying patients to apply for and (if they meet the legal requirements) receive, either the drugs to kill themselves, or a lethal injection administered by a physician or a nurse practitioner. This action, facilitated by health care professionals is a medical intervention that has no medical indication; it is not required by standards of care for any medical condition (Lemmens 2021) nor guided by scientific evidence. MAID is designed to be an individual's personal choice, defined by, and enabled by Canadian law.

This acceptance of radical autonomy, trumping all other valuable considerations, resulted in a high number of MAID cases being undertaken in Canada, not only for the comfort of the dying and relief of suffering but also for non-health related reasons, such as loss of dignity, being a burden, and asserting one's control over the when, and how to die (Heath Canada 2022). When the law (albeit, within legally defined boundaries) hands over to individuals, the power to end their own lives at a time and manner of their choosing, and it conscripts (without criminal indictment) the medical profession to carry out the wishes or demands of the qualifying applicants, it follows that the content, meaning and application of the PPR must, by necessity, change.

The principle of radical autonomy presented in the MAID encounter is defined and initiated by the patient's legal right, and the physician's duty to serve the law, where both are entering into a contract that does not require the formation of a PPR. Each actor becomes a means to an end, and as such, each can slip into conscious objectification of the other. For example, MAID puts a demand on the physician to comply with the patient's request, even if that request will later entail the physician's assessment, as required by law, that curiously enough, will qualify or disqualify the patient from exercising their autonomy. The completion of the application for MAID leaves no room for the nuances that are the requisite for an actual relationship to take place. If the MAID assessment and procedure does create some type of encapsulated relationship on the strength of the party's personalities, it still would not qualify as what is meant by a PPR.

Hence, what can be discerned is that the practice of MAID has introduced a new way for the concept of autonomy to operate. In pre-MAID, autonomy is aligned with the encounter in a patient-physician relationship and facilitates consent, fosters trust and strengthens the relationship. Conversely, during MAID, autonomy is in allegiance with the law, facilitates a legal precedent, and imposes on the physician a duty to obey the law via the patient's request for MAID. Therefore, the word, 'autonomy' serves two diverse purposes.

25.4.3 Shifting Ground for Patients

In current health care settings, patients will experience some periods of uncertainty, confusion and concerns (Sweeney 2018). When MAID is available, many patients, who must accept the care of a new physician, will have concerns that were unknown to past generations, and will beg to know whether the physician can be both a caring executioner for one person and a caring healer and protector for another. Patients contemplating their future vulnerabilities, and their inability to advocate, will wonder how a medical decision will be made about them. Will their wishes be respected if their goals are different from those of the physician and health care organization? Or will the other's power prevail? If the MAID encounter is to be contextualized within the PPR, there must be well defined ethical, legal boundaries to protect patients, who do not want MAID, even though they would qualify.

Consider the case of a young patient, Carl, in hospital with incurable advanced malignancy, who wishes to live his life to its natural end and who does not want MAID. His "old" physician retired, so he enters a new PPR to receive care from a physician who was assigned to look after him. Carl will immediately perceive that his personal autonomy is threatened because he does not know if the new physician will support his goal of natural death, or judge, now or at a later time, that he would be better off served by MAID. Physicians could nudge the patients toward MAID, or compromise their care, e.g., inadequate pain control, and influence patients to believe that MAID is their best option. Because the profession, currently, has no method to identify physicians who are MAID-willing or MAID-averse, if Carl wishes to know, he would have to start a conversation with his new physician on this topic and this is a demanding task for Carl who is tired and has difficulties to concentrate. Carl likely does not know that counselling a person to die by suicide remains, in Canada, a criminal offence (Criminal Law 2022), even for physicians and nurse practitioners who are permitted to provide medical assistance in dying. Some medical authorities also advise that physicians should never propose MAID to a patient; if it is done, it could represent an "external pressure" that would invalidate a person's application for MAID. In contrast, the Canadian Association of MAID assessors and providers recommends that every potentially eligible patient should be informed about the possibility of applying for MAID (CAMAP 2022) and some authors suggested that this information should be provided always when planning end-of-life care with a patient. Therefore, it is possible that the new physician may have some personal experience with providing MAID, and could propose it to Carl and share a glowing report about MAID. At this point Carl's previous autonomous decision to wait for natural death would be severely challenged. He may even feel that he has an obligation, at least a weak one, to accept what the doctor proposes. Carl could possibly comply with the doctor's preference for MAID and spend the last remaining days—while waiting for a MAID date—disappointed and frustrated because his well-considered plan for his remaining life was thwarted. Or he could resist the gentle nudging or perceived pressure from the physician and live in constant fear that, somehow, the physician will do something to get his or her way.

Because most patients, over 90% of those who are at the end of life, do not wish to have MAID, there is now a greater chance of patients being frustrated with feeling that their autonomy for life is now more compromised than pre-MAID time because the promise to enhance a person's autonomy by ending one's life figures large in arguments in favour of MAID. Descriptive and exploratory research would be required to obtain the data on this aspect of PPR, but there are reasons to suspect that the promise of enhanced autonomy as a way to improve the balance of power in PPR at the end of life is not achievable.

25.4.4 Shifting Ground for Medical Professionals

The practice of MAID has placed a physician as a professional in a precarious situation. As in ancient times, the god one served demanded its own particular sacrifices and bestowed its own unique rewards. Will a physician have to make a choice between serving a patient to improve health outcomes, or, serving an abstract concept dictated by law to terminate a life by MAID?

Many physicians, who have provided MAID have reported very rewarding experiences for themselves, the family members of the deceased, and the patients themselves prior to their death. These physicians, in their humanity, feel that they are stopping the suffering of those who are receiving MAID. However important this justification is in principle, Canada's 2022 report points to the fact that the majority of people who are accessing MAID cite non-medical reasons. But even if the MAID provider's full reward would be the gratitude felt by a suffering patient, and within this short engagement, a relationship develops, this relationship would not qualify as a PPR. The patient's legally enforced autonomy commands the conscripted/willing physician to act, first as the information gatherer to determine whether or not, by law, the patient qualifies for MAID, and then, as an agent in facilitating or causing a death. One could argue that information gathering also takes place in a traditional PPR, however, there is a significant difference. In a PPR, the questions are posed in order to identify the patient's health problem, establish the treatment proposals and obtain consent in order to achieve improvement in the patient's health outcomes and/or prolong life. Within the MAID protocol, there is a legal form with pre-set questions that are asked in order to establish whether or not the patient legally qualifies for MAID. The sought-after outcome is the termination of the patient's life. In each scenario, both, the reason and purpose of the questions asked are different. The relationship established by a MAID provider and patient could be called humanitarian, consoling, or anything else, but it would not qualify as what has been traditionally defined as physician–patient relationship. Ending a patient's life does not involve the art of medicine, it involves succumbing to an outside interference that is neither art nor medical science, but a legal dictate. The position statement of Canadian neurosurgeons stated: "Although in certain situations death may be seen as favourable or a desired outcome, it has never been accepted as the goal of treatment." (Barry et al. 2017). Similarly, the American Psychiatric Association asserted: "Protection of life is paramount…there is a profound ethical duty of a doctor that cannot be reconciled with participation in the killing of a patient" (APA 2022).

The physician must choose between being a creative agent in a healing profession that is in service to the living, or merely participating in a legally mandated activity. Ultimately, physicians will choose the goals and rewards that suit them as they participate within the dictates of their profession and the law; to enjoy the fruits of creative labor, be they sweet or sour, or settle in the comfort of having pleased and appeased the new rights-based autonomy.

The termination of human life by agency of a physician or a nurse practitioner raises important concerns about the goal and very purpose of the healing profession and therefore about the goal and purpose of the PPR. The members of the medical profession must ask and answer critical questions. Does it matter if the actual practice of MAID is not to relieve suffering but to facilitate individuals' legally framed autonomy, so they can control and determine the time and place of their death? How can the profession continue to be involved in a medical intervention that by definition does not qualify as medical care? Surely, even if by a long stretch of the imagination, MAID could be called medical care, then, because it is so radically different from the medical care that has historically been provided, would it not be best to see it as a separate goal from that which is already in existence? To accommodate this situation, the profession could formally assign two separate goals of medicine served by two distinct categories of physicians—the first to attend to the health of the individual, the second, to serve the newly defined autonomy of the individual. Another possibility is to explore a PPR that includes the moral interests of both physicians and patients similar to the Window-Mirror model mentioned earlier.

25.5 Views from the Trenches

What is the experience of Canadian physicians working in a MAID environment, in regards to the PPR and inter-professional relationship? The published experience is limited but several valuable studies must be noted.

The experience of 23 palliative care physicians and nurses in Ontario working for six months after MAID deliveries were started was characterized by a high level of distress for a variety of reasons. They discovered that the patient-caregiver relationship changed. Some patients and families now entering the unit were fearful that the palliative care unit would bring about their death, while other patients came in to pursue assisted death. If it became obvious to this group that a palliative professional had reservation about MAID, they would become uncommunicative. It was difficult for the staff to provide optimal palliative care as they had previously done. Staff became hesitant to identify themselves as conscientious objectors because that would strain the relationship with some patients. Participants in the study described how MAID provision consumed staff time and resources, which they needed for palliative care patients, and how MAID created "difficult conversations". Staff was concerned that just introducing the topic of MAID could be interpreted as an invitation to apply for it and that would erode trust of patients and families. Staff also attempted to reconcile the tensions and strained relations between patients and their dissenting families. There were questions about exact prognosis that were difficult to answer when MAID had to be scheduled. The situation somewhat improved after two separate streams were established in the unit, one for MAID and the other for palliative care (Mathews et al. 2021).

A smaller qualitative study involving hospice care workers in a medium-sized city revealed difficult changes in PPR. Despite a desire to support patient's autonomy, some interpreted a patient's choice for MAID as a rejection of a natural death experience that the hospice strove to perfect. They were uncertain how to support patients, who chose MAID. Many patients aiming for a MAID death were in good general condition and did not look like persons, who were close to natural death. Staff struggled to make sense of these patients' imminent death in spite of their normal daily functioning. MAID came to be understood as a disturbing element of normal hospice care. For some, a patient's choice of MAID left them questioning the success of their care, others felt that these patients missed out on their ability to experience a comfortable natural dying process and the full spectrum of hospice provisions. Staff were often left on their own to make sense of the motivations for MAID. Staff not only experienced the grief of others, but their own grief as a sense of loss due to patients' choice to received MAID. Combined with feelings of rejection, this produced troublesome emotional and moral distress (Freeman et al. 2021).

We know from Annual Reports on MAID (Health Canada 2022) that only about 3% of Canadian physicians and a similar percentage of nurse practitioners participate in the provision of MAID. What influences physicians to participate or not to participate, and has it anything to do with PPR? One recent qualitative study analyzed interviews with 35 physicians and nurse practitioners in western Canada, who identified themselves as non-participants and focused on exogenous factors. The study showed that the exogenous factor that influenced the non-participation included the health care system they worked in, the communities where they lived, the current practice content, and as they saw the risk of participation. But, a number of factors were related to patient-professional relationship. These clinicians were concerned how the public would view them and how their patients would feel about their participation in MAID. They were concerned that their participation would be viewed as "giving up" on patients, and this would complicate their mental health and suicide prevention conversations. How can a doctor advise a suicidal patient not to commit suicide when the patient knows that the doctor assisted in the death of his grandmother? (Brown et al. 2021).

Finally, a study was published focusing on inter-professional relationships of 16 MAID providers across the country. MAID providers, in interviews, noted enhanced relationships with other MAID providers while relationship with objecting colleagues were "somewhat strained". This had both emotional and practical impact on physicians. A colleague would no longer feel comfortable with a physician who was a MAID provider and would not share patients with them. One physician stated that the introduction of MAID had forever changed the mood for her in the unit. One MAID provider reported being forcibly attacked by palliate care colleagues (Khoshnood et al. 2018).

25.6 Conclusions and Recommendations

Only a small proportion of Canadian physicians, estimated to be about 3%, are actively involved in the delivery of MAID. Yet this new program appears to already be changing and redefining the patient-physician relationship. This phenomenon could be examined in a variety of ways. This chapter has focused on two important elements that have contributed to the formation of a healthy PPR—trust and respect for autonomy.

In pre-MAID time, patients could trust, knowing that all physicians are bound by professional and ethical codes that entails them to provide relief of suffering, possible cure and prolongation of life. Harm was always to be avoided, and a patient's death was the highest level of harm. When a patient created a PPR with a new doctor, the patient could always rely on this unwritten social contract, hence trust was present from the start. During MAID time, physicians have the privilege to terminate the life of a patient who asks for it, and meets the legal requirements. Some physicians adhere to the original intention of the code and would not kill, others would accept a patient's request, asses the patient for eligibility and terminate their life if eligible. Presently, a patient who needs to create a new PPR with a physician cannot be entirely certain to what the physician subscribes. If a patient would never want to have MAID, then trusting a new physician may be problematic, because distrust could be present even before the PPR is initiated. Respect for a patient's autonomy in pre-MAID time had been expressed chiefly in the process of informed consent, where the patient has the right to accept or reject the physician's proposal for a test or treatment, which must be based on science-based standards of care for the particular patient's condition. This autonomy was shaped by the goals of medicine and the physician's ethical commitment. In MAID time, patients have acquired a new radical autonomy that allows them to apply for self-death. This radical autonomy is not limited by standards of care or science or ethical responsibility of the physician, but only by law and the regulations of the state. At the moment, we do not have any comprehensive empirical studies of PPR development in Canada, since the introduction of MAID. However, recent Canadian studies about the experience of the patient-professional relationship from the point of view of professional caregivers, suggest that this relationship is in turmoil. It is necessary to initiate research projects that would thoroughly examine the PPR in the presence of MAID in our health care system. We have to rely on this future research to give us an indication on what needs to be done to rescue that rich and beneficial PPR that is so essential to health and healing.

References

ABIM Foundation. 2021. Poll: Physician's trust in health system leadership. https://abimfoundation.org/pressrelease/poll-physicians-trust-in-health-system-leadership-declines-during-covid-19-pandemic

APA. 2022. Principles of Medical Ethics. American Psychiatric Association. https://www.psychiatry.org/File%20Library/Psychiatrists/Practice/Ethics/Opinions-of-the-Ethics-Committee.pdf

Barry, Sean, Chris Ekong, Brian Wheelock, Richard Moulton, and Peter Gorman. 2017. AID and the neurosurgeon: Position statement of the Canadian Neurosurgical Society. *Canadian Journal of Neurological Sciences* 44 (6): 744.

Beauchamp, T.L., and J.F. Childress. 2021. *Principles of biomedical ethics*. New York, NY: Oxford University Press.

Brown, Janine, Donna Goodridge, Lilian Thorpe, and Alexander Crizzle. 2021. 'I am okay with it, but I am not going to do it': The exogenous factors influencing non-participation in medical assistance in dying. *Qualitative Health Research* 31 (12): 2274–2289.

Buetow, S., and G. Elwyn. 2018. The window-mirror: A new model of the patient-physician relationship. *Open Medicine* 2 (1): e20–e25.

CAMAP. 2022. Bringing up Medical Assistance in Dying (MAiD) as a clinical care option. https://camapcanada.ca/wp-content/uploads/2022/02/Bringing-up-MAiD.pdf

Coggon, John, and Jose Miola. 2011. Autonomy, liberty and medical decision-making. *Cambridge Law Journal* 70 (3): 523–547. https://www.ncbi.nlm.nih.gov/pmc/articles/PMC3535760/

Criminal Law. 2022. R.S.C, 1985, c. C.46, current to 2022-01-16.

Emanuel, Ezekiel J., and Linda L. Emanuel. 1992. Four models of the physician-patient relationship. *The Journal of the American Medical Association* 267 (16): 2221–2226.

Freeman, Shannon, Davina Banner, and Valerie Ward. 2021. Hospice care providers experiences of grappling with medical assistance in dying in hospice setting: A qualitative study. *BBC Palliative Care* 2021: 20–25.

Government of Canada. 2022. Criminal Code (R.S.C., 1985, c. C-46) Section 241.1. Medical assistance in dying. Last amended 2022-01-16. https://laws-lois.justice.gc.ca/eng/acts/C-46/index.html

Health Canada. 2022. Third annual report on medical assistance in dying in Canada 2020. https://www.canada.ca/en/health-canada/services/medical-assistance-dying/annual-report-2021.html

Hebert, Philip C. 1996. *Doing right. Chapter 5*. Don Mills, Ontario: Oxford University Press.

Ipsos. 2021. Doctors are the most trusted profession in Canada and across the world. https://www.ipsos.com/en-ca/news-polls/doctors-most-trusted-profession-in-canada

Jonsen, Albert R., Mark Siegler, and William Winslade. 1982. *Clinical ethics*. Macmillan Publishing, New York.

Khoshnood, Narges, Marie-Clare Hopwood, Bhadra Lokuge, Alison Kurahashi, Anastasis Tobin, Sarina Isenberg, and Amna Husain. 2018. Exploring Canadian physician's experience providing medical assistance in dying: A qualitative study. *Journal of Pain and Symptom Management* 56 (2): 222–229.

Lemmens, Trudo. 2021. How bill C-7 will sacrifice the medical profession's Standard of Care: https://policyoptions.irpp.org/magazines/february-2021/how-bill-c7-will-sacrifice-the-medical-professions-standard-of-care/

Marcum, J.A. 2008. Patient-Physician Relationships. In: Humanizing Modern Medicine: An Introductory Philosophy of Medicine. Springer Science and Business Media.

Mathews, J.J., D. Hausner, J. Avery, B. Hannon, C. Zimmermann, and A. Al-Awamer. 2021. Impact of medical assistance in dying on palliative care: A qualitative study. *Palliative Medicine* 35 (2): 447–454.

Olaisen, Henry R., Mark D. Schluchter, et al. 2020. Assessing the longitudinal impact of physician-patient relationship on functional health. *Annals of Family Medicine* 18 (5): 422–429.

Pearson, Steven D., and Lisa H. Raeke. 2000. Patients' trust in physicians: Many theories, few measures, and little data. *Journal of General Internal Medicine* 15 (7): 509–513.

Pellegrini, Carlos. 2017. Trust: The keystone to the patient-physician relationship. *Journal of American College of Surgeons* 224 (2): 95–102. https://doi.org/10.1016/j.jamcollsurg.2016.10.032.

Steward, Michael A. 1995. Effective physician-patient communication and health outcomes: A review. *Canadian Medical Association Journal* 152 (9): 11423–11433.

Stiggelbout, A.M. 2012. Shared decision making: Really putting patients at the centre of healthcare. *British Medical Journal* 344: e256. https://doi.org/10.1136/bmj.e256.

Sweeney, James F. 2018. The eroding trust between patients and physicians: Why the relationship is strained and what doctors can do to strengthen the connection. *Health and Medical Economics* 95 (7): 38.

Louisa Pedri, BA, B. Ed. (Lakehead University), MA (Ethics, King's College, University of London, UK), was a founding member and served as Education Coordinator of the Centre for Health Care Ethics Lakehead University and is now a Research Affiliate and member of its Executive Committee. She served as President of Via Vitae Palliative Care Association and member of the Research Ethics Board of Lakehead University. Her work has been published in the Formosan Journal of Medical Humanities and in HEC Forum.

Chapter 26
The Importance of Conscience as an Independent Protection

Derek Ross and Deina Warren

Abstract Freedom of conscience has received little attention as a distinct protection in Canadian case law to date. Where it has been judicially considered, it has largely been conflated with freedom of religion. While these two freedoms share similarities, there are important distinctions between them. This article examines some of these distinctions, and how the need for greater attention to conscience has come into sharp focus in the debate surrounding physician participation in medical assistance in dying (MAID). Building on a growing body of scholarship, this article makes the case for a more purposive and generous interpretation of freedom of conscience as an independent right with independent content, as required by established principles of Charter interpretation. First, the authors summarize how claims of conscience in the context of MAID have thus far been assessed primarily through the lens of freedom of religion. Second, they explain how conflating freedom of religion and conscience results in the former subsuming the latter, in problematic ways. Third, they propose some ideas on how freedom of conscience can be more robustly understood, calling for (1) a clearer articulation of the purpose for and necessity of protecting conscience; and (2) an analytical framework that recognizes the complexity of conscientious convictions, and the severe social, public, and individual harms of violating conscience.

Keywords Conscience · Conscientious objection · Effective referral · Freedom of conscience · Freedom of religion · Medical assistance in dying (MAID) · Medical ethics · Professional independence

D. Ross
Christian Legal Fellowship, London, Canada
e-mail: execdir@christianlegalfellowship.org

D. Warren (✉)
Canadian Centre for Christian Charities, Elmira, Canada
e-mail: deina.warren@cccc.org

J. Kotalik and D. W. Shannon (eds.), *Medical Assistance in Dying (MAID) in Canada*, The International Library of Bioethics 104,
https://doi.org/10.1007/978-3-031-30002-8_26

26.1 Introduction

Freedom of conscience is, along with freedom of religion, a cornerstone of liberal democracy. This was emphasized by the Supreme Court of Canada in *R v Big M Drug Mart*, the Court's first ruling interpreting the guarantee of freedom of conscience and religion in the *Canadian Charter of Rights and Freedoms* (Charter).[1] The Court noted that "an emphasis on individual conscience and individual judgment" sits at "the heart of our democratic political tradition" and sustains the "legitimacy, acceptability, and efficacy of our system of self-government".[2] The "rights associated with freedom of individual conscience" possess a "centrality" to "basic beliefs about human worth and dignity" and to the nature of a free and democratic society.[3] This centrality, the Court concluded, renders these rights the "*sine qua non*"—the essential and indispensable elements—of "the political tradition underlying the *Charter*."[4]

Yet, despite its recognized importance to the flourishing of the individual and of society at large, freedom of conscience has received little attention in Canadian jurisprudence to date. Where it has been judicially considered, it has largely been conflated with freedom of religion.[5] While these two protections share similarities and are located together in section 2(a) of the Charter, there are important distinctions between them, calling for recognition of conscience as an independent freedom.[6] Building on a growing body of scholarship, this article examines a number of reasons why freedom of conscience ought to be treated as an "independent right with independent content", not as a "derivative" of religious freedom.[7] As Supreme Court Justice Bertha Wilson observed: "as a matter of statutory interpretation, 'conscience' and 'religion' should not be treated as tautologous if capable of independent, although related, meaning."[8]

This is not merely a technical or academic point. The practical challenges of an under-developed conception of freedom of conscience have come into sharp focus

---

[1] *R v Big M Drug Mart*, [1985] 1 SCR 295 at para 122.

[2] Ibid.

[3] Ibid.

[4] Ibid.

[5] See discussion in Brian Bird, Dwight Newman, and Derek Ross, "The Charter's forgotten fundamental freedoms", Policy Options (16 June 2020), online: https://policyoptions.irpp.org/magazines/june-2020/the-charters-forgotten-fundamental-freedoms/.

[6] See Brian Bird, "The Call in *Carter* to Interpret Freedom of Conscience" (2018) 85 SCLR (2d) 107-141; Mary Ann Waldron, "Conscientious Objections to Medical Aid in Dying: Considering How to Manage Claims of Conscience in a Pluralistic Society" (2018) 85 SCLR (2d); Barry W. Bussey, "Blazing the Path: Freedom of Conscience as the Prototypical Right" (2020) 98 SCLR (2d); Brian Bird, "The Reasons for Freedom of Conscience" (2020) 98 SCLR (2d); and Howard Kislowicz, Richard Haigh, Adrienne Ng, "Calculations of Conscience: The Costs and Benefits of Religious and Conscientious Freedom" (2011) 48:3 Alta L Rev 679.

[7] The Supreme Court made this point with respect to s. 2(d) freedom of association in *Mounted Police Association of Ontario v Canada (Attorney General)*, 2015 SCC 1 at para 49.

[8] *R v Morgentaler* [1988] 1 SCR 30 at 179.

in the debate surrounding physician participation in euthanasia and assisted suicide, collectively described in Canada as "medical assistance in dying" (MAID).

This article makes the case for a more purposive and generous interpretation of freedom of conscience, as required by established principles of Charter interpretation,[9] and proceeds as follows:

- First, we summarize how claims of conscience in the context of MAID have thus far been assessed primarily through the lens of freedom of religion;
- Second, we explain how conflating freedom of religion and conscience results in religion subsuming conscience in problematic ways; and
- Third, we propose some ideas on how freedom of conscience can be more robustly understood and purposively interpreted as a distinct protection that rightly stands on its own, as equal to, not lesser than, other constitutional entitlements.

26.2 Conscience and MAID in Canadian Jurisprudence

At the outset of our discussion on MAID and conscience, it is important to note that no Canadian court has interpreted the Charter as requiring physician participation in procedures which violate their conscience.

In *R v. Morgentaler,* the Supreme Court decision that led to the decriminalization of abortion, Justice Beetz acknowledged that the law could not "force a physician to perform an abortion."[10] The unanimous Court in *Carter v Canada* approvingly cited this passage, stressing that nothing in its decision, which led to the decriminalization of euthanasia, "would compel physicians to provide assistance in dying."[11] To the contrary, the Court emphasized that "a physician's decision to participate in assisted dying is a matter of conscience and, in some cases, of religious belief."[12]

The Supreme Court did not suggest that conscience is automatically subjugated to access to MAID—instead, conscience rights were to be considered and *reconciled* with patients' access by "physicians' colleges, Parliament, and the provincial legislatures."[13] In other words, legislatures and policymakers were directed to accommodate *both* medical conscience *and* patient access.

[9] *R v Big M Drug Mart*, [1985] 1 SCR 295 at paras 117–118.

[10] *R v Morgentaler* [1988] 1 SCR 30 at 96.

[11] *Carter v Canada*, 2015 SCC 5 at para 132.

[12] Ibid.

[13] Ibid.

26.2.1 Effective Referrals, Conscience, and Medical Ethics

One response to this directive, adopted by the College of Physicians and Surgeons of Ontario (CPSO), was an "effective referral" requirement: physicians with a conscientious objection are not required to perform MAID, but must make "a personal and good faith effort to ensure that the patient is connected to the service they are seeking."[14] This requires "positive action" such as an "effective referral" to another willing and available physician.[15] However, this approach became problematic for a number of physicians because it explicitly requires taking "positive action" to facilitate and cooperate in a procedure they view as unethical and harmful: purposefully ending another person's life.[16]

The normative significance of a medical referral should not be overlooked.[17] For example, the World Medical Association' (WMA) position is that that no physician should be "obliged to make referral decisions" for euthanasia or assisted suicide (MAID), and recently re-affirmed its position that these practices are contrary to the principles of medical ethics.[18] The WMA also recently rejected a proposed change to its International Code of Medical Ethics requiring mandatory referrals in cases of

---

[14] The CPSO imported the "effective referral" requirement previously developed as part of its Professional Obligations and Human Rights Policy into its MAID Policy. See *Christian Medical and Dental Society of Canada v College of Physicians and Surgeons of Ontario*, 2019 ONCA 393 at para 78.

[15] Ibid. at para 24.

[16] While this article was at review stage, the CPSO issued a new draft policy, "Human Rights in the Provision of Health Service," and a companion "Advice to the Profession" document for public consultation. The Advice document provides an additional example to satisfy effective referral requirements that is arguably less onerous, in which the conscientiously-objecting physician may provide a patient with contact information for "a non-objecting, available, and accessible physician, other health-care professional, or agency". While this may allow some flexibility in specific circumstances, the document also enhances follow-up obligations requiring the physician to confirm that the patient was connected with a MAID assessor/provider and to "take a more active step" if the patient was not. Ultimately, it seems that the primary concern for conscientious objectors still remains, which is taking "positive action" in facilitating euthanasia or assisted suicide. See http://policyconsult.cpso.on.ca/?page_id=14240 for the draft documents.

[17] Indeed, while the context raises other complexities, the CPSO itself recognizes that, in the case of female genital cutting/mutilation, referrals have medical implications and are prohibited; see College of Physicians and Surgeons of Ontario, Statements & Positions, "Female Genital Cutting (Mutilation)", online: https://www.cpso.on.ca/Physicians/Policies-Guidance/Policies/Female-Genital-Cutting-Mutilation: "Physicians will be subject to disciplinary measures if they perform, assist in or refer patients for FGC/M procedures" (emphasis added). For further discussion on the moral implications of referral, see Dr. Ewan Goligher and Simon Czajkowski's chapter in the present volume.

[18] World Medical Association, "WMA Declaration on Euthanasian and Physician-Assisted Suicide: Adopted by the 70th WMA General Assembly, October 2019" (23 November 2021), online: https://www.wma.net/policies-post/declaration-on-euthanasia-and-physician-assisted-suicide/. See also "WMA Declaration of Venice on End of Life Medical Care" (revised by the 73rd General Assembly, October 2022), online: https://www.wma.net/policies-post/wma-declaration-of-venice/.

conscientious objection.[19] Likewise, the Canadian Medical Association has taken the position that "physicians must be able to follow their conscience without discrimination when deciding whether or not to provide or participate in assistance in dying", including whether to "refer the patient to a physician or a medical administrator who will provide assistance in dying to the patient".[20]

26.2.2 *Judicial Review of the "Effective Referral" Requirement*

Shortly after the CPSO's effective referral policy was introduced, a group of individual physicians and medical organizations commenced a legal challenge, arguing that the policy violated the freedom of conscience and religion of physicians who conscientiously object to euthanasia, as well as their right to equal treatment without discrimination based on their religion. Both the Divisional Court and the Court of Appeal of Ontario accepted that the policy infringed freedom of religion, as it would compel some religious physicians to leave their practice.[21] However, both courts found the infringement was justified and upheld the policy on the basis that it struck a reasonable balance between patients' interests and physicians' religious freedom in advancing the CPSO's goal of ensuring "equitable access to health care."[22]

It is important to note that the Ontario court decisions did not specify what the Charter *requires* in reconciling patient access with physicians' freedom of religion (or their freedom of conscience, which, as discussed below, was not substantively addressed). The Ontario courts concluded that the CPSO's referral policy fell "within the range of reasonable alternatives."[23] Thus, effective referrals were deemed *one* acceptable approach, but, as the Divisional Court noted, authorities elsewhere have adopted policies that are "arguably less restrictive."[24] In other words, the Charter was interpreted to *permit* an effective referral policy in a specific factual and evidentiary

---

[19] Instead, "The physician must immediately and respectfully inform the patient of this objection and of the patient's right to consult another qualified physician and provide sufficient information to enable the patient to initiate such a consultation in a timely manner." See WMA International Code of Ethics (revised by the 73rd General Assembly, October 2022), online: https://www.wma.net/policies-post/wma-international-code-of-medical-ethics/

[20] Canadian Medical Association, "CMA Policy: Medical Assistance in Dying" (2017), online: https://policybase.cma.ca/viewer?file=%2Fmedia%2FPolicyPDF%2FPD17-03.pdf.

[21] *Christian Medical and Dental Society of Canada v College of Physicians and Surgeons of Ontario,* 2019 ONCA 393 at para 79.

[22] Ibid. at paras 107, 112–113, 160; *Christian Medical and Dental Society of Canada v College of Physicians and Surgeons of Ontario,* 2018 ONSC 579 (Div Ct) at para 177.

[23] *Christian Medical and Dental Society of Canada v College of Physicians and Surgeons of Ontario,* 2018 ONSC 579 (Div Ct) at para 177; *Christian Medical and Dental Society of Canada v College of Physicians and Surgeons of Ontario*, 2019 ONCA 393 at para 158.

[24] *Christian Medical and Dental Society of Canada v College of Physicians and Surgeons of Ontario,* 2018 ONSC 579 (Div Ct) at para 174.

context, but it was not found to *mandate* such a policy. Different approaches (such as the implementation of a centralized co-ordination service in place of effective referrals) may also comply with the Charter.[25]

In addition, it is important to underline that the Ontario litigation was ultimately decided on the ground of freedom of religion; freedom of conscience, which the claimants also raised, was not substantively addressed. The Court of Appeal concluded that, "[t]o the extent the individual appellants raise issues of conscience, they are inextricably grounded in their religious beliefs", and explained:

> It is not appropriate to explore the contours of freedom of conscience in a case that does not have a robust evidentiary record. Like the Divisional Court, given my conclusion that the Policies infringe the appellants' s. 2(a) religious freedom, I find it unnecessary to consider the appellants' alternative argument that the Policies infringe the appellants' s. 2(a) freedom of conscience.[26]

Thus, whether Ontario's effective referral policies might violate freedom of *conscience* in certain circumstances remains an unresolved issue.[27] To answer that question, courts will need to closely examine the multi-faceted nature of MAID and conscience. This is not merely a question of legality, where legality alone determines whether a procedure is medically ethical. As the trial judge in *Carter* noted: "Actions may be ethical but not legal, and, conversely, may be legal but not ethical."[28] This must not be overlooked in the task of accommodating ethically diverse approaches to MAID.

MAID—the act of intentionally ending another person's life via the administration or provision of a lethal substance[29]—engages human rights considerations, as well

---

[25] As Chief Justice McLachlin observed (in a passage repeated in *Christian Medical and Dental Society of Canada v College of Physicians and Surgeons of Ontario,* 2018 ONSC 579 (Div Ct)): "The test must not be applied in a manner that amounts to identifying the Canadian province that has adopted the 'preferable' approach to a social issue and requiring that all other provinces follow suit"; see *Quebec (Attorney General) v A*, 2013 SCC 5 at para 440. See also the discussion in Christian Legal Fellowship, "Submissions of Christian Legal Fellowship re: The College of Physicians and Surgeons of Ontario's (CPSO) policies concerning *Medical Assistance in Dying* and *Professional Obligations and Human Rights*" (14 May 2021), online: https://static1.squarespace.com/static/57503f9022482e2aa29ab3af/t/609ef13c76d9926869ed4fec/1621029180353/CLF+submission+to+CPSO+re+MAID+and+POHR+policies+-+May+14%2C+2021.pdf.

[26] *Christian Medical and Dental Society of Canada v College of Physicians and Surgeons of Ontario*, 2019 ONCA 393 at para 85.

[27] The courts also dismissed the physicians' s. 15 equality claim, primarily based on their view that physicians could easily "transition to other areas of medicine in which these issues of faith or conscience are less likely to arise, if at all" (*Christian Medical and Dental Society of Canada v College of Physicians and Surgeons of Ontario*, 2019 ONCA 393 at para 94). This could potentially be distinguished with different evidence in the future (especially if MAID eligibility expands and becomes applicable in more areas of medicine), but at a fundamental level, it seems to ignore the principle that "the fact that a person could avoid discrimination by modifying his or her behaviour does not negate the discriminatory effect"; see *Lavoie v Canada* [2002] 1 SCR 769 at para 5 (per McLachlin CJC, dissenting, but not on this point).

[28] *Carter v Canada*, 2015 BCSC 886 at para 173.

[29] *Criminal Code of Canada*, RSC 1985, c C-64, s. 241.1, "medical assistance in dying".

as health, medical, clinical, social, and cultural factors.[30] All of these considerations are relevant for professional ethics (and, therefore, professional *conscience*), yet they are largely absent from the current s. 2(a) analytical framework, which focuses on *religious* matters, such as one's "system of faith and worship" and practices "which allow individuals to foster a connection with the divine or with the subject or object of that spiritual faith".[31]

Thus, a more specific, extensive, and discrete analysis of *conscience* is merited, as discussed in the following section.

26.3 The Need to Consider Conscience as a Distinct Freedom

As we consider conscience, it is helpful to place it in context. Conscience is not a mere psychological abstraction; it is significant and tangible in its impact. Consider a psychiatrist whose patients endured extended periods of hopelessness, persistent and long-standing desires to die, but who, with support, ultimately found meaning and fulfillment in their lives. Based on her medical training and this recurrent patient care experience, the psychiatrist has concluded it is impossible to determine with certainty whether a particular case of mental illness (e.g. depression) is "irremediable". It is her clinical position that introducing death as treatment option for patients with depression is inappropriate and would exacerbate the very symptoms for which they seek help. She cannot, therefore, in good conscience offer or refer for MAID in cases of clinical depression, as it defeats the very purpose of her psychiatry practice.[32]

There are a number of variations to this hypothetical; for example:

- A pediatrician who concludes, after years of practice, that children need significant time and experience with treatments and supports before it can be determined whether their disability is grievous and irremediable, and therefore is unable to participate in any pediatric MAID regime[33];
- A physician who refuses to participate in ending the life of an infant (which, given the infant's lack of capacity, would always be non-consensual), based on his

[30] See Suresh Bada Math and Santosh K Chaturvedi, "Euthanasia: Right to live vs right to die" (2012) 136:6 Indian J Med Res, online: https://www.ncbi.nlm.nih.gov/pmc/articles/PMC3612319/. This paper considers the "complex medical, social and legal dilemma" of euthanasia in court decisions in India.

[31] *Syndicat Northcrest v Amselem*, 2004 SCC 47, [2004] 2 SCR 551 at para 39.

[32] Currently, MAID is not available in cases where the sole underlying condition is a mental disorder, however, Bill C-7 will remove this exclusion as of March 2024. See also Coelho et al. (2022), citing research highlighting "the known risk of providing psychiatric MAD to suicidal individuals who would otherwise benefit from suicide prevention strategies" and that a significant number of psychiatrists oppose MAID solely on the grounds of mental illness.

[33] The federal government is exploring the expansion of MAID to "mature minors". See also "Normalizing Death", ibid.: "Evidence related to new illness or injury shows that suicidality is often present at the outset…but it is not enduring in the long run".

conviction that doing so is contrary to the basic principles of ethics and medicine, no matter how "severe" the infant's disability[34];

- A physician who is concerned about offering MAID in contexts where the patient's request to die is primarily driven by factors external to the underlying illness and/or disability, such as poverty, inadequate housing, psychosocial suffering, and/or prison conditions[35]—although the patient is legally eligible, the physician views the request as indistinguishable from suicide.[36]

In each of these scenarios, the physician's position is informed by a combination of his/her medical experience, ethical framework, and professional judgment. But where do "clinical reasons" end and "ethical/conscientious reasons" begin? In cases where a physician's objection is framed as "religious", attempts have been made to segregate these categories (discussed below), but in practice the issue of conscience is far more complex. And it will become even more complex as MAID expands and becomes a legal option in more areas of medicine.

These examples reveal something else about conscience that is often overlooked or misunderstood. One follows one's conscience not simply to shield oneself from moral discomfort, although it may have that effect, but to prevent the infliction or causation of harm—in these cases, harm to another person. Far from "abandoning" their patients, these conscientious physicians want to support and care for them, even at the risk of professional sanction; this is hardly an act of self-interest.[37] One may agree or disagree with the conscientious objector's conception of harm, but that makes it no less their sincere position and motivation.

As discussed below, the conflation (or, perhaps more accurately, the *substitution*) of conscience with religion has removed a number of these significant considerations from the legal analysis. When a commitment is seen as "religious" it is more likely to be perceived as personal, inward-focussed, and out of place in the context of a professional setting such as medicine. Religious belief is also more readily seen as *metaphysical* in nature, extraneous to science or reason, and perhaps even dogmatic or irrational.[38]

Therefore, assessing physicians' conscientious concerns solely as *religious beliefs* can obscure the complexity and nature of their position. For example, in the *CMDS v CPSO* decisions, the physicians' objections to MAID were not described in

[34] The Collège des médecins du Québec is advocating for the legalization of pediatric euthanasia for infants born with "severe deformities" in certain circumstances—see note 81, *infra*.

[35] See e.g. Simpson et al. (2022).

[36] Numerous media reports have highlighted examples of such cases. See also Alexander Raikin, "No Other Options", The New Atlantis (Winter 2023), online: https://www.thenewatlantis.com/publications/no-other-options.

[37] This aspect of conscience was ignored in *Christian Medical and Dental Society of Canada v College of Physicians and Surgeons of Ontario*, 2019 ONCA 393, where the court instead accepted arguments that conscientious objectors would effectively "abandon their role as patient navigators" (see paras 102 and 160).

[38] Of course, this characterization of religious thought is problematic in its own right and must similarly be resisted in freedom of religion claims; religious beliefs are often profound, sophisticated, and well-considered by the claimant.

connection with their clinical experience, medical opinion, professional judgment, or concern for their patients. Rather, their positions were framed as "religious beliefs", motivated by a desire to avoid "sinful" behaviour.[39] In this way, the physicians were perceived as seeking to practice medicine according to personal "religious" beliefs, and therefore as focusing on their "own interests" rather than those of their patients or the public.[40]

This approach has problematic implications when it comes time to balance the limitation of a claimant's (religiously-informed) conscience with competing interests under s. 1 of the Charter.[41] Instead of examining the inextricable connectivity between ethics, clinical practice, and one's (religiously informed) conscience, and how respect for these considerations might *enhance* the interests of patients and the public, only the claimant's "personal" religious element is considered. And if a claim is effectively characterized as a self-centred demand,[42] the outcome will be all but pre-determined.

This is not to suggest that the Ontario courts' framing of the claim as religious was intended to sideline the physicians' concerns. The courts could not ignore the factual and evidentiary context in which the case arose and was pled, which understandably emphasized the physicians' sincere religious beliefs. However, the physicians also raised a freedom of conscience claim, and in future cases, it will be important to recognize that religious beliefs may paint only part of the conscience picture.

If religiously-informed conscience claims are assessed *only* through the lens of freedom of religion, the implication is that freedom of conscience may be relevant as a distinct protection only for non-religious claimants. This ignores the reality that all physicians—whether religious or non-religious—possess a comprehensive worldview which informs their understanding of what is good (ethical) and bad (harmful) and which guides them in their work. For some doctors, religion may be an integral aspect of their ethical understanding, but it is not the *only* aspect.[43] Yet, unlike

---

[39] *Christian Medical and Dental Society of Canada v College of Physicians and Surgeons of Ontario*, 2019 ONCA 393 at paras 66, 73, 77.

[40] Ibid. at paras 102, 160, 187. See also *Christian Medical and Dental Society of Canada v College of Physicians and Surgeons of Ontario*, 2018 ONSC 579 (Div Ct) at paras 181, 197.

[41] Once it is established that government action or policy limits a Charter freedom, the government has the opportunity to convince the court that the limitation is reasonable and "demonstrably justified in a free and democratic society" under section 1 of the Charter.

[42] The claimants were perceived as seeking to "preserve <u>their</u> rights" and not the "interests of vulnerable patients": *Christian Medical and Dental Society of Canada v College of Physicians and Surgeons of Ontario*, 2019 ONCA 393 at paras 129, 160 (emphasis in original).

[43] *Chamberlain v Surrey School District No. 36*, 2002 SCC 86 at para 59: "...the fact that some parents and Board members may have been motivated by religious views is of no moment." See also para 137 (Gonthier J in dissent, but not on this point): "According to the reasoning espoused by [the lower court], if one's moral view manifests from a religiously grounded faith, it is not to be heard in the public square, but if it does not, then it is publicly acceptable. The problem with this approach is that everyone has 'belief' or 'faith' in something, be it atheistic, agnostic, or religious. To construe the 'secular' as the realm of 'unbelief' is therefore erroneous. **Given this, why, then, should the religiously informed conscience be placed at a public disadvantage or disqualification?** To do so would be to distort liberal principles in an illiberal fashion and would provide only a feeble notion of pluralism." (emphasis added).

"non-religious" medical frameworks which might be viewed more holistically and as informed by a variety of factors (including clinical experience and medical judgment), the religious doctor's ethic is viewed essentially as a *theological* objection. She may then be perceived as a medical outsider and as disrupting the accepted approach to clinical practice.[44]

This framing is problematic in another way: it suggests that religious claimants are seeking "special" treatment or protection not available to others, when, in fact, they are not. Conscientious dissenters—religious or otherwise—are not seeking a right to convert medicine into a religious exercise. They are simply seeking the same fundamental freedom enjoyed by everyone else—that is, the freedom not to be forced to participate in an act (in this case, the ending of a human life) that they have concluded to be harmful, based on a variety of well-considered factors (including clinical and ethical ones, informed by a sincere commitment to "do no harm" and to serve their patients with care and compassion).

Here again, conflating conscience with religion can create confusion. Freedom of religion protects many outward expressions, and is often conceived as protecting the *right* to do something without undue state interference (e.g. to worship, pray, engage in religious practices, etc.). Conscience is primarily engaged as a protection against being *forced* to do something in violation of one's deeply held convictions (e.g. participation in war, or, as some now argue should be the case, obliging physicians to *initiate* discussions about MAID even if a patient has not inquired about it but *might* be eligible).[45] Breaching conscience not only denies people the autonomy to do something they view as right, but it can empower the state to *compel* them to do something they view as wrong. This can involve coercion of a different nature than

---

[44] For further discussion on how religious claimants can be perceived as "the other", see Derek Ross and Deina Warren, "Religious Equality: Restoring Section 15's Hollowed Ground" (2019) 91 SCLR (2d).

[45] See Canadian Association of MAID Assessors and Providers, *Bringing up Medical Assistance in Dying (MAiD) as a clinical care option,* (Victoria: CAMAP, 2022), online: https://camapcanada.ca/wp-content/uploads/2022/02/Bringing-up-MAiD.pdf. The CAMAP document is referenced by the CPSO in its new draft "Advice to the Profession: Medical Assistance in Dying", online: http://policyconsult.cpso.on.ca/wp-content/uploads/2022/08/Medical-Assistance-in-Dying_Draft-Advice.pdf. Physicians, however, might conscientiously object to this requirement, concluding that initiating a discussion about MAID with a patient who is not seeking it could (1) send the message that the physician is of the view that the patient's life might not be worth living, (2) lead the patient to internalize this view herself. Again, this position is both an ethical conclusion and one rooted in medical considerations. A number of disability organizations have raised similar concerns about the devastating impact and trauma that can be experienced by a patient having MAID suggested to them as a "treatment option"; after its hearings on Bill C-7 in 2021, the Standing Senate Committee on Legal and Constitutional Affairs made the following observation: "**Requests for MAiD should be strictly patient-initiated**. The committee heard from many witnesses that the inherent power imbalance between patients and their medical practitioners constitutes a big challenge for ensuring that patients are making free and informed choices and are not inappropriately coerced into seeking access to MAiD." See Senate of Canada, "Observations to the fourth report of the Standing Senate Committee on Legal and Constitutional Affairs (Bill C-7") (43rd Parliament, 2nd Session), online: https://sencanada.ca/en/committees/LCJC/Report/90681/43-2 (emphasis added).

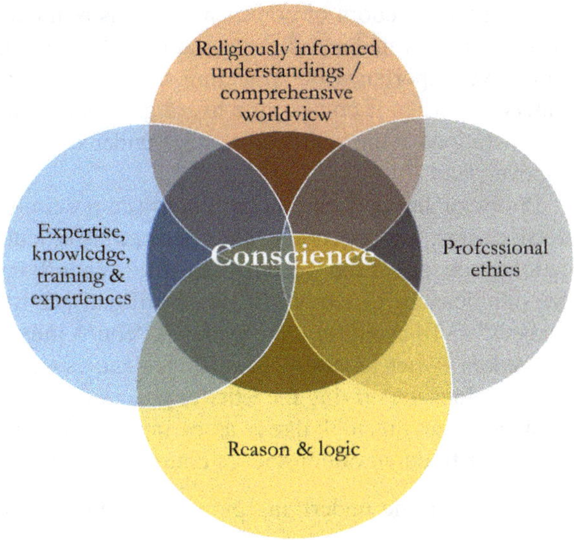

Fig. 26.1 Factors (non-exhaustive) informing conscience

may be associated with claims informing Canada's religious freedom jurisprudence to date.[46] The severity of coercion in such a breach of conscience is unique and requires greater attention and weight; this in turn must inform the analysis of whether such a breach can be "demonstrably justified in a free and democratic society" under s. 1 of the Charter.

26.3.1 Factors Informing Conscience

Every person has a conscience and is guided by it.[47] Accepting this fact at the outset of a legal analysis rightly levels the balancing scales. Conscience claims can be recognized as informed by religious beliefs where appropriate, but the legal analysis must also account for the conscience claimant's comprehensive worldview, professional ethics, reason and logic, life experiences, training, knowledge, and expertise, all of which also inform conscience. In other words, conscience may be grounded in and informed by religious beliefs (among other factors), but that does not mean that religion is the sum of conscience, as illustrated in Fig. 26.1.

[46] Of course, freedom of religion also protects freedom from coerced action, and freedom of conscience may also engage external manifestations. But this does not mean that the two protections are identical, nor are they substitutes for one another.

[47] This is a foundational principle of the *Universal Declaration of Human Rights*, which affirms that all human beings are "endowed with reason and conscience and should act towards one another in a spirit of brotherhood". See United Nations General Assembly, "Universal Declaration of Human Rights", GA/Res 217A (10 December 1948).

Assessing freedom of conscience claims requires careful consideration of each of these factors under a distinct framework,[48] one that must also account for the normative importance of protecting conscience in a free and democratic society. Otherwise, when it comes time to weigh a conscientious claim against competing considerations at the justification stage under s. 1 of the Charter, the analysis will be impoverished.[49]

Professor Jamie Cameron has expressed precisely this concern in the context of freedom of expression, noting that courts rarely articulate a robust theory for the foundational concept of the freedom, nor for its underlying values, thereby "discounting the significance [of expressive freedom] and leaving it barren in the justification analysis".[50] To avoid this, she explains, "courts must examine and explain the insult to freedom" when assessing its infringement, as well any consequences of its breach for the values underlying the Charter guarantee.[51]

What might this look like in assessing an infringement of conscience? Our thesis is that, for freedom of conscience claims to be properly assessed, courts should:

(1) articulate the underlying purpose for and necessity of protecting conscience; and
(2) develop an analytical framework that recognizes, at minimum (i) the complexity of conscientious convictions, and (ii) the severe social and public—not just individual—harms of violating conscience.

---

[48] See, for example, Justice Linden's comments in *Roach v Canada (Minister of State for Multiculturalism & Culture)* (1994), 113 DLR (4th) 67 (Fed CA): "It seems, therefore, that freedom of conscience is broader than freedom of religion. The latter relates more to religious views derived from established religious institutions, whereas the former is aimed at protecting views based on strongly held moral ideas of right and wrong, not necessarily founded on any organized religious principles. These are serious matters of conscience."

[49] This highlights another concern with considering religion to the exclusion of conscience: the s. 1 analysis may be very different, depending on the freedom in question. Certain interests protected by conscience, which might not be protected by religion, would not be considered at all in the proportionality analysis.

[50] Jamie Cameron, "Resetting the Foundations: Renewing Freedom of Expression under Section s. 2(b) of the Charter" (2022) 105 SCLR (2d) 121 at 143.

[51] Ibid.

26.4 Towards a More Purposive Interpretation of Freedom of Conscience

26.4.1 Recognizing the Ethical Complexity of a Conscientious Conviction

It is important that a claimant's conscientious position be assessed comprehensively. Sometimes, this will require a more explicit recognition that it is not an abstract "spiritual" or "dogmatic" claim, but a nuanced, well-considered, rationally defensible, ethical position, informed by moral philosophy, reason, logic, experience, and expertise.

This is not to say that courts must assess the veracity or accuracy of a claimant's position. However, the analysis must not presume (implicitly or explicitly) that a conscientious position is about advancing one's own interests or is irreconcilable with (for example) professional obligations, medical ethics, patient care, or clinical best practices, simply because it is a dissenting one or is one informed by moral reasoning.[52] In fact, as discussed below, the opposite may be true.

As a practical matter, this requires vigilance in avoiding assumptions about a conscience claimant's *motivations*. In the MAID context, for example, conscientious physicians have sometimes been characterized as invoking conscience to impose their beliefs on others, or to re-litigate the legalization of MAID. Although the Ontario courts, appropriately, did not adopt these characterizations, their reasons did seem to reflect discomfort with the moral aspects of the physicians' conscientious position, suggesting that it could exacerbate MAID's "stigmatizing legacy of several centuries of criminalization grounded in religious and secular morality".[53] This raises further questions. If it is objectionable to ground a criminal law in moral considerations, then wouldn't that render a great body of criminal law objectionable?[54] And if exposure to moral disagreement risks the harm of stigmatization, then what right does any person ultimately have to express a dissenting viewpoint on moral issues?[55]

Of course, regulators are justified in taking to steps to ensure that health care professionals treat patients with utmost respect and dignity, and it is always *possible* that an individual could assert a conscience claim for disingenuous and/or ulterior reasons. However, there was no evidence of this in the Ontario litigation. Regardless,

---

[52] See, for example, *Christian Medical and Dental Society of Canada v College of Physicians and Surgeons of Ontario*, 2019 ONCA 393 at para 187, which seems to view the physician claimants as prioritizing their personal interests over their professional obligations and/or the interests of their patients.

[53] Ibid. at para 123.

[54] Addressing moral concerns is one of the criminal law's core purposes: "Every generation faces unique moral issues. And historically, every generation has turned to the criminal law to address them. [...] Morality has long been recognized as a proper basis for the exercise of the criminal law power" (*Reference re Assisted Human Reproduction Act*, 2010 SCC 61 at paras 1, 49, per McLachlin CJC). See also *Reference re Genetic Non-Discrimination Act*, 2020 SCC 17 at para 90.

[55] See, for example, *Chamberlain v Surrey School District No. 36*, 2002 SCC 86 at paras 65–66.

the *potential* for questionable claims should never disqualify or undermine a Charter protection at the outset; otherwise, all Charter rights could be discounted, as all are susceptible to misuse. The constitutional orchard mustn't be abandoned over the hypothetical bad apple.

Relatedly, it is important for decisionmakers to avoid viewing conscientious claims as strictly "personal" and never "professional" in nature.[56] One's personal conscience has professional implications, and vice versa. In the MAID context, for example, a physician may take the position—informed by professional experience and ethics—that certain procedures are not clinically appropriate, such as MAID for mental disorders or for children.[57] Categorizing medical conscience as purely "personal" and never "professional" or "clinical" is a false dichotomy. It is also a dangerous delineation, because it separates ethical considerations from clinical practice, as several interveners argued in the *CMDS v CPSO* litigation:

> [A]s the [Ontario] Court of Appeal has recognized, 'ethical considerations form an essential part of medical decision-making'. Any exercise of professional judgment is inherently and necessarily holistic; it integrates clinical experience, education, and—critically—a morally-informed ethical framework. [...] **Attempts to bifurcate the 'moral' from the 'clinical' undermine the very idea of professional judgment, which necessarily integrates both elements.**[58]

This does not mean that every professional (or personal) opinion is protected by freedom of conscience. More work is needed to further define the threshold for what

[56] For example, the CPSO's current policy seems to assume that a physician's conscientious objection "is due to personal and not clinical reasons", and requires physicians to inform patients of same. This appears to overlook the fact that a physicians' ethical framework and approach to MAID may also be informed by medical/clinical considerations. See College of Physicians and Surgeons of Ontario, "Medical Assistance in Dying" (June 2016, last updated April 2021), online: https://www.cpso.on.ca/Physicians/Policies-Guidance/Policies/Medical-Assistance-in-Dying. While the CPSO's new, draft policy (see note 16, *supra*) does not include this phrase, the implication remains that conscientious objections are of a separate and distinct category from clinical considerations.

[57] Christian Legal Fellowship, "Submissions of Christian Legal Fellowship re: The College of Physicians and Surgeons of Ontario's (CPSO) policies concerning *Medical Assistance in Dying* and *Professional Obligations and Human Rights*" (14 May 2021), online: https://static1.squarespace.com/static/57503f9022482e2aa29ab3af/t/609ef13c76d9926869ed4fec/1621029180353/CLF+submission+to+CPSO+re+MAID+and+POHR+policies+-+May+14%2C+2021.pdf. Even where physicians are legislatively granted a "gatekeeping" role in connection with a particular regime, "they remain bound by their own ethics and codes of conduct" and may decline to participate (such as in providing medical marijuana): see *R v Mernagh*, 2013 ONCA 67 at para 88 (per Doherty JA, concurring).

[58] Factum of the Interveners The Evangelical Fellowship of Canada, The Assembly of Catholic Bishops of Ontario, and The Christian Legal Fellowship (whom the authors represented as litigation counsel in *CMDS v CPSO*) (Court of Appeal file no. C65397) at para 15 (references omitted; emphasis added), quoting *Flora v Ontario Health Insurance Plan*, 2008 ONCA 537 at para 75. See also *AC v Manitoba*, 2009 SCC 30 at para 143, where the majority observed: "Judgment is a function, not only of intellectual understanding of treatment and the consequences of refusing it, but of experience and independence [...] it requires 'ethical, emotional maturity'". While this was in the context of a patient exercising medical judgment, the same principles, in our view, apply to physicians.

freedom of conscience protects, a topic which is beyond the scope of this paper.[59] But as a preliminary consideration, some analogous guidance may be drawn from the Supreme Court's decision in *Amselem*, which refers to religious beliefs as sincere, "deeply held personal convictions" that are "integrally linked to one's self-definition and spiritual fulfillment".[60] In the context of conscience, "spiritual fulfillment" may be substituted for something connected to one's core moral compass or ultimate conception of the good. Convictions, in other words, that are so profound and central to one's moral identity that abiding by them is not a "desire" so much as a "duty"—one that a claimant cannot breach without doing harm to her (or others') "essential humanity".[61]

The Charter's text makes it clear that conscience, like the other freedoms protected by s. 2(a), engages interests more "fundamental" than most exercises or pursuits.[62] And while it is not a license to do whatever one *wants* for any reason one *chooses*, neither is it a "liberty full of threats to all"[63] (which seems to be the way fundamental freedoms are sometimes viewed and, therefore, more readily limited). Freedom of conscience must not be devalued in this manner. To borrow from Lord Acton, it protects not "the power of doing what we like, but the right of being able to do what we ought".[64]

Thus, in addition to adopting a more nuanced understanding of the multi-faceted nature of a claimant's conscientious position, courts must also consider the severe harms potentially caused when conscience is violated, as discussed in the following section.

---

[59] For an insightful doctrinal discussion on "what must be proven to conclude that the state has limited freedom of conscience" see Brian DN Bird, *The Forgotten Fundamental: Freedom of Conscience in Canada* (DCL Thesis, McGill University, 2021), online: https://escholarship.mcgill.ca/concern/theses/4j03d4180, at pp. 158–170. Bird identifies some threshold criteria, including that the limitation (1) be more than "trivial or substantial" (drawing from religious freedom jurisprudence), and (2) require the claimant to materially cooperate, at a requisite degree of proximity, in an act they sincerely view as unethical.

[60] *Syndicat Northcrest v Amselem*, 2004 SCC 47 at para 39.

[61] *R v Morgentaler* [1988] 1 SCR 30 at 165 and 179.

[62] Freedom of conscience is listed as a "fundamental freedom" in s. 2, and like all other fundamental freedoms, is separately delineated from the "right to liberty" in s. 7.

[63] William Shakespeare, *The Tragedy of Hamlet, Prince of Denmark* at Act 4, Scene 1.

[64] Gary M Galles, "Lord Acton on the Meaning of Freedom", Fee Stories (24 September 2017), online: https://fee.org/articles/lord-action-on-the-meaning-of-freedom.

26.4.2 Recognizing the Importance of Conscience and the Harms of Its Violation, Especially in Professional Disciplines

As Dr. Brian Bird explains, conscience is central to one's identity and integrity, both of which also engage one's inherent dignity.[65] This speaks powerfully to the significant weight that should be afforded to freedom of conscience. In addition to harming the *individual* (and those the individual serves), violating conscience harms *society* by depriving it of the benefits of diverse voices and ideas (among other goods).

This is especially so in fields where progress, advancement, and the search for truth depend on professional independence, academic freedom, and intellectual diversity. For example, compelling all physicians, or lawyers, or academics to affirmatively advance a particular ideological viewpoint leads to dangerous groupthink and suppresses ideas that could advance better solutions or correct faulty presuppositions. As the Supreme Court has recognized, "No one has a monopoly on truth."[66] "No one" includes public institutions, lawmakers, and regulators.

Legal scholarship has called for a renewed understanding of the relationship between the fundamental freedoms (including conscience) and the search for truth.[67] As Professor Cameron explains:

> The value of freedom, in this account, is that it provides a counterpoint to a '[c]onformity, deference to authority, stasis, [and] passivity in the realm of beliefs' that is 'not just unfortunate or unwise but *dangerous*' […] this principle of freedom offers the goal and opportunity to seek the 'truth of things generally', mindful and hopeful of achieving progressive objectives along the way.[68]

Society benefits immensely when the fundamental freedoms—and, with them, the search for truth—are advanced. Professional disciplines such as medicine and law are fields where conscience must be especially protected, because they intrinsically depend on it. Conscience is a prerequisite for, and an expression of, professional independence.[69] Professional independence, in turn, is an essential ingredient—a "hallmark"—of a free society.[70]

Much like academic freedom, professional independence requires freedom from external/state interference *as well as* freedom from majoritarian pressures *within*

---

[65] Brian Bird, "The Reasons for Freedom of Conscience" (2020) 98 SCLR (2d).

[66] *Reference re Secession of Quebec*, [1988] 2 SCR 217 at para 68.

[67] Kathleen Brady, "Religious Group Autonomy: Further Reflections About What is at Stake" (2006/2007) 22:1 J.L. & Religion 153; and Derek Ross, "Truth Seeking and the Unity of the Charter's Fundamental Freedoms" (2020) 98 SCLR (2d).

[68] Jamie Cameron, "Resetting the Foundations: Renewing Freedom of Expression under Section s. 2(b) of the Charter" (2022) 105 SCLR (2d) at 140–141 (references omitted, emphasis added by Cameron, citing V. Blasi, "Holmes and the Marketplace of Ideas" (2004) Sup. Ct. Rev. 1 at 29).

[69] See Julia Laffranque, "Dissenting Opinion and Judicial Independence" (2003) 8 Juridica Intl, 162 at 169, discussing this idea in the context of judicial independence.

[70] *A.G. Can. v. Law Society of B.C.* (1982), 335–336 (per Estey J, discussing the independence of the bar).

professions and institutions themselves. As Supreme Court Justice Suzanne Côté explains:

> I think that [dissent] is healthy, and for me judicial independence requires that. [...] And judicial independence is not only independence from the outside, from external pressure. It **is also independence within your institution.** I feel when I write my opinion, majority or dissent, that **I'm acting in accordance with my conscience.**[71]

Just as "legal conscience" is necessary for an independent bar and judiciary,[72] so too is "medical conscience" necessary for a just and equitable health care system: one that is open to and reflective of the diverse public it serves, and that is responsive to new, helpful ideas which can correct harmful ones.

Without minimizing the significance of individual integrity, dignity, and identity as underlying principles for protecting conscience, this broader context demonstrates that conscience is not a mere "personal" exercise of moral self-preservation in which professionals may be (unfairly) characterized as prioritizing their own moral comfort over the interests of those they serve. Rather, it is the exact opposite: a professional's conscience provides the very means by which they serve others and advance the public interest.

26.5 Conclusion

So what might all of this mean for future decisions? First, it is important to note that there remains room for legislatures and physicians' colleges to adopt different policies than those currently in place in Ontario.[73] It remains open to a legislature or regulatory college to conclude that patients are best served by a medical community that includes diverse ethical viewpoints and accommodates physicians and patients alike who, along with the WMA,[74] do not view MAID as a medical solution to suffering.

---

[71] Cristin Schmitz, "Litigators' champion celebrates five years at SCC, looks forward to five more years", Lawyers Daily (8 May 2020), online: https://www.thelawyersdaily.ca/articles/18911/litigators-champion-celebrates-five-years-at-scc-looks-forward-to-five-more-years?article_related_content=1 (emphasis added).

[72] Ibid. See also the comments of Justice Bernice B. Donald (U.S. Court of Appeals for the Sixth Circuit), who emphasizes the importance of the right to dissent and the "enormous value it can have as an expression of legal conscience", describing it as a "pillar of judicial independence"; see Bernice B. Donald, "Judicial Independence, Collegiality, and the Problem of Dissent in Multi-Member Courts" (2019) 94:3 NYU L Rev, 317 at 339.

[73] See, for example, Bill 34, *The Medical Assistance in Dying (Protection for Health Professionals and Others) Act* (assented to 10 November 2017) CCSM c M92.

[74] World Medical Association, "WMA Declaration on Euthanasian and Physician-Assisted Suicide: Adopted by the 70th WMA General Assembly, October 2019" (23 November 2021), online: https://www.wma.net/policies-post/declaration-on-euthanasia-and-physician-assisted-suicide/.

A decisionmaker may also determine that the public is best served when physicians are not forced to violate their ethical framework,[75] recognizing, as the trial judge did in *Carter*, that "thoughtful and well-motivated people can and have come to different conclusions about whether physician-assisted death can be ethically justifiable."[76] Decisionmakers may conclude that accommodating such diverse, thoughtful, and well-motivated viewpoints amongst medical professionals *contributes* to the public good by allowing a range of options for patients. As one legal organization has explained, accommodating conscience strengthens the health care system by fostering greater inclusivity and accessibility to health care for marginalized patients who seek support from physicians who share their commitments and approach to care.[77] For example, physicians who affirm that MAID is not a "medical solution" for non-life-threatening disabilities and/or for mental illness may be uniquely equipped to support patients who hold the same view, but who may feel unsafe or unsupported in our health care system.[78]

Second, as discussed above, there remain some unanswered questions about the constitutionality of coercing health care professionals to participate in procedures they view as unethical through mandatory referrals, as it pertains to their freedom of *conscience*. The importance of reviving conscience and understanding it as a distinct and unique freedom is accentuated now that MAID may be expanded in more areas of medicine and to a much larger group of people than initially contemplated. Since the *CPSO v CMDS* decisions were released, Bill C-7 has expanded MAID eligibility to include those who are *not* dying or near death and will further expand eligibility to those whose sole underlying condition is mental illness at a date to be confirmed.[79] Indeed, a Parliamentary Committee recently heard evidence that

---

[75] See, for example, Michael Quinlan, "When the State Requires Doctors to Act Against their Conscience: The Religious Freedom Implications of the Referral and the Direction Obligations of Health Practitioners in Victoria and New South Wales", (2017) 2016:4 BYU L Rev 1237 at 1271: "Health practitioners who consistently act against their conscience can also become desensitized to it. They are at greater risk of developing indifference to patients and 'doubling' or 'compartmentalization,' leading to a weakened ability to make the types of ethical decisions critical for health practitioners."

[76] *Carter v Canada*, 2012 BCSC 886 at para 323.

[77] Christian Legal Fellowship, "Submissions of Christian Legal Fellowship re: The College of Physicians and Surgeons of Ontario's (CPSO) policies concerning *Medical Assistance in Dying* and *Professional Obligations and Human Rights*" (14 May 2021), online: https://static1.squarespace.com/static/57503f9022482e2aa29ab3af/t/609ef13c76d9926869ed4fec/1621029180353/CLF+submission+to+CPSO+re+MAID+and+POHR+policies+-+May+14%2C+2021.pdf.

[78] Ibid.

[79] Bill C-7, *An Act to amend the Criminal Code (medical assistance in dying)*, 2nd Sess, 43rd Parl, 2021 (assented to 17 March 2021), SC 2021, c 2. MAID for mental illness as a sole underlying condition was originally legislated to be introduced in March 2023. In response to concerns that proper safeguards are not yet in place, Parliament voted to delay introducing this next step of expansion until March 2024; See Department of Justice Canada, "Statement by Ministers Lametti, Duclos and Bennett on medical assistance in dying in Canada" (15 Dec 2022), online: https://www.canada.ca/en/department-justice/news/2022/12/statement-by-ministers-lametti-duclos-and-bennett-on-medical-assistance-in-dying-in-canada.html.

Canadian veterans were offered MAID for post-traumatic stress disorder (PTSD).[80] Still broader expansion to mature minors, to those who are no longer competent to consent but made known a prior request, and even to newborn infants[81] might also be possible.[82] Careful reflection and scrupulous review of such regimes is absolutely necessary. The Supreme Court required no less in its *Carter* ruling.[83] Further, legal, medical, and ethical views vary widely on these subjects.[84] Therefore, not only will conscientious concerns about participating in these procedures become more complex, the diversity of viewpoints and positions will also likely increase.[85]

Perhaps a place where theses varied views can meet is in the concept of conscience, the basis for our shared "essential humanity".[86] To return to Justice Dickson's remarks in *Big M Drug Mart*:

> What unites enunciated freedoms in [...] s. 2(a) of the *Charter* and in the provisions of other human rights documents [...] is the notion of the centrality of individual conscience and the

---

[80] Standing Committee on Veterans Affairs, *Evidence,* 44-1, No. 27 (24 November 2022).

[81] Special Joint Committee on Medical Assistance in Dying, *Evidence*, 44-1, No 19 (7 October 2022) at 10:00 (Dr. Louis Roy, on behalf of the Collège des médecins du Québec): "With support from the board of directors, the committee recommended that emancipated minors be able to apply for MAID in conjunction with their parental authority or guardian. [...] The same is true for babies from birth to one year of age who come into the world with severe deformities and very serious syndromes for which the chances of survival are virtually nil..."

[82] The Special Joint Committee on Medical Assistance in dying is currently exploring the issue of MAID for mature minors and MAID by advance requests.

[83] *Carter v Canada*, 2015, SCC 5 at para 117. Even MAID providers have testified that the current regime is so ill-defined, it "can render the legislative safeguards impotent." Special Joint Committee on Medical Assistance in Dying, *Evidence*, 44-1, No 19 (18 October 2022) at 19:40 (Dr. Madeline Li).

[84] See, for example.: Catalina Devandas-Aguilar, "Visit to Canada: Report of the Special Rapporteur on the rights of persons with disabilities", Human Rights Council, 43rd Sess, A/HRC/43/41/Add.2 (19 December 2019), online: https://www.un.org/en/ga/search/view_doc.asp?symbol=A/HRC/43/41/Add.2; The Expert Panel Working Group on Advance Requests for MAID, "The State of Knowledge on Advance Requests for Medical Assistance in Dying", Council of Canadian Academies (2018) Ottawa (ON), ISBN: 978-1-926522-51-7; The Expert Panel Working Group on MAID for Mature Minors, "The State of Knowledge on Medical Assistance in Dying for Mature Minors", Council of Canadian Academies (2018) Ottawa (ON, ISBN: 978-1-926522-46-3; The Expert Panel Working Group on MAID Where A Mental Disorder Is the Sole Underlying Medical Condition, "The State of Knowledge on Medical Assistance in Dying Where a Mental Disorder Is the Sole Underlying Medical Condition", Council of Canadian Academies (2018) Ottawa (ON), ISBN: 978-1-926522-48-7; Gerard Quinn, Claudia Mahler, and Olivier De Schutter, "Disability is not a reason to sanction medically assisted dying—UN experts" (25 January 2021), online: https://www.ohchr.org/en/press-releases/2021/01/disability-not-reason-sanction-medically-assisted-dying-un-experts?LangID=E&NewsID=26687.

[85] See, for example, the recent comments of the Expert Panel on MAiD and Mental Illness: "If there are cases whose characteristics do not correspond with an assessor's own values, objecting on conscience grounds may be appropriate.": "Final Report of the Expert Panel on MAiD and Mental Illness", Health Canada (May 2022), online: https://www.canada.ca/content/dam/hc-sc/documents/corporate/about-health-canada/public-engagement/external-advisory-bodies/expert-panel-maid-mental-illness/final-report-expert-panel-maid-mental-illness/final-report-expert-panel-maid-mental-illness.pdf.

[86] See *R v Morgentaler* [1988] 1 SCR 30 at 165 and 179.

> inappropriateness of governmental intervention to compel or to constrain its manifestation. In *Hunter v. Southam Inc., supra,* the purpose of the *Charter* was identified [...] as 'the unremitting protection of individual rights and liberties'. It is easy to see the relationship between respect for individual conscience and the valuation of human dignity that motivates such unremitting protection.[87]

Agreeing to the importance of conscience and dignity does not eliminate cognitive dissonance, but it can provide a framework of understanding. As the Supreme Court has emphasized, cognitive dissonance is "neither avoidable nor noxious", but "is simply a part of living in a diverse society".[88]

In the case of the most basic of human instincts—to not end another person's life—it is essential to guard against eliminating cognitive dissonance on the ethics of these matters. These are, after all, questions of life and death. If freedom of conscience has no role to play here, where does it?

Acknowledgements We are grateful to Jamie Cameron and Brian Bird for their thoughtful comments on earlier drafts of this paper, and to Garifalia Milousis for her editorial review. Any errors are solely our own.

References

AC v Manitoba (Director of Child and Family Services), 2009 SCC 30, [2009] 2 SCR 181.

AG Can v Law Society of BC [1982] 2 SCR 307.

Bill C-7, *An Act to amend the Criminal Code (medical assistance in dying)*, 2nd Sess, 43rd Parl, 2021 (assented to 17 March 2021), SC 2021, c 2.

Bill 34, *The Medical Assistance in Dying (Protection for Health Professionals and Others) Act* (assented to 10 November 2017) CCSM c M92.

Bird, Brian. 2018. The Call in *Carter* to interpret freedom of conscience. *SCLR (2d)* 85: 107–141.

Bird, Brian. 2021. *The forgotten fundamental: Freedom of conscience in Canada* (DCL Thesis, McGill University). Online: https://escholarship.mcgill.ca/concern/theses/4j03d4180

Bird, Brian. 2020. The reasons for freedom of conscience. *SCLR (2d)* 98.

Bird, Brian, Dwight Newman, and Derek Ross. 2020. The Charter's forgotten fundamental freedoms. Policy Options (16 June 2020). Online: https://policyoptions.irpp.org/magazines/june-2020/the-charters-forgotten-fundamental-freedoms/

Brady, Kathleen. 2006. Religious group autonomy: further reflections about what is at stake (2006/2007). *J.L. & Religion* 22 (1): 153.

Bussey, Barry W. 2020. Blazing the path: Freedom of conscience as the prototypical right. *SCLR (2d)* 98.

Cameron, Jamie. 2022. Resetting the foundations: Renewing freedom of expression under Section s.2(b) of the Charter. *SCLR (2d)* 105: 121.

Canadian Medical Association. 2017. CMA policy: Medical assistance in dying. Online: https://policybase.cma.ca/viewer?file=%2Fmedia%2FPolicyPDF%2FPD17-03.pdf

Carter v Canada, 2012 BCSC 886.

Carter v Canada, 2015 SCC 5.

Chamberlain v Surrey School District No. 36, 2002 SCC 86.

[87] *R v Big M Drug Mart*, [1985] 1 SCR 295 at 346.

[88] *Chamberlain v Surrey School District No. 36*, 2002 SCC 86 at para 65.

Christian Legal Fellowship. 2021. Submissions of Christian Legal Fellowship re: The College of Physicians and Surgeons of Ontario's (CPSO) policies concerning *Medical Assistance in Dying* and *Professional Obligations and Human Rights* (14 May 2021). Online: https://static1.squarespace.com/static/57503f9022482e2aa29ab3af/t/609ef13c76d9926869ed4fec/1621029180353/CLF+submission+to+CPSO+re+MAID+and+POHR+policies+-+May+14%2C+2021.pdf

Christian Medical and Dental Society of Canada v College of Physicians and Surgeons of Ontario, 2018 ONSC 579 (Div Ct).

Christian Medical and Dental Society of Canada v College of Physicians and Surgeons of Ontario, 2019 ONCA 393.

Coelho, Ramona, K. Sonu Gaind, Trudo Lemmens, and John Maher. 2022. Normalizing death as a 'treatment' in Canada: whose suicides do we prevent, and whose do we abet? *World Medical Journal* 70 (3): 27–35.

College of Physicians and Surgeons of Ontario. Female genital cutting (mutilation). Online: https://www.cpso.on.ca/Physicians/Policies-Guidance/Policies/Female-Genital-Cutting-Mutilation

College of Physicians and Surgeons of Ontario. Medical assistance in dying. (June 2016; last updated April 2021). Online: https://www.cpso.on.ca/Physicians/Policies-Guidance/Policies/Medical-Assistance-in-Dying

College of Physicians and Surgeons of Ontario. Human rights in the provision of health services. (Draft policy; accessed December 2022). Online: http://policyconsult.cpso.on.ca/wp-content/uploads/2022/08/Human-Rights-in-the-Provision-of-Health-Services_Draft-Policy.pdf

College of Physicians and Surgeons of Ontario. Advice to the profession: Human rights in the provision of health services. (Draft policy; accessed December 2022). http://policyconsult.cpso.on.ca/wp-content/uploads/2022/08/Human-Rights-in-the-Provision-of-Health-Services_Draft-Advice.pdf

Criminal Code of Canada, RSC 1985, c C-64, s. 241.1.

Department of Justice Canada. 2022. Statement by Ministers Lametti, Duclos and Bennett on medical assistance in dying in Canada (15 Dec 2022). Online: https://www.canada.ca/en/department-justice/news/2022/12/statement-by-ministers-lametti-duclos-and-bennett-on-medical-assistance-in-dying-in-canada.html

Devandas-Aguilar, Catalina. 2019. Visit to Canada: Report of the Special Rapporteur on the rights of persons with disabilities. Human Rights Council, 43rd Sess, A/HRC/43/41/Add.2 (19 December 2019). Online: https://www.un.org/en/ga/search/view_doc.asp?symbol=A/HRC/43/41/Add.2

Donald, Bernice B. 2019. Judicial independence, collegiality, and the problem of dissent in multi-member courts. *NYU L Rev* 94 (3): 317–339.

Expert Panel on MAiD and Mental Illness. 2022. *Final report of the Expert Panel on MAiD and Mental Illness*. Health Canada (May 2022). Online: https://www.canada.ca/content/dam/hc-sc/documents/corporate/about-health-canada/public-engagement/external-advisory-bodies/expert-panel-maid-mental-illness/final-report-expert-panel-maid-mental-illness/final-report-expert-panel-maid-mental-illness.pdf

Expert Panel Working Group on Advance Requests for MAID. 2018. *The state of knowledge on advance requests for medical assistance in dying*. Ottawa, ON: Council of Canadian Academies. ISBN 978-1-926522-51-7

Expert Panel Working Group on MAID for Mature Minors. 2018. *The state of knowledge on medical assistance in dying for mature minors*. Ottawa, ON: Council of Canadian Academies. ISBN 978-1-926522-46-3

Expert Panel Working Group on MAID Where a Mental Disorder Is the Sole Underlying Medical Condition. 2018. *The state of knowledge on medical assistance in dying where a mental disorder is the sole underlying medical condition*. Ottawa, ON: Council of Canadian Academies. ISBN 978-1-926522-48-7

Galles, Gary M. 2017. Lord action on the meaning of freedom. Fee Stories (24 September 2017). Online: https://fee.org/articles/lord-action-on-the-meaning-of-freedom

Kislowicz, Howard, Richard Haigh, and Adrienne Ng. 2011. "Calculations of Conscience: The Costs and Benefits of Religious and Conscientious Freedom" (2011) 48:3 Alta L Rev 679.

Laffranque, Julia. 2003. Dissenting opinion and judicial independence. *Juridica Intl* 8: 162–172.
Lavoie v Canada [2002] 1 SCR 769.
Math, Suresh Bada, and Santosh K. Chaturvedi. 2012. Euthanasia: Right to live vs right to die. *Indian J Med Res* 136(6). Online: https://www.ncbi.nlm.nih.gov/pmc/articles/PMC3612319/.
Mounted Police Association of Ontario v Canada (Attorney General), 2015 SCC 1.
Quinlan, Michael. 2016. When the state requires doctors to act against their conscience: The religious freedom implications of the referral and the direction obligations of health practitioners in Victoria and New South Wales. *BYU L Rev* 2016 (4): 1237–1287.
Quinn, Gerard, Claudia Mahler, and Olivier De Schutter. 2021. Disability is not a reason to sanction medically assisted dying—UN experts (25 January 2021). Online: https://www.ohchr.org/en/press-releases/2021/01/disability-not-reason-sanction-medically-assisted-dying-un-experts?LangID=E&NewsID=26687
Quebec (Attorney General) v A, 2013 SCC 5.
R v Big M Drug Mart [1985] 1 SCR 295.
R v Mernagh, 2013 ONCA 67.
R v Morgentaler [1988] 1 SCR 30.
Raikin, Alexander. 2023. No other options. *The New Atlantis* (Winter 2023). Online: https://www.thenewatlantis.com/publications/no-other-options
Reference re Assisted Human Reproduction Act, 2010 SCC 61.
Reference re Genetic Non-Discrimination Act, 2020 SCC 17.
Reference re Secession of Quebec, [1988] 2 SCR 217.
Roach v Canada (Minister of State for Multiculturalism & Culture) (1994), 113 DLR (4th) 67(Fed CA).
Ross, Derek. 2020. Truth seeking and the unity of the charter's fundamental freedoms. *SCLR (2d)* 98.
Ross, Derek, and Deina Warren. 2019. Religious equality: Restoring Section 15's hollowed ground. *SCLR (2d)* 91.
Schmitz, Cristin. 2020. Litigators' champion celebrates five years at SCC, looks forward to five more years. *Lawyers Daily* (8 May 2020). Online: https://www.thelawyersdaily.ca/articles/18911/litigators-champion-celebrates-five-years-at-scc-looks-forward-to-five-more-years?article_related_content=1
Senate of Canada. Oberservations to the fourth report of the Standing Senate Committee on Legal and Constitutional Affairs (Bill C-7) (43rd Parliament, 2nd Session). Online: https://sencanada.ca/en/committees/LCJC/Report/90681/43-2
Shakespeare, William. *The Tragedy of Hamlet, Prince of Denmark* at Act 4, Scene 1.
Simpson, Alexander, Jason Tran, and Roland Jones. 2022. Ethical considerations regarding mental disorder and medical assistance in dying (MAiD) in the prison population. *Medicine, Science and the Law*, 1–3.
Syndicat Northcrest v Amselem, 2004 SCC 47, [2004] 2 SCR 551.
United Nations General Assembly. 1948. Universal Declaration of Human Rights. GA/Res 217A (10 December 1948). Online: https://www.un.org/sites/un2.un.org/files/2021/03/udhr.pdf
Waldron, Mary Ann. 2018. Conscientious objections to medical aid in dying: considering how to manage claims of conscience in a pluralistic society. *SCLR (2d)* 85.
World Medical Association. 2021. WMA Declaration on Euthanasian and Physician-Assisted Suicide: Adopted by the 70th WMA General Assembly, October 2019 (23 November 2021). Online: https://www.wma.net/policies-post/declaration-on-euthanasia-and-physician-assisted-suicide/
World Medical Association. 2022. WMA Declaration of Venice on End of Life Medical Care (revised by the 73rd General Assembly, October 2022), online: https://www.wma.net/policies-post/wma-declaration-of-venice/
World Medical Association. 2022. WMA International Code of Ethics (revised by the 73rd General Assembly, October 2022). Online: https://www.wma.net/policies-post/wma-international-code-of-medical-ethics/

Derek Ross, LL.B., LL.M. is Executive Director & General Counsel for Christian Legal Fellowship (CLF), Canada's national association of Christian lawyers, jurists, and law students. He has appeared before all levels of court, including the Supreme Court of Canada, and has acted for public interest interveners in a number of cases involving the *Canadian Charter of Rights and Freedoms.* He has also appeared before legislative and Parliamentary committees to present on legal and constitutional issues, including medical assistance in dying. In 2017, Derek helped found CLF's annual *Academic Symposium on Religion, Law & Human Rights,* and has edited/co-edited four resulting academic collections published as volumes of the *Supreme Court Law Review.* Derek is a centre associate with the University of British Columbia's Centre for Constitutional Law and Legal Studies, and currently serves as an executive member of the Constitutional & Human Rights Law Section of both the Canadian Bar Association and the Ontario Bar Association.

Deina Warren, LL.B., LL.M. is the Director of Legal Affairs at the Canadian Centre for Christian Charities. She advises on a broad range of charity, tax and regulatory matters, religious freedom and human rights law and regularly engages in government consultations and advocacy efforts. She has served as counsel in a variety of trial and appeal courts, including interventions before the Supreme Court of Canada in *Trinity Western* and *Wall v Highwood Congregation*, and on matters at all levels of court in Ontario. She has been part of a number of charity boards and is active in the Charity and Not-for-Profit and Constitutional, Civil Liberties and Human Rights sections of the Ontario Bar Association. Much of her spare time is spent attempting to garden, baking, exploring provincial parks with her family and walking their rambunctious rescue dog.

Chapter 27
Freedom of Conscience and Medical Assistance in Dying—Clinical Perspective

Simon Czajkowski, Sean Murphy, and Ewan C. Goligher

Abstract In *Carter v. Canada (Attorney General)*, the Supreme Court emphasized that a physician's decision to participate in assisted dying is a matter of conscience or religious belief, and that nothing in *Carter* would compel physicians to provide assistance in dying. Legislators and regulators of medical practice have responded by affirming, in principle, the importance of upholding the freedom of conscience of physicians. In practice, however, varying policies have been implemented across Canada, with some regulators compelling physicians to refer patients to available providers who would be willing to administer medical assistance in dying (MAID). We begin this chapter by defining 'conscience' and discussing how it operates in the context of medical practice vis-à-vis the goals of medicine and its supporting virtues; we argue that professional obligations can sometimes necessitate conscientious objection. We proceed to explain how reasonable disagreement can exist regarding the consistency of MAID with the goals and virtues of medicine. We then discuss and critique the concept of the 'conventional compromise'—a proposed justification for mandating referrals from conscientiously objecting physicians. Finally, we argue that referrals are not morally neutral acts and that long-standing professional obligations can prevent physicians from providing them for MAID.

Keywords Medical assistance in dying (MAID) · Conscience · Conscientious objection · Dignity · Goals of medicine · Professional obligations · Virtues · Access · Referral · Conventional compromise

---

S. Czajkowski
Division of Urology, Department of Surgery, University Health Network, Toronto, Canada
e-mail: Simon.Czajkowski@uhn.ca

S. Murphy
Protection of Conscience Project, Powell River, Canada

E. C. Goligher (✉)
Interdepartmental Division of Critical Care Medicine, University of Toronto, Toronto, Canada
e-mail: ewan.goligher@uhn.ca

Department of Medicine, Division of Respirology, University Health Network, Toronto, Canada

J. Kotalik and D. W. Shannon (eds.), *Medical Assistance in Dying (MAID) in Canada*, The International Library of Bioethics 104,
https://doi.org/10.1007/978-3-031-30002-8_27

27.1 Introduction

When the Supreme Court of Canada struck down the absolute prohibition on physician-assisted death in *Carter v. Canada (Attorney General)*, it indicated that its decision did not necessarily entail an obligation for individual physicians to participate in the provision of this intervention. Bill C-14, the legislation amending the criminal code to permit medical assistance in dying (MAID) also suggested that the freedom of conscience and religion of all Canadians remained protected under the *Charter*. In principle, provincial medical colleges affirm the value of physicians' right to conscientious objection in the context of MAID. In practice, however, substantially different policies and guidelines outlining the concrete positive and negative obligations[1] owed by conscientiously objecting physicians to MAID-seeking patients exist across Canada. Table 27.1 outlines the key specific obligations for physicians with a conscientious objection to MAID defined by each of the various provincial or territorial regulators of medical practice in Canada. Of note, some provinces specifically stipulate that an objecting physician is required to refer a patient who is requesting MAID to a willing provider. This requirement, while respecting conscience to some extent insofar as it does not actually obligate the objecting physician to personally administer lethal medications, has occasioned considerable legal and ethical conflict in Canada and has been specifically criticized by professional organizations such as the Canadian Medical Association.[2] We evaluate the impact of MAID legislation and regulation on conscientious practice of medicine in Canada. Because of widespread confusion and misunderstanding regarding the concept of conscience, we give considerable effort to discussing how conscience generally operates in the context of medical practice vis-à-vis the goals of medicine and its essential virtues, and we show that professional obligations can sometimes necessitate conscientious objection. We examine whether there can be reasonable disagreement about the consistency of MAID with the goals and virtues of medicine. We then evaluate the 'conventional compromise'—the rationale for mandating physicians with a conscientious objection to refer patients to practitioners who are willing to provide the objectionable intervention. We argue that such referrals are not morally neutral acts and that long-standing professional obligations apply to acts of referral.

---

[1] In this case, positive and negative do not refer to moral valence, but rather to performing an action in order to uphold another person's *positive right* or refraining from an action in order to uphold another person's *negative right,* respectively. For a comprehensive discussion about positive and negative rights and obligations, as well as their origins, see *Basic Rights: Subsistence, Affluence, and U.S. Foreign Policy* by Henry Shue (1980) and *Two Concepts of Liberty* by Isaiah Berlin (1958).

[2] Canadian Medical Association, "CMA's Submission to the College of Physicians and Surgeons of Ontario (CPSO): Consultation on CPSO Interim Guidance on Physician-Assisted Death."

Table 27.1 Access-related obligations of physicians with a conscientious objection to 'medical assistance in dying' (MAID) according to the policies and guidelines of medical regulators in Canada

Medical regulator	Obligations of physicians with a conscientious objection
College of Physicians and Surgeons of Alberta (CPSA)	"When Charter freedom of conscience and religion prevent a regulated member from providing or offering access to information about a legally available medical or surgical treatment or service, the regulated member must ensure that the patient who seeks such advice or medical care is offered timely access to: (a) a regulated member who is willing to provide the medical treatment, service or information; or (b) a resource that will provide accurate information about all available medical options"[3] Referring the patient to the AHS Care Coordination Service, which can guide patients and practitioners through the MAID process, would satisfy the above requirement[4]
College of Physicians and Surgeons of British Columbia (CPSBC)	"Registrants who object to MAID on the basis of their values and beliefs are required to provide an effective transfer of care for their patients by advising patients that other practitioners may be available to see them, suggesting the patient visit an alternate physician or service, and if authorized by the patient, transferring the medical records as required. Where needed, registrants must offer assistance to the patient and must not abandon the patient. Any registrant receiving a written request for MAID who transfers the care of the patient to another provider or care coordinator for any reason must complete the provincial form to report details about this transfer of care"[5]
College of Physicians and Surgeons of Manitoba (CPSM)	"A physician who refuses to refer a patient to another physician or to personally offer specific information about MAID on the grounds of a conscience-based objection must: ... provide the patient with timely access to a resource that will provide accurate information about MAID, including how a patient can make a request for MAID or to be assessed for eligibility for MAID … and make available the patient's chart and relevant information (i.e., diagnosis, pathology, treatment and consults) to the physician(s) providing MAID to the patient when authorized by the patient to do so …"[6]

(continued)

[3] "CPSA Standards of Practice: Conscientious Objection."

[4] "CPSA Advice to the Professional: Medical Assistance in Dying."

[5] "CPSBC Practice Standard: Medical Assistance in Dying."

[6] "The College of Physicians & Surgeons of Manitoba Standard of Practice: Medical Assistance in Dying (MAiD)."

Table 27.1 (continued)

Medical regulator	Obligations of physicians with a conscientious objection
College of Physicians and Surgeons of New Brunswick (CPSNB)	"When moral or religious beliefs prevent a physician from providing or offering access to information about a legally available medical or surgical treatment or service, that physician must ensure that the patient who seeks such advice or medical care is offered timely access to another physician or resource that will provide accurate information about all available medical options"[7]
College of Physicians and Surgeons of Newfoundland & Labrador (CPSNL)	"The College recommends that a physician who declines to participate in medical assistance in dying offer the patient timely access to another medical professional (or appropriate information resource, clinic or facility, care provider, health authority, or organization) who is: (a) available to assist the patient; (b) accessible to the patient; and (c) willing to provide medical assistance in dying to a patient who meets the eligibility criteria"[8]
College of Physicians and Surgeons of Nova Scotia (CPSNS)	"The physician unable or unwilling to participate or to continue to participate in medical assistance in dying must complete an effective transfer of care for any patient requesting or eligible to receive medical assistance in dying"[9] An "'effective transfer of care' means a transfer made by one practitioner in good faith to another practitioner who is available to accept the transfer, who is accessible to the patient, and willing to provide medical assistance in dying to the patient if the eligibility criteria are met"
College of Physicians and Surgeons of Ontario (CPSO)	"... Physicians who decline to provide MAID due to a conscientious objection: ... must provide the patient with an effective referral".[10] An 'effective referral' entails "taking positive action to ensure the patient is connected to a non-objecting, available, and accessible physician, other health-care professional, or agency" It is important to note that the 'effective referral' is a transfer of care that is specific to the care to which the physician objects[11]

(continued)

[7] "College of Physicians and Surgeons of New Brunswick: Moral Factors and Medical Care."

[8] "College of Physicians & Surgeons of Newfoundland and Labrador Standard of Practice: Medical Assistance in Dying."

[9] "College of Physicians & Surgeons of Nova Scotia Professional Standard Regarding Medical Assistance in Dying (MAiD)."

[10] "CPSO Medical Assistance in Dying."

[11] "CPSO Advice to the Professional: Medical Assistance in Dying."

Table 27.1 (continued)

Medical regulator	Obligations of physicians with a conscientious objection
College of Physicians and Surgeons of Prince Edward Island (CPSPEI)	"A physician who for conscientious reasons declines to provide medical assistance in dying must: … provide, or arrange to be provided, the patient's chart and sufficient medical information, with the patient's consent, to the patient or to other physicians or nurse practitioners involved in the process … [and] … enable access to another physician, nurse practitioner or service"[12] "A physician may delegate the responsibility for communication of information regarding medical assistance in dying to another person (who is competent to do so and for whom the physician is responsible), or to another agency"
Collège des médecins du Québec (CMQ)	The physician "must tell [the patient] about the range of available medical options and, by notifying the authorities, help him find a colleague who is willing to provide such aid provided the legal and medical requirements are met"[13] "The authorities to be notified vary depending on the physician's place of practice" (e.g. physicians practicing at a centre operated by an institution, must notify the President and CEO of that centre, or their delegate)
College of Physicians and Surgeons of Saskatchewan (CPSS)	"The physician is expected to provide sufficient information and resources to enable the patient to make his/her own informed choice and access all options for care. This means arranging timely access to another physician or resources, or offering the patient information and advice about all the medical options available"[14] "The obligation to inform patients may be met by delegating this communication to another competent individual for whom the physician is responsible"
Yukon Medical Council (YCM)	"Objecting physicians must provide the patient with timely access to another non-objecting physician or resource with accurate information about all available medical options. The objecting physician must document, in the medical record, the date on which the referral was made, and the physician, nurse practitioner and/or agency to which the referral was directed"[15] "When moral or religious beliefs prevent a physician from providing or offering access to information about a legally available medical or surgical treatment or service, that physician must ensure that the patient who seeks such advice or medical care is offered timely access to another physician or resource that will provide accurate information about all available medical options"[16]

(continued)

[12] "The College of Physicians and Surgeons of Prince Edward Island Policy on Medical Assistance in Dying."

[13] "Medical Aid in Dying: Practice and Pharmacological Guidelines."

[14] "College of Physicians and Surgeons of Saskatchewan POLICY—Medical Assistance in Dying: Patient's Death Is Not Reasonably Foreseeable."

[15] "Yukon Medical Council Standard of Practice: Medical Assistance in Dying (MAID)."

[16] "Yukon Medical Council Medical Practice: Moral or Religious Beliefs Affecting Medical Care."

Table 27.1 (continued)

Medical regulator	Obligations of physicians with a conscientious objection
Health and Social Services (Government of the Northwest Territories)	"Other than providing the Central Coordinating Service contact card to a patient who requests information on Medical Assistance in Dying, no part of the *Medical Assistance in Dying Interim Guidelines* compels a Practitioner to provide Medical Assistance in Dying or another Practitioner, Registered Nurse, or Pharmacist to aid a Practitioner in providing Medical Assistance in Dying to a patient"[17]

Table contents updated as of 31 March 2022
No relevant information was available for the Department of Health and Social Services (Government of Nunavut)

27.2 Defining Conscience

Philosophers of medicine have sought to specify the notion of conscience in the context of medical practice. Daniel Sulmasy defines conscience as "a conjunction of will and judgment"—the willful assent to moral truths and the judgment that arises in particular moral deliberations.[18] In a similar way, Anthony Fisher suggests that it is helpful to think of conscience as an activity oriented toward actualizing universal values and mediating universal norms rather than an inbuilt satellite navigator or "a sort of angelic voice distinct from our own reasoning".[19] Nevertheless, he emphasizes that conscience cannot be conceived as absolutely autonomous by pointing out, for instance, that the Latin root word *conscientia* can be broken down into two parts: 'scire' which means 'to know', and 'con' which means 'with'. Fisher is alluding to more communitarian conceptions of conscience advocated by such philosophers as Alasdair MacIntyre and Charles Taylor. Both MacIntyre and Taylor critique modern conceptions of the self as overemphasizing autonomy and failing to appreciate the extent to which community tradition and social values shape personal identity, of which conscience can be considered a corollary. In *After Virtue,* MacIntyre contends that encouraging people "to separate themselves and their values from the roles they perform, is a recipe for the dissolution of character."[20] It is not difficult to appreciate how a separation of personal values from roles might precipitate a decline in ethical conduct.

Other authors emphasize the connection between conscience and personal integrity or self-concept. Mark Wicclair, for example, asserts that moral integrity is based on the principles, virtues, and/or paradigm-based maxims that lie at the core

[17] "Government of Northwest Territories Medical Assistance in Dying Interim Guidelines for the Northwest Territories."

[18] Sulmasy, "What Is Conscience and Why Is Respect for It so Important?"

[19] Fisher, "The Voice of Conscience."

[20] Macintyre, *After Virtue: A Study in Moral Theory.*

of a person's identity or self-concept.[21] Moral integrity allows a person to function in a state of unity and detect conflicts between internal and external values and/or norms. Moral integrity therefore closely resembles the idea of conscience. This is especially apparent when considering the concepts of *preservative conscience* and *perfective conscience.* According to Sean Murphy and Stephen Genuis, a person exercises their *preservative conscience* when freely refraining from participating in wrongdoing, which preserves their moral integrity.[22] Conversely, a person is said to exercise their *perfective conscience* when they freely choose a moral good in order to contribute to human flourishing or perfection to the extent that what is chosen is truly good. Martin Benjamin conceptualizes conscience in three ways: as a general inner capacity to distinguish right act from wrong, as internalized parental or broader social norms, and as "the exercise and expression of a reflective sense of integrity."[23] Others emphasize this reflective aspect of conscience as an ability to scrutinize past, present, and future moral actions, while maintaining that conscience also has a reflexive component that "provides feedback in the form of internal dissonance or discomfort" when a person engages in immoral action.[24]

In sum, conscience is that element of our persons that combines the desire to do good and to be good with the innate capacity to critically and rationally evaluate the morality of intentions and actions. Insofar as humans can be held morally responsible for their actions, conscience is the element of our humanity that gives rise to that moral responsibility. Conscience, in other words, is essential to our humanity.

27.3 The Role of Conscience in Medical Practice

What then does conscience have to do with medical practice? The conscience is continuously engaged in the practice of medicine because every action of the physician carries moral significance. Medical practice is an innately and unavoidably moral activity because physicians are entrusted with the care of persons possessed of intrinsic and incalculable value, and because of the covenant of trust into which physicians enter with their patients. The physician is expected to do good for the patient. Every judgment and action, even the most technical matters, are subject to moral valuation from this vantage.

Medicine derives its true moral authority from the greatness of the patient's value. If the patient were of little or no objective value, medicine would also be of little or no objective value as well. Belief in the value and importance of health care is predicated on the belief that patients matter, and that the health of patients is a fundamental moral good. The drama and glory of medicine arises from the goodness

[21] Wicclair, "Is Conscientious Objection Incompatible with a Physician's Professional Obligations?"

[22] Murphy and Genuis, "Freedom of Conscience in Health Care: Distinctions and Limits."

[23] Benjamin, "Conscience."

[24] Genuis and Lipp, "Ethical Diversity and the Role of Conscience in Clinical Medicine."

of health, the tragedy of illness, and the power of medicine to uphold and display the fundamental dignity of those with illness, even when cure is impossible. Even when cloaked in the garb of a scientist (white coat, etc.), the good physician does not regard the patient as a mere artefact of intellectual or scientific curiosity. Rather, the patient is honoured, treated with respect and care, patience and kindness. Immanuel Kant insightfully explained the essential nature of respect for persons by stipulating that persons must always be treated as ends rather than as mere means to ends.[25] Medicine regards the patient not as a mere means to an end, but rather as an end in herself. Where doctors are found to treat patients as mere means to ends (e.g., for financial gain, sexual pleasure, scientific research without appropriate informed consent, etc.), they are censured. Thus, medicine is innately moral because patients matter. The patient's intrinsic moral worth becomes an essential guidepost for the conscience of the physician.

Medicine is also innately and unavoidably moral because of the nature of the relationship between physician and patient. Edmund Pellegrino and Daniel Sulmasy describe the essence of this relationship beautifully:

> When patients seek help they are anxious, often in pain, fearful, dependent, and therefore vulnerable and exploitable. In that state, physicians ask then, "How can I help you?" By that act physicians invite the patient's confidence and trust that they are competent and will use that competence primarily in the patient's interests. The relationship is therefore not a contract but a covenant of trust, to which physicians must be faithful even if it means some suppression of their own self-interest. The good of the patient is thus a moral compass.[26]

In undertaking to care for patients, we make a promise to help. All of medical practice is subject to moral evaluation in light of faithfulness to this promise. In an exposition of the patient-physician covenant published in the *Journal of the American Medical Association*, Ralph Crawshaw and colleagues write: "Medicine is, at its centre, a moral enterprise grounded in a covenant of trust. This covenant obliges physicians to be competent, and to use their competence in the patient's best interests. Physicians, therefore, are both intellectually and morally obliged to act as advocates for the sick wherever their welfare is threatened and for their health at all times."[27] It is interesting to note that under this covenant even the technical competence of physicians is a matter of ethical concern; physicians owe it to their patients to be good at what they do—accordingly, their competence is subject to both technical and moral evaluation. Thus, the professional technical judgments and actions of physicians cannot be divorced from their moral judgment. This covenant of trust between patients and physicians constitutes another essential guidepost for the conscience of physicians.

If medical practice is inspired by the value of the patient and the covenant of trust between patient and physician, medicine is also constrained by those same moral foundations. These constraints give moral direction to the physician's activity *qua* physician and define what it means to practice medicine conscientiously.

---

[25] Johnson and Cureton, "Kant's Moral Philosophy: The Humanity Formula.".

[26] Pellegrino and Sulmasy, "Medical Ethics."

[27] Crawshaw et al., "Patient-Physician Covenant."

First, the covenant of trust gives medicine a particular goal. The goal of medicine is to help the patient, to advance the patient's health, and to do them good vis-à-vis their health. Philosophers of medicine have made careful and thoughtful attempts to expand on this basic notion to specify more precisely the ways in which physicians are to help the patient. A consensus statement issued by the Hastings Center identified four goals for medicine[28]:

1. The prevention of disease and injury and the promotion and maintenance of health;
2. The relief of pain and suffering caused by maladies;
3. The care and cure of those with a malady, and the care of those who cannot be cured;
4. The avoidance of a premature death and the pursuit of a peaceful death.

These goals form the basis for the physician's professional obligations. We may reasonably expect doctors to always act in accordance with these goals.

Second, the covenant of trust, the intrinsic value of the patient, and the goals of medicine entails that medicine must be practiced in accordance with certain virtues, virtues that engender trust, reflect the value of the patient, and promote the end of medicine. This implies that the goals of medicine cannot be pursued in any manner deemed merely effective. So, for example, doctors ought not to employ deception as a means of achieving the goals of medicine—we ought not to give placebos or lie to patients about our intentions or actions. We are not to coerce or force patients to co-operate with our treatment plans. We are not to have sexual relations with a patient, even if the patient consents and seeks such relations.[29] Such actions are forbidden because they risk devaluing the patient; they treat patients as mere means to ends, rather than as ends in themselves, and they fail to recognize and respect patients' moral agency. The virtuous physician practices with diligence, attention to detail, kindness, mercy, altruism, trustworthiness, and self-sacrifice. These virtues reflect the worth and significance of the patient with whom the physician enters into a covenant of trust.

This moral framework for medicine serves to define the professional obligations of physicians. Physicians are obliged to promote the ends of medicine in a manner consistent with the virtues that engender trust and reflect the value of patients. Professional obligations do not appear arbitrarily and at random. Rather, they arise from the moral foundations of medicine—the moral value of the patient and the special nature of the covenantal relationship between patient and physician.

Conscience enters this equation as the "conjunction of will and judgment" (in Sulmasy's terms)—the will to be faithful to one's professional obligations as physician and the judgment as to what it means to be so faithful, in any given clinical situation. Thus, the Physician's Pledge (Declaration of Geneva), a medical oath ratified by the World Medical Association, calls upon physicians to practice with 'conscience and dignity', to dedicate themselves to 'the service of humanity,' and to

[28] "The Goals of Medicine."

[29] Barilan, "Of Doctor-Patient Sex and Assisted Suicide."

make the health of their patients their 'first consideration'.[30] To be a conscientious physician is to strive always and with care to fulfill one's professional obligation to advance the goals of medicine in a manner consistent with the virtues of medicine. A properly formed conscience recognizes the patient's value and directs the physician to behave in a manner that comports with that value.

27.4 Professional Obligations Can Require Conscientious Objection

Critically, the same commitment to the goals and virtues of medicine prevent physicians from acting in ways that hinder those goals or violate those virtues. Sometimes the physician's professional obligations require her to refuse a patient's request. This refusal may be respectful, gentle, kind, and accompanied by a sincere explanation; but refuse she must, or risk being unfaithful to her professional obligations as a physician.

This analysis provides a basis for defining justifiable cases of conscientious objection. Conscientious objections offered by physicians without regard for the patient's health or well-being seem difficult to justify. Conscientious objections offered by physicians out of legitimate concern for the patient's health and well-being are salutary and even obligatory. The former kind of objection violates the goals and virtues of medicine; the latter kind honours the goals and virtues of medicine. Castigating conscientious objection as an expression of self-interest without regard for the patient's welfare attacks a straw man and misses the point. Properly formed conscientious objection derives from the physician's professional obligations; it does not set aside those obligations.

In fact, the claim that a physician can legitimately object to certain medical practices on the basis of professional obligation is unobjectionable. Even strong self-identified opponents of conscientious objection in medicine like Udo Schuklenk and Julian Savalescu seem to accept this claim. They cite the case of a military nurse who refused to force-feed patients at the Guantanamo Bay detention camp. The nurse's objection, they note, was "justified on professional grounds".[31] They write, "Why do we applaud this conscientious objector, but reject others, who may hold as strong views about abortion as this nurse did about torture? It is trivially true that torture is not a medical practice which is in the interests of the patient or desired by the patient." They recognize that the nurse was justified in objecting to the demand to participate in torture because it ran contrary to professional obligations. They draw a distinction, however, between objection to abortion and objection to participating in torture because in their personal view the latter is self-evidently neither in the interests of the patient, nor desired by the patient. This demand for torture was forced

[30] WMA Declaration of Geneva, "The Physician's Pledge."

[31] Savulescu and Schuklenk, "Doctors Have No Right to Refuse Medical Assistance in Dying, Abortion or Contraception."

upon the nurse apart from the patient's will or concern for the patient's interests; thus objection was obligatory.

We have now made considerable progress in isolating the issues at stake. The question is not whether doctors may offer conscientious objections in light of their professional obligations, but whether any particular request by a patient is believed to run counter to the professional obligations of physicians. The question then is whether requests for MAID could run counter to the professional obligations of physicians. More to the point in Canada, the question is whether a request for a referral to an available physician who is willing to provide MAID could contravene the professional obligations of the referring physician and, thereby, justify a conscientious objection.

27.5 MAID and the Goals and Virtues of Medicine

Is MAID consistent with the goals and virtues of medicine? If so, then a physician ought to provide it. If not, then a physician ought not to provide it. Here we consider both points of view.

Proponents of euthanasia argue that MAID is consistent with the goals of medicine. They reason that if a patient faces grievous and irremediable suffering, it might be better for that patient to be dead than to remain alive. In his seminal tome on the ethics of assisted death, philosopher L. W. Sumner writes, "The harm of death lies in the value of the life thereby foregone",[32] and that our "moral status...is not independent or logically prior to the prudential value of [life]; it is instead conditional on that value".[33] Thus, if the patient does not value life, then death ought not to count as harm but rather as a benefit. On these grounds, MAID is held to be consistent with the goals of medicine. The physician would be obliged to facilitate the patient's death because the patient's death would be consistent with the purported goals of medicine.

Opponents of MAID do not accept this line of reasoning. First, they reject the assumption that the value of life depends on the value subjectively ascribed to it by patients. A subjective self-valuing of life would raise serious concerns about the value of the lives of many human persons, including small children who by virtue of limited cognitive capacity have little or no ability to hold or express articulate value judgments. Second, opponents of MAID would point out that attempts to calculate the relative benefits and harms of continued life versus death face an insurmountable problem, for existence after death is empirically unverifiable. Viewing MAID as an effective treatment for suffering implicitly assumes that death is a state relatively free from suffering. While one may hope so, such claims are beyond the pursue of medical science and necessarily constitute a kind of blind faith in death. Third, it seems strange to claim that one's health and well-being can be advanced by non-existence. How

[32] Sumner, "Suicide and the Sanctity of Life."

[33] Sumner, "The Ethics of Assisted Death."

does one quantify the well-being, or lack thereof, attached to being dead? Fourth, high quality palliative care can work wonders for pain relief and symptom management, such that it is possible for dying patients to have a peaceful death without resorting to extinguishing their existence. Other arguments against MAID include a concern that the limits on MAID gradually erode over time with legalization, thereby further endangering patient welfare, as well as the conviction that participation in MAID ultimately undermines the trust patients place in physicians as healers and the health care enterprise more generally.[34]

Proponents of MAID argue that it is consistent with the virtues of medicine to intentionally cause a patient's death. They hold that respect for the patient's worth primarily manifests in respect for the patient's autonomous preferences, and that if the patient prefers to be dead, then that preference ought to be honoured.

Opponents of MAID disagree. They hold that one cannot manifest respect for the patient's intrinsic and incalculable value by extinguishing the patient. Certainly, physicians ought not to force patients to undergo medical treatment and must invite patients to share responsibility for medical decision-making. Such respect for the patient's autonomy, however, does not entail treating patients as mere means to ends by ending them.

The point here is not to conclusively make the case one way or the other. The point, rather, is to show that there is fundamental but reasonable disagreement about the moral legitimacy of MAID and whether it comports with the goals and virtues of medicine, properly understood. Because of the fundamental nature of the disagreement, there is little hope for broad consensus, at least in the foreseeable future. At the very least, we must recognize that, in this context, the conscientious objector's concerns can be properly located in a desire to remain faithful to the goals and virtues of medicine. Objection to participating in MAID therefore does not necessarily reflect mere self-interest or neglect of the patient's good. Rather, it has everything to do with a fundamental commitment to the patient's value and to remaining faithful to the covenant of trust entered into with the patient.

27.6 Conscientious Objection and Ensuring Access to MAID

Conscientious objection to a legally available and professionally endorsed intervention raises questions about access for patients. If patients request MAID and that request is denied, are they to be denied access to an otherwise legally available intervention? Canadian provinces and medical regulators have resolved this problem in different ways. No physician in Canada is required to personally administer MAID—an important (but often unrecognized) manifestation of respect for the physician's conscience. Rather, controversy surrounds the question whether physicians should

[34] Sulmasy et al., "Non-Faith-Based Arguments against Physician-Assisted Suicide and Euthanasia."

be expected to refer a patient who is seeking MAID to an available provider who is willing to administer MAID.

Many seem to regard a referral as a reasonable compromise between respecting the conscience of physicians and ensuring access for patients. Dan Brock argues that this 'conventional compromise' represents a reasonable accommodation for the conscientiously objecting physician. He contends that the medical profession is obligated to "ensure that the service/product is available to the patients/public."[35] Since this obligation rests with the profession as a whole, but not necessarily with individual members of the profession, physicians may fulfill their professional obligations by referring the patient to another physician who is willing to provide the service in question. Ronit Stahl and Ezekiel Emmanuel express the expectation for referral bluntly: "Conscientious objection still requires conveying accurate information and providing timely referrals to ensure patients receive care."[36]

Different versions of this compromise have been implemented by medical regulators in Canada (Table 27.1). The policy of the *College of Physicians and Surgeons of Ontario* (CPSO) coheres most closely to Brock's conventional compromise in that it compels physicians with a conscientious objection to provide an 'effective referral' to an available and non-objecting provider specifically for the intervention to which the physician objects.[37] This 'compromise' resulting in referral mandates merits close scrutiny. The question is whether the moral foundations of medicine, such as the goals and virtues of medicine that shape long-standing professional obligations, apply to the act of referral, broadly or narrowly understood. Stated yet another way, if a physician believes that her professional obligations would make it wrong to personally administer MAID, do those same obligations also apply to the act of referral to a willing provider?

Several considerations suggest that the answer is "yes". First, a formal referral is an act of the physician *qua* physician. A referral is not something that a non-physician can offer or provide—it is something that only the professional can provide by virtue of her role as a professional. Because she is acting as a professional, all the moral considerations that bear on her professional practice also apply to the act of referral. If a physician is convinced that performing MAID violates her professional obligations to the patient, it is difficult to see how specifically facilitating access to MAID by an act of referral can be consistent with the same obligations.

Second, it is widely accepted that physicians are morally responsible for the referrals they make. If a physician knowingly and deliberately refers a patient to another provider who is known to be incompetent or malicious, the referring physician can be found partly culpable for any harm incurred by the patient. Such findings of culpability are consistent with the policies of the same medical regulators who compel physicians with a conscientious objection to MAID to refer for it. In its policy on

---

35 Brock, "Conscientious Refusal by Physicians and Pharmacists: Who Is Obligated to Do What, and Why?"

36 Stahl and Emanuel, "Physicians, Not Conscripts—Conscientious Objection in Health Care."

37 "CPSO Medical Assistance in Dying"; "CPSO Advice to the Professional: Medical Assistance in Dying."

female genital mutilation, the CPSO states that physicians "will be subject to disciplinary measures if they perform, assist in or refer patients for FGC/M [female genital cutting/mutilation] procedures."[38] On this matter, there is clearly no apparent moral distinction between performing the procedure and referring for the procedure. Brock himself recognizes that the 'conventional compromise' underpinning referral mandates makes the referring physician complicit in the eventual performance of the action by the accepting physician: "the conventional compromise do[es] leave the objecting physician complicitous to some degree." He believes that this degree of complicity is insufficient to justify refusal to refer. Yet if the conventional compromise purports to protect and uphold individual moral integrity of physicians while also admitting to making them complicit in the administration of MAID, then it fails to be any sort of meaningful compromise at all.

Third, the arguments put forward in favour of requiring referrals from conscientiously objecting physicians seem to beg the question: what if the conscientious objection is appropriately grounded in a concern for the patients' well-being, as we have outlined above? Consider Stahl and Emmanuel's position: "To invoke conscientious objection is to reject the fundamental obligation of health care—the primary duty to ensure patients' continued well-being."[39] Here they take for granted that the requested referral comports with good health care, yet that is exactly the issue being debated. When they stipulate that conscientious objectors are required to provide "timely referrals to ensure patients receive care", they take for granted that the procedure for which the referral is required constitutes "care". The duty to provide access is necessarily predicated on the claim that the proposed intervention is good. There are many potential therapies to which Canadians do not have access, and this is because the medical community considers such therapies to be harmful, not sufficiently beneficial, or even not sufficiently cost-effective to fund in a public health care system. The recent controversies over access to venoplasty for chronic cerebrospinal venous insufficiency in multiple sclerosis or ivermectin for COVID-19 are salient examples.[40] Access to a given intervention is not primarily conditioned on respect for patient autonomy (as many seem to think), but rather on the medical community's beliefs about the appropriateness of the intervention. Unless physicians are reasonably persuaded of the benefit and appropriateness of an intervention, they should not be obliged to offer or facilitate it.

At bottom, then, the question of effective referral comes down to a basic understanding of how 'care' and 'professional obligations' should be defined. More specifically, the question comes down to 'who' should define 'care' and 'professional obligations.' From Brock's standpoint, professional obligations seem to derive from legal standards as well as consensus of medical professionals at large—i.e. "the law, professional norms, and a consensus in medical ethics." This account of professional obligations fails to consider the possibility that segments of the medical profession

---

[38] "Female Genital Cutting (Mutilation)."

[39] Stahl and Emanuel, "Physicians, Not Conscripts - Conscientious Objection in Health Care."

[40] The New York Times, "What's Behind the Ivermectin Frenzy?"; Zamboni et al., "Efficacy and Safety of Extracranial Vein Angioplasty in Multiple Sclerosis: A Randomized Clinical Trial."

can sometimes be seriously mistaken. Elie Wiesel poignantly describes the moral failings of the medical establishment in Nazi Germany.[41] Brock, Stahl, Emmanuel and others who would deny freedom to object to an effective referral have an idealistic and unrealistic vision of the moral perfections of established consensus, a vision that is undermined by repeated failures of the medical establishment to be faithful to the goals and virtues of medicine (cf., Tuskegee syphilis project). Each individual physician has a responsibility to be formed, not by momentary majoritarian consensus, but by the fundamental and long-standing moral principles of medical practice, as discussed above. If conscientious objectors are required to exit the profession in order to avoid violating their conscientious commitments to patients by making referrals, it is difficult to see how Brock's ideal of consensus underpinning the 'conventional compromise' can ever be meaningfully criticized or effectively revised. Respect for the reasonable and well-formed conscientious convictions of virtuous doctors who are committed to the good of patients can only benefit the quality of medical care, even if it means that professional consensus cannot always be achieved.

27.7 Policy Recommendations

In light of these considerations, we would propose the following statements as an appropriate policy on respecting conscientious objection in healthcare:

1. Physicians may decline to recommend, encourage, provide, collaborate in or help to arrange for MAID if they reasonably believe that to be inconsistent with the good of their patients, and may refuse to affirm what they believe to be false, misleading or harmful in that regard. In so doing, physicians must acknowledge patient concerns, treat them kindly, compassionately and respectfully, and must not attempt to convince patients to adopt their ethical, moral or religious convictions.
2. In declining to recommend, encourage, provide, collaborate in or help to arrange for MAID, physicians must not discriminate against persons by reason of their race, national or ethnic origin, religion, sex, sexual orientation, gender, age, marital status, mental or physical disability or political views.
3. Physicians must give reasonable notice to patients of ethical, moral or religious beliefs that prevent them from providing MAID so that patients may, if they wish, consult other physicians or health care personnel. Notice is reasonable if it is given before or at the time physicians assume primary responsibility for a patient's care, or as soon as practicable after becoming aware that a patient who is potentially eligible for MAID probably wishes to discuss it. Notice should also be given to colleagues and institutions likely to be affected.
4. Physicians who decline to provide or refer for MAID must advise affected patients that they may seek the services elsewhere and explain how affected patients can

[41] Wiesel, "Without Conscience."

contact other physicians or health care personnel. When appropriate, they must promptly communicate to a person in authority a patient's request for a transfer of care and cooperate in a transfer of care arranged by the patient or a person in authority. They must continue to provide treatment or care unrelated to MAID unless the patient requests otherwise.

27.8 Conclusion

Conscientious objection is a logical corollary of conscientious practice. Conscientious objection in medicine is justifiable and even obligatory when it is based on the professional obligations arising from the goals and virtues of medicine. Physicians who object to MAID insist that, contrary to certain perceptions, they are deeply concerned for the health and well-being of patients. They hold the value of the patient to be so great that they are unwilling to negate that value by causing the patient's death, or by facilitating the causing of the patient's death through an effective referral. They reasonably regard MAID as a violation of their long-standing professional obligations. Of course, those same obligations require physicians to treat patients with respect, to be honest with patients, and to avoid obstructing or interfering with patients' pursuits of controversial procedures. A culture of medical practice that prioritizes conformity to establishment consensus over individual persuasion and reasoned conviction as to whether any given medical act comports with the goals and virtues of medicine makes patients vulnerable to harm: the establishment consensus is fallible. Conscientious devotion to the health of the patient, motivated by respect for the intrinsic and incalculable worth of the patient and enacted through well-honed technical skill and moral judgment, is an essential characteristic of the good physician—the kind of physician that patients deserve.

References

Barilan, Michael Y. 2003. Of doctor-patient sex and assisted suicide. *The Israel Medical Association Journal* 5 (6): 460–463.

Benjamin, Martin. 2004. Conscience. In *Encyclopedia of bioethics*. Macmillian Reference.

Brock, Dan W. 2008. Conscientious refusal by physicians and pharmacists: who is obligated to do what, and why? *Theoretical Medicine and Bioethics* 29 (3): 187–200. https://doi.org/10.1007/S11017-008-9076-Y.

Canadian Medical Association. 2016. CMA's Submission to the College of Physicians and Surgeons of Ontario (CPSO): Consultation on CPSO Interim Guidance on Physician-Assisted Death.

College of Physicians & Surgeons of Newfoundland and Labrador Standard of Practice: Medical assistance in dying (2022). https://cpsnl.ca/wp-content/uploads/2020/11/Standard-of-Practice-Medical-Assistance-in-Dying-2017.pdf

College of Physicians & Surgeons of Nova Scotia Professional Standard Regarding Medical Assistance in Dying (MAiD) (2021). https://cpsns.ns.ca/resource/medical-assistance-in-dying/

College of Physicians and Surgeons of New Brunswick: Moral Factors and Medical Care (2017). https://cpsnb.org/en/medical-act-regulations-and-guidelines/guidelines/445-moral-factors-and-medical-care.

College of Physicians and Surgeons of Saskatchewan POLICY—Medical Assistance in Dying: Patient's Death Is Not Reasonably Foreseeable (2021). https://www.cps.sk.ca/imis/CPSS/CPSS/Legislation__ByLaws__Policies_and_Guidelines/Legislation_Content/Policies_and_Guidelines_Content/MAiD-Patient_Death_NOT_Reasonably_Foreseeable.aspx

CPSA Advice to the Professional: Medical Assistance in Dying (2021). https://cpsa.ca/wp-content/uploads/2020/06/AP_Medical-Assistance-in-Dying.pdf

CPSA Standards of Practice: Conscientious Objection (2016). https://cpsa.ca/wp-content/uploads/2020/05/Conscientious-Objection.pdf

CPSBC Practice Standard: Medical Assistance in Dying (2022). https://www.cpsbc.ca/files/pdf/PSG-Medical-Assistance-in-Dying.pdf

CPSO Advice to the Professional: Medical Assistance in Dying (2021). https://www.cpso.on.ca/Physicians/Policies-Guidance/Policies/Medical-Assistance-in-Dying/Advice-to-the-Profession-Medical-Assistance-in-Dyi.

CPSO Medical Assistance in Dying (2021). https://www.cpso.on.ca/Physicians/Policies-Guidance/Policies/Medical-Assistance-in-Dying.

Crawshaw, Ralph, David E. Rogers, Edmund D. Pellegrino, Roger J. Bulger, George D. Lundberg, Lonnie R. Bristow, Christine K. Cassel, and Jeremiah A. Barondess. Patient-Physician Covenant. *JAMA* 273 (19) (May 17, 1995): 1553–1553. https://doi.org/10.1001/JAMA.1995.03520430089054

Female Genital Cutting (Mutilation) (2021). https://www.cpso.on.ca/en/Physicians/Policies-Guidance/Statements-Positions/Female-Genital-Cutting-(Mutilation)

Fisher, Anthony. 2012. The voice of conscience. In *Catholic bioethics for a new millennium*, 38–52. Cambridge (UK): Cambridge University Press.

Genuis, Stephen J., and Chris Lipp. Ethical diversity and the role of conscience in clinical medicine. *International Journal of Family Medicine* 2013 (December 12, 2013): 1–18. https://doi.org/10.1155/2013/587541

Government of Northwest Territories Medical Assistance in Dying Interim Guidelines for the Northwest Territories (2021). https://www.hss.gov.nt.ca/sites/hss/files/resources/interim-guidelines-medical-assistance-dying.pdf

Johnson, Robert, and Adam Cureton. 2022. Kant's moral philosophy: The humanity formula. *Stanford Encyclopedia of Philosophy.*

Macintyre, Alasdair. 1984. *After virtue: A study in moral theory*, 2nd ed. Notre Dame, IN (USA): University of Notre Dame Press.

McLeod, Carolyn. 2006. Demanding referral in the wake of conscientious objection to abortion. In Comparative Program on Health and Society Working Papers Series 2004–2005, ed. J.C. Cohen and J.E. Keelan, 130. University of Toronto: Munk Centre for International Studies at Trinity College.

Medical Aid in Dying: Practice and Pharmacological Guidelines (2018). https://numerique.banq.qc.ca/patrimoine/details/52327/3438228

Murphy, Sean, and Stephen J. Genuis. 2013. Freedom of conscience in health care: Distinctions and limits. *Journal of Bioethical Inquiry* 10 (3): 347–354. https://doi.org/10.1007/S11673-013-9451-X

Pellegrino, Edmund D., and Daniel P. Sulmasy. 2003. Medical ethics. In *Oxford textbook of medicine*, ed. David A. Warrell, Timothy M. Cox, John D. Firth, and Edward J. Benz, 4th ed., 14. New York (NY): Oxford University Press.

Savulescu, Julian, and Udo Schuklenk. 2017. Doctors have no right to refuse medical assistance in dying, abortion or contraception. *Bioethics* 31 (3): 162–170. https://doi.org/10.1111/BIOE.12288

Stahl, Ronit Y., and Ezekiel J. Emanuel. 2017. Physicians, not conscripts—Conscientious objection in health care. *The New England Journal of Medicine* 376 (14): 1380–1385. https://doi.org/10.1056/NEJMSB1612472

Sulmasy, Daniel P. 2008. What is conscience and why is respect for it so important? *Theoretical Medicine and Bioethics* 29 (3): 135–149. https://doi.org/10.1007/S11017-008-9072-2

Sulmasy, Daniel P., John M. Travaline, Louise A. Mitchell, and E. Wesley Ely. 2016. Non-faith-based arguments against physician-assisted suicide and Euthanasia. *The Linacre Quarterly* 83 (3): 246. https://doi.org/10.1080/00243639.2016.1201375

Sumner, Leonard Wayne. 2011a. Suicide and the sanctity of life. In *Assisted death: A study in ethics and law*, 77. New York (NY): Oxford University Press.

Sumner, Leonard Wayne. 2011b. The ethics of assisted death. In *Assisted death: A study in ethics and law*, 85. New York (NY): Oxford University Press.

The College of Physicians and Surgeons of Prince Edward Island Policy on Medical Assistance in Dying (2019). https://www.cpspei.ca/wp-content/uploads/2019/11/Medical-Assistance-in-Dying-Policy-Nov-2019.pdf

The College of Physicians & Surgeons of Manitoba Standard of Practice: Medical Assistance in Dying (MAiD) (2021). https://cpsm.mb.ca/assets/Standards%20of%20Practice/Standard%20of%20Practice%20Medical%20Assistance%20in%20Dying%20(MAID).pdf

The Goals of Medicine. *Hastings Center Report* 26 (6): S1–27. https://doi.org/10.1002/J.1552-146X.1996.TB04777.X

The physician's pledge. WMA Declaration of Geneva (2017). https://www.wma.net/policies-post/wma-declaration-of-geneva/

What's Behind the Ivermectin Frenzy? The New York Times (2021). https://www.nytimes.com/2021/10/04/podcasts/the-daily/ivermectin-misinformation-covid19-drugs.html

Wicclair, Mark R. 2008. Is conscientious objection incompatible with a physician's professional obligations? *Theoretical Medicine and Bioethics* 29 (3): 171. https://doi.org/10.1007/S11017-008-9075-Z

Wiesel, Elie. 2005. Without conscience. *The New England Journal of Medicine* 352 (15) (April 14, 2005): 1511–1513. https://doi.org/10.1056/NEJMP058069

Yukon Medical Council Standard of Practice: Medical Assistance in Dying (MAID) (2021). http://www.yukonmedicalcouncil.ca/pdfs/MAID.pdf

Yukon Medical Council Medical Practice: Moral or Religious Beliefs Affecting Medical Care (2021). http://yukonmedicalcouncil.ca/pdfs/Moral_or_Religious_Beliefs_Affecting_Medical_Care.pdf

Zamboni, Paolo, Luigi Tesio, Stefania Galimberti, Luca Massacesi, Fabrizio Salvi, Roberto D'Alessandro, Patrizia Cenni, et al. 2018. Efficacy and safety of extracranial vein angioplasty in multiple sclerosis: A randomized clinical trial. *JAMA Neurology* 75 (1): 35–43. https://doi.org/10.1001/JAMANEUROL.2017.3825

Simon Czajkowski (chai-kov-ski), HB.Sc., M.Sc. has an Honours Bachelor of Science degree (majors in Psychology and Human Biology) and a Master of Science degree (in Physiology), both from the University of Toronto. Simon also completed a Master of Bioethics degree at the University of Leuven (KU Leuven) in Belgium, where we wrote a thesis in which he applied a framework similar to the *Oakes* test for reconciling physician conscientious objection with patient access to health care related services. Simon currently works as a Clinical Research Analyst at the University Health Network, in Toronto. He continues to write and publish in the field of bioethics and hopes to pursue further education in bioethics and/or medicine.

Sean Murphy has been the Administrator of the Protection of Conscience Project since its inception in 1999, responsible for its day-to-day management and operations. He has been described as "a leading advocate of conscience protection laws by health care providers" (McLeod 2006). The Project is a non-profit, non-denominational initiative with international interests. It does not take a position on the morality of contentious procedures, but critiques policies of coercion, encourages accommodation, and promotes clarification and understanding of the issues involved. It advocates for freedom of conscience among health care workers, supporting those under pressure, and encouraging and facilitating collaboration among freedom of conscience advocates.

Ewan C. Goligher, M.D., Ph.D., FRCPC is Assistant Professor of Medicine in the Interdepartmental Division of Critical Care Medicine at the University of Toronto. He trained in biochemistry and medicine at the University of British Columbia. Following medical school, he pursued postgraduate residency training in internal medicine and critical care medicine at the University of Toronto. He obtained a doctoral degree in Physiology from the University of Toronto. He is appointed as a clinician-scientist at Toronto General Hospital Research Institute and he practices critical care medicine at Toronto General Hospital. His CIHR-funded research program spans clinical and translational aspects of respiratory pathophysiology and mechanical ventilation, and he develops novel clinical trial approaches to accelerate discovery and precision medicine in respiratory failure. Dr. Goligher has published on ethical aspects of end-of-life care in the intensive care unit and the role of conscience in medical practice.

Chapter 28
Spirit at the Gateway: Religious Reflections on Medical Assistance in Dying

Christine Jamieson, Kathleen Absolon, S. Zaki Ahmed, Rabbi Catharine Clark, Sister Nuala Kenny, and Nita Kotiuga

Abstract This chapter provides religious and spiritual reflection on medical assistance in dying (MAID) from five perspectives: Indigenous, Muslim, Jewish, Roman Catholic, and Protestant. Each contributor was asked the following basic questions to direct their reflection. What would be the guidelines that you would offer people of your faith tradition? What would people from your faith tradition be concerned about? How would they make a decision regarding MAID? What guidance, as a leader in your faith, would you offer? Each contributor brought their own religious tradition to bear on addressing these questions in various manners. While each offers unique points, there are common threads or themes that emerge, for example, our lives are a gift and not something we necessarily "own" and MAID is not a solitary decision, but others and communities are implicated. Each contributor pushes the reader into the religious realm of meaning and value and so raise important points not considered in the secular realm.

Keywords Indigenous · Muslim · Judaism · Catholic · Protestant · Religious · Spiritual · Euthanasia

28.1 Introduction

This chapter deals with the challenging theme of medical assistance in dying (MAID) from a religious/spiritual perspective. How are religious or spiritual traditions considering the issue of MAID? Does MAID challenge faith traditions in relation to questions concerning human dignity, the meaning of human suffering, and the trajectory of one's human journey. Does seeking medical assistance in dying contravene what

---

C. Jamieson (✉) · K. Absolon · S. Z. Ahmed · R. C. Clark · S. N. Kenny · N. Kotiuga
Department of Theological Studies, Concordia University, Montreal, Canada
e-mail: christine.jamieson@concordia.ca

J. Kotalik and D. W. Shannon (eds.), *Medical Assistance in Dying (MAID) in Canada*, The International Library of Bioethics 104,
https://doi.org/10.1007/978-3-031-30002-8_28

many consider the will of God or the Creator? Does making a decision to end one's life because one experiences overwhelming suffering betray the gift of life and one's responsibility to seek meaning and value despite one's challenges? In other words, does suffering have the final word on the worth of life or is there something "redemptive' beyond one's suffering? These are religious questions, questions that are not considered anywhere expect within religious or spiritual perspectives.

We are privileged to have short reflections from five different religious/spiritual perspectives. While not exhaustive, it is a beginning to explore this important dimension of MAID. Each contributor was asked the following basic questions to direct their reflection. What would be the guidelines that you would offer people of your faith tradition? What would people from your faith tradition be concerned about? How would they make a decision regarding MAID? What guidance, as a leader in your faith, would you offer? Each contributor brought their own religious tradition to bear on addressing these questions in various manners. While each offers unique points, there are common threads or themes that emerge, for example, our lives are a gift and not something we necessarily "own" and MAID is not a solitary decision, but others and communities are implicated. Each contributor pushes the reader into the religious realm of meaning and value and so raise important points not considered in the secular realm.

We begin with Dr. Kathy Absolon's reflection from an Indigenous perspective. Dr. Absolon's piece is particularly valuable as there has been little reflection on MAID from an Indigenous standpoint. Dr. Absolon speaks from the perspective of her own Anishinaabe tradition. She does not propose a generic Indigenous response, yet her reflections provide important insights, helping us to understand something about an Indigenous worldview in relation to MAID. Dr. Kathy Absolon, Anishinaabekwe from Flying Post First Nation, is Director of the Centre for Indigegogy in the Faculty of Social Work at Wilfrid Laurier University in Kitchener, Ontario where she is full professor. Dr. Absolon is the author of numerous articles, book chapters, and an important book that is in its second edition, *Kaanodosowin: How We Come to Know*.

Following Dr. Absolon, we have Dr. S. Zaki Ahmed's reflections on MAID from a Muslim perspective. Drawing on important Muslim teachings and texts, Dr. Ahmed is clear and concise about a Muslim response to MAID. It is valuable to read this decisive reflection that draws so carefully on the sacred texts of Islam. It is striking to note that while Islam is clear in prohibiting MAID, Dr. Ahmed points to the importance of each person's right to choose. Dr. Ahmed is from Thunder Bay Regional Health Sciences Centre, where he is Chief of Staff. He is an Associate Professor at Northern Ontario School of Medicine. He is an internist, respirologist and intensivist by training and completed a Masters Certificate in Physician Leadership from Schulich School of Business/York University and a Masters in Health Sciences in Health Administration from the Institute of Health Policy Management and Evaluation at University of Toronto.

The Jewish perspective on MAID is taken up by Rabbi Catharine Clarke. Speaking as a rabbi in the Conservative Movement, Rabbi Clarke points to a "narrow circumstance" that would allow for MAID with careful oversight. While Jewish teaching does not advocate MAID, exceptional cases demand compassion and careful reflection. Still, no exception is warranted for MAID based on financial reasons. Rabbi Clarke is the rabbi at Congregation Or Shalom in London, Ontario. She graduated from the Jewish Theological Seminary in 2012 and holds an M.A. and a J.D. from the University of Chicago.

Dr. Nuala Kenny draws deeply on papal documents reflecting on the value of life and the importance of human dignity. Encyclicals, pastoral letters, and other teachings reflecting on the moral tradition of the Roman Catholic church provide the foundation of Dr. Kenny's presentation on the Roman Catholic position on MAID. Dr. Kenny is currently Emeritus Professor in the Department of Bioethics at Dalhousie University. She is Health Policy Advisor for the Catholic Health Alliance of Canada. She is a prolific author; her most recent book is *Rediscovering the Art of Dying: How Jesus' Experience and Our Stories Reveal a New Vision of Compassionate Care.*

The final piece by Dr. Nita Kotiuga is a reflection on the questions posed from a Protestant perspective. Drawing mainly on the Judeo-Christian Bible as her authoritative source, Dr. Kotiuga offers a pastoral approach. She pays attention to the concrete situations of those considering MAID. The final word is left to those who are suffering yet, with an important reminder that one's suffering is not in isolation, but within a community affected by individual decisions. Dr. Kotiuga is Pastor of Connection and Theology Study at Westview Church in Pierrefonds, Quebec. She is Director of Spiritual Formation at Bakke Graduate University and online Mentor at Okanagan Bible College.

28.2 An Indigenous Perspective by Dr. Kathy Absolon, PhD

Boozhoo. Minogiizhigokwe n'dishnaukauz. Flying Post First Nation Treaty 9 n'doonjibaa. Greetings from my spirit to yours. My name is Shining Day Woman (the one that brings goodness and beauty to the day) and I am from Flying Post First Nation Treaty 9. My English name is Kathy Absolon. Miigwech for the invitation to include my perspective as an Anishinaabekwe (Ojibway woman). I qualify that my perspective is mine as an Anishinaabe woman and does not reflect that of all other Indigenous Nations or teachings. My perspective is informed by Anishinaabe teachers and Elders who have informed my knowing. I have been informed by beautiful Anishinaabe Elders, Grandmothers, and teachers. I also carry teachings informed by Midewewin Lodges across both Canada and the United States. What I am about to share has been informed by our Anishinaabe Creation Story.[1] Much of my knowledge I acquired orally by attending ceremony, being in circle, and listening in the sacred lodges for hours, days and years. What I am presenting here is only a brief glimpse of how life and death is informed by our oral traditions, teachings, and understanding of the spirit world. My brief summary is limited by the space herein and I acknowledge this topic is deeper that I can humanly offer.

[1] Benton-Banai (1998); Nabigon (2006).

In our Anishinaabe teachings we are taught about the spirit and sacredness of life. All of Creation has a spirit and that life itself has a spirit. All of life is sacred. I was in a Sweat Lodge ceremony with an Anishinaabe Elder (*Pathfinder*), who shared that we are spiritual beings having a human experience. Prior to my spirit taking its place in my human vessel, my spirit lived in the Spirit World alongside of the Creator and all other spirit beings. We believe that we exist within multiple realms and that this earthly humanly realm is only one stop along the way. We also believe that All Indigenous peoples have a Creation story and worldviews that inform their relationship to one's humanly journey in this earthly lodge.

As an Anishinaabe kwe, our creation story teaches of a spirit that journeys into this earth lodge and takes its place through our humanly vessel. This vessel is our body. We believe that who we are is comprised of a spirit, heart, mind, and body. The physicality of our being is only one element of our existence and that the totality of our being includes our spirit, heart, and mind. In our Creation Story, a spirit comes from a sacred realm, the Creator's realm, and takes its place in this earthly lodge with intention of having a human experience. Each spirit having a human experience is here with purpose and each life has a meaning. Sometimes we are not ever certain what the nature of our existence is, however, when one attends to their spiritual existence the answers to our purpose and existence become clearer. One begins to attend to their relationship with their spirit. We have spirit names, clans, helpers, and guides during our time in this earthly lodge. Throughout our humanly life it is the responsibility of our families and communities to ensure we know our clans, kinships systems and spirit names. As we grow and age in our earthly body, we learn more about what it means to be who we are. Our spirit names, clans, guides, and relationship to all of Creation help us navigate our human experience, along with our human families and kinship systems. When a person passes on from this humanly world we see that person is beginning their spirit journey back to the spirit world. In a sense, the spirit being having a human experience is simply returning back home. Sometimes you hear death being referred to as a person is beginning their spirit journey and our ancestors are preparing for their arrival. Death, to me, is a birth in the spirit world. Birth in this humanly world, may perhaps be a spirit leaving the spirit world to come to this earthly world. The cycle of life comes from spirit and when the spirit returns home, it leaves the body and human vessel.

From this Anishinaabe perspective, there is a real difference between the idea of MAID and deciding on when one's death or spirit journey begins and a spirit journey that happens naturally (accidental or natural). Another example is death from suicide. There are teachings and ceremonies that I have witnessed and been a helper for that are specific to helping a spirit on its journey when the death was not necessarily the persons time but was forced by suicide. Is MAID a forced death? I don't really know and cannot say for sure. These are matters that are sacred. From a traditional point of view we would not construct death or predetermine the time when a spirit is done having their humanly experience. We would not typically interfere with such a sacred journey. The responsibility of humans is to support each other and to do ceremonies to help prepare the spirit for its journey home. We often don't talk about medically assisted dying and link that to one's spirit journey because when a person reaches the end of their human journey it is more of in the hands of the Creator. Honestly, I have been a helper to many funerals and helped carry the lantern for those passing onto their spirit journey. I have little experience with those who are passing onto their spirit journey because of a medically assisted death. While some people may consider MAID, I do not believe that within a traditional knowledge system, we believe we have the right to aid in ending a human life. With that said, if such a situation presented itself, I would definitely consult with ceremonial leaders who work with passing on ceremonies.

If an Indigenous person is considering MAID and wants to understand and receive support from their traditional community, I would encourage the family to seek out their Elders and traditional leaders. I would suggest they seek their guidance and advice. I also believe that the Elders and ceremonial leaders, would generally determine the course of action based on their knowledge and teachings. This determination would likely happen in ceremonies and that guidance from spirit would be searched out through various sacred ceremonies. Only after engaging in ceremony themselves would Elders and ceremonial leaders be in a position to offer guidance about such a situation as MAID.

The concern around death is always about the spirit journey of the one passing on to the spirit world. It is the responsibility of the ceremonial family, kinship system and community to ensure the passing on ceremonies, protocols and spirit journey teachings are given. There is much to do to prepare one for such a journey and our funerals and protocols center around preparing the spirit for its journey home. The humanly vessel has done its work and is now reached a death. While the humanly vessel is shutting down, much of the preparations are for the spirit journey. People need to have their spirit names and there are protocols and preparations to attend to. All of this work is for the spirit journey, and in this work, the human also receives comfort as they prepare to leave this realm. It is truly a sacred time.

28.3 An Islamic Perspective by Dr. S. Zaki Ahmed, MD. FRCP(C), FCCP, CCPE

Islam is the second largest religion in the world with over two billion Muslims spread across the globe.[2] The religion is based on the concept of monotheism, Allah being the only deity, and Muhammad as the prophet of Islam. The Islamic law, Shariah, is based on the Quran, the holy book and the Hadith which are the ways and sayings of Muhammad.[3,4] Topics like artificial ventilation, cardiac resuscitation, supplemental artificial nutrition, organ donation and Medical Assistance in Dying (MAID) are topics that arose centuries after the Quran and the Hadith were completed. Hence

[2] World Population Review, Muslim Population by Country 2022, World Population Review. https://worldpopulationreview.com/country-rankings/muslim-population-by-country.

[3] Smith (2001).

[4] Nasr (2002).

these topics are decided upon by rulings known as Fatwa. Fatwas are arrived upon by scholars based on the above sources, scientific insight, and cultural and societal context.[5]

There are two large sects in IslamSunnis and Shias with multiple groups of varied beliefs within them.[6] Islam is not monolithic and there is diversity in views based on culture and individuality. Each individual's and their family's wishes and beliefs should be considered when deciding on providing, withdrawing, or not offering medical therapies. There is significant variation in the religiosity of an individual and their personal belief regarding MAID.[7] The following section is based in published literature and represents the popular dogma within the Islamic scholars.

Health is considered to be a blessing in Islam and self-harm is not allowed as per the Quran—"do not throw yourselves into destruction with your own hands." (2:195)[8] The prophet advised Muslims to seek and accept medical treatments for illnesses. Afflictions and pain are ways of expiating sins and is often considered to be a test for Muslims.[9] Giving or taking away life is a task reserved for God and for no one else—"Nor can a soul die except by Allah's leave, the term being fixed as by writing." (3:145)[10] Hence murder, suicide and Medical Assistance in Dying are not permitted in Islam—"and do not kill yourselves (or one another)." (4:125).[11,12,13,14]

There is however, a considerable difference between active euthanasia and passive euthanasia in Islam. Where active euthanasia and MAID are not allowed, life-sustaining therapies like intubation and ventilation as well as resuscitation can be withheld if death is inevitable. This should be discussed in conjunction with the family and providers.[15] Often, this may require three clinicians to agree on the prognosis of outcomes.[16] Although not universally agreed upon,[17] pain management and psychological comfort should not be denied to the patient by the family or the

---

[5] Kersten (2019).

[6] Zubaida (2009).

[7] Aghababaei (2013).

[8] Ali (2011).

[9] Nasafi (2021).

[10] Ali (2011).

[11] *Topics of Quran*. http://www.topicsofquran.com/toq/quran-translations/display?tid=664&s=4&f=29&t=29 Accessed February 28, 2022.

[12] Madadin et al. (2020).

[13] Izgandarova (2015).

[14] Freeland (1997).

[15] Sachedina (2005).

[16] Saeed et al. (2015).

[17] Malek (2018).

provider.[18,19] The provision of palliative care and comfort measures is allowed and often encouraged by Islam.[20,21,22]

None of the 50 Muslim countries in the world allow MAID and the provision of MAID by a healthcare provider is considered to be equivalent to murder in most of those countries.[23] There is a general disdain against suicide that causes significant strife amongst healthcare providers across the world.[24,25,26,27,28,29,30,31] In countries where MAID is permissible, healthcare providers may have difficulty in providing these services due to their religious beliefs. In Canada, physicians may decide not to perform medical assistance in dying but are required to make a formal referral on behalf of the patient in Ontario and Nova Scotia.[32,33] They must continue caring for the patient until transfer of care is possible even when they are not obligated to make the referral.[34] This ruling upheld by the Supreme Court can be morally challenging for Muslim and other physicians who have conscientious objections to Medical Assistance in Dying and even the process of a referral may equate to asking a colleague to murder their patient.[35]

Almost all the literature sources agree that MAID is not permissible in Islam. Islam, however, has always established the greatest importance to free will and autonomy to its followers.[36]

There are many patients who self-identify as Muslims living in jurisdictions where MAID is permitted by law. It is important to value each individual's right to choose based on their own individual beliefs grounded in their religiosity and circumstances.

---

[18] Canadian Council of Imams. Senate of Canada https://sencanada.ca/content/sen/committee/432/LCJC/Briefs/CanadianCouncilofImams_e.pdf Accessed, February 28, 2022.

[19] Chakraborty et al. (2017).

[20] Al-Shahri and Al-Khenaizan (2005).

[21] Baddarni (2010).

[22] Leong et al. (2016).

[23] Sachedina (2005).

[24] Rathor et al. (2014).

[25] Aramesh and Shadi (2007).

[26] Jafary (1996).

[27] Ayuba (2016).

[28] Ahaddour et al. (2018).

[29] Izgandarova (2015).

[30] Damghi et al. (2011).

[31] Razban et al. (2016).

[32] Kelsall (2018).

[33] College of Physicians and Surgeons Ontario. Medical Assistance in Dying. https://www.cpso.on.ca/Physicians/Policies-Guidance/Policies/Medical-Assistance-in-Dying Accessed February 28, 2022.

[34] College of Physicians and Surgeons of British Columbia. Practice Standard—Medical Assistance in Dying https://www.cpsbc.ca/files/pdf/PSG-Medical-Assistance-in-Dying.pdf Accessed February 17, 2022.

[35] Kelsall (2018).

[36] Rizvi (2019).

28.4 A Judaic Perspective by Rabbi Catharine Clarke, M.A., J.D.

In Judaism,[37] God is understood to have created the universe and everything in it, including every human being. As Creator, God owns our bodies and has granted us normal use of them during our lifetimes. We have a duty to care for our bodies[38] and are not permitted to inflict injuries on ourselves, especially injuries likely to lead to death.[39] In Judaism, God also is understood to be infinitely compassionate.[40] Psalm 145:9 declares "God's compassion embraces all God's creations." Moreover, we are to "walk in all God's ways," including responding to suffering with compassion.[41]

The tension between these two principles yields permission for aid in dying in very narrow circumstances. Specifically, aid in dying is permitted for a patient who "has a terminal, incurable disease, which despite available treatments, will, in the opinion of at least two independent physicians, bring about the person's death within six months or shortly thereafter… and that involves pain or suffering that cannot be quelled."[42] Moreover, unlike with current Canadian assistance in dying laws, healthcare professionals must provide no more assistance in dying than supplying drugs for the patient to administer himself.[43]

Permission to use aid in dying is tightly restricted because of several concerns that have not been, and are not likely to be, adequately addressed by law or society. One of the weightiest of these concerns is the financial burden of illness and disability.[44] In Canada, we are fortunate that most of the costs of healthcare do not fall on the individual already burdened with poor health, but we must resist the perverse incentives brought about by the fact that providing medical assistance in dying is cheaper for the public health system than long-term treatments and pain relief or other forms of

---

[37] Jews are a small but diverse people with many ways to understand the proper relationship between God and humankind. The author of this essay is a rabbi in the Conservative Movement, and she relies on the rulings of Jewish law (*halachah*) made by the Committee on Jewish Law and Standards, the leading decision-making body on matters of *halachah* for the Conservative Movement. Other branches of Judaism take different approaches to this matter of life and death.

[38] Judaism does not, however, require one to begin or continue machines and medications that do not benefit the patient in that patient's evaluation. In addition, pain medication should be used to relieve suffering even if that medication may hasten death, so long as the intent is to alleviate pain. Dorff, Elliot N., "Assisted Suicide/Aid in Dying Reconsidered: "God's Compassion Embraces All God's Creations (Psalms 145:9)"," https://www.rabbinicalassembly.org/sites/default/files/2021-04/Assisted%20Suicide%20Revisited%20final.pdf (accessed 5/10/2022) 4, 33, 42.

[39] Dorff, Elliot N. "Assisted Suicide," https://www.rabbinicalassembly.org/sites/default/files/public/halakhah/teshuvot/19912000/dorff_suicide.pdf (accessed 5/10/2022) 380–381.

[40] Dorff, Elliot N. "Aid in Dying Reconsidered," 28.

[41] *Sifrei Devarim* 49:1, Ekev, on Deuteronomy 11:22.

[42] Dorff, "Aid in Dying Reconsidered," 43.

[43] Dorff, Aid in Dying Reconsidered," 43–44. A Jewish healthcare professional may provide aid in dying under the circumstances specified in this essay, but Jewish law does not obligate them to do so, even to patients in uncontrollable pain.

[44] Dorff, "Aid in Dying Reconsidered," 6.

end-of-life care, such as hospice care. Especially troubling are reports of individuals with chronic conditions who seek to die because they cannot afford an apartment that accommodates their chronic conditions.[45] Budgetary considerations must not make us callous to the value of a life created by God.

Another concern behind the strict limits Judaism places on aid in dying is inadequate attention to mental health.[46] Someone who has received a terminal diagnosis or who suffers from a chronic condition understandably may suffer from depression. Rather than aid in dying, our response to this emotional ache must be manifested in measures that counteract depression, such as medication and therapy from professionals and visits and compassion from family, friends, and community members. We better practice the value of compassion when we improve access to mental health care and put as many resources into developing effective treatments for mental health conditions as we do into, say, cancer research.

Jews reading this essay will be interested in one more question. Judaism includes many laws and customs concerning burial and mourning, including burial in a cemetery with other Jews, a seven-day period of formal mourning (*shivah*), and the recitation of memorial prayers (*Kaddish* and *Yizkor*). The traditional understanding is that an individual who died by suicide would not receive these rites, although such a rule is rarely enforced.[47] Few rabbis would enforce it in the situation described here, where an individual in extreme uncontrollable pain received aid in dying, and many would not deny burial and mourning rites even to a Jew who died under the much broader provisions of MAID. Such a death should be met with the compassion God requires of us.

28.5 A Catholic Perspective by Dr. N. P. Kenny O.C., M.D., FRCP (C)

Catholic teaching on MAID is rooted in the belief that we have a God who so loved us that He sent His only Son to suffer and die for us.[48] In Gethsemane, Scriptures portray Jesus experiencing the natural human desire to avoid pain and suffering:

> For it is not as if we had a high priest who was incapable of feeling our weakness with us; but we have one who has been tempted in every way that we are, though he is without sin. (Hebrews 4:15)[49]

---

[45] CTV News. "Woman with Disabilities Nears Medically Assisted Death after Futile Bid for Affordable Housing," (April 30, 2022, updated May 4, 2022) https://www.ctvnews.ca/health/woman-with-disabilities-nears-medically-assisted-death-after-futile-bid-for-affordable-housing-1.5882202 (accessed May 10, 2022).

[46] Dorff, "Aid in Dying Reconsidered," 6–7.

[47] Dorff, "Aid in Dying Reconsidered," 34.

[48] Kenny (2017).

[49] *New Jerusalem Bible: Standard Edition* (1999). All bible quotations are from the *New Jerusalem Bible:Standard Edition.*

Jesus prayed intensely before surrendering to trust in the Father's love no matter how difficult things might become:

> Nevertheless, let your will be done, not mine. (Luke 22:43)

The dying Jesus is not passive but actively living out his mission to His last breath.

There is a long Catholic moral tradition guiding decision making in illness and dying based on belief that "The dignity of man (*sic*) rests above all on the fact that he is called to communion with God."[50] All share in this God-given inherent dignity which is not an attribute lost in weakness and dependence.

The church also teaches that "Life and physical health are precious gifts entrusted to us by God. We must take *reasonabl*e care of them, taking into account the needs of others and the common good."[51] Biologic life is not an absolute good. It follows that "To forego extraordinary or disproportionate means is not the equivalent of suicide or euthanasia; it rather expresses acceptance of the human condition in the face of death."[52] Medically assisted death is rejected because "Euthanasia is a false solution to the drama of suffering…the true response cannot be to put someone to death, however 'kindly,' but rather to witness to the love that helps people to face their pain and agony in a human way."[53]

Catholics are concerned that MAID is a rejection of the suffering, death and resurrection of Jesus Christ and a medicalization of suffering and death. Christians do not seek suffering but believe that "In suffering there is concealed a particular power that draws a person interiorly closer to Christ, a special grace" in which there is "above all a call."[54]

Our "witness to the love" of the dying requires an evangelizing community which "gets involved by word and deed in people's lives; it bridges distances, it is willing to abase itself if necessary, and it embraces human life, touching the suffering flesh of Christ in others."[55] We also advocate for hospice and palliative services because "Palliative care is an expression of the properly human attitude of taking care of one another, especially of those who suffer… when life becomes very fragile and the end of the earthly existence approaches."[56] It addresses medical care for pain and other symptoms. "The use of painkillers to alleviate the sufferings of the dying, even at

---

[50] *Catechism of the Catholic Church* (1993).

[51] *Catechism of the Catholic Church*, no. 27.

[52] Pope John Paul II. *Evangelium Vitae*, no. 66. Located at: https://www.vatican.va/content/john-paul-ii/en/encyclicals/documents/hf_jp-ii_enc_25031995_evangelium-vitae.html.

[53] Pope Benedict XVI, Angelus, February 1, (2009). Located at: https://www.vatican.va/content/benedict-xvi/en/angelus/2009/documents/hf_ben-xvi_ang_20090201.html.

[54] Pope John Paul II. Salvific Doloris, (1984). Located at: https://www.vatican.va/content/john-paul-ii/en/apost_letters/1984/documents/hf_jp-ii_apl_11021984_salvifici-doloris.html.

[55] Pope Francis. Evangelium Gaudi, no. 24. Located at: https://www.vatican.va/content/francesco/en/apost_exhortations/documents/papa-francesco_esortazione-ap_20131124_evangelii-gaudium.html.

[56] Pope Francis. Address to the Pontifical Academy of Life, March 5, 2015. Located at: https://www.vatican.va/content/francesco/en/speeches/2015/march/documents/papa-francesco_20150305_pontificia-accademia-vita.html.

the risk of shortening their days, can be morally in conformity with human dignity if death is not willed as either an end or a means, but only foreseen and tolerated as inevitable."[57]

Responding to Catholics who are considering MAID demands a re-assessment of medical and spiritual care. It presents the pastoral challenge of balancing compassion and accompaniment in a time of crisis with education of the serious moral consequences.

Catholic bioethicists respond from our spiritual and moral tradition not from procedural principle-based reasoning. They can help individuals consider that MAID is not limited to dying but for "intolerable suffering" as the person determines it, including psychological distress, uncertainty about care, desire to control the death, fear of dependence, feelings of loss of dignity, and guilt at being a burden to others. There is high risk for the vulnerable. MAID has been rapidly normalized in Canada, which has the most liberal access in the world. Preaching and teaching should promote and support prayerful reflection on values and beliefs regarding health decisions and dying long before times of crisis.

28.6 A Protestant Perspective by Nita Kotiuga, MA, DTL

In reflecting on medical assistance in dying (MAID) from a Protestant perspective, there are several guidelines I would offer people from my faith tradition. Life is precious. Human beings value and enjoy life. We procreate for a varied number of reasons but some include forming the next generation, enjoying having little people who were made by 2 people, we want to have people somewhat made in our image. Having future generations live on beyond us is a human value and now society is asking that we consider how we depart from this life as well.

From the Christian scriptures we are taught that we are created beings, made not by our own hands but by a Creator. We have extrapolated that our lives are not our own but belong to our Creator, who lovingly made us in his image and breathed his life into us. This is unique in the creation story in Genesis. The Genesis story, nevertheless, is not about us, the created, but about Who created us. The *Enuma Elish*[58] is the basis for our story and our story declares one God and points to *El*/God as creator.

In my tradition, some people believe it is an offence against God to end life, and a crime against the person. Only God is allowed to end a life because he is the originator of that life. Yet, as science and medical support has lengthened human life, and many times this lengthened life is spent in a hospital bed, one might ask, when is it time to say goodbye? Families and loved ones can push the medical system

[57] *Catechism of the Catholic Church*, no. 2279.

[58] The *Enuma Elish* is an ancient Babylonian creation story with both similarities and differences from the creation story in the Hebrew bible in Genesis. It sheds light on Biblical scholars' understanding of the creation story in the book of Genesis.

to extend life through artificial support and it is difficult to know when to end this support. Age does play a factor as we understand that a life lived does end, and it is acceptable to let an older person die without heroics.

The difficulty for many is when the person is young and could have the potential of an active contributive life if they were to be healed from their illness. The temptation to hope for a new breakthrough for a disease keeps many from choosing to allow death or even choose to end life. We say "I don't ever want to live like this" when we see people struggle but wouldn't choose to let them die or let them choose death.

How would we make a decision regarding MAID? The apostle Paul wrote in Philippians 1:21 that for him to live is Christ and to die is gain. He saw death as a better choice because he would be reunited with Christ, the very Christ whom he met on the road to Damascus. (Acts 22: 6–11) He understood that life was an opportunity to serve humankind and learn to love God, but that death would be the reward for a life well-lived. My question would be: have you or are you living such a life? Is your life dedicated to being a blessing? If it is, then dying for you means being with Christ. If it is not, then death may not be the gain you are hoping for. If you are looking to end your current or potential suffering, first ask yourself what are your various thoughts about an afterlife?

In responding to the question of what guidance I, as a leader in my Church, would offer, I direct my response to a younger person looking at a gradual diminishment of their life and perhaps a painful end. I believe this is a personal choice to be made between God and the person because God holds us each accountable for the choices we make. The question each person needs to ask of themselves is whether this is truly the way they wish to end their life. Having established that life is Christ and death is gain, are they truly feeling that their contribution to life is finished? If so, how have they determined this? Is there anything they have overlooked? Are they in a place where the future is slowly narrowing and life diminishing? If so, how would they, in their own words, describe what and when an end to life should be?

What would I suggest they need to consider in contemplating MAID? Closure would be needed before one can say goodbye to life. Relationships that are unfinished, loved ones who need reassurance, and other forms of reconciliation need to be considered in order to leave well. Dying is never a solo endeavour but a consequence felt by many and so others need to be prepared as well. We were born into community and need to leave our community as well as we can given our circumstances.

28.7 Conclusion

A common thread among all five responses to the question of MAID is that human beings are relational beings and MAID is not a solo decision nor a solo endeavour. The importance of relationships to others, of community and communal living, and of the common good are raised in different manners in each contribution. That the Creator, God, Allah, is a compassionate Creator comes through clearly for all and so, in the midst of terrible, unbearable suffering, compassion and solidarity must be offered. All contributors point to the belief that human beings are not the owners of their lives but are created and have a responsibility to our Creator. Unlike some who may consider MAID a gift and even a legacy that one leaves to others,[59] in none of the contributions to this chapter is MAID seen as a good to be sought after. The importance of religious reflection on end-of-life issues such as medical assistance in dying is that it raises questions that are generally left out of the discussion in a secular context. Such questions might be: What is my responsibility in face of life and the possibility of death? What am I called to live in face of suffering? Is there something I am still called to give in this life that I have been given? How might my relationship with my Creator be lived in the midst of suffering and nearness of death? These are religious questions, questions concerning ultimate value and meaning that push one into a different realm of existence, a religious realm that can shift our experience in a dramatic and graced manner.

Acknowledgements I would like to thank Matthew R. Anderson for suggesting the title of this chapter.

References

Aghababaei, Naser. 2013. The euthanasia-religion nexus: Exploring religious orientation and euthanasia attitude measures in a Muslim context. *OMEGA—Journal of Death and Dying* 66 (4): 333–341.

Ahaddour, Chaïma, Stef Van den Branden, and Bert Broeckaert. 2018. God is the giver and taker of life: Muslim beliefs and attitudes regarding assisted suicide and euthanasia. *AJOB Empirical Bioethics* 9 (1): 1–11.

Ali, Abdullah Yusuf. 2011. *The Meaning of the Holy Qu'ran*. Beltsville, MD: Amana Publications.

Al-Shahri, M.Z., and Abdullah Al-Khenaizan. 2005. Palliative care for Muslim patients. *Journal of Supportive Oncology* 3(6): 432–436.

Aramesh, Kiarash, and Heydar Shadi. 2007. An Islamic perspective on Euthanasia. *The American Journal of Bioethics* 7 (4): 65–66.

Ayuba, Mahmud Adesina. 2016. Euthanasia: A Muslim's perspective. *Scriptura: Journal for Contextual Hermeneutics in Southern Africa* 115(1): 1–13.

Baddarni, Kassim. 2010. Ethical dilemmas and the dying Muslim patient. *Asian Pacific Journal of Cancer Prevention* 11 (1): 107–112.

Benton-Banai, Edward. 1998. *The Mishomis book*. Hayward, Wisconsin: Indian Country Communications Inc.

Canadian Council of Imams. Senate of Canada. https://sencanada.ca/content/sen/committee/432/LCJC/Briefs/CanadianCouncilofImams_e.pdf.

Catechism of the Catholic Church. 1993. Liberia Editrice Vaticana, Vatican City.

Chakraborty, Rajshekhar, Areej R. El-Jawahri, Mark R. Litzow, Karen L. Syrjala, Aric D. Parnes, and Shahrukh K. Hashmi. 2017. A systematic review of religious beliefs about major end-of-life issues in the five major world religions. *Palliative and Supportive Care* 15 (5): 609–622.

[59] I am referring to the May 25, 2017 *New York Times* article about John Shields, titled "At His Own Wake, Celebrating Life and the Gift of Death," by Catherine Porter, where Shields, who requested and received MAID, communicated to Porter that he "believed that dying openly and without fear could be his most meaningful legacy."

College of Physicians and Surgeons of British Columbia. Practice standard—Medical assistance in dying. https://www.cpsbc.ca/files/pdf/PSG-Medical-Assistance-in-Dying.pdf.

College of Physicians and Surgeons of Ontario. Medical assistance in dying. https://www.cpso.on.ca/Physicians/Policies-Guidance/Policies/Medical-Assistance-in-Dying.

CTV News. Woman with disabilities nears medically assisted death after futile bid for affordable housing, (April 30, 2022, updated May 4, 2022). https://www.ctvnews.ca/health/woman-with-disabilities-nears-medically-assisted-death-after-futile-bid-for-affordable-housing-1.5882202. Accessed 10 May 2022.

Damghi, Nada, Jihane Belayachi, Badria Aggoug, Tarek Dendane, Khalid Abidi, Naoufel Madani, Aicha Zekraoui, Abdellatif Benchekroun Belabes, Amine Ali Zeggwagh, and Redouane Abouqal. 2011. Withholding and withdrawing life-sustaining therapy in a Moroccan emergency department: An observational study. *BMC Emergency Medicine* 11(1): 11–12.

Dorff, Elliot N. Assisted suicide/Aid in dying reconsidered: God's compassion embraces all god's creations (Psalms 145:9). https://www.rabbinicalassembly.org/sites/default/files/2021-04/Assisted%20Suicide%20Revisited%20final.pdf. Accessed 5 October 2022. 4, 33, 42.

Freeland, Richard. 1997. Euthanasia and Islamic law. *Medico-Legal Journal* 65 (4): 196–198.

Izgandarova, Nazila. 2015. Physician-assisted suicide and other forms of Euthanasia in Islamic spiritual care. *Journal of Pastoral Care and Counselling* 69 (4): 215–221.

Jafary, Maqbool H. 1996. Euthanasia: Changing trends. *Journal of Pakistan Medical Association* 46(12): 269–270.

Kelsall, Diane. 2018. Physicians are not solely responsible for ensuring access to medical assistance in dying. *Canadian Medical Association Journal* 190 (7): E181.

Kenny, Nuala. 2017. *Rediscovering the art of dying: How Jesus' experience and our stories reveal a new vision of compassionate care.* Toronto, ON, Canada: Novalis.

Kersten, Carool. 2019. *The Fatwa as an Islamic legal instrument concept, historical role.* Berlin and London: Gerlach Press.

Leong, Madeline, Sage Olnick, Tahara Akmal, Amanda Copenhaver, and Rab Razzak. 2016. How Islam influences end-of-life care: Education for palliative care clinicians. *Journal of Pain and Symptom Management* 52 (6): 771–774.

Madadin, Mohammed, Houria S Al Sahwan, Khadijah K Altarouti, Sarraa A Altarouti, Zahra S Al Eswaikt, and Ritesh G Menezes. 2020. The Islamic perspective on physician-assisted suicide and euthanasia. *Medicine, Science and the Law* 60(4): 278–286.

Malek, Mohammad Mustaqim. 2018. Islamic considerations on the application of patient's autonomy in end-of-life decision. *Journal of Religion and Health* 57 (4): 1524–1537.

Nabigon, Herb. 2006. *The hollow tree. Fighting addiction with traditional native healing.* Montreal and Kingston: McGill-Queen's University Press.

Nasafi, Abdullah. 2021. *Translation of Sahih Al Bukhari.* Independently Published.

Nasr, Syed Hossein. 2002. *Islam: Religion, history and civilization.* San Francisco: Harper Collins Publisher.

New Jerusalem Bible: Standard edition. 1999. New York: Doubleday.

Pope John Paul II. *Evangelium Vitae*, no. 66. Located at: https://www.vatican.va/content/john-paul-ii/en/encyclicals/documents/hf_jp-ii_enc_25031995_evangelium-vitae.html.

Pope Benedict XVI, Angelus, February 1, 2009. Located at: https://www.vatican.va/content/benedict-xvi/en/angelus/2009/documents/hf_ben-xvi_ang_20090201.html.

Pope John Paul II. Salvific Doloris, 1984. Located at: https://www.vatican.va/content/john-paul-ii/en/apost_letters/1984/documents/hf_jp-ii_apl_11021984_salvifici-doloris.html.

Pope, Francis. Address to the pontifical academy of life, March 5, 2015. Located at: https://www.vatican.va/content/francesco/en/speeches/2015/march/documents/papa-francesco_20150305_pontificia-accademia-vita.html.

Pope, Francis. *Evangelium Gaudi*, no. 24. Located at: https://www.vatican.va/content/francesco/en/apost_exhortations/documents/papa-francesco_esortazione-ap_20131124_evangelii-gaudium.html.

World Population Review, Muslim Population by Country 2022, World Population Review. https://worldpopulationreview.com/country-rankings/muslim-population-by-country.

Porter, Catherine. At his own wake: Celebrating life and the gift of death. New York Times, May 25, 2017.

Rathor, Mohammad Yousuf, Mohammad Fauzi Abdul Rani, Mohammad Arif Shahar, A. Rehman Jamalludin, Shahrin Tarmizi Bin Che Abdullah, Ahmad Marzuki Bin Omar, and Azarisman Shah Bin Mohamad Shah. 2014. Attitudes toward euthanasia and related issues among physicians and patients in a multi-cultural society of Malaysia. *Journal of Family Medicine and Primary Care* 3(3): 230–237.

Razban Farideh, Sedigheh Iranmanesh, Hasan Eslami Aliabadi, and Mansooreh Azzizadeh Forouzi. 2016. Critical care nurses' attitude towards life-sustaining treatments in South East Iran. *World Journal of Emergency Medicine* 7(1): 59–64.

Rizvi, Syed Abbas. 2019. *The house of Muhammad: The sectarian divide and the legacy about Aya Tatheer*. Bloomington, Indiana: Author House.

Sachedina, Abdulaziz. 2005. End-of-life: An Islamic view. *The Lancet* 366 (9487): 774–779.

Saeed, Fahad, Nadia Kousar, and Sohaib Aleem. 2015. End-of-life care beliefs among muslim physicians. *American Journal of Hospice and Palliative Care* 32 (4): 388–392.

Sifrei Devarim 49: 1, Ekev, on Deuteronomy 11: 22.

Smith, Huston. 2001. *Islam: A concise introduction.* San Francisco: Harper Collins Publishers.

Topics of Quran. http://www.topicsofquran.com/toq/quran-translations/display?tid=664&s=4&f=29&t=29.

Zubaida, Sami. 2009. Sects in Islam in *The Oxford handbook of the sociology of religion*, ed. Peter B. Clarke, 545–561. Oxford: Oxford University Press.

Christine Jamieson is Associate Professor in the Department of Theological Studies at Concordia University in Montreal, Canada. Her areas of expertise are Christian social ethics, bioethics, and Indigenous spirituality.

Chapter 29
MAID Practice and Impact

Jaro Kotalik and Louisa Pedri

Abstract Medical Assistance in Dying (MAID) in Canada is an ongoing project. Required by a court order in 2015, legalized in 2016 and amended in 2021, it was expected to have further expansion of eligibility criteria in 2023 but after much public debate and extensive print media engagement, it was postponed until March 2024. From the beginning, MAID attracted a large number of patients, so that by the years 2019–2021 its approximate annual increase was 30%, unrivalled in other jurisdictions that provide voluntary euthanasia or assisted suicide. We suggest that this extraordinary growth in Canada is due to MAID's genesis. Within the matrix of the Canadian Charter of Rights and Freedoms, MAID was declared a constitutional right enabling the exercise of one's autonomy. The medical profession bore the responsibility of performance, with physicians and nurse practitioners remaining free to object, but still legally bound to provide a referral. With the powers conferred upon them by the federal government, the provinces and territories' health authorities dispatched orders to all publicly funded, government-run health institutions to set up MAID access points. Having assumed responsibility for providing MAID, these health care institutions have been expeditious in educating staff and building an abiding infrastructure for MAID's delivery. We can only get a glimpse of the impact of this very liberal law and its subsequent managerial success because Canadian federal statistics lag behind and independent research studies in critical areas of concern are few. However, that which is available reveals a very complex interrelated web of enablers that not only fuel the high rate of government sanctioned deaths, but also will cause self-inflicted damage. The casualties will be found within the medical

J. Kotalik (✉) · L. Pedri
Centre for Health Care Ethics Lakehead University, 955 Oliver Road, Thunder Bay, Ontario P7B 5E1, Canada
e-mail: jkotalik@nosm.ca

L. Pedri
e-mail: lpedri@lakeheadu.ca

J. Kotalik
Northern Ontario School of Medicine University, Thunder Bay, ON, Canada

Department of Philosophy, Lakehead University, Thunder Bay, Canada

J. Kotalik and D. W. Shannon (eds.), *Medical Assistance in Dying (MAID) in Canada*, The International Library of Bioethics 104,
https://doi.org/10.1007/978-3-031-30002-8_29

profession, the health care system and the population in general, but especially the vulnerable and neglected groups in our society.

29.1 Introduction

This chapter has two objectives. (1) to extend the story of MAID'S debates and development in the law and policy as presented in the first chapters of the book, up to February 2023. (2) to report and comment on the practice of MAID as it is expressed in publications from health care institutions, and from health care professions. The conclusion will attempt to draw a unique composite picture of MAID practice in Canada.

29.2 Parliamentary Debates and Professional Positions

The first sections of this book reported on Canada's MAID law up to its latest development, the passing of Bill C-7 of March of 2021 (House of Commons 2021; Lemmens et al. 2021). This controversial Bill amended the original MAID bill, C-14 by modifying the procedural safeguards and expanding the eligibility criteria to persons whose death is not reasonably predictable and also to persons with mental disorder as a sole condition, with the latter condition to take effect two years later, on March 17, 2023. The prior controversies did not abate after the law was enacted, in fact the struggle intensified; its key issue, "if and how" to make eligible for MAID, those persons suffering solely from mental disorders (discussed also in Chaps. 17–19). On one side of the debate, the upholders of "radical autonomy" (discussed in Chap. 1) defended the position of an individual's right to access MAID, even if that decision is compromised by the presence of ulterior motives (e.g., poverty, isolation, angst, etc.), and even if such acquiescence to MAID would contravene the intent of the law's eligibility criteria. On the other side, pointing out the law's many flaws and potential dangers, the upholders of a broader, classical view of autonomy, which accepts restrictions on individual action as required by common interest, are more concerned with the safety and value of the individual and society. The intensity of this debate as it unfolded in the public square between parliament, the public and numerous professional groups, and recorded in professional journals or popular print media will illustrate the inherent challenges facing law development and policy-making in this domain.

To deal with unresolved issues after the adoption of Bill C-7, the Parliament re-established the Special Joint Committee on MAID, which allowed many witnesses and experts an opportunity to give their, often contradictory, views. The committee's interim report endorsed governmental plans for MAID expansion for people with sole mental illness, for mature minors, and for people not capable to decide but who wrote an advanced request. However, three members of the committee gave a dissenting report, claiming that the hearings were rushed and asked the federal government to pause the expansion of MAID (Nelson 2022). Concurrently, an Expert Panel on MAID and Mental Illness was set up to provide guidelines, safeguards and protocols for the implementation of MAID for those with mental illness. Again, two members of the panel, including the panel's ethicist, resigned, citing serious flaws in the process. In May 2022, this panel reported that it was impossible to provide a scientific rule on "incurability" and "irreversibility" of mental illness, the two conditions that the Criminal Law requires for eligibility. The Panel admitted that it will be impossible to differentiate between a MAID request and a suicidal ideation that is sometimes associated with mental illness, but evaded the problematic issue by claiming that Canadian society already made the "ethical choice" to enable certain patients to receive MAID, regardless. The panel subsequently formulated 19 recommendations, which they concluded could be adopted without any change to the law, believing the existing safeguards were sufficient, and no additional safeguards were recommended for allowing MAID to persons with solely mental disorders. The panel concluded that the expansion of eligibility for MAID can proceed as scheduled with assessors making eligibility decisions on a case-by-case basis (Government of Canada 2022).

This panel's recommendation raised a storm of objections from parliamentarians, various professional and community groups as well as the media. Parliamentarians challenged the Expert Panel to actually review the compliance and effectiveness of the safeguards (The Canadian Press 2022), but their request, as reasonable as it was, was ignored. The Society of Canadian Psychiatry called for a delay in the implementation and provision of MAID for mental illness until 2024 to allow a proper consultation, with a review of the safety and development of evidence-based recommendations regarding MAID for sole mental illness (Call for Action 2022). Also calling for a delay was the Association of Chairs of Psychiatry in Canada, who pointed out that the current problem of poor access to psychiatric care was exacerbated by an opioid epidemic and mental health pandemic (CTV 2022). The Canadian Association for Suicide Prevention issued a statement, asking that the expansion of MAID be postponed until a consensus could be achieved on the issues of 'irremediability' and 'suicidality', as was requested also by the Canadian Association of Mental Health (CASP 2022). Finally, the Commissioner of the Canadian Human Rights Commission remarked that, "Leaving people to make this choice [to die], because the state is failing to fulfill their fundamental human rights, is unacceptable" (BBC 2023).

29.3 Views from the Media and Political Actions

An understanding of MAID policy development would be incomplete without appreciating the role played by print media. Until 2022, the usual stories published were about patients being grateful for the opportunity to obtain relief of their suffering via MAID, and often with their families' enthusiastic encouragement. Media were initially effective promoters of this new civil right, but gradually more and more stories emerged, showing the darker side of this program, about people who applied or receive MAID because of their poverty, lack of access to health care or housing. Stories and commentaries definitely changed in 2022 as opposition to governmental plans for the expansion of MAID intensified. Major Canadian and foreign print media began reporting on serious issues. A national columnist from the popular daily *Toronto Star* published an interview with Justice Minister David Lametti, where she concluded that the minister's comments are "shocking because they fly in the face of suicide prevention efforts" and it sounded like he's encouraging suicide (Raj 2022). In November, the *Toronto Star* published an editorial entitled, "We need to put the brakes on expanding access to MAID" (Star Editorial Board 2022). A major national daily paper, the *Globe and Mail*, published an editorial stating that, "In Canada, people can die with dignity, they should be able to live that way, too" (The Editorial Board 2022 (1)), and two days later an opinion piece "Euthanasia without real mental health is a moral failure" (The Editorial Board 2022 (2)). A more recent article from the same newspaper describes the case of a woman who received MAID without her family's knowledge, even if they were in daily contact with her. The journalist described several other similar cases and questioned the "impoverished autonomy" that seems to be driving this approach. The writer was concerned of how a proper assessment can be made of a person applying for MAID, without also gathering information from people who know the patient best; her close family (Anderssen 2023). The *Canadian Lawyer* magazine published an opinion piece stating that, "Expanding Medical Assistance in Dying will have devastating effects on Canadians with disabilities" (MacDonald 2022).

Abroad, a bombshell among numerous reports was an article in the *New Atlantis*. An investigative reporter criticized various aspects of the MAID program, but he was mostly concerned about the failure of the program to protect the vulnerable, in fact encouraging vulnerable people, including war veterans, to seek medical death. He quoted Dr. Madeline Li, a psychiatrist and clinical scientist in Toronto, who had extensive experience with MAID, as saying that, "legislated safeguards are impotent." He discovered that the Association of MAID Assessors and Providers, at their internal gathering, had shared evidence that many applicants were driven to MAID because of their crushing poverty, difficulties to get medical care, poor housing, credit card debt and other social troubles. At a meeting of the Association, it was stated that, "(I)t's not that they are terminally ill and truly beyond help, but that they can't get the help they need in the system right now, so euthanasia seems like the only way out." If a MAID assessor wished to withdraw from such a case, it was suggested that it would be appropriate to refer the patient to another assessor, who would be willing to do it (Raikin 2022). During the same month, a columnist for the *New York Times* wrote, "The rules of civilization necessarily include gray areas … It is barbaric, however, to establish a bureaucratic system that offers death as a relief of suffering and enlists the healing profession in delivering this 'cure'" (Douthat 2022). Another US paper, *Washington Examiner*, gave voice to a Canadian living with autism, who wrote that Canada "creates a society in which decency for the most vulnerable people is substituted by something as radical and inhuman as euthanasia."

A journal of religion and public life stated that the Canadian assisted suicide regime, is a "crime against humanity" (Radner 2023). Also in December 2022, the British weekly magazine *Spectator* used similarly strong words, declaring that, "Canada's assisted dying catastrophe is a warning to Britain" (Ashworth-Hayes 2022). In addition to these stated opinions, newspapers also published numerous stories of people who received, or, were about to receive MAID because they were under duress of personal economic and social conditions, or lack of care. It seems that by the end of the year, there were very many loud voices, perhaps even signs of an emerging national consensus, asking the Canadian government to change their course.

On December 15, 2022, Justice Minister David Lametti indeed announced that the government will seek to delay the expansion (no date was set) of Canada's assisted dying regime that includes people whose sole underlying conditions are mental disorders (Major 2022). By February 2, 2023, Lametti announced that the Liberal government is introducing legislation to delay this expansion by one year.

29.4 How MAID works in practice; reports from institutions

To appreciate how MAID is practiced in Canada, it is necessary to know that it began as a court order. In order to legalize the practice of assisted suicide and voluntary euthanasia, the Supreme Court of Canada declared that the absolute prohibition of these acts in the Criminal Code infringes on the rights to life, liberty and security of persons as guaranteed by the Canadian Charter of Rights and Freedoms (Supreme Court of Canada 2015). By declaring the prohibition unconstitutional, the Court, in fact, established access to MAID as a constitutional right of Canadians. 'Rights,' by definition, require the means by which individuals can exercise them, in other words, agents are needed, whose responsibility is to fulfill those rights. The Court also ruled that assisted dying can be provided lawfully only by physicians, yet, no one can be compelled to provide it. After Bill C-14 was adopted to enshrine in law the Court judgement (Parliament of Canada 2016), it became the obligation of the government to assure that all eligible persons had access. This quandary of how to solicit physicians (most of whom are independent contractors) and nurse practitioners (who were included in Bill C-14) to fulfill the stated obligation of the government was resolved when the federal government entrusted the job, of providing access, to the provincial and territorial governments, who in turn conscripted the publicly funded hospitals, health sciences centres and regional health authorities to provide access to MAID. This high-level authoritative decision firmly pushed the provision of MAID into already overloaded acute care facilities. If the organization of MAID had been open to public debate, then, agents, other than medical professionals, could have been identified, and other forms of delivery, such as independent MAID clinics, could have been discovered. This commandeering approach unfortunately gave rise to the many procedural and ethical difficulties that are now being faced.

Nonetheless, these institutions, within the parameters of their established practices, were largely successful in recruiting physicians and nurse practitioners, who were inclined to take on the role of MAID providers and assessors.

29.4.1 British Columbia

In the province of British Columbia, for example, a province wide committee was created with representation from the Ministry of Health, Ministry of Justice, colleges of health professions, regional health authorities, First Nation Health Authority and Coroners Service. The committee worked to standardize all aspects of provision to MAID, such as forms, procedures and practice standards. To implement the program on Vancouver Island (part of the BC province), the Island Health Authority established a working group and started to issue MAID privileges to physicians, who provided evidence of training in Europe or self-training. In just six months after legalization, there were eight physicians providing MAID, and 30 others who had completed training. A large group of willing MAID providers, the demographics of the province (many affluent retirees), and a history for the legalization of MAID (strong advocates for legalization; Mrs. Rodriguez and Mrs. Carter came from this province), may explain why this province has the highest number of MAID deaths in Canada (Robertson et al. 2017).

29.4.2 Toronto, Ontario

Most other institutions could not rely on a strong provincial framework as that of British Columbia. There are several accounts of how Canadian institutions coped with the task of making MAID available. The first report from the University Health Network in Toronto, which consists of four large, tertiary care teaching hospitals, covers a period of one year after MAID was legalized. To "meet the institution's legally mandated obligations" an institutional framework was created that consisted of clinical, assessment, and intervention teams. In addition, a multidisciplinary quality committee was established. The Network recorded, in one year, 74 MAID inquiries that resulted in 29 formal assessments for eligibility; the approval of 25 persons for MAID; and 19 receiving MAID. Natural death, or a change in a patient's decision were the reasons for six of the approved individuals not receiving MAID. Some initial organizational problems were noted. No clinical departments were willing to be formally associated with MAID and so the psychosocial oncology and palliative care services of the oncology section assumed the operational responsibility. Delivering MAID in a single location would have been an operational advantage, but no clinical unit was willing to be identified as the de facto "MAID unit." The fears of staff that there would be "over-zealous" deliveries of MAID was reduced by staff education on all the checks and balances in the operational framework.

This is a well written report, but raises a number of questions. Not surprisingly, of those patients presenting for assessment of MAID eligibility, 87% were receiving palliative care services, but only 52% were receiving specialized psychosocial care. Given that, as the report indicates, the main reason why these patients were seeking MAID was psychosocial, not pain or physical symptoms, then psychosocial counselling and assistance should have been offered to all MAID applicants. Another concern is that at the onset, the organization decided that they would only offer MAID in a hospital setting, by IV injection, administered by a physician or nurse practitioner. Patients did not have the opportunity to ask for a drug prescription and then consume the medication at will. Keeping MAID intramural was likely an operational advantage, because, if necessary, a second dose of drugs could be readily injected; the pronouncement of death could be readily done on the ward; there would be no need for house visits; and no need to keep track of drugs that would be taken home. But there are important issues at stake. The law provides for both options of voluntary euthanasia and assisted suicide, and to exclude one option by institutional policy could frustrate some patients and be open to legal challenge. From an ethical perspective, we need to note that MAID's legitimacy is the patients' right to exercise their autonomy at the end of life, so it is troublesome that these patients do not have the option to choose between an assisted suicide (patient self-administers a lethal medication at the time and place of their own choosing) and voluntary euthanasia (the physician administers the lethal medication to the patient), but instead must accept the latter and be fitted into an institutional schedule for an injection. Another concern with the institutional delivery of MAID deals with use of resources. The Network did not receive any designated funding for introducing MAID, yet the program was described as "resource intensive." MAID delivery teams had to be created by workload redistribution, and in Canadian institutions where manpower and resources are always scare, this could have led to compromising care for patients in other parts of the hospital. Finally, the authors need to be congratulated for establishing a MAID quality committee, unfortunately, no data are provided in their document about the experiences of this committee. Given that MAID was an entirely new program, unlike anything else the institution had ever undertaken, and that errors in MAID delivery may not be reversible, then, real-time quality control would have been highly desirable. The report ends with the authors raising a question as to "whether this 'brave new world' will ultimately be regarded as enlightened or dystopian" (Li et al. 2017).

More recently, the leading author of the above report who spearheaded the rollout of the program six years ago expressed her significant concerns about the expansion of the program beyond the terminally ill and noted that, "Making death too ready a solution, disadvantages the most vulnerable people, and actually lets society off the hook." She added that, "The MAID law is very Canadian. It was left so vague it could offend nobody," and, "is not specific enough to protect people" (Honderich 2023).

29.4.3 Hamilton, Ontario

The McMaster University Medical Centre, with the Hamilton Health Sciences teaching hospital of McMaster University, is another organization that reported on their effort to create a hospital-based MAID program. In that institution, the Director of the Office of Ethics and Care Ecologies became the Director of MAID. She and her group published 10 papers in a single issue of HEC Journal that describes the great effort which was required to start providing MAID, and it demonstrates the strengths and weaknesses of MAID as an institution-based program. The introductory paper points out that in six years there were over 30,000 Canadians who received MAID, over 10,000 of those in 2021 alone and suggests that this "may be seen as a triumph of patients' advocacy" and represents an "incredible success." This is, of course, a narrow managerial perspective, and the authors do not provide any alternative explanation for these remarkable figures. A critique may interpret this as evidence that society failed to provide care and support to over 30,000 people so that they, in spite of diseases, disability or aging, might have been able to live a comfortable and meaningful life to the end of their natural lifespan. A greater part of this paper deals with a concern that, "The patients' demands for MAID is outstripping the supply of willing and available providers." It was noted that oncologists, neurologists and cardiologist were underrepresented as MAID providers and these professionals are seen as "an untapped resource for future recruitment." This introductory paper ends with an ambitious goal to practice MAID "in a principled and sustainable manner" and "promote transparency, integrity, quality and respect for moral diversity" (Frolic and Oliphant 2022).

Another paper in this series refers to a survey and focus group to determine the wishes, concerns and hopes of 'stakeholders'; the creation of an Assisted Dying Resource and Assessment Service; a volunteer group with representation from 10 different professions with responsibility for a centralized consultation; and a referral service to manage patients' requests and to support clinical teams caring for these patients (Frolic and Oliphant 2022). The most promising paper from this institution is on the evaluation of the hospital's MAID program, which made an assertion that the hospital is delivering "high-quality MAID care" (Frolick et al. 2022). This conclusion is problematic because collecting staff's opinions, perspectives, views and attitudes concerning the interventions they provide is not a sufficient measure of quality for those interventions. Also, it was not clear what the phrase, "MAID care" involved, other than the lethal injection of drugs. Researchers did not report on any evaluation of the documentation of their MAID recipients, so we do not know, for example, if all MAID recipients provided a written, signed and dated request for MAID; and whether or not it was also signed and dated by two independent witnesses; whether the assessment was done by two independent physicians or nurse practitioners, and so on. Proof of strict adherence to eligibility criteria and procedural safeguards would be the first indicators of a high-quality MAID program.

29.4.4 London, Ontario

Another large academic institution, London Health Sciences Centre reported on an entirely different operational structure for the provision of MAID. In that institution, a high-level Internal Resource Group (IRG) had been established in 2015, consisting of senior leadership, departmental leadership, and director of Patient Safety and Experience, and chaired by the hospital ethicist. After some effort, the IGR created a MAID Provider Group. Seven physicians were recruited to provide assessment, but it was more difficult to recruit MAID providers. The authors stated that, "(I)n at least two cases, despite being strong MAID advocates, physicians have been unwilling to step forward as providers out of fear of potential adverse personal and professional consequences. Professionally, the concern was primarily related to judgment by colleagues and peers (which was experienced by at least one of our members)." The IRG initiated staff and public education, addressed non-compliance with the law and regulation of physicians and the staff, advised on difficult cases, triaged referred cases if required. The group also was receiving referrals from the community and through the hot line established by the province. Patients frequently complained that it required many requests addressed to their physicians before they were referred. However, once a patient was assessed, MAID was provided very rapidly, 11 days on average, except for cases where reflection time was shortened to three days average, as permitted by law (Ball et al. 2019).

29.4.5 Profile of MAID recipients

Another report comes from Sunnybrook Health Sciences Centre, a teaching hospital, also in Toronto, and is based on a retrospective review of patients' charts who had submitted a formal request for MAID over a period of 24 months. Of the 107 applicants, 97 were found to be eligible, and 80 received MAID. Most patients were receiving palliative care before or after the requests. Physical symptoms were rarely the main reason for seeking MAID. Reviewing the demographics of these patients, investigators noted that 48% were living alone, comparing with 28% in the adult population, and cited just a few close relationships. Investigators suggested that this finding is in keeping with reports from Switzerland and Oregon, in which euthanasia requestors are often characterized by "dismissive attachment style." Consistent with this hypothesis, they thought, is also the observation that in their group of patients, the primary reasons for MAID, besides loss of ability to engage in desired activities, were the loss of independence and dignity. Their cohort was asked several times about the reason for seeking MAID, and researchers reported that most held a persistent and longstanding philosophical belief about death, and had decided prior to their illness to control the circumstances of their deaths. From this description, MAID emerges as a lifestyle choice for some patients, rather than a means to provide comfort in the presence of "enduring physical and psychological suffering that is intolerable to them," as the law requires. It appears that some MAID seekers are taking advantage of the permissiveness of this law, and with the indulgence of MAID assessors are using MAID for reasons diametrically opposed to the law's intent (Selby et al. 2020).

This chart review would have been a good opportunity to determine, for example, if the documentation confirmed patients' eligibility criteria of "incurable illness, disease or disability" and if the applicants were indeed in "advanced state of irreversible decline." It was stated that 77% of these patients had a diagnosis of cancer, but this information is not helpful in determining legal eligibility because about half of cancer patients at the time of diagnosis have a curable disease, and furthermore the stage of the cancer is not routinely collected for MAID reporting.

29.5 How MAID works in practice; Physicians' reports

The Canadian Medical Association called MAID "the most profound change in medical practice in modern times" (Kotalik 2020). Physicians and nurse practitioners are the only individuals who can legally provide MAID. There are no special qualifications or training required to assess the eligibility of applicants for MAID and, when they are found eligible, to terminate their life, which in 99% of cases is done by intravenous injection. It is actually a very small fraction of the physicians' population who provide MAID. The federal government's Third Annual Report indicated that 10,064 deaths in Canada in 2021 were assisted by 1577 physicians and 89 nurse practitioners (Third Annual Report 2022). Given that Canada has a little less than 100,000 physicians, this means that 1.5% are MAID providers. This providers' group has been increasing about 17% per year.

29.5.1 MAID Assessors and Providers

The Canadian Association of MAID Assessors and Providers (CAMAP) supports its members, provides continuing education and has issued a number of position statements, protocols and guides. One of their controversial guides states that physicians have a professional obligation to raise the topic of MAID as a treatment option when it is "medically relevant" and the patient is likely eligible (CAMAP 2019). This guide may be contrary to the Criminal Code, which states that anyone (and this includes physicians), who counsels, abets or aids a person to die by suicide is guilty of an indictable offence. CAMAP justifies this guide by insisting that counselling is not what doctors do, and the prohibition is not relevant. However, there are also strong ethical reasons against this recommendation. In any case, the extensive policies of the Ontario College of Physicians and Surgeons for MAID provision and for the end-of-life care, do not mention any such obligations (CPSO 2019, 2021a). An extreme case of such an attitude was apparent when a Veteran Affairs caseworker responded to a veteran's request for help, by saying that MAID is a better option than "blowing your brain out" (Brewster 2022). In New Zealand and Victoria, Australia, doctors are explicitly prohibited from raising the topic. There is no legal prohibition in Belgium and Netherlands, but it is considered inappropriate (Kirkey 2022). An indicator of existing tension between MAID assessors and providers, who consider that assisted dying is 'health care,' and other health care professionals who disagree, is that CAMAP has recently issued a Position statement on Professionalism and Harassment. The statement is a reminder that there is legislation that prohibits violence and harassment in the workplace. It also points out that it is a criminal offence to obstruct a health care worker from performing their duties and to use fear to stop a health care worker or those who assist from performing their duties, or to prevent a person from obtaining health services (CAMAP 2022). CAMAP issued a number of other position statements, protocols, interpretations and guides on matters of MAID administration, capacity assessment, MAID and dementia, and palliative care, which are available on their website.

CAMAP policy is not to promote MAID expansion, but some of their very visible members do. Dr. Wiebe, a Vancouver physician, when interviewed by CTV News, strongly endorsed expansion of MAID to include people with solely a mental disorder (Favaro 2022). Dr. Wiebe, a board member of Dying with Dignity Canada, speaking at a seminar for physicians described her experience with an unnamed man, who was found to be ineligible for MAID by an assessor because he did not have the capacity to make an informed consent decision about his health. This man contacted Dr. Wiebe, who assessed him remotely, declared him MAID eligible, asked him to fly to Vancouver and later, in her clinic, administered MAID to him by injection. In a video, Dr. Wiebe commented that, "it's the most rewarding work …" and revealed that she helped 400 people to die via the MAID program (Reyes 2023). At one point, Dr. Wiebe stated that the main risk for MAID providers are angry families. Two of the first MAID providers, each wrote a book about their experiences (Green 2022; Marmoreo and Schneller 2022). Yet, a Belgian physician titled her paper about assisted dying in her country as "The doctor turned into an instrument" (Dopchie 2021). A large study in Netherlands showed that 72% of physicians experienced feelings of discomfort after performing the act and consider it an emotional burden (Evenblij et al. 2019). A literature survey confirmed these finding (Stevens 2006). So, it is somewhat surprising that no such reports have been published in Canada. A report based on the interviews of 21 MAID providers (1.5% of CAMAP members), who volunteered to respond to a call on the CAMAP listserv, indicates that all responders found the experience "profoundly positive" (McKee 2018). No large study was done as yet on this topic. But it may be difficult to write and have accepted for publication any data that is seen to discourage physicians and nurse practitioners from becoming MAID providers because of the stress it may pose on the assured access of this intervention as a constitutional right in our political agenda.

As concerns the physician's role in MAID delivery, it appears, increasingly, the procedure will be provided by those physicians, who elect to become 'functional specialists' and that this group may even strive for recognition as specialists. This probability needs to be carefully considered. There are possible advantages to this arrangement (e.g., advance technical skill of providers, avoiding burden of MAID on physicians, who do not wish to get involved with it), but there are also possible disadvantages (e.g., MAID providers will not know their candidates, yet they are expected to rapidly make a life and death decision).

29.5.2 MAID Critics

On the other end of the spectrum are physicians, who without exception are opposed to MAID, and then there are some, who would accept the procedure only under extreme and rare situations. By and large they would agree that, "(G)iving a lethal injection to a patient which results in his or her immediate death, implies an inversion of attitude a doctor has towards his or her patient in a normal therapeutic treatment" (Lemmens et al. 2021).

These physicians are not as publicly visible as MAID providers, but their observations and arguments deserve consideration. A group of seven Canadian Physicians, engaged in family medicine, palliative care, hematology and psychiatry published a high-spirited defence of Hippocratic medicine, which they felt is fundamentally incompatible with MAID. They pointed out that, besides many other problems, the state-run health insurance system is supposed to pay only for 'medically necessary' health services. If the insurance also pays for MAID, it gives the impression that it also is medical necessary. These physicians stressed that in Belgium and the Netherlands, euthanasia is permitted only if a physician is personally convinced that there is not a reasonable alternative and that a person's suffering is intolerable and enduring. In Canada, in contrast, physicians are directed not to question the suffering and to defer entirely to the patient, who can reject any alternative intervention, even if it could potentially eliminate all suffering. The authors published their paper in the journal of *World Medical Association* as a "cautionary tale" and a warning to colleagues in other countries (Leiva et al. 2018).

Surveying the developments in Canada four years later, a family physician, two psychiatrists and a health law professor were alarmed at how death is becoming normalized as a "treatment" option. They point to the broad and imprecise legislation, the problematic process of assessing eligibility, the removal of some safeguards and the current crisis in health care and social systems that caused unacceptably long waiting times for health care, especially mental health consultation and treatments. They concluded that, "Canadians are increasingly opting to receive medical state-funded death, not because they no longer want to live, but because our society has failed them" (Coelho et al. 2022). On a similar note, a family physician attempted to stop euthanasia and assisted suicide by appealing to the Canadian courts (Saba 2021). Analysis of monitoring and reporting of MAID cases was found to be very deficient (Kotalik 2020).

A qualitative research project focused on physicians' experiences in Northwestern Ontario suggests that some of the concerns that have been expressed above are justified. Rural physicians, of which about half were MAID providers or assessors, were concerned about the lack of access to palliative care, the lack of which forced some patients to ask for MAID, instead of them choosing MAID only if and when palliative care would no longer keep them comfortable. They noted that MAID is "an easier shortcut than trying to deal with complexities of someone's existential suffering." They commented that a request to hasten death is first and foremost a cry for help. Of these 23 physicians, many provide care for Indigenous patients. They noted that MAID seems incongruous with the Indigenous person's fundamental spiritual beliefs. Indigenous patients would rarely seek MAID, and physicians do not feel comfortable to speak about MAID with them (Durant and Kortes-Miller 2020). This is the only publication out of about 130 reviewed for this chapter, which mentions Indigenous people.

This is one indicator, which shows that the needs and concerns of Indigenous populations of Canada, about 5% of our residents, have been largely ignored during the development of the MAID program (See also Chap. 25). During the early days of parliamentary discussions of Bill C-14, a Cree member of Parliament, Robert-Falcon Ouelette (2016) spoke about the necessity for him to oppose the Bill because it was in conflict with native spirituality and because of his concerns about the current suicide pandemic in Indigenous communities. Representatives of 15 Indigenous Nations sent an open letter to members of Parliament and regulators and pointed out that "Bill C-7 goes against many of our cultural values, beliefs and sacred teachings. The view that MAID is a dignified end for terminally ill or for those living with disabilities should not be forced on our peoples" (CPSO 2021b). They stated that the expansion of MAID would have a lasting impact on their vulnerable populations. Anderson (2019) from Northwester Territories, similarly opposed the MAID bill pointing out that in the Territories the right to live is already compromised by inadequate housing, food and health care. The Report of the Centre for Suicide Prevention pointed out that suicide and self-inflicted injuries are the leading cause of death of the First Nations' youth and adults up to age 44. For First Nations, the suicide rate is three times the national average, and for Inuit it is nine times the national rates (Centre for Suicide Prevention 2021).

29.5.3 Problematic Assessment for MAID Eligibility

A case reported in a medical journal by a physician illustrates that its occurrence is more than just an anecdote (i.e., rare, single case that cannot be generalized). As the case will show, this incident appears to be an indicator that assessments for MAID are being completed without sufficient knowledge of the patient and therefore likely not taking into account that the MAID candidate could still enjoy a prolonged and symptom free life. A physician started an elderly patient, who had developed metastatic carcinoma of breast, on hormonal therapy, which had never been offered to her. On follow-up, the patient started to show an early response to the therapy and her general condition was very good. To the physician's astonishment, she applied for and received MAID. This woman had informed her doctor about the procedure being scheduled, yet, no MAID assessor or provider approached him to ask for a consultation or opinion about his patient. The doctor felt that this was a premature decision and that the patient had a good chance to enter a long-term remission (Rayson 2022). This is not likely an unusual situation. The Third Annual Report indicates that consultation with primary care physicians is sought by MAID assessors only in about 34% of cases (Third Annual report 2022, Sect. 6.3). Another federal report for previous years indicated that in about 55% of cases, MAID assessors did not seek the opinion of any other health care professional. The emerging picture is that in a typical case, a patient is assessed and then injected by a MAID provider, who had no contact with the patient before the MAID request, and that, that provider frequently proceeds without making contact with the patient's own physician or other professionals. There is, of course, a second physician or a nurse practitioner, who by law has to confirm that the patient is MAID eligible, but according to the current federal monitoring regulations, the only information collected about them is their profession, not their specialty. The law requires that a MAID candidate is made aware of "all available treatment options." Is it realistic to think that this requirement can be fulfilled, for example, by a practitioner of family medicine, who is the assessor in about 68% of cases (Third Annual report 2022, Sect. 5.4)? Given that oncology consultation has been requested by only 12% of patients undergoing assessment for MAID eligibility (Third Annual Report 2022, Sect. 6.3), can one trust the knowledge base of the involved family physician—or any other assessor for that matter—to know and tell the patient, the most current treatment options for the 100 plus variants of cancer?

29.5.4 Hastening death

The perception that health care staff is getting accustomed to the hastening of death is supported by the following study. Intensivists from adult and pediatric Intensive Care Units across Canada were asked to fill in a questionnaire. The majority (86%) of the 150 physicians, who responded were familiar with MAID law; 71% were in favour of MAID; 14% were conflicted and 11% were opposed. Only eight of them (5%) thought that MAID influenced their approach to the withdrawal of life-sustaining therapies. Yet, 41% of Intensivists had observed medication being given in disproportionately high doses during withdrawal, with 13% having personally administered such doses. 80% had experienced explicit requests from families to hasten death and 47% thought that it was ethically permissible to hasten death following withdrawal of life sustaining therapies (Andersen et al. 2022). Given that a great majority of these patients die within 24 h., (Cooke et al. 2010) what is the purpose of hastening death? Are these clinicians actually providing MAID outside the legally permitted context? Another report from a physician, which was published in a medical journal reveals related concerns. An emergency physician was concerned that death is being much more readily accepted in hospitals now, than in pre-MAID times. As an example, he presents a case of an elderly patient in hospital, who suddenly became unresponsive. Because she was DNR, her monitor was simply discontinued and family was told she may be dying. The physician, who knew that she was well in the morning, took notice, assessed her, found a readily correctable problem, and resumed her treatment. The patient, who had been left to die, was discharged to her nursing home 10 days later (Waghmare 2022). So, it seems likely that the practice of MAID is having a covert—if not overt—effect on clinicians' relationship to patients. Hastening of a patient's death without consent, is conflated with the normalization of MAID.

29.5.5 MAID patients with cancer

Because over 60% of patients who received MAID were patients with cancer, it is important to know the oncologists' views on MAID. A recent survey of radiation and medical oncologists, with responses from 224 clinicians, revealed that about 60% of oncologists judged that their knowledge of MAID is very good or excellent, and only 11% reported it as poor. Most oncologists worked in large institutions, and if their patients inquired about MAID, they were referred to MAID management teams. Under some circumstances, slightly more than half of the oncologists would initiate a conversation about MAID with patients. The main reasons for not initiating a discussion, were concerns that it could be construed as coercion. Oncologists also maintained that the role of the oncologist is to offer hope, to improve and extend life. To the question of what they would be prepared to do if surgery, radiation therapy and chemotherapy could no longer offer anything to their patients, 40% indicated that they would be willing to be assessors and 23%, under certain circumstances, would be willing to prescribe a lethal dose of drugs, 18% would possibly give the lethal injections (Chandhoke et al. 2020). The picture that emerges is that oncologists are cautious about discussing MAID and getting involved in MAID, but would consider it in symptomatic patients, who no longer can benefit from conventional therapy. There was only a 32% response rate to this survey after two reminders, and it is possible that oncologists who are opposed to MAID did not respond to the survey. Significantly, only 1.3% of MAID deaths in 2021 involved an oncologist (Third Annual Report 2022). Oncologists' lack of interest in MAID for their patients is possibly not limited to Canada. A major US oncology journal published an editorial which listed some perceived implications of physician assisted suicide practice for oncologists and their patients. It noted, decreased professionalism, the possibility of error, a diminished physician–patient relationship, and a slippery slope toward the practice of euthanasia (O'Rourke et al. 2017).

Another report about MAID for patients with cancer deals largely with issues of process in assessment and delivery of MAID. The report is from the province of Alberta, which established a provincial MAID Coordinating Service with three zones. Anticipating that a substantial proportion of requests for MAID would come from people with cancer, the tertiary cancer centre in the central zone created two part-time oncology positions to support their MAID Coordinating Service. If their formal assessment determined the patient to be eligible, one of the oncologists would administer the lethal injection (Wu et al. 2018). It may be beneficial for patients with cancer, to be assessed for MAID by oncologists. On the other hand, for every oncologist who becomes an assessor, there is one less oncologist available to work in the active healing practice.

29.5.6 MAID instead of cancer treatment

At least in one province, it was the inability to receive oncology care at a reasonable time that drove patients to MAID. A typical case reported by a physician to a journalist was that of a British Columbia farmer, diagnosed with carcinoma in June, who at that time was still otherwise healthy and active. By the time he was given his first opportunity to speak with an oncologist in September, the man was bedridden and too weak to travel to the cancer centre to be assessed for treatment. He subsequently received MAID. The journalist reported that in the fall of 2022, only 5% of patients referred to BC Cancer, which has a monopoly on comprehensive cancer care in the province, had their first consultation within the recommended period of four weeks. Doctors told the journalist that many patients have to wait months before they are seen by an oncologist, and when pain and anxiety becomes unbearable, they turn to MAID. At least one other province, Nova Scotia, failed to meet the benchmark on the start of radiation therapy (Woo 2022a, b). It is unethical and likely illegal if patients are approved for MAID on the basis of a tentative diagnosis of incurable cancer by a non-specialist, a diagnosis which may be incorrect. Even if it were correct, a consultation with a cancer specialist is important because it would give the patient an alternate choice, one to partake in effective palliative treatment.

A group of researchers in Ottawa examined the files of MAID recipients in their region spanning about two years and identified patients with lung cancer (Moore et al. 2022). This diagnosis, until recently was associated with a high mortality rate, but during the last few years, with biomarker testing and treatment by targeted therapy and immunotherapy, these patients' survival rate is rapidly improving. These researchers discovered that 13% of those 45 patients with 'cancer' did not have a biopsy of the tumor, only clinical and imaging impressions, which resulted in a non-definitive diagnosis. Further, 10 out of 45 patients did not have a medical oncology consultation, so testing for the best therapy was not offered to them. They signed the consent for MAID without being informed "of the means available to relieve their suffering," as the law requires (Bill C-14 241.2(1)). It seems that in those situations, MAID becomes a solution for a severe lack of access to cancer care, a situation, which in human terms, is appalling and if investigated, the practitioners would likely be found to knowingly fail to comply with the Criminal Code.

29.6 Conclusion

MAID in Canada is set to become ever more inclusive. Since the original law of 2016, the eligibility criteria have been modified to include people, whose death is not foreseeable. An additional change, scheduled to take effect March 2024, will make eligible, people with a mental disorder as a sole underlying condition. The federal government is poised to continue to expand eligibility to children and people who provide advanced directives.

Presently, MAID death's surge of approximately 30% per year is unprecedented in the world. The information gleaned from publications reviewed for this study leads the writers to conclude that multiple, complex factors are contributing to MAID's excessive yearly increases. It is impossible to cover the whole spectrum of influences (e.g., philosophical, sociological, etc.), but one, decidedly identifiable factor of impact is related to the genesis and evolution of MAID's architecture, which is markedly different from all other jurisdictions where euthanasia and/or assisted suicide are permissible.

To put things in perspective, it is known that in other jurisdictions, the law makes it possible for health professionals to end a patient's life as long as they follow certain legal and regulatory directions. The procedure generally takes place as an arrangement between a patient and their physician (most often the patient's long term personal physician). In contrast to the Canadian way, access to euthanasia or assisted suicide in other jurisdictions, is not defined as a person's constitutional or civil right. Furthermore, hospitals, public health authorities and governments are not involved in its delivery, education, or assuring its access. It was the Supreme Court of Canada, in 2015, which concluded that denying citizens access to a medical assisted death was in conflict with the Canadian Charter of Rights and Freedoms, consequently, MAID was indirectly introduced as a constitutional right for Canadians.

To assure availability, provincial ministries of health made 'access to MAID' the responsibility of hospitals, health sciences centres and regional health authorities, all governed and funded by governments. These organizations in turn established committees and MAID teams and had to recruit (sometimes aggressively) physicians, or nurse practitioners, as well as other professionals and support staff. Creating this infrastructure for the delivery of MAID was accompanied by extensive surveys of knowledge and attitudes of staff, as well as engaging staff in a continuous, multimedia education blitz dealing with MAID law, regulations and practice. Staff was expected to provide information to patients and, if patients were interested, to direct them to MAID eligibility assessment teams. Considerable administrative effort, focused on access and efficiency, was invested in setting up the MAID delivery systems, which operated quickly and efficiently, and, in most places, is coping well with an increasing patient demand.

But there is likely an unintended and undesirable side of such managerial success. It is possible that such a high profile, high priority institutional program can consciously or subconsciously affect the philosophy of care. A seemingly routine encounter with a patient in a "MAID-soaked" hospital, where MAID has been normalized and trivialized can potentially affect a health care worker to evaluate a patient, any patient, as a potential MAID candidate. To verify or dismiss this hypothesis, which presents a serious concern about the moral environment of our MAID-delivering institutions would require a dedicated research program. If the hypothesis would be confirmed, MAID-delivering institutions would need to develop, in parallel with the MAID operation, a comprehensive program that would operationalize the charge given to us by Parliament, when approving Bill C-14: "…it is important to affirm the inherent and equal value of every person's life and to avoid encouraging negative perceptions of the quality of life of persons who are elderly, ill or disabled" (Parliament of Canada 2016). Other ethical and operational issues emerging in the program, besides those already discussed in this book are the problematic assessment of eligibility for MAID, and providing MAID when access to treatments have very long waiting times, or are unavailable to the patient.

Given the government's enthusiasm to expand eligibility for MAID, and the ultra-efficient institutional infrastructure, plus, the 'rights' driven autonomy to access MAID, it is most likely that we will continue to witness a steady increase in MAID deaths. Again, this situation requires study. MAID is a very large program affecting thousands of people across Canada. It has been established quickly, with priority given to providing efficient access to the intervention, and focused on expansion of eligibility and sustainability. Little attention has been paid to any type of quality control process, and to the assessment of MAID long-term impacts on vulnerable individuals, on families and communities, on health care professionals, on the ethos of health care, and on the shared understanding among Canadians of what we owe to each other. It is time for adjustment of priorities.

References

Anderson, M. 2019. Canadian Senator calls for strengthening Inuit right to live. *Canadian physicians for life*. https://www.physiciansforelife.ca/strenghten-inuit-right-to-live/.

Andersen, S.K., S. Stewart, B. Leier, L.E. Alden, D.R. Townsend, and D. Garros. 2022. Hastening death in Canadian ICUs: End of life care in the era of medical assistance in dying. *Critical Care Medicine* 50(5).

Anderssen, E. 2023. A complicated grief: Living in the aftermath of a family member's death by MAID. *Globe and Mail,* January 20. https://www.theglobeandmail.com/canada/article-maid-death-family-members-privacy/.

Ashworth-Hayes, Sam. 2022. Canada's assisted dying catastrophe is a warning to Britain. *The Spectator*. December 11, 2022. https://www.spectator.co.uk/article/canadas-assisted-dying-catastrophe-is-a-warning-to-britain/.

Association of Chairs of Psychiatry in Canada Dec. 1. 2022. https://www.ctvnews.ca/health/canada-should-delay-maid-for-people-with-mental-disorders-psychiatrists-1.6177908.

Ball, I.M., B. Hodge, and R.W. Sibbald. 2019. A Canadian academic hospital's initial MAID experience: A health-care system review. *Journal of Palliative Care* 34(2). https://journals.sagepub.com/doi/abs/10.1177/0825859718812446.

BBC News. 2023. January 18. https://www.bbc.com/news/world-us-canada-6400432.

Brewster, M. 2022. Former Paralympian tells MP veterans department offered her assisted death. CBC News. December 1.

Call for Action. 2022. Society of Canadian psychiatry. https://www.socpsych.org/calltoaction#:~:text=CALL%20to%20ACTION%3A%20SoC%CE%A8%20calls,2.

CAMAP. 2019. *Canadian association of MAID assessors and providers bringing ups MAID as a clinical care option*. https://camapcanada.ca/wp-content/uploads/2022/02/Bringing-up-MAiD.pdf.

CAMAP. 2022. *Canadian association of MAID assessors and providers.* Position statement on professionalism and harassment. https://camapcanada.ca/wp-content/uploads/CAMAP-Position-Statement-on-Professionalism-and-Harassment.pdf.

Canadian Press. 2022. Staff. Liberal government seeking delay to expanding medically assisted dying program. December 15, 2022. https://www.cbc.ca/news/politics/maid-deadline-extension-mental-illness-1.6687429.

Canadian Press, Staff. The Canadian Press. 2022. MPs, Senators' debate requirements for medically assisted dying with mental disorders. September 23, 2022.

Carter v. Canada (Attorney General). 2012. BCSC 1587.

CASP Issues statement about MAID for mental illness, Statement December 14, 2022. https://suicideprevention.ca/media/casp-issues-statement-about-maid-for-mental-illness/#:~:text=%E2%80%9CAs%20we%20have%20made%20clear,definitions%20of%20irremediableness%20and%20suicidality.%E2%80%9D.

Centre for Suicide Prevention. 2021. Indigenous people, trauma, and suicide prevention. June 20, 2021. Indigenous people, trauma, and suicide prevention—Centre for Suicide Prevention (suicideinfo.ca).

Chandhoke, G., G. Pond, O. Levine, and S. Oczkowski. 2020. Oncologists and medical assistance in dying: Where do we stand? Results of a national survey of Canadian oncologists. *Current Oncology* 27 (5): 263–269.

Coelho, R., T. Lemmens, K.Sonu. Gaind, and J. Maher. 2022. Normalizing death as treatment in Canada: Whose suicides do we prevent, and whose do we abet? *World Medical Association Journal* 70(3): 27–35.

College of Physicians and Surgeons of Ontario. 2021. Indigenous people should not be compelled to provide or facilitate medical assistance in dying. Declaration, February 2021. Indigenous-Peoples-Should-Not-Be-Compelled-to-Provide-or-Facilitate-MAiD_Redacted.pdf (cpso.on.ca).

Cooke, C.R., D.L. Hotchkin, R.A. Engelberg, L. Rubinson, and J.R. Curtis. 2010. Predictors of time to death after withdrawal of mechanical ventilation in the ICU. *Chest* 138 (2): 289–297.

CPSO. 2019. Planning for and providing quality end-of-life care. Policy of Ontario College of Physicians and Surgeons.

CPSO. 2021a. Medical assistance in dying. Advice to the profession. Ontario College of Physicians and Surgeons.

CPSO. 2021b. Indigenous peoples should not be compelled to provide or facilitate medical assistance in dying. http://policyconsult.cpso.on.ca/wp-content/uploads/2021/05/Indigenous-Peoples-Should-Not-Be-Compelled-to-Provide-or-Facilitate-MAiD_Redacted.pdf.

CTV. 2022. *C. Bains.* The Canadian Press. December 2, 2022. https://www.ctvnews.ca/health/canada-should-delay-maid-for-people-with-mental-disorders-psychiatrists-1.6177908.

CTV Health News. Canada should delay MAID for people with mental disorders: psychiatrists. https://www.ctvnews.ca/health/canada-should-delay-maid-for-people-with-mental-disorders-psychiatrists1.6177908#:~:text=Canada%20is%20not%20ready%20to,the%20specialists%20across%20the%20country.

Dopchie, C. 2021. The doctor turned into instrument. In *Euthanasia: Searching for the full story*, ed. Devos, T. Springer, Cham. https://doi.org/10.1007/978-3-030-56795-8_2.

Douthat, Ross. What euthanasia has dome to Canada. *New York Times*, December 3, 2022. https://www.nytimes.com/2022/12/03/opinion/canada-euthanasia.html.

Durant, K-L., and K. Kortes-Miller. 2020. Physician snapshot: The forming landscape of MAID in northwestern Ontario. *Palliative Care and Social Practice* 14: 1–14.

Editorial Board. 2022. (1). *Globe and mail.* December 5, 2022. https://www.theglobeandmail.com/opinion/editorials/article-in-canada-people-can-die-with-dignity-they-should-be-able-to-live-that/.

Editorial Board. 2022. (2). *Globe and mail.* December 7, 2022. https://www.theglobeandmail.com/opinion/editorials/article-euthanasia-without-real-mental-health-care-is-a-moral-failure-fund-it/

Evenblij, K., H.R.W. Pasman, J.J.M. van Delden, A. van der Heide, S. van de Vathorst, D.L. Willems and B.D. Onwuteaka-Philipsen. 2019. Physicians' experiences with euthanasia: a cross-sectional survey amongst a random sample of Dutch physicians to explore their concerns, feelings and pressure. *BMC Family Practice* 20(1): 177. https://pubmed.ncbi.nlm.nih.gov/31847816/

Favaro, A. 2022. The death debate: why some welcome the Canada's move to assist dying for mental illnesses and other fear it. CTV News, December 10. https://www.ctvnews.ca/w5/the-death-debate-why-some-welcome-canada-s-move-to-assisted-dying-forental-illness-and-others-fear-it-1.6109646.

Frolic, A., and A. Oliphant. 2022. Introducing medical assistance in dying in Canada: lessons on pragmatic ethics and the implementation of a morally contested practice. *HEC Forum* 34:307–319. https://doi.org/10.1007/s10730-022-09495-7

Frolic, A., M. Swinton, A. Oliphant, L. Murray and P. Miller. 2022. Access isn't enough: evaluating the quality of a hospital medical assistance in dying program. *HEC Forum* 34: 429-455.

Government of Canada. 2022. Final report of the expert panel on MAID and mental illness. 2023. https://www.canada.ca/en/health-canada/corporate/about-health-canada/public-engagement/external-advisory-bodies/expert-panel-maid-mental-illness/final-report-expert-panel-maid-mental-illness.html.

Green, S. 2022. This is assisted dying. Simon and Schuster Canada.

Honderich, Holly. Who can die? Canada wrestles with euthanasia for mentally ill. BBC News, March 26, 2023. https://www.bbc.com/news/world-us-canada-64004329.

House of Commons-Bill C-7. 2021. https://canada.justice.gc.ca/eng/trans/bm-mb/other-autre/c7/c7.html.

Justice Minister David Lametti under fire for 'unbelievable' comparisons between euthanasia and suicide. The Star.

Kirkey, S. 2022. Canadian doctors encouraged to bring up medically assisted death before their patients do. *National Post*, November 2. https://nationalpost.com/news/canada/canada-maid-medical-aid-in-dying-consent-doctors.

Kotalik, J. 2020. Medical assistance in dying: challenges of monitoring the canadian program. *Canadian Journal of Bioethics* 3 (3): 202–209. https://www.erudit.org/en/journals/bioethics/2020-v3-n3-bioethics05693/1073799ar/.

Leiva, R., M.M. Cottle, C. Ferrier, S. Rutledge-Harding, T. Lau, T. McQuiston, and J. Scott. 2018. Euthanasia in Canada: A cautionary tale. *World Medical Journal* 64 (3): 17.

Lemmens, T., M. Shariff, and L. Herx. 2021. How bill C-7 will sacrifice the medical profession' standard of care, *Policy Options*, February 11, 2021.

Lemmens, W. When conscience wavers. Some reflections on the normalization of euthanasia in Belgium. In *Euthanasia: Search for the full story*, ed. Devos, T. Springer Nature, Switzerland AG. Open access.

Li, M., S. Watt, M. Escaf, M. Gardam, A. Heesters, G. O'Leary and G. Rodin. 2017. Medical assistance in dying – implementing a hospital-based program in Canada. *The New England Journal of Medicine* 376(21): 2082–2088.

MacDonald, Lorin. Expanding medical assistance in dying will have devastating effects on Canadian with disabilities. *Canadian Lawyer*. 16 December 2022. https://www.canadianlawyermag.com/news/opinion/expanding-medical-assistance-in-dying-will.

Major, D. 2022. Government seeking delay to MAID expansion that would cover mental illness. *CBC News*. https://www.cbc.ca/news/politics/maid-deadline-extension-mental-illness-1.6687429. Accessed 15 Dec 2022.

Marmoreo, J., and J. Schneller. 2022. *The last doctor: lessons in living from the front lines of medical assistance in dying*. Viking Press. New York, NY.

McKee, M. (Lakehead University, School of social work). Self-care/burnout study results. Presented at the 2018 CAMAP Conference.

Moore, Sarah, Thabet, Chloe, and Wheatley, Paul. 2022 February. Medical assistance in dying in patients with lung cancer. *JTO Clinical Research Report*. 3(2): 100283. Published online 2022 Jan 21. https://doi.org/10.2016/j.jtocrr.2022.1000283.

Nelson, Jessica. Northern Alberta MP announces dissenting report on assisted death. *RMOToday.com.* July 1, 2022.

O'Rourke, A.A., M.C. O'Rourke, and M.F. Hudson. Reasons to reject physician assisted suicide/physician aid in dying. *Journal of Oncology Practice* 13(10): 683–687.

Parliament of Canada. Bill C-14. Statues of Canada 2016. Chapter 3. Assented to June17, 2016.

Ouelette, R.F. Parliament of Canada. House of Common debates, 42 Parl, 1st sess, 2 May 2016, 1830–1835.

Radner, E. 2023. Slippery slopes. *First Things.* January 2023. https://www.firstthings.com/article/2023/01/slippery-slopes.

Raikin, Alexander. 2022. No other option. *The New Atlantis*, December 16. https://www.thenewatlantis.com/publications/no-other-optionshave-devastating-effects-on-canadians-with-disabilities/372424.

Raj, A. Justice minister David Lametti under fire. *Toronto Star,* November 29, 2022.

Rayson, D. 2022. An end too soon? *Canadian Medical Association Journal* 194: E262–E263. https://doi.org/10.1503/cmaj.211748.

Reyes, Ronny. It's the most rewarding work we've ever done… *Daily Mail*, January 8, 2023. https://www.dailymail.co.uk/news/article-11611095/Canadian-doctor-who.

Robertson, W.D., J.A. Pewarchuk, J. Reggler, S. Green, T. Daws, and Trouto, K. Case review of medically assisted deaths on Vancouver Island. *BC Medical Journal* 59(6): 305–309.

Saba, P. 2021. *Made to live: a physician's journey to save life.* Word Alive Press, Montreal. https://wordalivepress.ca/

Selby, D., S. Bean, E. Isenberg-Grzeda, B. Henry, and Nolen, A. 2020. Medical assistance in dying (MAID): A descriptive study from a Canadian tertiary care hospital. *American Journal of Hospice and Palliative Care* 27(1): 58–54.

Simson-Tirone, Martha, Samatha, Jansen, and Marilyn, Swinton. 2022. MAID care coordination: Navigating ethics and access in the emergence of a new health professions. *HEC Forum* 34(4): 457–481.

Society for Canadian Psychiatry. Call to Action, November 10, 2022. tatic1.squarespace.com/static/61db373a8e4e00423c117825/t/637bef81233c136bd0eb6d6d/1669066626054/SocPsych+Call+to+Action.pdf Seeking better understanding of assessment of eligibility for MAID.

Star Editorial Board. 2022. https://www.thestar.com/opinion/editorials/2022/11/25/medically-assisted-deaths.html#:~:text=We%20need%20to%20put%20the%20brakes%20on%20expanding,from%20terminal%20illness%2C%20but%20from%20poverty%20and%20despair.%3C/div%3E

Stevens, K.R., Jr. 2006. Emotional and psychological effects of physician-assisted suicide and euthanasia on participating physicians. *Issues in Law and Medicine* 21 (3): 187–200.

Supreme Court of Canada. 2015. Carter v Canada (Attorney General). https://scc-csc.lexum.com/scc-csc/scc-csc/en/item/14637/index.do. Accessed 06 Feb 2015.

Third Annual Report on MAID of Health Canada. 2022. S.5.3. https://www.canada.ca/en/health-canada/services/medical-assistance-dying/annual-report-2021.html.

Waghmare, R. 2022. DNR. *The Medical Post*, 36–39. December 2022.

Woo, A. 2022a. Cancer care system in B.C. Bucklin as staff shortages lead to soaring wait times. *The Globe and Mail* September 22, 2022. https://www.theglobeandmail.com/canada/article-british-columbia-cancer-care-staff-shortages/.

Woo, A. 2022b. Lengthy waiting times have some cancer patients dying before their first consultations. *The Globe and Mail*, November 16, 2022. https://www.theglobeandmail.com/canada/british-columbia/article-cancer-patients-treatment-wait-times/.

Wu, J.S.Y., J. Pinila, S. Watsin, S. Verma, and I. Olivotto. 2018. Medical assistance in dying for cancer patients one year after legalization: A collaborative approach at a comprehensive cancer centre. *Current Oncology* 25 (5): e486–e489.

Jaro Kotalik MD (Charles University, Prague), DMRT (University of Toronto), M.A. (Medical Ethics and Law, Kings College, London, UK), FRCPC, is Professor, Northern Ontario School of Medicine University; Adjunct Professor, Department of Philosophy, Lakehead University; Associate Member of the World Medical Association. He served as a radiation oncologist, CEO of Thunder Bay Regional Cancer Centre, Vice President of Cancer Care Ontario and Clinical Professor of McMaster University. He founded the Centre for Health Care Ethics at Lakehead University, was its director for 12 years and is now Chair of its Executive Committee. He was appointed to the National Council for Ethics in Human Research and to several committees of the Royal College of Physicians and Surgeons. He served as a bioethics consultant to many organizations, including Health Canada and Swiss National Bioethics Committee.

Louisa Pedri BA, B.Ed. (Lakehead University), MA (Ethics, King's College, University of London, UK), was a founding member and served as Education Coordinator of the Centre for Health Care Ethics Lakehead University and is now a Research Affiliate and member of its Executive Committee. She served as President of Via Vitae Palliative Care Association and member of the Research Ethics Board of Lakehead University. Her work has been published in the Formosan Journal of Medical Humanities and in HEC Forum.

Chapter 30
MAID: Pasts, Present, and Futures

Tom Koch

> Even after 2000 years of traditional medical ethics we can be swept off our medical moorings in the medical profession in the face of rapidly changing social mores.
> Dr. Mark Komrad 2020.

Abstract Others have carefully described the history of MAID in Canada and its ramifications, legal and social. This short piece attempts to consider the possible future of MAID in Canada. Those speculations are set in a brief history of MAID and how it came to be a 'right' in the Canadian legal nad medical system.

Keywords Canada · Euthanasia · legal history · Holland · Medical termination

30.1 Introduction

Because "that's the way it is," does not mean "that's the way it will be," forever. This book documents the legal and social histories that have resulted in progressively permissive legal policies permitting, and some would say today encouraging, medical termination in Canada. It is tempting to think, "Well, that's … done so, live with it." But history is not a spectator sport. We parse the past to identify old wrongs in an attempt to forge a more acceptable future. To do this it's helpful to remember that the movement for what we Canadians politely call Medical Aid in Dying (MAID) only began in earnest in Holland in the 1970s. Perceived, at first, as a radically unacceptable departure from traditional medical and social prohibitions, its acceptance—contested at first—spread over subsequent decades to other countries in Europe and Canada. The adoption of permissive legislation, and an expansion everywhere of the categories

T. Koch (✉)
University of British Columbia, Vancouver, BC, Canada
e-mail: tomkoch@kochworks.com
URL: http://kochworks.com

Alton Medical Centre, Toronto, Canada

J. Kotalik and D. W. Shannon (eds.), *Medical Assistance in Dying (MAID) in Canada*, The International Library of Bioethics 104,
https://doi.org/10.1007/978-3-031-30002-8_30

of eligible patients, has been either extraordinarily rapid or arduously slow, depending on one's point of view. What's next?

30.2 The Dutch Beginnings

An answer requires, first, a brief review of its beginnings described to me in the early 1990s by Dr. Cornelisse-Claussen, then director of the Dutch Society for Voluntary Euthanasia. It was founded in the 1970s to advocate for medical termination as a way to shorten or eliminate the agony experienced by unpalliated, end stage cancer patients in an era in which palliative medicine had few options other than morphine (Koch 1996). The crusade was personal for Cornelisse-Claussen who remembered her mother's painful death from end-stage, metastatic cancer. At the same time, she and her society sought to promote a libertarian ideal of individual choice unimpeded by the niceties of medical control or judgment. End-stage cancer treatment was the springboard for a drive to de-medicalizing legislation until any who wished might simply apply to the local pharmacist for a supply of drugs that would result in a painless and certain end.

The progression from, first decriminalization to legalization with medical supervision, took place over the next decade. While few were truly alarmed at the medical deaths of patients in the last days of an irremediably painful disease, the euthanasia in 1993 of Boudewign Chabot was met with public alarm. After an unpleasant divorce followed by, in rapid succession, the death of her two sons, she was understandably depressed and saw no reason to continue living. Through the Dutch Voluntary Euthanasia Society, she was referred to a sympathetic psychiatrist who, after 12 unproductive counseling sessions, defined her condition as chronic and hopeless. Violating regulations requiring a second opinion, he unilaterally decided medical death was appropriate and proceeded to assist in her requested termination.

"A Bridge Too Far," declared the country's largest daily newspaper, *De Telegraaf.* Euthanasia in the case of painful, terminal disease is one thing, it argued. Killing a twice bereaved, recently divorced, severely depressed woman was something else. Officials in the Dutch Physicians Union which had opposed liberalizing physician assisted-suicide were not surprised, however. This was what they expected. "Once you start looking at killing as a means to solve problems," its director, Karl gunning, told reporters, "You'll find more and more problems where killing can be the solution" (Toronto Star 1994).

And, certainly, that has been the case both in both Holland and Canada. Amidst an ever expanding class of legally eligible patients, medical termination has become a default response for many treating non-fatal, non-progressive chronic conditions causing anxiety, depression, and fear. As one Dutch physician reported, when a patient refused medical termination for a bowel obstruction he was forced to seek advice on 'alternate treatments', those commonly employed by others who did not reflexively advise termination in non-terminal cases (Zylicz 2002). "Usually I solve this kind of problem with euthanasia," he said, "but this patient seems to not be pleased by this." In describing this case and the physician's surprise Zylicz wrote that, "The physician's remark that he usually solves such problems with euthanasia … illustrates how euthanasia becomes a substitute for learning how to relieve the suffering of dying patients" (Zylicz 2002). The same has been occurring in Canada.

We find that, today, in Canada.

30.3 Canada: Futures

At this writing progressive permissiveness encourages a continually expanding class of potential deaths from medical termination for reasons other than irremediable pain or physical distress. As one oncologist put it: "Loss of autonomy, control, and dignity are the primary drivers for seeking medical assistance in dying rather than uncontrolled symptoms, and the decision to pursue medical assistance in dying may occur pre-illness" (Susman 2021). From the 1993 court case of Sue Rodriguez through *Carter* et al. *v Canada* to the Royal Assent in 2016 for Bill C-14 few if any of those seeking legal permission in seminal court cases involved patients suffered irremediable, untreatable physical pain (Fig. 30.1).

In the early 1990s, Sue Rodriguez was well palliated at the time of her Supreme Court petition seeking an exception to laws then prohibiting medical assistance in dying (Rodriguez 1993). Instead, it was her fear that ALS (amyotrophic lateral sclerosis) would reach a point of respiratory crisis that would leave her gasping for breath and in physical distress. She, and her supporters, ignored the ability of palliative care specialists to treat this medically, reducing consciousness, and then, if

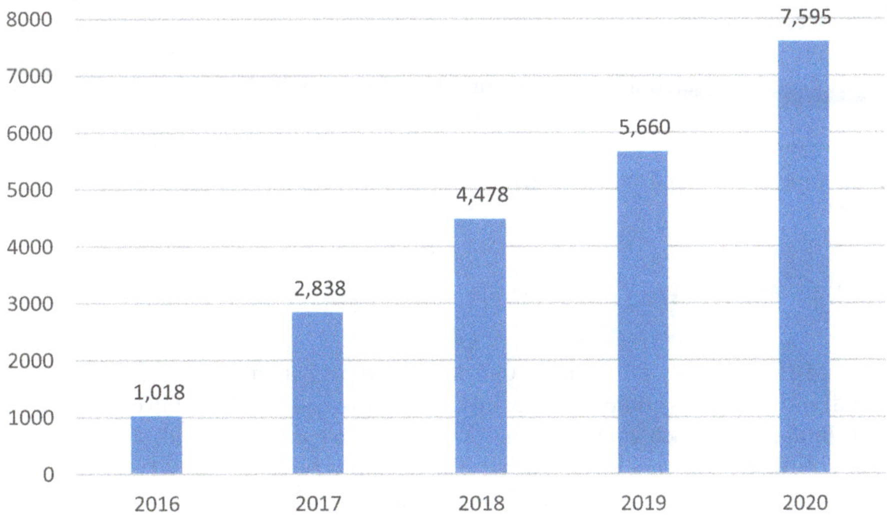

Fig. 30.1 The annual increase of medical terminations from 2016, when enabling legislation was passed, through 2020. *Source* Health Canada 2021

desired, providing continued respiratory support. And, too, she refused to accept the possibility of a future with, if she chose to survive with respiratory assistance, near complete paralysis. With that she rejected the example of popular figures like actor Christopher Reeves, paralyzed in a riding accident, physicist Steven Hawking, jurist Mr. Justice Same Filer, and others with high spinal paralysis who had lived for years with respiratory assistance and despite paralysis (Koch 1996).

Future fears rather than the lack of palliative care is a theme that runs through many of the seminal Canadian cases. In 1997 philanthropist Natverlal Thakore traveled to the United States to be terminated by Dr. Jack Kevorkian soon after being diagnosed with Parkinson Disease (Rinehart 1997). His condition was well controlled and while he might have lived comfortably for years, with treatment, he worried about the possibility of late stage future incapacities. In 1993, Saskatchewan farmer Robert Latimer killed his daughter Tracy because he believed procedures proposed by physicians to *eliminate* the pain caused by repeated hip dislocations would be painful and disfiguring (Butts 2021). Unremarked either in legal or popular review of this case was the potential for intrathical delivery of Baclafen as a non-surgical, less radical treatment of her condition. It is useful to note that at that time it was already in use in British Columbia but not in Saskatchewan.

Gloria Taylor, the primary petitioner in *Carter v Canada* (Attorney General, 2015), testified she was not suffering from untreated or untreatable pain resulting from her Amyotrophic Lateral Sclerosis (ALS). Rather, her desire for termination was like Rodriguez's based on fear of the disease's future progression, of possible incontinence and increased mobility limits. That fear of an unpleasant future was, as well, the rationale for the MAID request in a well publicized case in 2019 by an elderly couple in their early 90's with relatively minor chronic illnesses. Living in a comfortable seniors' setting, well-to-do and relatively healthy they sought, and received, MAID because they feared the possibility of future incapacities. The physician assessing their request described the medical rational as a 'terminal diagnosis' of … age (Grant 2018). Ironically, their story was juxtaposed in one newspaper with an article on the celebration of a group of centenarians living in an assisted living facility in Quebec (Koch 2018). Reviewing many cases in which MAiD was sought and granted it was similarly a fear of future limits rather than discomfort and limits at the time of the request that underlay many other requests for premature death.

30.3.1 MAID Abuses

Were MAID still illegal it is likely that some or most patients today seeking MAID could, with sufficient clinical and social support, live full if different lives adjusted to their limitations (Koch 2000a). Because the life one has lived is ended by accident, illness or injury does not mean that a meaningful life need be lost. "Quality of Life" is as much a result of clinical and social support as it is a measurable and firm diagnosis based on limit or immediate loss (Koch 2000b). Concerns of burden on others, for example, reflects the limits of social support in cases of physical fragility.

It is that mix of social and clinical in the context of patient choices that, under current legislation ha been ignored in the face of Corneliesse-Claussen's libertarian ideals. Talk of MAID and patient 'autonomy' but forget that patient choices occur within the fabric of family needs and the presence or absence of social support.

More fundamentally, perhaps, at this writing, a series of reports suggest MAID is being performed without attention to accepted procedures and safeguards and in the absence of programs of care (diagnostic and rehabilitative) that might have provided the necessities of a full, continued life. These "hastened deaths" resulted from a patient's sense of hopelessness based not simply in their illnesses but a lack of sustaining resources and treatments (Gallagher et al. 2020). In cancer care, for example, "Biomarker-driven targeted therapy and immunotherapy offer effective and tolerable new treatments, but a subset of patients undergoes medical assistance in dying without accessing—or, in some cases, without being assessed for—these treatment options" (Susman 2021). In that study approximately 20 percent of the patients surveyed did not have access to a medical oncologist or a radiation oncologist who might best advance treatment options. Decisions to terminate on the basis of these cancers thus were made in the absence of information on treatments that might have changed patient choices from termination to new treatment modalities.

The basic problem was well presented in a 2018 paper by David Baker and Geoff Cross in The Supreme *Court Law Review*: "The Court referenced the need for adequate safeguards in any constitutionally valid system of medical assistance in dying (MAID), but left unanalyzed the constitutional consequences of creating a MAID scheme without robust access to PC [Palliative Care]." They argue for one of two correctives under either the Canada Health Act, in which palliative care would be a stated right, or as part of the Criminal Code. What was significant, here, was the careful legal analysis of the law as it stands today, it's documented failures and the presentation of potential correctives.

More critically, or at least more publicly, testimony in Parliament by the Special Joint Committee on Medical Assistance in Dying heard reports from family members whose deceased relatives were not simply urged but in some cases bullied into accepting medical termination, sometimes on the basis of hasty and incorrect diagnoses (CPAC 2022). In one example, presented at a news conference, a patient's daughter said her father was encouraged to MAID by hospital staff using a patently false diagnosis. He accepted their recommendation after days of non-care (CPAC 2022). In another well publicized example, in 2022 a 51-year-old Quebec women with extreme sensitivities to chemicals chose medical termination after a search for affordable housing free of to her toxic cigarette smoke and chemical cleaners failed (Favaro 2022a). It was not that she did not want to live, supporters said. Rather, the failure of the provincial social services to assure her a safe living environment drove her to accept death as an answer.

A 31-year-old Toronto woman has applied for and at this writing is near approval for MAID. "I've applied for MAID essentially … because of abject poverty," she told CTV reporter Avis Favaro. Because she requires a wheelchair and as limited financial resources she has been unable to find suitable accommodations. That and a general bias on the part of some to hire those with disabilities has driven her to this extreme.

Finally, and in the same vein, Roger Foley has remained in a London, Ontario, hospital bed, for over four years, refusing to leave until he is assured adequate home care support. Unprofessional and dangerously inadequate support by caregivers assigned by a company on contract to the province had, he charged, resulted in repeated injuries requiring numerous hospitalizations. "Going home" would only mean more of the same without a system assuring well-trained home care personnel. In hospital he was politely "informed" that he was eligible for MAID as a way of ending his "suffering" (CTV 2018). Refusing to leave until his demands are met, or at least a court application for relief is heard, the hospital has chosen to bill him for his stay against their advice. To date they claim he owes over $1.2 million (US) for his intransigent insistence and resulting tenure in their bed.

All these issues were summarized in the United Nations (UN) *Report of the Special Rapporteur on the rights of persons with disabilities* in 2020. After reviewing the limits available for those with disabilities in various provinces the Rapporteur wrote, in the third person:

> She has learned that there is no protocol in place to demonstrate that persons with disabilities deemed eligible for assistive dying have been provided with viable alternatives. Moreover, she has received worrisome information about persons with disabilities in institutions being pressured to seek medical assistance in dying and of practitioners not formally reporting cases involving persons with disabilities (UN 2020).

30.4 Another Future

Canada might continue on the current path toward the original, libertarian ideal of the Dutch Voluntary Euthanasia Society: Termination on demand at your local pharmacy. This would be a natural extension of the demedicalization of care and the provision of MAID for an ever expanding class of non-terminal patients. Here is Henk ten Have on the general future of a depersonalized, commercial, medicine:

> Future medicine will be personalized and democratized because patients will manage their own care and make decisions about medical interventions with the assistance of artificial intelligence and robots (Ten Have 2022).

That would be the logical end point of the medicine promised in todays TV commercials touting any one of a number of powerful drugs (Just ask your doctor!) for cancer, Parkinson Disease, and memory loss that may or may not signal a dementia. The complexities of the individual case can be ignored with an algorithms based on patient symptoms but without regard to the complex issues of family support or the necessity for rehabilitative, psychiatric, or social services. In this future MAID would be a de factor response built into the overreaching if necessarily limited algorithms of ten Have's imagining.

Or, Canada may choose a different path. It begins with the recognition that the original rationale for medical termination is rarely the reason given today for its inaction. Palliative medicine has, in recent decades progressed to the degree that unpalliated, end-stage pain is very rare. Canada may draw away form the reflexive expansion of MAID on the basis of a growing class of reported cases of abuse accompanying police investigations of patient deaths resulting from questionable applications of MAID (Favaro 2022b). Some such reports refer to the questionable medical termination of the medically ill (Maddeaux 2022).

And, too, with greater publicity official attention will necessarily be paid to the types of medical abuse reflected in the stories of those who testified in the Parliamentary committee investigations, previously cited. As this type of public reportage increases, public sentiment may begin to question the legislation as it exists today.

With these public reports there are, in addition, legal challenges slowly working their way toward judicial review. These include, importantly, the Roger Foley amended petition in Ontario Supreme Court claiming the failure to assure adequate, safe, and professional home support violates his Charter rights (Foley v Attorney General of Canada 2021). In that petition he names both the home care agency that provided what he alleges has been dangerously inadequate care, the hospital he refuses to leave and has billed him for his stay, the province and the federal health agencies.

In the same vein, and more broadly, an application to the Ontario Supreme Court (Preston et al. v Attorney General of Canada 2021) alleges that the failure to assure comprehensive palliative care while enshrining a legal right to MAID creates a context of inequality violating specific sections of the Canadian Charter of Rights. Simply, the argument is that without a predicate right of appropriate palliative care and support that a right to medical termination denies the predicate necessities by which a person might choose honestly between living with dignity, despite restrictions, and a "death with dignity" through medical intervention.

These cases, while critical, will be a slow path toward legal and social change. And, in Canada, they sit between what is today a division between provincial authority in health care allocation and federal guarantees under the Canadian Charter of Rights and Freedoms. They will be heard within a public context in which MAID has become an accepted conclusion in almost all cases where ennui and fear are motive causes rather than irremediable end-stage disease. Since the 1990s and 'the Myth of Loving Murder" (Koch 1995) there has been a de factor sense in most stories that death is preferred in the face of most clinical challenges.

But if public opinion enforced by medial reportage changed from the 1980s into this century it may change again as reports of MAID abuses and, as the Foley case represents, the failure of rehabilitative, psychiatric and social supports becomes the focus.

30.5 Conclusion

Because something is legal does not mean it is ethical, or morally appropriate. History is replete with the dire result of what occurs when we are swept from "our medical moorings in the medical profession in the face of rapidly changing social mores" (Komrad 2020). I would like to think that the growing evidence of abuses and the emphasis on a rapid death rather than assisted live will result in both a growing public awareness of the need to focus the communal (and thus legal) will. I would like to think that a renewed legal, popular, and professional insistence on assistance to those who now fear dependence will result in a 'right' to MAID only in exceptional circumstances.

Terminating a patient is easy. Caring for the person is more difficult. And if it "takes a village" to raise a child it will take that same village to assure the fragile among us can continue to be with us rather than medicated out of our presence.

References

Baker, D., Cross, G. 2018. Establishing the right to palliative care in Canada. *Supreme Court Law Review (2d)* 293: 296–332.

Bill C-14 (Royal Assent). 14 June 2016. https://www.parl.ca/DocumentViewer/en/42-1/C-14/royal-assent.

Boudewign, Chabot. 16 September 1993. Court Docket No. 478–1993.

Butts, Edward. 2021. Robert Latimer case. Canadian encyclopedia. https://www.thecanadianencyclopedia.ca/en/article/robert-latimer-case?gclid=CjwKCAjw9-KTBhBcEiwAr19ig1R2-leaV0Fd4pkAqHmZBzs5Afdt7HXJ8GMgRJCO0hkVxZAbjprLJRoCAF0QAvD_BwE.

Carter v Canada (Attorney General), 2015 SCC 5, [2015] 1 SCR 331 [cited 27 Apr 2015]. Available from: http://canlii.ca/t/gg5z4.

CPAC. 10 May 2022. Conservative MPs on medical assistance in dying safeguards. YouTube. *Video*, 31: 23. https://www.youtube.com/watch?app=desktop&v=LY5obUDerQY.

CTV. 18 August 2018. Chronically ill man releases audio of hospital staff offering assisted death. *CTV News*. https://www.ctvnews.ca/health/chronically-ill-man-releases-audio-of-hospital-staff-offering-assisted-death-1.4038841.

Favaro, Avis. 14 April 2022a. Woman with chemical sensitivities chose medically-assisted death after failed bid to get better housing. *CTV News*. https://www.ctvnews.ca/health/woman-with-chemical-sensitivities-chose-medically-assisted-death-after-failed-bid-to-get-better-housing-1.5860579.

Favaro, Avis. 26 April 2022b. Police investigating medically-assisted death of B.C. woman." *CTV National News*. https://www.ctvnews.ca/health/police-investigating-medically-assisted-death-of-b-c-woman-1.5877288.

Foley v. Attorney General of Canada et al. 2021. Ontario Superior Court, Amended Claim. Court File No. CV-18–00603786.

Gallagher, Romayne, Michael J. Passmore, and Caroline Baldwin. 2020. Hastened death due to disease burden and distress that has not received timely, quality palliative care is a medical error. *Medical Hypotheses* 142: 109727.

Grant, Kelly. 1 April 2018. Medically assisted death allows couple married almost 73 years to die together. *Globe and Mail*. https://www.theglobeandmail.com/canada/article-medically-assisted-death-allows-couple-married-almost-73-years-to-die/.

Health Canada. 2021. *Second annual report on medical assistance in dying in Canada 2020*. Ottawa: Health Canada. https://www.canada.ca/en/health-canada/services/medical-assistance-dying/annual-report-2020.html.

Koch, Tom. 1996. Living vs dying 'with dignity' a new perspective on the euthanasia debate. *Cambridge Quarterly of Healthcare Ethics* 5 (1): 50–61. https://doi.org/10.1017/S0963180100006721.

Koch, Tom. 2000a. Future states: Testing the axioms underlying prospective, future-oriented, health planning instruments. *Social Science and Medicine* 52 (4): 453–466. https://doi.org/10.1016/S0277-9536(00)00154-4.

Koch, Tom. 1995. The myth of loving murder: The constructed tale of Jean and Cecil Brush. *Toronto Life Magazine* 61–63.

Koch, Tom. 2000b. Life quality vs the 'quality of life': Assumptions underlying prospective quality of life instruments in health care planning. *Social Science and Medicine* 51(3): 419–428. https://doi.org/10.1016/s0277-9536(99)00474-8. https://repository.library.georgetown.edu/handle/10822/516459.

Koch, Tom. 9 April 2018. Old age alone shouldn't be considered a justification for physician-assisted death. *Globe and Mail.* https://www.theglobeandmail.com/opinion/article-old-age-alone-shouldnt-be-considered-a-justification-for-physician/.

Komrad, Mark. 16 November 2020. Vulnerable ethics: The role of psychiatrists in the US eugenics movement and the Nazi Holocaust. *Yale Psychiatry Grand Rounds.* New Haven, CT. Yale University. Video, 1: 16:08. https://www.youtube.com/watch?v=cji13_m5Bfg.

Maddeaux, Sabrina. 2 May 2022. Euthanizing the mentally ill without providing proper supports is reprehensible. National Post. https://nationalpost.com/opinion/sabrina-maddeaux-euthanizing-the-mentally-ill-without-providing-proper-supports-is-reprehensible.

Preston, Wong-Ward et al. 2021. Attorney General of Canada. Application under Rule 140.05. Ontario Superior Court. Court File No. CV-21–00674049–0000.

Rinehart, Dianne. 23 September 1997. Natverlal Thakore goes to the U.S. for a Dr. Kevorkian death. *Vancouver Sun*, A1.

Rodriguez V. British Columbia, 3 SCR 519. Case No. 23476. (1993).

Susman, ed. 21 September 2021. Study sheds light on physician-assisted suicide in lung cancer patients. *Med Page.* https://www.medpagetoday.com/meetingcoverage/iaslc/94429.

Ten Have, Henk. April, 2022. Medicine and machines. *Medicine Health Care and Philosophy.* https://doi.org/10.1007/s11019-022-10080-5.

Toronto Star. 23 June 1994. Dutch woman's aided death rekindles euthanasia debate. *Toronto Star.*

United Nations (UN). February 24–March 2020. Report of the Special Rapporteur on the rights of persons with disabilities. *Human Rights Council Forty-third session* H-69: 13.

Zylicz, Zbigniew. 2002. Palliative care and Euthanasia in the Netherlands: Observations of a Dutch physician. In *The case against assisted suicide: For the right to end-of-life care*, ed. Kathleen Foley and Herbert Hendin, 122–143. Baltimore, Maryland: The John Hopkins University Press.

Tom Koch, PhD Prof. Tom Koch PhD is a medical ethicist, geographer and historian who has written both academic articles and popular investigations of medical termination since 1995. A comprehensive history of that work is available at kochworks.com/.

Part VII
Overview and Conclusions

Chapter 31
Death on Demand?: Canada's Culture and Values at a Crossroads

Harvey Schipper

Abstract The 2015 Supreme Court of Canada decision *Carter v. Canada* marks an inflection point in legal and medical thought and practice. For the first time since capital punishment was abolished in Canada in 1976, active euthanasia has become a legal and clinical reality. More than 31,000 Canadians have sought and received euthanasia. What was initially a narrowly circumscribed set of end of life circumstances, characterized by 'intolerable suffering' looks to be expanded to include chronic non-terminal conditions, mental health disorders and potentially conditions affecting adolescents considered capable of giving consent. The intended liberalization, if fully expressed, would be the broadest in the world. More than 30 authors, with expertise crossing medicine, ethics, law, psychology, sociology and anthropology, and politics have explored the implications of the Court judgments and legislation which followed and have raised many concerns, including:

- Does the particular concept of autonomy upon which this edifice is built represent the views of Canadians? Are there other constructs of equal or greater validity and resonance?
- Do we fully understand, as a practical matter, what consent means in conditions of intolerable suffering?
- Is there reasonable consensus about the natural history of mental health disorders, and post-traumatic distress that we can safely and reliably end someone's life without error?
- Have we heard and taken into account the concerns of the disabled?
- Do we understand the knock-on effects medical euthanasia has on the practice of medicine, and the health care system?
- Do we have in place fundamental safeguards about the selection of MAiD assessors and providers, as well as the independent rigorous oversight of their conduct which are required for public confidence?

H. Schipper (✉)
University of Toronto, Toronto, Canada
e-mail: hschipper@mindenschipper.com

J. Kotalik and D. W. Shannon (eds.), *Medical Assistance in Dying (MAID) in Canada*, The International Library of Bioethics 104,
https://doi.org/10.1007/978-3-031-30002-8_31

This concluding chapter seeks to bring together contentious themes to encourage thoughtful deliberation of a fundamental rethinking of our societal values which are the foundation of public policy.

Keywords Active euthanasia · Medically assisted suicide · MAiD · Oversight · Autonomy · Suffering · Medical ethics · Legal reasoning · Suicide vs. euthanasia · Palliative care

In June of 2014, before the Supreme Court of Canada (2015) *Carter v. Canada* (Attorney General)[1] decision was rendered, I wrote a piece for the *Globe and Mail* anticipating that decision (Schipper 2014). As an oncologist, with considerable experience in the evolution of palliative care as well as in related public policy here and abroad, I made the following statement: *Put as a dichotomy (unresolved suffering or active euthanasia), I'm prepared that a few might suffer more than they can bear, rather than countenance in the name of some kind of generosity of spirit the active taking of a life. I know from history, and have seen too much of the slippery slope of convenience, to find confidence in any permissive legislative process* (Schipper 2014).

Not surprisingly the opinion piece drew condemnation, not all of it polite, from advocates for what has become known, in Canada, as MAiD. I was criticized as an uncaring doctor who didn't respect the wishes of patients. Perhaps most worrisome, advocates insisted that there was no slippery slope. The argument was made that the depreciation of human life, the marginalization of the vulnerable, the utilitarian economics, as happened in other places like Nazi Germany, could not happen here. Advocates declared that to even raise the comparison with that horrible moment in human history was, in itself, an outrage. Notwithstanding that euthanasia discussion and the Nazi experience is, across the Western world and beyond, part of every bioethics course.

There is a philosophic construct sometimes called 'the fence within the fence'. Some values cannot be abrogated directly because they are so deeply embedded in the fabric of civilization that to puncture the veil, as it were, would be to unleash unforeseeable consequences, many potentially dire. So, a moral and philosophical second fence is drawn around the issue, where one might consider the limits of the principle, or how adherence might be determined or adjudicated. State sanctioned euthanasia meets that criterion.

The experience of the years since *Carter* is that the veil around a fundamental societal value has indeed been torn, and already the ramifications extend far beyond those anticipated by the Court, or dismissed by advocates, or feared by those opposed. What we are seeing is not a disciplined, evidence driven, incremental integration of this radical moral departure into our community, but rather a rapid expansion absent of even the consideration mandated by legislation.

[1] Hereinafter "*Carter*".

While this has been posited as an ultimate means of relieving suffering, and as an expression of one's individual autonomy with respect to the body, the reality is different. The implementation of MAiD has broadly reframed our societal understanding of the value of life, how we address illness, the changes in our physical and mental capacity that may result, and how as a community we deal with death.

Consider this analogy. We are confronted by a new technology. Perhaps it is nuclear power, which offers the prospect of providing the energy our society requires without the health, economic and environmental downsides of fossil fuels. It can be argued that fossil fuels lead to more illness and death *each year* than all the nuclear episodes since its modern inception, including Hiroshima and Nagasaki. It is a plausible hypothesis. Our response has been to study it intensely and move very cautiously. We consider the immediate and obvious hazards, such as leaks and waste disposal, and invite considered speculation about the unpredictable, the unfathomable. We want fail safes, because the stakes are so high. We do this because our ultimate concern is the integrity and stability of our community. This critical engagement does not have a time limit. For some, the process seems too slow, as lives continue to be lost. But the debate drives innovation and commitment toward a better solution. It broadly engages the community and works toward a consensus. It permeates our work environment. It becomes part of the education of the next generation, those who will live with the consequences of our decisions.

The path toward the liberalization of medical euthanasia has followed a very different path. That path has been ideological. It is *go forward absent of evidence of adverse effect or consequence,* the exact opposite of what we expect for new interventions. This approach confuses, potentially lethally, the absence of evidence with the evidence of absence. Absence of evidence thinking is comfortable with not looking for evidence, something that is a notable feature of Parliamentary debate about both Bill C-14 and Bill C-7. It also tends to narrow the discussion, making the assumption that satisfaction under one discipline of analysis (legal for example) can safely be generalized, obviating the need for multidisciplinary analysis. Viewed from this perspective, the current state of affairs for medical euthanasia in Canada is only at its starting point. We simply do not know enough to make so profound a change in the legal, medical, scientific and cultural ethos of this country.

Advocates for medical euthanasia did not engage in broad societal discussion. They did not engage the medical and scientific communities. They chose instead a narrowly legal approach, made possible by Canada's *Charter of Rights and Freedoms* (1982) which holds a fundamental tension between the broadest feasible expression of individual autonomy (Section 7) and the aggregate welfare of the society (Section 1). Supporters would argue that ultimately this was to be a legal issue because it meant creating a specific exemption in the *Criminal Code* for what otherwise would be murder. Critics would argue that it was advocacy by stealth.

Attempting suicide was removed from the *Criminal Code of Canada* (1985) in 1972. However, counselling suicide remained a criminal offense. Thus, the legal proceedings had to walk a fine line. If medical euthanasia were to be made legal, then assisting in that procedure had to be made legal as well. The balance would be that the patient was to initiate the discussion and not be subjected to advocacy or coercion, direct or indirect.

The *Rodriguez v. British Columbia* (Attorney General)[2] (1993) case was the first instance, wherein a close and anguished decision by the Supreme Court of Canada considered and ultimately refused medical euthanasia, then termed assisted suicide. Sue Rodriguez was a woman with an irreversible neurodegenerative disease with no prospect of recovery.

> Sopinka J, writing for the majority, found there was no violation of section 7. He first considered whether the prohibition on ending one's life engaged the right to security of person. He found the prohibition had sufficient connection with the justice system by its impact on an individual's autonomy and right to life by causing physical and psychological pain.
>
> However, Sopinka J found the provision did not violate any principles of fundamental justice. He examined the long history of the prohibition of suicide and concludes that it reflects part of the fundamental values of society and so could not be in violation of fundamental justice.
>
> He also rejected the claim that the provision violated the section 12 right against cruel and unusual treatment or punishment as a mere prohibition did not fall within the meaning of treatment.
>
> Lastly, he considered the section 15 equality challenge. He noted the issue is best not resolved under this right, but in assuming it did violate section 15, he found it was clearly saved under section 1. He found the objective was pressing and substantial, rational and that there was no lesser means to achieve the goal.

In the end, she took her own life with the assistance of a physician, and with a parliamentarian present. No charges were laid. It was moral and legal decision-making on the knife-edge.

In contrast, the 2015 *Carter* decision allowed the exemption from the criminal prohibition against murder under a set of narrow and specific conditions. The Court accepted the plaintiffs' position that the possibility that a patient would be forced to seek an earlier death at his or her own hands for fear that their suffering might not be relieved when they were more debilitated violated their *Charter* right, under Section 7, and that right was not such a threat to the greater societal obligation set out in Section 1. The Court cited evidence that some other jurisdictions had found a pathway and safeguards, enough to relieve the justices' concern about safety.

> Section 241(b) and s. 14 of the *Criminal Code* unjustifiably infringe s. 7 of the *Charter* and are of no force or effect to the extent that they prohibit physician-assisted death for a competent adult person who (1) clearly consents to the termination of life and (2) has a grievous and irremediable medical condition (including an illness, disease or disability) that causes enduring suffering that is intolerable to the individual in the circumstances of his or her condition.

In *Carter*, the Court gave the legislature 12 months, subsequently extended owing to an election, to come up with law giving effect to its decision. That led to Bill C-14 which viewed this as an end-of-life issue.

---

[2] Hereinafter "*Rodriguez*".

The person has a grievous and irremediable medical condition (new section 241.2(1)(c)).

The phrase "grievous and irremediable medical condition" is defined in new section 241.2(2) as requiring all of the following criteria:

- the person has a serious and incurable illness, disease or disability;
- the person is in an advanced state of irreversible decline in capability;
- the illness, disease or disability or the state of decline causes enduring physical or psychological suffering that is intolerable and cannot be relieved under conditions that the person considers acceptable; and
- natural death has become reasonably foreseeable, taking into account all of the medical circumstances, although a prognosis as to the specific length of time remaining is not necessary.

Parliament, as its right and role, interpreted the limitation in Carter to "only the facts of this case," to mean end of life. The legislation then mandated a five year review of what actually happened, as well as a specific study of exclusions to the law: advance directives (which would serve to transfer agency in certain circumstances), euthanasia when mental health was the sole eligibility criterion (addressing both suicidal thinking as a manifestation of illness, and its prognosis, as well as capacity to consent), and mature minors (confronting the relationship between parents and child, in the context of what is known about neurodevelopmental maturation in relation to age). The Council of Canadian Academies Expert Report (2018), (I was part of that process) was specifically not empowered to make recommendations, only to distill the state of knowledge. Nonetheless it made one explicit recommendation: more knowledge and experience was required before broadening eligibility. The federal ministers of Health and Justice at the time have commented that Bill C-14 appeared to balance conflicting views in that neither side was entirely satisfied.

With the passage of Bill C-14 the federal government punted operational responsibility to the provinces, given that what by then had come to be known as MAiD was considered a health matter. The provinces predictably assigned responsibility to the health professional colleges who struggled to provide operational regulations, and some elements of oversight. The issue of conscientious objection on the part of health providers provided an early warning that the regulatory role would not be simple. The provinces, with input from the courts, have established different rules, often parsing the meaning of 'effective referral' to another provider should the responsible physician be morally opposed to being part of actively ending a patient's life. By December of 2021, more than 31,000 Canadians had availed themselves of the procedure (Government of Canada 2022). We were beginning to position ourselves for the mandated review, when a lower court decision from Quebec rocketed the topic to the top of the national political agenda.

The *Truchon* decision of the Quebec Superior Court (Truchon 2019), held that the 'death reasonably foreseeable' limitation in Bill C-14 was unconstitutional in that it discriminated against those who wished to have their lives ended, but whose medical condition was not foreseeably terminal. The Quebec Superior Court decision was explicit in its position that the individual autonomous right over-rode Section 1 of the *Charter*, the section upholding the societal balance of the tension. In the face of an impending federal election, the federal government chose not to appeal. While the decision is only directly binding in Quebec, the federal government chose to introduce legislation effectively extending its application across the country.

Bill C-7 was the result. The 'reasonably foreseeable death' restriction was removed. A two-track system was introduced, one for those whose death was imminent, and another where it was not. Waiting periods were shortened or eliminated. Perhaps most striking, active euthanasia for mental health as the sole malady was to become legal by the Spring of 2023 *unless evidence to the contrary were to come forward.* In other words, the absence of evidence would be taken as tantamount to the evidence of absence. Moreover, the review of prior experience mandated by Bill C-14 would only take place *after* the Bill was enacted. An expert committee was to review the evidence regarding mental health. The mandate was to articulate the issues to consider in the course of implementing the expansion of MAiD with respect to mental health.

With the *Truchon v. Canada*[3] decision by the Quebec Superior Court, and the subsequent promulgation of Bill C-7 (2021), we reach a watershed in Canadian social policy. With the *reasonably foreseeable death* limitation of the predecessor Bill C-14 (2016) removed, and the impending eligibility of mental health as the sole criterion, the legislation is close to sanctioning state provided death on demand. While several Western jurisdictions have enacted permissive legislation for physician assisted suicide, none will have gone as far as Canada.

The *Final Report of the Expert Panel on MAiD and Mental Illness* was released in May 2022. What is striking about the 19 recommendations is both their breadth and the detail with which their concerns are expressed. Yet no additional safeguards were proposed. For example, with respect to capacity assessment, Recommendation 5:

> MAiD assessors should undertake thorough and, where appropriate, serial assessments of a requester's decision-making capacity in accordance with clinical standards and legal criteria. These assessments should be consistent with approaches laid out in standardized capacity evaluation tools.

A review of capacity assessment in the context of 'intolerable suffering' conducted by the Law Commission of Ontario (Cartagena 2016) revealed that there were no such reliable or validated capacity assessment tools.

As another example, Recommendation 10:

> The requester should be assessed by at least one assessor with expertise in the condition(s). In cases involving MAiD MD-SUMC, the assessor with expertise in the condition should be a psychiatrist independent from the treating team/provider. Assessors with expertise in the person's condition(s) should review the diagnosis, and ensure the requester is aware of all reasonable options for treatment and has given them serious consideration.

However, as yet, there are no independent, formalized training programs for assessors other than those organized by groups of self-selected MAiD providers. Presumably this is a role the professional colleges, such as the Royal College of Physicians,

[3] Hereinafter "*Truchon*".

might assume, as they have for other medical specializations. Meticulous assessment of the qualifications, motivations and personal characteristics of potential providers is a foundational requirement, perhaps more so than in any other area of qualification because this role is directed to ending a life…a unique perspective in all of health practice.

This report appears limited by its mandate to simply catalogue issues (and even then it seems there was division to the point where there were resignations from the panel prior to publication), and turn responsibility over to the legislature. There were more than a dozen recommendations documenting shortcomings at every step in the process. The panel may have assumed that legislators would do what they have done in aircraft safety, nuclear energy, smoking, new drugs and devices and medical procedures: understand the issue fully, document and explain the risks, and move prudently, learning at every step.

This chain of legal events brings us to an uncomfortable moment. We would appear to be in the process of considerably expanding state-supported euthanasia without adequate evidence of its ramifications. If MAiD, as the procedure has come to be known, is to find an acceptable place in Canadian society, there must be a full and nuanced understanding that it is a form of ending one's life. Such a departure from centuries of principle echoes across the entire society: in medicine, economics, public policy, law and jurisprudence, faith and education. The process thus far has been almost entirely legal, circumventing all other concerns under the rubric that the law is supreme, and having litigated a rationale, all other concerns will find a subsidiary space.

In these early days we already have inklings of painful unresolved issues. The review of MAiD for mental illness conducted by a Parliamentary Committee in May of 2022 has elements of partisanship and polarization that are uncommon, to some unseemly… because the issue is so contentious. In addition, case stories are emerging in the press citing improper process, and even subtle if not overt coercion. Some of these are coming before the Courts. They are countered by other stories of grace, dignity and peace with the successful provision of MAiD. Which is to say this is not simple. It is not settled law. It is not, to this point, something about which the community has even full knowledge, let alone reached consensus.

Where to go from here? Having deeply studied medical euthanasia for a decade or more, I have learned that it will find a place in our society. This is not what I would have wished back in 2014, and as I've written and spoken since, we will only find that space by slow evolution based on deliberate discussion and debate. Conviction must not debase evidence. Passion cannot ride roughshod over doubt.

This book is an attempt to engage that learning. More than anything else, what these more than 30 chapters make clear, is that the sages were right. Having rent the inner fence we are discovering an issue much more complex than anticipated. Though most of the writers are cautionary, there is a range of views, from fine tuning the current process to urging the use of the Not-Withstanding Clause to invalidate Bill C-7.

If we are to achieve a reasonable balance in legislation and implementation, the starting point must be discussions that are frank and clear. What follows is something of a starting point list of important issues raised by contributors to this volume.

31.1 Language

The first issue is language. Actively ending another person's life in the face of medical futility, or unbearable unrelievable suffering can be described in several ways: murder, mercy killing, active euthanasia, assisted suicide, or as more recently euphemized, medical aid in dying (MAiD). Each word has nuance and meaning that may differ among users. There are few polls which directly address the language issue, but it would appear that when people are asked if it is permissible to kill someone at their request, the answer is different than when the word 'killing' is not used. The Supreme Court decisions refer to 'medically assisted suicide'. The legal carveout is with respect to murder, which we have always called the intentional ending of someone else's life. Somehow, without much consideration, our legislators and regulators have accepted the euphemism "Medical Assistance in Dying", implying that it is simply an addition to what physicians have done through recorded history… easing the transition from life to death. That terminology offends and insults many physicians. Moreover, it sanitizes a difficult reality: we are talking about killing people.

There is a further confusion at the bedside. The lay public, even many interested in this issue, do not understand that this has nothing to do with their long standing, unequivocal right to decline further intervention. Withdrawing intervention is not the same as actively ending life.

31.2 Through Legal Eyes

The first section of the book explores the legal framework and assumptions underlying cases leading to *Rodriguez, Carter, Truchon* and the consequent legislation. The first presumption is that this is a legal matter, yet when Quebec first ventured into the thicket it considered active euthanasia to be a medical procedure. Lost in translation was a fundamental disconnect between medical thinking and epistemology, and legal reasoning. Medicine is at root Hippocratic: forward thinking, dedicated to the preservation of life and in sole obligation to the patient. Over the past century is has become more and more an art deeply invested in science and objective evidence. It fundamentally accepts that biology rules, and we advance ourselves by knowing and working within those rules. Legal reasoning is different. It is post hoc based on principles and precedent. Legislatures make the rules. Canute can order the waves to cease.

A second presumption concerns the concept of autonomy. Several scholars explore it. The highly individualistic concept of autonomy which lies at the root of the medical euthanasia debate is but one of several formulations of autonomy. This 'radical autonomy', as discussed by Dylan McGuinty Jr, is 'defined by freedom from any comprehensive moral anthropology'. It derives from Sartre's notion of existential freedom. The argument is made that such a view of autonomy also prioritizes the autonomy of those who wish to die over the well-being of those who wish to live on, but for the hardship of their circumstances, including things like pain control and lack of social support. It is a position of moral neutrality. There is another notion of autonomy, relational autonomy, that is more deeply rooted in the collective welfare of the community. It is certainly much closer, and likely derives from moral principles of faith that are common to both Western and Eastern belief systems. One can speculate on how the Supreme Court came to embrace this individualistic, arguably libertarian view but that is beside the point. What is pertinent is whether this view, when clearly articulated vis-à-vis other concepts truly represents a Canadian ethos. Moreover, it leads directly to a third presumption.

Over the past decades, the courts have gone from values-based judgements, to harms-based judgements, and from there to risk of harms based judgement. This aligns with a more self-centered autonomy. Translated: What once was proscribed on principle, became legal if there was no harm to others. The next step was it became legal if the risk of harm was tolerable. It is values free and transactional. Should this be the legal foundation for legislation permitting the active ending of life?

Of course, this begs a fourth question. Should this be a legal matter at all? What business do the Courts have at the bedside? It becomes a cumbersome and vexing question. We don't employ legal reasoning to ascertain any other medical procedure. We accept that severe injury and death may ensue from medical judgement. Imperfect but pretty reasonable informed consent procedures work well within the constraint that physicians are not historically or culturally the agents of a priori ending life. And equally historically, as in but not exclusively in Nazi Germany, when that role has been added to the medical lexicon, the results have been disastrous. Other sections of the book raise the issue of conscientious objection, a direct consequence, which is a serious matter.

Fifth, the late legal scholar Peter Hogg devoted much of his career to articulating nuances of the relation between the Legislature and the Courts. He spoke of a dialogue, wherein the elected legislature articulated the views of Canadians in laws which went forward unless the Courts found they violated fundamental principles, in this instance those laid out in the *Charter of Rights and Freedoms.* All the pertinent Court decisions are based on an interpretive balance between the welfare of the community (Section 1) and the rights of the individual (Section 7) of the *Charter.* Taking into account the divergent modes of reasoning and evidence between medicine and law, and the very particular concept of autonomy, could the Courts have overreached?

Canada does not exist in isolation from the rest of the world. The evolution of our medical euthanasia framework has also drawn the attention of the United Nations, specifically with respect to the *International Covenant on Civil and Political Rights*

(United Nations 1966), and the *Convention on the Rights of Persons with Disabilities* (United Nations 2010), among others. There is concern that any medical euthanasia framework discriminates against those who are different, those with disabilities. Several chapters, and a whole section on disabilities invite a serious consideration of the issue. The more so because Canada's medical euthanasia is among, if not the most liberal in the world.

31.3 Implementation—Making It Work

Chapter 7 by Bagheri and Dholakia titled "Medical Assistance in Dying Step-by-Step: Processes and Challenges" opens with a quotation from Felix Frankfurter, the late eminent US Supreme Court Justice: "The history of liberty has largely been the history of the observance of procedural safeguards." That is the leitmotif of this section of the book.

As a physician, one starts from the bedside. Who selects the assessors and providers? What personal qualifications must they possess, and perhaps uniquely in this exceptional and counter-cultural setting, ought we be especially careful when we allow people to intentionally end lives?

What should be the elements of the training programs? Is there a review process for each and every case, as in the long-established practice of morbidity and mortality rounds in hospitals? How do we respond, and who is responsible for the response when concerns are raised? Nowhere in medicine are our decisions infallible. Why should it be different here, especially when the recipient of the erroneous treatment is dead?

Is there something that distinguishes those MAiD providers who process hundreds of cases, representing the majority of a practice, from those, who by current statistics provide a handful of services? It is the case in every other medical discipline.

Every criterion for MAiD, across all the bills and regulations uses terms, such as 'intolerable', which though legally precise, are clinically ambiguous. With time and experience, can we sharpen our understandings so as to provide consistency and reassurance?

Moving from the bedside to the aggregate impact, what questions should be asked and what information collected about each case? Clearly process information is important, but the critical data is the detailed documentation of the particular medical condition, trajectory of interventions and supports offered and actually provided, family and social context, and reasons for requesting MAiD.

Our contributors note that a start has been made. They are concerned that the right questions may not have been posed, that there is inconsistency both within and across jurisdictions, and that what should be deep learning is at present only a gathering of gleanings. As they point out, it is not merely a question of whether the decision to provide MAiD and its administration was correct within the law. The more important learning is how requests for, and the provision of MAiD reflect on our health care system as a whole, and from there how our society values and supports those who are suffering.

31.4 Palliative Care in the Realm of MAiD

The moral and clinical foundation of palliative care is the belief that human dignity and connectedness can be maintained despite terminal illness. The relief of suffering is holistic practice. Easing the end of life, rather than accelerating it, is a foundational value that derives from the history of palliative care. The epochal work of Elizabeth Kubler Ross in articulating the stages of coming to terms with one's impending end of life, and Dame Cicely Saunders' focus on a holistic, connected caring for those at the end of their lives mark the foundations of palliative care. In part the rise of the palliative care movement was a corrective to the then almost unitary focus on treating the disease, rather than the person. The WHO Cancer Unit, led by the Swedish oncologist Dr. Jan Stjernsward, focused on adequate pain management, at a time when the misguided, if well intentioned, fear of addiction left many patients to suffer unnecessarily. Saunders from England, and Balfour Mount and Neil Macdonald from Canada were leaders in expanding the ambit of palliative care to address dignity, connectedness, and a focus on making every day worthwhile. They brought to life Saunders' theme, *"You matter because you are you, and you matter to the end of your life. We will do all we can not only to help you die peacefully, but also to live until you die."* We quickly learned that when patients knew they could receive hospital-based care when they needed it, they felt liberated to continue with their lives with palliative care support which extended *into* the community. Several studies have shown that people who are provided palliative care lived longer. They remained connected.

Thus, it is not surprising that palliative care providers have serious misgivings about medical euthanasia. Mary Shariff in Chap. 12 of this book takes umbrage with the BC Supreme Court's *factual* finding that the:

> Preponderance of evidence from ethicists is that there is no ethical distinction between physician-assisted death and other end-of-life practices whose outcome is highly likely to be death.

She does so because it unequivocally posits medical euthanasia as simply one choice on a menu of treatments, rather than as a last resort. The issue cuts to the heart of palliative care. It has caused divisions in a previously tightly knit clinical specialty.

All through the debates about medical euthanasia, palliative care has been an elephant in the room. Advocates for medical euthanasia advocate for better palliative care, from their position that euthanasia is co-equal with palliative care, rather than a last resort. Two questions follow:

How readily must we accede to a patient's request to die if we have not done our best to understand and relieve their suffering? David Shannon and his colleagues in Chap. 17 point to our societal and legal inconsistency, if not confusion. Bill C-7 contains a sunset clause, effective in 2023 which will permit medical euthanasia for mental illness, on demand, albeit with some perhaps enhanced safeguards. At the very same moment, laws in Ontario and elsewhere permit involuntary incarceration for treatment of persons who are considered suicide risks.

Second, do we Canadians believe that state sanctioned euthanasia should be a *right*, while the ability to have our suffering relieved to the extent our medical and social knowledge allows is only a *wish*?

There is a historical analogy worth considering. In the 1970s, pain-wracked patients suffered unnecessarily because we were afraid to use adequate doses of pain relievers, particularly the opioids. The science of pain relief was being advanced within palliative care, but the practice was in its infancy and we had rather few pharmacologic options. Dr. Kenneth Walker, an obstetrician and gynecologist writing as W. Gifford Jones, became an outspoken advocate for the legalization of heroin for the relief of cancer pain. That served to galvanize public perception. The governmental response was insightful. Heroin was legalized, *in concert* with intensive health provider education about the safe use of larger doses of morphine. The change in practice was remarkably rapid and substantial. More effective drugs and procedures were developed, ushering in the evolution of contemporary pain management. Effectively what had been a *wish* for better pain relief, became a practical possibility; a reasonable expectation, if not formally a *right*. It is not an empty speculation that had medical euthanasia been an option then, many who might have made that choice instead went on to live their remaining days with a much enhanced quality of life.

There is a stark question at the interface of palliative care and medical euthanasia. While comprehensive community centered end-of-life care is cost effective, medical euthanasia is cheaper. When Bill C-14 was under consideration, estimates were made of its additional cost to the system. It may give pause to some that when the analysis was done, in Ontario for example, the *savings* were estimated to be in excess of $60 M per year (Parliamentary Budget Office Canada 2020; Trachtenberg 2017). The well documented examples of institutional pressure to advocate for medical euthanasia is an unavoidable reality that follows. What are our core social values? How are they to be reflected on those who are disabled, vulnerable or disenfranchised?

31.5 The Concerns of the Disabled

All of us are destined to experience disability, unless we were to die suddenly and unexpectedly. The limitations may be physical, intellectual, sensory or mental illness. The disability may be transient or permanent, stable or progressive. These limitations powerfully interact with social determinants of health.

Ableism is discrimination and social prejudice against people with disabilities and/or people who are perceived to be disabled. In some respects our community, through law, regulation and habit is attentive, accommodating and respectful of those perceived as disabled. However, the medical euthanasia debate has brought a spotlight to some serious deficiencies. The MAiD deaths of Archie Rolland (Fidelman 2013) and Sean Taggart are cited as examples of an implicit view that those who are disabled, in some manner and degree have lives less worth living than the rest of us. Both had serious neurodegenerative diseases requiring extensive support. Both had those services withdrawn for administrative reasons. What does this say for the rest of us, as the economic and social consequences of our ageing population become apparent? There is a real danger that disability may be construed as having a life of lesser value, a flawed understanding of great consequence.

When is the desire to end one's life a consequence of an illness or injury, or painful loss? How long does that dark period last? There is copious evidence which Marini and Villareal cite in Chap. 23 of this volume, that environment, the experience of people, place and purpose, is a powerful determinant. It is also clear that the grieving period can last many months, even years before we accommodate to the new reality. How do we ensure that a safeguard against inadvertent euthanasia takes into full and transparent account of both the determinants of post-traumatic grief, and of its duration?

Between the *Rodriguez* and *Carter* Supreme Court cases, was another legal test relevant to the medical euthanasia debate, that of Robert Latimer (Butts 2021). He was tormented by the severe and interminable suffering, as he perceived it, of his severely mentally and physically disabled child. He chose to end that life when Tracy was 13 years of age. He was imprisoned as a consequence. Projecting some years from now, in the context of the more liberalized criteria for medical euthanasia being considered today, might we view Latimer differently? Would we weigh the paucity of supports in the family's rural community? Would we buy the argument that Robert Latimer restored his daughter's dignity by ending her life? Would we say that given Tracy's severe disability, her life was of lesser value? That was the basis of the eugenics movement in Germany in the 1930's, concepts notably borrowed from jurisdictions in both the US and Canada. And what would we say about consent for those adolescents considering ending their lives? As we consider allowing medical euthanasia for mature minors it will be important to take into account the differences between the binary legal notion of capacity (you have it or you do not) on one hand, and the much more nuanced clinical reality. Capacity is context dependent. Someone in the throes of anguish has less capacity to make a nuanced and consequential decision than the same person in more normal circumstances. The mature expression of capacity is a frontal lobe process which does not become functional until the third decade of life. The Latimer story tells us that the agency relationship between parent and child is a deeply held social value. Robert Latimer spent years in jail despite the great sympathy his plight engendered. The science tells us clearly that children and adolescents are not simply smaller adults.

31.5.1 In the Broader Social Context

The concluding section of this book explores some of the broader social issues which flow from the medical euthanasia discussion.

Does this change the nature of medical practice? If the timeless and expressed goal of medicine is to maintain dignity in our lives, how do we reconcile that with the central notion of medical euthanasia, namely that it restores dignity by ending it? How does the relationship between health providers change when medical euthanasia is part of the clinical armamentarium? Does it further compromise the already strained Hippocratic obligation to the patient first and foremost? As a distinguished physician colleague has phrased it, "Will patients trust us in the same way if they know it is within our bounds to end their lives rather than continue to be there for us when the going gets difficult?".

Have we fairly balanced the likely irreconcilable competing rights to medical euthanasia for patients, as against the right of conscientious objection by health providers, as required by the *Carter* decision? Would it be best to create a separate specialty, outside of medicine and nursing for this singular role… death doulas?

One of my regrets as a member of the Federal Expert Panel which arose out of Bill C-14 was the absence of faith-centered voices around the table. Submissions were invited and received, but that is different from having a participatory voice. In the end, a group of Expert Panel members met with our Indigenous communities. Within a range of views, there was a common theme: these are not our values. It turns out that across faith communities there are struggles to come to terms with this momentous social change. Is it sufficient to say those values will be respected when it is becoming clear that the echoes of legalized medical euthanasia are more far reaching than most people anticipated? Several authors raise similar questions. This is not to argue for a faith driven veto. It is a recognition that, whatever their failings as institutions, that is where our moral values take their root. As Dostoevsky wrote in *The Brothers Karamazov* (2003), where there are no morals, there is no law.

31.6 So Where Does This Leave Us?

The largest lesson of our medical euthanasia experience over the years echoes that editorial of 2014. We have gone down the slippery slope. By the end of 2021, more than 31,000 Canadians had chosen to end their lives by MAiD, the largest number in any country thus far. In other countries, where some form of medical euthanasia is permitted, growth in prevalence of the procedure has been gradual. In Canada it appears rapid. We have chosen to favour one group of Canadians with an epochal right to state-provided death, at the expense of others who might not have really wanted it. We can't hear from them directly because they are dead. But we are hearing reports of uncertainty and perplexity from families and troubling case stories.

Moreover, by overturning what advocates will call one tiny element of the fundamental Western, if not civilized global prohibition on killing, we have indeed punctured the inner fence. And as predicted we have unleashed issues and forces much broader and deeper than we could have imagined. The contributors to this volume have given clear expression to some of those concerns.

Supreme Court decisions can change society, for better and worse. In Canada, the case of *The Queen v. Drybones* (1970) advanced our First Nations rights before the law by striking down the *Indian Act*. In contrast, the legal case of *Dred Scott v. John F.A. Sandford* (1856) in the United States depreciated the value of many Americans. One is a decision we applaud. The other is our neighbour's disgrace. Where will *Rodriguez, Carter and Truchon* find their place?

In December of 2022 the federal government announced that the March 2023 expansion of eligibility criteria, in particular relating to mental health, would be delayed to allow for further reflection. Hopefully this is a starting point, not a political tactic. This volume contains many suggestions for addressing the almost insurmountably difficult problem. Surely, fundamentally and morally we need to proceed with clear and frank discussion, full and continuing examination and understanding, and deliberate incremental process. The rush forward in full flight of ideology and passion in lieu of careful consideration is to invite very deep regret.

31 December 2022.

References

Bill C-7. 2021. *An Act to amend the Criminal Code (medical assistance in dying)*, S.C. c. 2.

Bill C-14, *Act to amend the Criminal Code and to make related amendments to other Acts (medical assistance in dying)*, S.C. 2016, c. 3

Butts, Edward. 2021. Robert Latimer case. *Canadian Encyclopedia.* https://www.thecanadianencyclopedia.ca/en/article/robert-latimer-case?gclid=CjwKCAjw9-KTBhBcEiwAr19ig1R2-leaV0Fd4pkAqHmZBzs5Afdt7HXJ8GMgRJCO0hkVxZAbjprLJRoCAF0QAvD_BwE.

Canadian Charter of Rights and Freedoms, Part I of the *Constitution Act, 1982*, being Schedule B to the *Canada Act, 1982* (U.K.), 1982, c. 11.

Canadian Council of Academies. 2018. *The State of knowledge on medical assistance in dying where a mental disorder is the sole underlying medical condition.* The Expert Panel Working Group on MAID Where a Mental Disorder Is the Sole Underlying Medical Condition. Ottawa, Ontario.

Cartagena, Rosario, Alison, Thompson, Kaveh, Catebian, Trudo, Lemmens, Rose, Geist, Harvey, Schipper, Sandy, Buchman, Heesoo, Kim, and Mark, Handelman. 2016. *Understanding the relationship between suffering and capacity at the end-of-life: A pilot study.* Toronto: ON: York University, Law Commission of Ontario.

Carter v. Canada (Attorney General), 2015 SCC 5, [2015] 1 S.C.R. 331.

Criminal Code of Canada, R.S.C. 1985. c. C-46

Dostoyevsky, Fyodor. 2003. *The brothers karamazov.* Penguin UK.

Fidelman, Charlie. 23 July 2013. Life in long-term hospital 'unbearable': Montreal man with ALS. *Montreal Gazette.* https://montrealgazette.com/news/local-news/life-in-long-term-hospital-unbearable-montreal-man-with-als.

Government of Canada. 2022. *Third annual report on medical assistance in dying in Canada 2021.* https://www.canada.ca/en/health-canada/services/medical-assistance-dying/annual-report-2021.html.

Health Canada. May 2022. *Final report of the expert panel on MAiD and mental illness.* https://www.canada.ca/en/health-canada/corporate/about-health-canada/public-engagement/external-advisory-bodies/expert-panel-maid-mental-illness/final-report-expert-panel-maid-mental-illness.html.

Parliamentary Budget Office Canada. October 2020. *Cost estimate for bill C-7 "medical assistance in dying"*. https://www.pbo-dpb.gc.ca/web/default/files/Documents/Reports/RP-2021-025-M/RP-2021-025-M_en.pdf.

Rodriguez v. British Columbia (Attorney General). 1993. 3 S.C.R. 519

Schipper, Harvey.6 June 2014. With assisted suicide, context is everything. *Globe and Mail*. https://www.theglobeandmail.com/opinion/assisted-suicide-context-is-everything/article19012073/.

Scott v. 1856. *Sandford*, 60 U.S. 19 How. 393.

The Queen v. Drybones [1970] S.C.R. 282; (1970), 71 W.W.R. 171.

Trachtenberg, Aaron J., and Braden Manns. 2017. Cost analysis of medical assistance in dying in Canada. *CMAJ* 189 (3): E101–E105. https://doi.org/10.1503/cmaj.160650.

Truchon v Canada (Attorney General), 2019 QCCS 3792.

United Nations. 1966. *International Covenant on Civil and Political Rights*, 999 UNTS 171, Can TS 1976 No 47 (entered into force 23 March 1976, Convention ratified by Canada May 1976). https://www.ohchr.org/en/instruments-mechanisms/instruments/international-covenant-civil-and-political-rights.

United Nations. 2010. *Convention on the rights of persons with disabilities*. Geneva: United Nations. https://www.un.org/development/desa/disabilities/convention-on-the-rights-of-persons-with-disabilities/convention-on-the-rights-of-persons-with-disabilities-2.html.

Dr. Harvey Schipper B.A.Sc.(Eng.), MD, FRCPC brings a broad multi-dimensional understanding of medicine as a whole, and end of life issues, including medical euthanasia, in particular. His experience in navigating sensitive health public policy issues, such as the tainted blood crisis of the late 1990's, is widely recognized and pertinent to the MAiD debate, about which he has written, spoken, and advised patients, professionals, institutions and governments.

He is an oncologist, Professor of Medicine and Adjunct Professor of Law at the University of Toronto. Initially trained as an engineer, he has been at the interface of bioscience, clinical medicine and public policy for half a century. Working with the WHO he helped shape pain control and palliative care programs around the world. He was at the forefront of patient centred cancer therapy, introducing quality of life benchmarks for treatment development and evaluation. He currently works at St. Michael's Hospital in Toronto.

Index

J. Kotalik and D. W. Shannon (eds.), *Medical Assistance in Dying (MAID) in Canada*, The International Library of Bioethics 104,
https://doi.org/10.1007/978-3-031-30002-8

D

Printed in the USA
CPSIA information can be obtained
at www.ICGtesting.com
LVHW082053051123
762769LV00020B/14

9 783031 300011